Statistical Case Studies for Industrial Process Improvement

ASA-SIAM Series on Statistics and Applied Probability

The ASA-SIAM Series on Statistics and Applied Probability is published jointly by the American Statistical Association and the Society for Industrial and Applied Mathematics. The series consists of a broad spectrum of books on topics in statistics and applied probability. The purpose of the series is to provide inexpensive, quality publications of interest to the intersecting membership of the two societies.

Editorial Board

Donald P. Gaver
Naval Postgraduate School, Editor-in-Chief

Alan F. Karr
National Institute of Statistical Sciences

John Lehoczky
Carnegie Mellon University

Robert L. Mason
Southwest Research Institute

Robert Rodriguez
SAS Institute

Andrew Solow
Woods Hole Oceanographic Institution

Werner Stuetzle
University of Washington

Grace Wahba
University of Wisconsin

Eric Ziegel
Amoco Corporation

Latouche, G. and Ramaswami, V., *Introduction to Matrix Analytic Methods in Stochastic Modeling*

Peck, R., Haugh, L. D., and Goodman, A., *Statistical Case Studies: A Collaboration Between Academe and Industry, Student Edition*

Peck, R., Haugh, L. D., and Goodman, A., *Statistical Case Studies: A Collaboration Between Academe and Industry*

Barlow, R. E., *Engineering Reliability*

Czitrom, V. and Spagon, P. D., *Statistical Case Studies for Industrial Process Improvement*

Statistical Case Studies for Industrial Process Improvement

Veronica Czitrom
Bell Laboratories, Lucent Technologies
Orlando, Florida

Patrick D. Spagon
Motorola University Southwest
Austin, Texas

Society for Industrial and Applied Mathematics
Philadelphia, Pennsylvania

ASA

American Statistical Association
Alexandria, Virginia

© 1997 by the American Statistical Association and the Society for Industrial and Applied Mathematics.

10 9 8 7 6 5 4 3 2

All rights reserved. Printed in the United States of America. No part of this book may be reproduced, stored, or transmitted in any manner without the written permission of the publisher. For information, write to the Society for Industrial and Applied Mathematics, 3600 University City Science Center, Philadelphia, PA 19104-2688.

No warranties, express or implied, are made by the publisher, authors and their employers that the data on the enclosed disk are free of error. They should not be relied on as the sole basis to solve a problem whose incorrect solutions could result in injury to person or property. If the data are employed in such a manner, it is at the user's own risk and the publisher, authors and their employers disclaim all liability for such misuse.

Library of Congress Cataloging-in-Publication Data

Czitrom, Veronica.
　　Statistical case studies for industrial process improvement / Veronica Czitrom, Patrick D. Spagon.
　　　　p. cm. -- (ASA-SIAM series on statistics and applied probability)
　　Includes index.
　　ISBN 0-89871-394-3 (pbk.)
　　1. Process control--Statistical methods--Case studies.
　　2. Semiconductor industry--Statistical methods--Case studies.
　　I. Spagon, Patrick D. II. Title. III. Series.
　　TS156.8.C95 1997
　　621.3815'068'5--dc21 　　　　　　　　　　　　　　97-12582

Understanding the Nature of Variability in a Dry Etch Process by Richard O. Lynch and Richard J. Markle reprinted with permission from SPIE Proceedings, Vol. 1185 (1989), ©1989 SPIE.

The introduction to this book was derived from Robert I. Scace, *Metrology for the Semiconductor Industry*, NISTIR 4653, National Institute of Standards and Technology, Gaithersburg, MD 20899 (1991).

Facts about SEMATECH based on the handbook *SEMATECH: Innovation for America's Future*.

IRONMAN is a registered trademark of Motorola.

siam is a registered trademark.

Contents

Foreword ix
Preface xi
Acknowledgments xv
Introduction xvii
Facts About SEMATECH xix
SEMATECH Qualification Plan xxi
    Veronica Czitrom and Karen Horrell

Part 1:   **Gauge Studies**

Chapter 1:   Introduction to Gauge Studies 3
    Veronica Czitrom

Chapter 2:   Prometrix RS35e Gauge Study in Five Two-Level Factors
    and One Three-Level Factor 9
    James Buckner, Barry L. Chin, and Jon Henri

Chapter 3:   Calibration of an FTIR Spectrometer
    for Measuring Carbon 19
    Peter C. Pankratz

Chapter 4:   Revelation of a Microbalance Warm-Up Effect 39
    James Buckner, Barry L. Chin, Todd P. Green, and Jon Henri

Chapter 5:   GRR Methodology for Destructive Testing
    and Quantitative Assessment of Gauge Capability
    for One-Side Specifications 47
    Teresa Mitchell, Victor Hegemann, and K. C. Liu

Part 2:   **Passive Data Collection**

Chapter 6:   Introduction to Passive Data Collection 63
    Veronica Czitrom

Chapter 7:   Understanding the Nature of Variability
    in a Dry Etch Process 71
    Richard O. Lynch and Richard J. Markle

Chapter 8:   Virgin Versus Recycled Wafers for Furnace Qualification:
    Is the Expense Justified? 87
    Veronica Czitrom and Jack E. Reece

Chapter 9:   Identifying Sources of Variation
    in a Wafer Planarization Process 105
    Arnon M. Hurwitz and Patrick D. Spagon

Chapter 10: Factors Which Affect the Number
    of Aerosol Particles Released by Clean Room Operators 115
    William Kahn and Carole Baczkowski

Chapter 11: A Skip-Lot Sampling Plan Based on Variance Components
    for Photolithographic Registration Measurements 133
    Dwayne Pepper

Chapter 12: Sampling to Meet a Variance Specification:
Clean Room Qualification 145
Kathryn Hall and Steven Carpenter

Chapter 13: Snapshot: A Plot Showing Progress
Through a Device Development Laboratory 155
Diane Lambert, James M. Landwehr, and Ming-Jen Shyu

Part 3: Design of Experiments

Chapter 14: Introduction To Design Of Experiments 171
Veronica Czitrom

Chapter 15: Elimination of TiN Peeling During Exposure to CVD Tungsten
Deposition Process Using Designed Experiments 199
James Buckner, David J. Cammenga, and Ann Weber

Chapter 16: Modeling a Uniformity Bulls-Eye Inversion 213
*James Buckner, Richard Huang, Kenneth A. Monnig, Eliot K. Broadbent,
Barry L. Chin, Jon Henri, Mark Sorell, and Kevin Venor*

Chapter 17: Using Fewer Wafers to Resolve Confounding
in Screening Experiments 235
Joel Barnett, Veronica Czitrom, Peter W. M. John, and Ramón V. León

Chapter 18: Planarization by Chemical Mechanical Polishing: A Rate
and Uniformity Study 251
Anne E. Freeny and Warren Y.-C. Lai

Chapter 19: Use of Experimental Design to Optimize a Process
for Etching Polycrystalline Silicon Gates 265
Fred Preuninger, Joseph Blasko, Steven Meester, and Taeho Kook

Chapter 20: Optimization of a Wafer Stepper Alignment System
Using Robust Design 283
Brenda Cantell, José Ramírez, and William Gadson

Part 4: Statistical Process Control

Chapter 21: Introduction to Statistical Process Control (SPC) 299
Veronica Czitrom

Chapter 22: Removing Drift Effects When Calculating Control Limits 315
Ray L. Marr

Chapter 23: Implementation of a Statistical Process Control Capability
Strategy in the Manufacture of Raw Printed Circuit Boards
for Surface Mount Technology 325
Ricky M. Watson

Chapter 24: Obtaining and Using Statistical Process Control Limits
in the Semiconductor Industry 337
Madhukar Joshi and Kimberley Sprague

Part 5: Equipment Reliability

Chapter 25: Introduction to Equipment Reliability 359
Veronica Czitrom

Chapter 26: Marathon Report for a Photolithography Exposure Tool 365
John T. Canning and Kent G. Green

Chapter 27: Experimentation for Equipment Reliability Improvement 387
Donald K. Lewis, Craig Hutchens, and Joseph M. Smith

Chapter 28:	How to Determine Component-Based Preventive Maintenance Plans .... 403 *Stephen V. Crowder*	
Part 6:	**Comprehensive Case Study**	
Chapter 29:	Introduction to Comprehensive Case Study .... 427 *Veronica Czitrom*	
Chapter 30:	Characterization of a Vertical Furnace Chemical Vapor Deposition (CVD) Silicon Nitride Process .... 429 *Jack E. Reece and Mohsen Shenasa*	
Part 7:	**Appendices**	
Appendix:	Introduction to Integrated Circuit Manufacturing .... 469 *Ronald J. Schutz*	

Glossary of Selected Statistical Terms 477

Table of Statistical Topics by Chapter 500

Index 501

Foreword

In 1987, a unique experiment was initiated between fourteen companies in the U.S. semiconductor industry and the U.S. government. The experiment, called SEMATECH, joined together more than 80% of the U.S. semiconductor industry, with the support of the federal government, to bring the U.S. semiconductor industry back to world leadership in semiconductor manufacturing technology. The premise behind the experiment was that market competitors working together with the federal government could perform certain kinds of research and development more effectively than they could working alone.

Today SEMATECH, its member companies, and the federal government have enjoyed several successes, returning the U.S. semiconductor and equipment industries to global market leadership. Because the cost of silicon integrated circuit manufacturing facilities approached $1 billion in the mid-1990s, and the major pieces of equipment and processes going into a fabrication facility can cost several million dollars each, one key to our success is the joint qualification of semiconductor manufacturing equipment. Here our focus has been to ensure that the development, qualification, and improvement of such equipment and processes is done jointly so that SEMATECH member companies see a major cost reduction through the elimination of redundant efforts.

The development of a suitable statistical data base for processes and equipment is vital to the qualification of semiconductor manufacturing equipment. An important element of this development is SEMATECH's Qualification Plan, which provides a statistical roadmap for equipment and process characterization and improvement. The Qualification Plan includes data collection and analysis, design of experiments, process optimization and sensitivity analysis, statistical process control and process capability, and equipment reliability. These topics and others are covered by the series of papers collected for this book. These papers represent the work of key scientists and engineers from SEMATECH member companies, universities, and national laboratories.

This broad participation by scientists and engineers indicates their strong interest in equipment qualification and the use of statistical methodology to improve that qualification process. As part of our Total Quality process, we at SEMATECH focus as much as possible on data-driven decision making, using statistical techniques for collecting, analyzing, and presenting this data as a key part of our programs. In turn, we hope that this collection of papers will be useful to equipment designers, semiconductor manufacturers, and people involved in semiconductor research and development. The development of standard processes and common ways to collect, analyze, and present data should provide a major step forward in the continuing rapid evolution of semiconductor technology.

W.J. Spencer
Austin, Texas

Preface

The primary objective of this book is to demonstrate how American industry would benefit from using statistical methods. Many industries already benefit from their use, especially in the semiconductor, medical, and agricultural fields. This book gives case studies of the use of statistics in the semiconductor industry, which can also be used by engineers and scientists in other industries.

Another important objective of the book is to enhance the characterization and improvement of industrial processes and equipment through the use of statistical methods. The broad selection of statistical case studies offers engineers, scientists, technicians, and managers numerous examples of best-in-class practices by their peers. The case studies enable engineers and technicians to use statistical methods, and give them the resources to do so by providing their managers with an understanding of the potential impact of the use of statistics.

Another major objective is to provide examples of successful industrial applications of statistics for use in industrial workshops and in academic courses. Engineering and science students can use the examples to learn how to use statistics to solve their problems, and statistics students can use the examples to learn how to apply statistics to other disciplines. It is important for students to see real examples with real data. The Harvard School of Business has very successfully used case studies to train MBAs. A table before the index shows statistical topics considered in each case study.

Engineers and scientists with a modest background in statistics can understand most of the case studies. The emphasis is on the use of mainstream statistical methods that can be understood and used by practitioners. The introductions to the units and a glossary give basic definitions of common statistical terms, and provide suggested reading.

The case studies illustrate applications of statistics to the semiconductor industry, though the applications are also relevant to other industries. The objectives of process and equipment characterization and improvement also apply to the chemical, automotive, steel, plastics, textile, and food industries, among others. Engineers, physicists, chemists, material scientists, equipment improvement personnel, statisticians, and many other scientists will benefit from applying the techniques discussed in this book. The case studies can be used in industry and in academia. Each case study describes the engineering process considered so that it can be understood by the lay person. An introduction to integrated circuit (semiconductor device) manufacturing by Ronald J. Schutz is given at the end of the book.

The case studies illustrate a wide variety of implementations of statistics. Some give examples from manufacturing, and others from research and development. They cover engineering areas such as chemical vapor deposition, furnace and implant, metallization, etch, and photolithography. They deal with topics such as process optimization, equipment reliability, detection of drift and warm-up effects, sampling plan reduction, tracking the progress of wafer lots through a device development laboratory, preventive maintenance plans, selection of bunny suits for use in clean rooms, use of cheaper recycled wafers instead of virgin wafers, and wafer planarization. Most case studies are straightforward applications of statistics that require little formal training, while some require a broader knowledge of statistics to fully understand the material. The unifying theme of the book is the SEMATECH Qualification Plan.

SEMATECH qualification plan

The SEMATECH qualification plan is a roadmap for process and equipment characterization and improvement that can be used in different industrial settings. The qualification plan is a versatile framework that has been successfully used by companies which produce semiconductor devices, by companies that make the equipment which makes the semiconductor devices, by companies that supply the material used in the fabrication of semiconductor devices, and by SEMATECH. A methodology similar to the SEMATECH qualification plan is employed in the chemical industry.

The main objectives of the SEMATECH qualification plan are to characterize and improve process and equipment performance. These engineering objectives are reached most efficiently by using statistical methods. Statistical methods are an integral part of the structured approach of the qualification plan. They provide a framework for thinking about an engineering problem and a vocabulary that helps engineers communicate their results across different engineering areas. The case studies in the book illustrate applications of these statistical methods.

The main objectives of the SEMATECH qualification plan and the statistical methods used to reach these objectives are:

Establish process and equipment baseline.
- Characterize the measurement systems to ensure adequate measurement capability (gauge studies, Part 1).
- Characterize the current process and equipment performance, and establish sources of variability (passive data collection, Part 2).

Optimize process and verify stability.
- Actively improve process performance by exploring and optimizing the process (design of experiments, Part 3).
- Establish and maintain process stability (statistical process control, Part 4).

Improve reliability and demonstrate manufacturability.
- Improve equipment reliability and process throughput and establish cost of ownership (reliability, Part 5).

Organization of the book

On page xxi you will find an introduction to the SEMATECH qualification plan. The first five parts of the book illustrate the main statistical methods that support the qualification plan, namely gauge studies, passive data collection (observational studies), design of experiments, statistical process control (SPC), and equipment reliability. Part 6 is a comprehensive case study. A glossary of selected statistical terms and an introduction to integrated circuit manufacturing are given at the end of the book. The data sets used in the case studies are included on the diskette.

Each part of the book begins with an introduction to the statistical methods considered in that part. The introduction includes definitions of the main statistical terms used in the chapters in that part, an introduction to each case study, and suggested reading. The definitions use simple non-technical language, reducing statistical jargon to a minimum. Very few definitions involve formulas. The typical engineer needs to understand the *meaning* of the statistical terms that appear in computer output, and will seldom need to use formulas directly.

After the introduction to each part, case studies illustrate the statistical method. The format of each case study is:

- Abstract
- Executive summary for a high level overview, with problem statement, solution strategy, and conclusions
- Description of the engineering process
- Data collection plan
- Data analysis and interpretation of results (the main body of the text)
- Conclusions and recommendations
- Acknowledgements
- References
- Appendices.

Most of the datasets in the book appear on the disks included with the book. Both a Macintosh disk and an IBM compatible disk are included with each book.

How this book came to be

This book was started in November of 1991, when Veronica Czitrom was on a two-year assignment at SEMATECH. It grew out of her and Pat Spagon's appreciation for an earlier book of case studies involving the application of designed experiments to various industries, *Experiments in Industry: Design, Analysis, and Interpretation of Results*, Ronald D. Snee, Lynne B. Hare, and J. Richard Trout, editors, American Society for Quality Control, 1985.

Veronica and Pat wanted the book to give a broad overview of the application of statistics in the semiconductor industry, and to use the SEMATECH qualification plan as the framework to do so. Accordingly, they issued a broad call for papers, and were gratified to receive a very enthusiastic response. Case studies were contributed by people working throughout the semiconductor industry.

To ensure that the case studies were technically correct, the editors asked for volunteers to review them. Many well-respected professionals contributed their time and expertise to review one or two, and sometimes more, case studies. The list of sixty-seven reviewers is given in the Acknowledgements section. Each case study was reviewed by between two and six independent reviewers, who followed a form provided by the editors to standardize the reviewing process. The editors also reviewed each case study. All the reviews were sent to the authors along with suggestions for improvement.

The authors made changes to the case studies and returned them to Bob Stevens, the capable production editor at SEMATECH who managed the production of the internal edition of the book, which included thirty-one case studies by sixty-seven authors. Distribution was limited to authors, reviewers, and SEMATECH member company libraries. A subset of the case studies was selected for the joint publication by ASA (American Statistical Association) and SIAM (Society for Industrial and Applied Mathematics).

Veronica Czitrom
Bell Laboratories, Lucent Technologies

Patrick D. Spagon
Motorola University Southwest

October 1996

Acknowledgments

The editors would like to thank the authors of the case studies for their excellent and varied contributions. Thanks are also due the reviewers who made invaluable comments and suggestions. The final form of the book owes much to Bob Stevens, who capably managed the production process of the SEMATECH edition of the book. We thank Susan Rogers of the SEMATECH Technology Transfer Department for providing editing and production resources. We also thank Patty Wertz and Karen Kelley of SEMATECH for their invaluable administrative support. In addition, Mike Adams, Curt Moy, Paul Tobias, and Chelli Zei provided legal and administrative assistance critical to the publication of this book.

Sixty-seven people volunteered to review case studies in response to requests through SEMATECH, at professional meetings, in publications, and by word of mouth. The reviewers generously contributed their time by reviewing one or two, and sometimes more, case studies or unit introductions. Again, many thanks to the reviewers. Their names follow.

Carl Almgren
NCR

Lesly Arnold
Lucent Technologies

Diccon Bancroft
W.L. Gore and Associates

Oren Bar-Ilan
National Semiconductor

Carol Batchelor
National Semiconductor

W. Michael Bowen
Intel

James Buckner
Texas Instruments

Michele Boulanger
Motorola

Youn-Min Chou
University of Texas at San Antonio

Stephen V. Crowder
Sandia National Labs

Veronica Czitrom
Bell Labs, Lucent Technologies

Vallabh Dhudshia
Texas Instruments

David Drumm
Lucent Technologies

Steven G. Duval
Intel

Itzak Edrei
National Semiconductor

Curt Eglehard
Intel

John Fowler
Arizona State University

Bruce Gilbert
Symbios Logic

Rob Gordon
Intel

Andy Ruey-Shan Guo
National Semiconductor

Kathryn Hall
Hewlett-Packard

Laura Halbleib
Sandia National Labs

Lynne Hare
NIST

Ian Harris
Northern Arizona University

Victor Hegemann
Texas Instruments

Richard E. Howard
Luxtron Corp.

Arnon Hurwitz
Qualtech Productivity Solutions

Peter W.M. John
University of Texas at Austin

Madhukar Joshi
Digital Equipment Corporation

William Kahn
Mitchell Madison Group

Behnam Kamali
Mercer University

Bert Keats
Arizona State University

Georgia Anne Klutke
University of Texas at Austin

Ramón V. León
University of Tennessee at Knoxville

Cathy Lewis
Lucent Technologies

Dennis K.J. Lin
Pennsylvania State University

Noam Lior
University of Pennsylvania

Jack Luby
AT&T

Richard O. Lynch
Harris Semiconductor

Brian D. Macpherson
University of Manitoba, Canada

Mike Mahaney
Intel

Ray L. Marr
Advanced Micro Devices

Robert E. Mee
University of Tennessee at Knoxville

Kyle Merrill
Hewlett-Packard

Teresa Mitchell
Texas Instruments

M.T. Mocella
Dupont

Ken Monnig
SEMATECH

Mike Pore
Advanced Micro Devices

Randall W. Potter
Lucent Technologies

José Ramírez
Digital Semiconductor

Jennifer Robinson
SEMATECH

Pegasus Rumaine
Sieman's

Edward L. Russell, III
Cypress Semiconductor

Eddie Santos
Lucent Technologies

Ronald D. Snee
Joiner Associates

Patrick D. Spagon
Motorola University Southwest

Lynne Stokes
University of Texas at Austin

William R. Sype
Intel

Paul Tobias
SEMATECH

Ledi Trutna
Advanced Micro Devices

Wayne Van Tassell
Ashland Chemical

Mario Villacourt
Intel

Esteban Walker
University of Tennessee at Knoxville

Ricky M. Watson
Tracor, Inc.

Skip Weed
Motorola

Steve Zinkgraf
Allied Signal

Introduction

The importance of semiconductor technology in today's world can hardly be overemphasized. Semiconductor devices are absolutely essential components of all electronic products. Without semiconductors, electronic products and systems cannot be made or operated. The world semiconductor industry produced $153.6 billion of products in 1995[a] and has been growing at an average compound annual growth rate of about 17.4 percent annually for the last decade. Electronic systems are integral parts of communication systems, transportation, medical care, banking, manufacturing, entertainment, etc. Electronics is one of the largest manufacturing segments of the world economy, amounting to some $778.7 billion in 1995 and growing at about 7.6 percent annually over the last decade.

Today's sub-micrometer integrated circuit technology is only achievable by pressing science and skill to very expensive limits. The breadth of science and technology used in semiconductor manufacturing is rapidly expanding. Problems of imperfect understanding and inability to measure what is being done abound. Integrated circuits today are built with great ingenuity by using very complex processes to deposit and remove surface films. The chemistry and physics of these processes are not well understood. The surfaces on which films are deposited affect both the deposition processes and the properties of the semiconductor devices in ways that are evident but not understood; these effects are becoming ever more significant. A clear view of what is being made requires electron microscopes, but accurate dimensional measurements with these are not now possible. Soon it will be necessary to create ultra-small patterns using x-rays or other new techniques that will require totally new measurements. Control of contamination from particles and stray impurities is pressing the limits of measurement and control. As device dimensions shrink, the details of their behavior change significantly. Previously negligible effects become important, and must be accounted for in engineering design. New device theory and computer models of device behavior are needed.

Many kinds of process equipment contain measurement systems to control their own operation. The strong present-day move to greater automation means that these measurements have a very great effect on the overall control of the entire manufacturing process. Even imprecise measurements of many process variables are difficult to perform with human aid. Obtaining accurate measurements without that aid is even more difficult. All of these factors add greatly to the challenge to support the processing equipment, materials, and device manufacturing industries with adequate measurements.

Semiconductor manufacturing is thus a measurement-intensive business. As semiconductor devices become smaller and smaller and chips become more complex, the demands on measurement technology increase. Better sensitivity, improved accuracy, totally new measurements, and the pressure to automate the manufacturing process are all needs that must be met. New materials and new processes bring with them new measurement problems that must be solved. The measurement imperative has been recognized by the industry in identifying metrology as an essential technology that is required for every aspect of semicon-

---

[a] Data are from VLSI Research; includes both production by captive and merchant manufacturers.

ductor device design and manufacture[b]. Statistical methods are essential for interpreting and understanding the data resulting from all of these measurements. Single measurements are seldom sufficient for many reasons. Because production techniques and the metrology needed to control them are both commonly just at the edge of what is technically and economically possible, statistical process controls must be used. Measured data often must be averaged to reduce the random uncertainties of individual measurements. Process control data are needed over long periods of time, raising questions about the stability of metrological instrumentation as well as of the process. Experiments to optimize process performance must be designed and the resulting data interpreted. These and other needs raise significant questions:

- Given a particular question, what data are needed to answer it? Is there a better question to pose or a more appropriate data set to obtain from an economic, technical, marketing, emotional, etc., point of view?
- Does the data have a normal distribution, or is it skewed, bounded, or censored?
- Does this experiment find a true operating optimum for this process or only a local optimum condition?
- How can the effects of differing materials, instruments, operators, and other variables be separated in analyzing these data?
- Does this piece of equipment produce the same effects (physical properties, chemical purity, reliability of the final product, etc.) on the material being processed as does a different tool made by another manufacturer?
- Does this piece of equipment reliably perform its functions as defined in the manufacturers specifications at least for the mean time between failures (MTBF) agreed upon between the purchaser and equipment manufacturer?
- Do these two instruments, made by different manufacturers, give comparable measurements of film thickness, line width, particle size and count, etc.?
- How can data from plant location A be reconciled with data from plant locations B, C, and D that all make the same products?

These are real-life issues in making integrated circuits, and no doubt for making many other types of products. This book is intended to be an aid, not only to engineers, technicians, and managers, but to statistical practitioners as well, in dealing with such situations.

Clearly, metrology and statistical data analysis go hand in hand. Measurements produce data, and statistics helps interpret data in ways that make its meanings clearer. Semiconductor manufacturing, as a metrology-intensive industry, is implicitly a statistics-intensive industry as well. I welcome the appearance of this book as a significant step in the exploration of statistical tools for answering some of the hard questions in semiconductor manufacturing, as a way of stimulating discussion on and improvement of the mathematical foundations for these tools, and as a medium for adding to the competence of engineers and scientists from other disciplines in the informed use of statistical methods in the regular course of their work. I am sure the authors will welcome constructive discussion and debate from readers interested in furthering those goals.

Robert I. Scace, Director
Microelectronics Programs, National Institute of Standards and Technology, Gaithersburg, Maryland

---

[b] See *Semiconductor Technology Workshop Working Group Reports* and the *National Technology Roadmap for Semiconductors*, published respectively in 1992 and 1994 by the Semiconductor Industry Association, 4300 Stevens Creek Boulevard, Suite 271, San Jose, CA 95129.

Facts about SEMATECH

Semiconductors are the foundation of the electronics industry, the largest industry in the United States, employing nearly 2.7 million Americans. Used in products ranging from television sets and automobiles to computers and aircraft, modern semiconductors were invented in the United States in the late 1950's. The United States enjoyed world leadership until the early 1980's, when it lost ground.

SEMATECH (SEmiconductor MAnufacturing TECHnology) is a unique example of industry/government cooperation. Because semiconductors are an essential component of electronic products, dependence on foreign suppliers for this technology has posed a serious threat to the United States economy and military security. To respond to this threat, the United States semiconductor industry worked with the government to found SEMATECH. It has become another example of American industry, government, and academia working together toward a common goal.

The SEMATECH Mission

SEMATECH is a consortium of United States semiconductor manufacturers (member companies) working with government and academia to sponsor and conduct research in semiconductor manufacturing technology. It develops advanced semiconductor manufacturing processes, materials and equipment. SEMATECH's emphasis is on manufacturing capability, not on device design. SEMATECH does not produce chips for sale.

SEMATECH's mission is simple: to create fundamental change in manufacturing technology and the domestic infrastructure to provide industry in the United States with the domestic capability for world leadership in semiconductor manufacturing.

The SEMATECH Organization

The SEMATECH operational concept is dependent upon knowledge from its member companies, American equipment and materials suppliers (SEMI/SEMATECH), the Department of Defense (Advanced Research Projects Agency, ARPA), American universities and government laboratories designated as SEMATECH Centers of Excellence, and the Semiconductor Research Corporation which administers research grants.

In 1996, the ten SEMATECH member companies were: Advanced Micro Devices, Inc., Digital Equipment Corporation, Hewlett-Packard Company, Intel Corporation, International Business Machines Corporation, Lucent Technologies, Motorola, Inc., NCR Corporation, National Semiconductor Corporation, Rockwell International Corporation, and Texas Instruments, Inc. Through 1996, member companies provided half of SEMATECH's funding, with the federal government providing the other half. After 1996, all SEMATECH funding will be provided by the member companies.

SEMATECH Qualification Plan

Veronica Czitrom and Karen Horrell

The SEMATECH Qualification Plan (Qual Plan) is a flexible guideline for effective process and equipment characterization and improvement. The Qual Plan uses statistical methods to efficiently reach its engineering goals. The engineering goals include gauge characterization, process and equipment characterization, process optimization, process stability control, and equipment reliability improvement. The statistical methods used to achieve these engineering goals include gauge studies, passive data collection (observational studies), design of experiments, statistical process control, and reliability.

Introduction

SEMATECH (SEmiconductor MAnufacturing TECHnology) is a consortium of major U.S. semiconductor manufacturers, founded with support from the U.S. government (Advanced Research Projects Agency). The consortium's mission is to create fundamental change in manufacturing technology and the domestic infrastructure to provide U.S. semiconductor companies the continuing capability to be world-class suppliers. The SEMATECH Qualification Plan (Qual Plan) is one of the tools SEMATECH uses to help create this fundamental change.

The Qual Plan is a framework for engineering process and equipment characterization and improvement. The Qual Plan uses statistical methods to reach its engineering goals efficiently. The engineering objectives include characterizing the measurement system, characterizing current process and equipment performance, optimizing the process and establishing the process window, maintaining process stability, and improving equipment reliability and throughput. The main statistical methods used to achieve these objectives are gauge studies, passive data collection (observational studies), design of experiments, statistical process control, and reliability. The statistical methods provide a framework for thinking about the engineering problem and a vocabulary that helps engineers communicate their results to others, including engineers in different areas and even in different companies.

The Qual Plan is a structured and flexible framework that can be adapted for different uses. The Qual Plan has been successfully used in the semiconductor industry by companies that produce semiconductor devices, by companies that manufacture the equipment that makes the semiconductor devices, by companies that supply the materials used in the fabrication of semiconductor devices, and by SEMATECH. SEMATECH uses the Qual Plan on its projects to improve semiconductor manufacturing equipment before it is introduced into a manufacturing facility. SEMATECH also used parts of the Qual Plan to convert its fab from the production of 150 mm wafers to the production of 200 mm wafers, and to characterize the 200 mm wafer equipment. SEMATECH member companies have adapted the Qual Plan for their own use. Segments of Hewlett-Packard have extended the Qual Plan to monitor semiconductor equipment during manufacturing. Intel Corporation has modified the Qual Plan for use in its packaging and testing areas.

The methodology of the Qual Plan is general and it can be adapted for use outside the semiconductor industry. The objectives of process and equipment characterization and improvement apply to other industries such as the chemical, automotive, plastics, and food processing industries, to name a few. A qualification plan similar to SEMATECH's is used in the chemical industry.

This section gives an overview of the Qual Plan. The next section describes its development, and the following section examines the Qual Plan in greater detail. The remaining parts of the book give case studies of the application of the five main statistical methods used in the Qual Plan.

Evolution of the Qual Plan

The Qual Plan evolved from the equipment burn-in methodology developed by Intel Corporation. Before the Qual Plan was used, new and upgraded equipment and processes underwent varying degrees of testing, from rigorous statistical tests and improvement efforts to limited acceptance testing where a piece of equipment was declared ready for production upon installation. Non-rigorous characterization can lead to months of work in an attempt to stabilize and optimize the equipment.

A group of SEMATECH engineers and statisticians began to modify the Intel burn-in methodology early in 1988. The first presentation of a strategy flow chart was in the Fall of that year. Equipment characterization was the focus of a SEMATECH Statistical Methods Workshop in February of 1990, where the Qual Plan was presented as a statistical methodology for equipment characterization. That initial methodology focused only on the process. The Qual Plan was subsequently expanded to include equipment utilization (marathon) to investigate the hardware and software reliability of the equipment.

During the summer of 1992, a task force of statisticians, engineers, and other topical experts was formed to incorporate management tasks into the existing plan. The result of that task force was the creation of a document on the Qual Plan, the *SEMATECH Qualification Plan Guidelines for Engineering*. This guideline was further expanded to include reliability improvement testing based on the Motorola "IRONMAN"™ procedure in 1995.

SEMATECH Qualification Plan

The objectives of the Qual Plan are to improve process and equipment performance. Statistical methods are used to support these objectives by describing and quantifying process and equipment behavior.

This section gives an overview of the Qual Plan, including a flow chart, a description of its three stages, and a description of the individual elements of each stage.

Flowchart of the Qual Plan

Figure A is a flowchart of the SEMATECH Qualification Plan. The Qual Plan consists of three stages: establish process and equipment baseline, optimize the process, and improve equipment reliability.

- **Stage I: Establish process and equipment baseline.** Gauge studies (Part 1) are performed to characterize the measurement systems to ensure they are capable of performing the required measurements. Passive data collection (Part 2), also called observational studies, are performed to characterize current process and equipment performance and sources of variability. If necessary, the process and/or the equipment hardware are improved.

- **Stage II: Characterize and optimize process and verify stability.** Designed experiments (Part 3) are performed to explore and optimize the process and establish a process window by simultaneously studying the effect of several process inputs on one or more process outputs. Statistical process control (Part 4) is used to establish and monitor long-term process stability.
- **Stage III: Improve equipment reliability and demonstrate manufacturability and competitiveness.** Equipment reliability (Part 5) is improved by using the IRONMAN™ methodology to cycle the equipment at night and carry out engineering improvement activities based on root cause analysis of failures during the day. Manufacturability and competitiveness are demonstrated by using a marathon to consider equipment reliability and process throughput, and by establishing cost of ownership metrics.

One of the strengths of the Qual Plan is its versatility. Modification of the basic structure of the Qual Plan accommodates the goals of individual projects. For example, if the goal is process improvement, stage III may be omitted. If the goal is equipment reliability, stage II may be omitted. Note that the flow chart and the stages of the Qual Plan emphasize engineering goals, and that statistical methods are used in support of those goals.

Process characterization is the description of the performance of a piece of equipment and its associated process, and optimization of the process recipe. Process characterization is performed during stages I and II of the Qual Plan. *Equipment qualification* covers the three stages of the Qual Plan, including the act of stressing the equipment in a manufacturing environment to measure its reliability in stage III. An equipment qualification prepares the process and the equipment for performance in an integrated manufacturing environment.

Elements of the Qual Plan

We will now consider the individual elements, or steps, of the Qual Plan flowchart given in Figure A.

Stage I: Establish process and equipment baseline. The objective of this stage is to establish the current performance of the equipment and its associated processes. Gauge studies of the metrology systems are used to ensure adequate measurements, and a passive data collection is used to establish current process and equipment performance. If process and equipment performance are not adequate, it may be necessary to improve the process and the hardware.

- **Plan the project and begin training.** The first element of the Qual Plan includes business aspects of the project and planning. The team members and customers determine the project goals and the resources required. It is important to define specification metrics, data collection methods, ways to report the results, and baseline metrics to evaluate the success of the qualification. Training on the equipment, metrology system, and related software can begin in this step, and can continue throughout the duration of the project.
- **Gauge capability study.** The objective of a gauge capability study is to ensure that the measurement (metrology) tool can perform the required measurements. This requires making sure that the measurement tool is stable over time, accurate so that the measurements are close to their target value, precise so that the variability of the measurements is small enough for the intended purpose, and under statistical process control (SPC) so that unexpected tool behavior is taken care of. A gauge study needs to be performed on each measurement tool used during the project. Part 1 gives an introduction to gauge

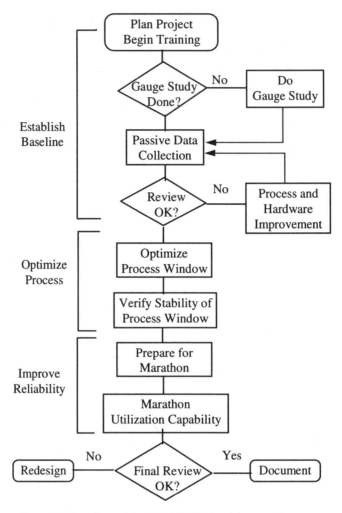

Figure A. *Flowchart of the SEMATECH Qualification Plan. Three stages of the Qual Plan: Establish process and equipment baseline, Optimize process, and Improve reliability. In the first stage, a gauge study ensures capable measurement systems, and a passive data collection establishes the process and equipment baseline.*

In the second stage, the process is optimized, and process stability is monitored.

In the third stage, equipment reliability is assessed and improved.

capability studies, and four case studies illustrate applications.

- **Passive data collection (PDC).** The objective of a passive data collection (PDC) (or observational study, or capability demonstration) is to establish the current process and equipment performance and to find areas for improvement. The PDC can be used to identify sources of variability, evaluate sampling plans, establish stability over time, determine whether the process variability is sufficiently small so that the process is capable of consistent performance, and set the direction of future experimentation for further process and equipment improvement. During a PDC, the current process and equipment performance is established by gathering information on some thirty independent process runs without any adjustments (tweaks). Normal maintenance procedures are allowed, but should be documented. Part 2 gives an introduction to passive data collection, and seven case studies.

- **Review.** A review assesses the results of the project at the end of stage I. Two major areas require review: the results of the gauge study and of the PDC, and plans for future activities. The project team presents the results of its activities for discussion by all concerned parties, including management. If the results are acceptable, the project can go on to stage II. If the results are not acceptable because of process instability or hardware reliability problems, the project team needs to correct the problems through process and/or hardware improvement.

- **Process & hardware improvement.** If hardware reliability problems keep the process equipment from attaining manufacturing requirements, hardware improvement must be considered. An equipment reliability improvement methodology based on the Motorola IRONMAN™ technique (Improve Reliability Of New Machines At Night) is described in stage III. If the process is unstable, a designed experiment, described in stage II, is typically performed. Once the reliability and/or stability issue has been resolved, another PDC is performed to confirm the improvement.

Stage II: Characterize and optimize process and verify stability. The objectives of this stage are to actively improve the performance of a stable process by exploring and optimizing the process, defining the process window (the region in which the process inputs may vary and still produce acceptable output), and then to ensure its long-term stability. Designed experiments are used to characterize and optimize the process, and statistical process control is used to monitor process stability.

- **Optimize process window.** The most efficient way to explore and optimize a process that depends on several process inputs is to use statistical design of experiments (DOE). The design of experiments methodology uses structured groups of experiments that change several process inputs simultaneously in order to determine cause-and-effect relationships between process inputs (factors) and process outputs (responses). As the process is optimized, the process window is established. The design of experiments methodology is used to identify key factors affecting performance, and to manufacture products that consistently meet specifications by making them robust to external variability. Part 3 gives an introduction to design of experiments, and six case studies.

- **Verify stability of process window.** The stability of the process window needs to be established and monitored. The stability can be verified by performing a passive data collection on the optimized process recipe. Statistical process control (SPC) is used to monitor process stability for an extended period of time. Statistical process control includes the use of control charts, a corrective action plan to handle unusual process or equipment behavior, and estimation of process capability. Part 4 gives an introduction to statistical process control, and three case studies.

Stage III: Improve equipment reliability and demonstrate manufacturability and competitiveness. The objectives of this stage are to improve equipment reliability, and to obtain reliability metrics that show that the process is manufacturable and competitive. An

IRONMAN™ is used to improve equipment reliability, and a marathon is used to obtain reliability metrics. This stage focuses on equipment reliability, not on the engineering process.

- **IRONMAN™.** The IRONMAN™ (Improve Reliability Of New Machines At Night) reliability improvement methodology typically consists of dedicating software and hardware engineering resources for one thousand or more hours of almost continual testing and improvement activities. The test is usually run at the equipment manufacturer's location. Hardware and software improvements based on root-cause analysis of all failures are incorporated during the day and proven at night. The testing cycle begins with a passive cycle debug where mechanical handling problems are investigated and eliminated without actually running the full process. After mechanical problems have been minimized, an active cycle debug simulates actual production, with observed failures driving the reliability improvement process. It is often possible to save time and improve efficiency by simultaneously optimizing the process and performing the IRONMAN (concurrent engineering).

- **Marathon-equipment utilization capability.** A marathon simulates equipment operation in a manufacturing environment. A marathon, or marathon run, consists of using the equipment 24 hours a day, seven days a week, for several weeks. The equipment reliability data generated during the marathon is used to calculate equipment utilization statistics such as mean time between failures and mean time to repair. The utilization statistics give the maximum possible utilization of the equipment, given its downtime characteristics. The cost of ownership model uses the utilization statistics to give the cost of processing one wafer on the equipment. Part 5 gives an introduction to reliability, and three case studies.

- **Final review and report.** At the end, the entire project is reviewed and documented. The results of the project, including identification of the goals that were not reached, are used to define future activities.

Summary

The SEMATECH Qualification Plan is a roadmap that uses statistical methods to efficiently characterize and improve process and equipment performance and reliability. The engineering goals of the Qual Plan include characterizing gauges, characterizing process and equipment performance, optimizing the process, establishing and monitoring process stability, and improving equipment reliability. The main statistical methods used in the Qual Plan to reach these engineering goals include gauge studies, passive data collection, design of experiments, statistical process control, and equipment reliability. These statistical methods, and case studies illustrating their use to reach engineering goals, are given in the remainder of the book. The Qual Plan has been successfully used in the semiconductor industry, and its versatility and generality allow its use in other industries.

References

All SEMATECH statistical publications that are available to the general public can be accessed at the following Internet URL: *http://www.sematech.org*

Biographies

Veronica Czitrom is a Distinguished Member of Technical Staff at Bell Laboratories, Lucent Technologies. She leads consulting and training efforts for the application of statistical methods, primarily design of experiments, to semiconductor processing. Dr. Czitrom was on a two year assignment at SEMATECH. She has published research papers on design of experiments and mixture experiments, and four engineering textbooks. She received a B.A. in physics and an M.S. in engineering from the University of California at Berkeley, and a Ph.D. in mathematics with concentration in statistics from The University of Texas at Austin. She was a faculty member at the University of Texas at San Antonio, and at the Universidad Nacional Autónoma de México.

Karen Horrell provides statistical support and coordinates the statistical training effort at SEMATECH in Austin, Texas. She has also worked as a statistician at Motorola's MOS13 semiconductor manufacturing facility in Austin, Texas. Karen provides statistical consulting support, develops statistical training materials and delivers classes teaching statistical methods. She led the task force that initially reviewed and documented the SEMATECH Qualification Plan in 1992 and was a member of the task force that updated the SEMATECH Qualification Plan in 1995. She graduated from Texas A&M with a B.S. in applied mathematics, a Secondary Education Certification and an M.S. in statistics.

PART 1
GAUGE STUDIES

CHAPTER 1

INTRODUCTION TO GAUGE STUDIES

Veronica Czitrom

INTRODUCTION

Gauges, or measurement tools, are the eyes with which we see engineering processes. As such, they are a critical link in the chain of events that occur as data is turned into information for decision-making. If we can't see clearly what a process is doing, we can't hope to manage it effectively. An inadequate gauge may yield process measurements that bear little relationship to what the process is actually doing. A gauge study is performed to characterize the measurement system in order to help identify areas of improvement in the measurement process, and to ensure that the process signal is not obscured by noise in the measurement.

Gauges are used for a variety of measurements. In Part 1, gauges are used to measure wafer sheet resistance, carbon contamination on a wafer, mass gain of a wafer after deposition of a thin film of tungsten, and pull strength of a wire bonding an integrated circuit to a lead frame.

A case study in Part 4 mentions difficulties in data analysis due to problems with gauges. Watson (Chapter 23) found that poor gauge resolution incorrectly gave observations of zero during a Statistical Process Control implementation.

The next section of this introduction to gauge studies defines some key terms, the following section gives a brief description of the case studies given in the remainder of Part 1, and the last section gives suggestions for further reading on gauge studies.

DEFINITION OF KEY TERMS

A *measurement system* consists of the measurement tool or gauge (including software), and of the procedure that is used to take a measurement (including operator handling, set-up techniques, off-line calculations and data entry, and calibration frequency and technique). A *gauge study*, or *Gauge Repeatability and Reproducibility (GRR) study*, can be used to estimate the total variability or error inherent in the measurement process, provide information on the magnitude and component sources of measurement error, estimate the closeness of the measurements to the "true" value they should have, determine the adequacy of the gauge for the intended purpose, and check gauge stability over time. All this is part of characterizing the measurement system.

Two major concepts of measurement process characterization are the *location* of the measurements and the *variability* of the measurements. It is important for the location (middle of the data, or central tendency) of the measurements to be close to the intended target value, since the measurements should not be consistently too high or consistently too low. On the other hand, measurements have inherent variability: ten measurements of film thickness at the same site on the same wafer will probably give ten slightly different readings. The variability in the observed measurements is the composite of two effects: variability in the engineering process being measured, and variability from the measurement process itself. The variability in the measurement process comes from a variety of sources such as operator technique, ambient temperature and humidity, and tool wear. A gauge study can be used to reduce the variability of the measurement process so that the engineering process can be adequately measured. A gauge study can also be used to estimate the variability of the measurement process, which can be used to estimate the *real* variability of the engineering process from the *measured* variability of the engineering process. Two passive data collection case studies in Part 2 (Czitrom and Reece, and Hall and Carpenter) use analysis of variance techniques to separate the total measurement error from variability in the process. The concepts of precision, repeatability, and reproducibility relate to the variability of the observations, and the concepts of accuracy, bias, and calibration relate to the location of the observations.

The *precision* of a gauge is the total variability in the measurement system, or total measurement error. Precision is given by the standard deviation $\sigma_{measurement}$ of the measurements. The smaller the standard deviation, the higher the precision. Precision can be separated into two components, repeatability and reproducibility: $\sigma^2_{measurement} = \sigma^2_{repeatability} + \sigma^2_{reproducibility}$.

Repeatability is the variability inherent in the measurement system. Repeatability, given by the standard deviation $\sigma_{repeatability}$, is short-term variability under identical measurement conditions (e.g., same operator, set-up procedure, test wafer, wafer site, and environmental conditions). Repeatability can be estimated using thirty or more repeated measurements taken back-to-back under identical measurement conditions.

Reproducibility is the variability that results from using the measurement system to make measurements under different conditions of normal use. Reproducibility, given by the standard deviation $\sigma_{reproducibility}$, is long-term variability that captures the changes in measurement conditions (e.g., different operators, set-up procedures, test wafers, wafer sites, and environmental conditions). Reproducibility can be estimated using an experiment where the potential sources of variation are changed systematically

Mitchell, Hegemann, and Liu present a typical gauge study in Chapter 5. The gauge study consists of *n* samples, with *k* readings per sample (repeatability) taken by each one of *m* operators (reproducibility). Repeatability and reproducibility are estimated using analysis of variance (ANOVA) techniques. Analysis of variance is considered in the glossary at the end of the book. This typical gauge study is considered by Buckner et al. in Chapter 4, with $n = 10$ wafers (samples), $k = 2$ readings per wafer, and $m = 3$ operators.

A measurement process is *accurate* if the measurements are close to a true or target value. The measurements should be in agreement with the "true" value of standards if standards are available, for example from the National Institute of Standards and Technology, NIST. If standards are not available, a reference product can be created and used throughout the project. The measurements should be close to the target value of the reference product. This allows tracking tool drift over time. *Bias*, which is the difference between the average value of all the measurements and the true or target value, expresses the extent to which the measurements deviate from the desired value. A gauge needs to be *calibrated* so that it is accurate (small bias). Calibration is the procedure used to adjust the measurements from a gauge so that they are close to the true or target values, which may cover a range of values. For example, Pankratz in his case study uses five wafer specimens with known carbon contaminations ranging in value from 0.0158 to 2.2403 ppma, to set the tool parameters so that the gauge reads the specimens correctly, and, by interpolation, the remaining values in the range that was considered.

Capability is the ability of the gauge to give measurements that are satisfactory for the intended purpose. Several indices have been developed to measure gauge capability. One frequently used index is the *precision over tolerance (P/T) ratio*, a capability ratio that compares the measurement error to the specification window of the process. The P/T ratio, which is the fraction of the process window that is used up by measurement error, is defined as

$$P/T = \frac{6\sigma_{measurement}}{USL - LSL},$$

where *USL* is the *Upper Specification Limit* for the process, *LSL* is the *Lower Specification Limit* for the process, the *process tolerance* is the difference USL − LSL, and the specification interval goes from LSL to USL. P/T is often expressed as a percentage by multiplying the above quotient by 100. It is desirable for P/T to be small, in which case only a small fraction of the tolerance of the process is used up by the measurement variability. As a rule of thumb, a measurement tool is judged capable of measuring a process if the corresponding P/T ratio is less than 0.3 (30%). If the gauge is not capable a different gauge can be used, or n measurements can be taken and averaged to increase the precision since the average has smaller variability than the individual observations (the standard deviation of the average of *n* independent measurements is $\sigma_{measurement}/\sqrt{n}$, which is smaller than the standard deviation $\sigma_{measurement}$ of one measurement). Pankratz uses the average of several measurements near the detection limit of the gauge to increase the precision of the reported value.

A measurement process is *stable* if it does not exhibit unusual changes over time, such as sudden shifts, drifts, and cycles. This means the location and the variability of the distribution of the measurements remains constant and predictable over time. Stability can be evaluated visually by constructing a trend chart, a plot of the observations in time sequence. For example, Buckner et al. (Chapter 4) detected a warm-up shift in a microbalance by plotting observations (in box plots, grouped by operator) in chronological order.

Once the tool is deemed adequate to perform the intended measurements, its behavior should be monitored using *control charts* (see Part 4, on Statistical Process Control). Pankratz detected an abrupt shift in carbon contamination measurements using a control chart.

Gauge studies should be carefully planned, including provisions for unexpected events. Several of the case studies point out examples where the gauge study did not proceed as expected. Pankratz selected a simple experimental design expressly to minimize mistakes in the execution of the experiment. He also indicates the importance of looking at the data as soon as possible, since looking at SPC charts earlier would have allowed him earlier detection of an abrupt shift in the measurements. Buckner et al. (Chapter 4) found that miscommunication led to a mistake in the data collection (repeatability measurements that should have been taken consecutively were separated by over half a day), which did not allow the intended separation of repeatability and reproducibility. In fact, the main result of their case study is the discovery and correction of a practical problem, the warm-up effect of a gauge.

INTRODUCTION TO CASE STUDIES

This section gives a brief overview of the case studies presented in Part 1 with applications of gauge capability studies.

Chapter 2. Prometrix RS35e Gauge Study in Five Two-Level Factors and One Three-Level Factor. *James Buckner, Barry Chin, and Jon Henri.*

This gauge study considers a four-point probe, a tool frequently used in the semiconductor industry to measure thin film sheet resistance. The case study illustrates the use of P/T ratios to determine the adequacy of the measurement tool for the intended purpose, in this case the measurement of tungsten in an Equipment Improvement Project at SEMATECH. The experimental design (a 16 run fractional factorial modified to accommodate a three-level factor as well as five two-level factors), is more complex than the standard full factorials commonly used for gauge studies. The results of the designed experiment were used to estimate the wafer-to-wafer and within-wafer sheet resistance uniformity, from which the corresponding P/T ratios were computed. These ratios were found to be acceptable. The measurements were not significantly affected by the six factors considered in the experiment. The case study gives a good engineering introduction to the use of four-point probes to measure sheet resistance, and to the reasons used to select the factors in the designed experiment.

Chapter 3. Calibration of an FTIR Spectrometer for Measuring Carbon. *Peter C. Pankratz.*

This case study considers the calibration of a Fourier transform infrared spectrometer used to measure carbon contamination on silicon wafers. The carbon contamination measurements from the spectrometer were calibrated to agree with the "true" carbon contamination values of five specimens with known and different amounts of carbon contamination. Design of experiments, statistical process control and regression are used in the calibration. An important issue considered is how to report measurements near the detection limit of a gauge. Pankratz uses the average of several measurements of a specimen, instead of one measurement, to increase the precision of the reported value.

Chapter 4. Revelation of a Microbalance Warm-Up Effect. *James Buckner, Barry Chin, Todd Green, and Jon Henri.*

This case study describes the discovery that the microbalance used to monitor the wafer-to-wafer uniformity of a tungsten thin film deposition needs to warm up before it will give accurate readings. Warm-up effects are encountered frequently in the semiconductor industry, and are not limited to gauges. The warm-up effect, which was found by displaying the data graphically using box plots, was eliminated by not turning the gauge off. When the observations taken while the gauge was warming up were removed, the gauge was found to be only marginally capable.

Chapter 5. GRR Methodology for Destructive Testing and Quantitative Assessment of Gauge Capability for One-Side Specifications. *Teresa Mitchell, Victor Hegemann, and K.C. Liu.*

This case study deals with a very important and common problem: estimation of gauge repeatability for a destructive measurement, which destroys the sample to be measured in the process of measuring it. Two related nested designed experiments are used to estimate repeatability. The case study also proposes a new estimate of measurement system capability in the case of one-sided specifications. The calculations of repeatability and system capability are illustrated for the bond pull test for the bond between a die and a lead frame.

REFERENCES AND FURTHER READING

All SEMATECH statistical publications that are available to the general public can be accessed from the following Internet URL: http://www.sematech.org/

Several topics in gauge studies are introduced by:
> Montgomery, Douglas C. *Statistical Quality Control*, second edition. New York: John Wiley and Sons, Inc., 1991. Section 9-5.

Introductory books on gauge studies are:
> Wheeler, Donald J. and Lyday, Richard W. *Evaluating the Measurement Process*, second edition. Knoxville, Tennessee: SPC Press, Inc., 1989.
> ASQC Automotive Industry Action Group. Measurement Systems Analysis Reference Manual. American Society for Quality Control, 1990.

The following books describe variance components:
> Box, George E.P., Hunter, William G., and Hunter, Stuart J. *Statistics for Experimenters*. New York: John Wiley and Sons, 1978. Pages 571 to 582.
> Leitnaker, Mary G., Sanders, Richard D., and Hild, Cheryl. *The Power of Statistical Thinking— Improving Industrial Processes*, Massachusetts: Addison–Wesley Publishing Co., 1996.
> Montgomery, Douglas C. *Design and Analysis of Experiments*, fourth edition. New York: John Wiley and Sons, Inc., 1997.
> Graybill, Richard K., and Graybill, Franklin A. *Confidence Intervals on Variance Components*. New York: Marcel Dekker, Inc., 1992.

CHAPTER 2

PROMETRIX RS35e GAUGE STUDY IN FIVE TWO-LEVEL FACTORS AND ONE THREE-LEVEL FACTOR

James Buckner, Barry L. Chin, and Jon Henri

The total measurement precision (variability) of a Prometrix RS35e automatic four-point prober tool was determined for two responses: wafer-to-wafer and within-wafer sheet resistance uniformity. Five two-level factors and one three-level factor were used. The precision of the tool was found to be adequate for monitoring the progress and completion of an improvement project driving to a goal of 2% for both types of uniformity. The measurement is not significantly affected by any of the six factors studied: operator, cassette type, slot loading, prior film, C2C station, and probe conditioning.

EXECUTIVE SUMMARY

Problem

Improvement of wafer-to-wafer and within-wafer sheet resistance uniformity is one of the goals of the SEMATECH Equipment Improvement Project (EIP) on the Novellus Concept One-W tungsten deposition tool. Because of the tight uniformity objectives (2% σ for each), it was necessary to determine whether the total measurement precision of the Prometrix RS35e four-point prober is adequate to monitor the progress of this project goal.

Solution Strategy

A gauge study, described in this case study, was performed on the Prometrix. The experimental design in 16 treatment combinations accommodated five two-level factors and one three-level factor which were identified as potential sources of variation in the measurement of sheet resistance. This experimental design was used to measure the effect of each factor on the responses, as well as the total measurement precision of the prober.

Conclusions

The Prometrix RS35e is adequate to monitor the progress and completion of the uniformity goals in the EIP. It displays precision-to-tolerance ratios much better than the requirement of 0.3 for both responses (wafer-to-wafer and within-wafer uniformity) when operated under normal conditions.

The measurement is not significantly affected by any of the six factors studied: operator, cassette type, slot loading, prior film, C2C station, and probe conditioning.

PROCESS

A number of critical processes in the manufacture of integrated circuits are monitored and controlled using measurements of thin film sheet resistance. Prometrix automatic four-point probers are ubiquitous instruments that perform this measurement task. These probers operate by setting down four collinear electrodes onto the surface of the film to be measured. An electric potential is applied to two of the electrodes, and the resulting electric current flowing through the other two is measured. The sheet resistance is extracted from the theoretical relationship between the two and adjusted by an empirically determined coefficient that is dependent upon the geometry of the probes and film.

Automatic four-point probers typically measure numerous sites on the sample wafer according to any of several standard patterns or by a pattern defined by the user. It is common to use this capability to measure uniformity of the thin film across the wafer. This is commonly referred to as within-wafer uniformity, and it is defined as the sample standard deviation of the site sheet resistance values divided by the sample mean multiplied by one hundred.

The particular four-point prober that is the subject of this study is a Prometrix model RS35e, located in the applications development laboratory at Novellus Systems, Inc. It is primarily used to measure the sheet resistance uniformity of tungsten thin films. A measurement capability study was required to determine whether the tool could be used to adequately assess the progress and completion of a SEMATECH Equipment Improvement Project (EIP) on the Novellus Concept One-W CVD tungsten deposition reactor.[1] The relevant EIP process tolerance goals are summarized in Table 2.1.

DATA COLLECTION PLAN

Since the measurements performed by the RS35e would be pivotal to the success of the EIP, the project team elected to explicitly include all suspected sources of variation as

---

[1] See Shah (1992) and (1993) for a full report on the Novellus Concept One-W Equipment Improvement Program.

factors in the experimental design. These are summarized in Table 2.2. The rationale for selecting these factors is explained as follows.

Table 2.1. *Novellus Concept One-W Equipment Improvement Project Goals for Sheet Resistance and Uniformity. The gauge study was performed to determine whether the measurement system was capable of monitoring progress toward these process tolerance goals.*

Response	Tolerance Goal
Sheet resistance	2%
Uniformity	2%

Table 2.2. *Factors and Levels. Note that the first factor has three levels.*

Factor	Levels
Operator	Larry, Curly, Moe
Cassette Type	A, B
Slot Loading	Top, Bottom
Prior Film	W, TiN
C2C Station	Front, Back
Probe Conditioning	Yes, No

Even though this tool is highly automated, it was still conceivable to the team that there could be some effect associated with the operator. All three operators who normally use the tool would be tested, requiring a three-level factor. Because of the variety of functions performed in this laboratory, it is common to use two different types of wafer holding cassettes. This appears as the second factor in Table 2.2, at two levels. The automated wafer handler, called the C2C, may pull a wafer from any of the 25 slots in the wafer cassette. In order to test for a possible positional effect, slot loading was included as a two-level factor, where the first level was to load the test wafers at the bottom of the cassette and the second level was to load the test wafers at the top of the cassette.

The tool is used to measure other films besides the tungsten films of interest in the EIP. The team believed it might be possible for prior measurement of one of these other films to affect the measurement of subsequent tungsten films, perhaps through the effect of residues of the prior film adhering to the probe tips. This possibility was included as a two-level factor by measuring both tungsten and titanium nitride, altering the order in different treatment combinations, and always taking the response from the tungsten film. In this way, half the tungsten measurements were preceded by another tungsten measurement, and half were preceded by a titanium nitride measurement.

Either of two cassette stations may be used on the C2C automated wafer handler. To test for a possible positioning effect, C2C Station was included as a two-level factor.

Finally, tool maintenance requires periodic conditioning of the probe tips by touching them down numerous times on a ceramic block to abrade away any residues of the measured films and to expose fresh electrode metal. Since this procedure is supposed to reset tool accuracy, the team decided to explicitly look for an effect by including pre-measurement probe conditioning as a factor. It was thought that there might be both a short lived effect and a long lasting effect of probe conditioning. Therefore, it was appropriate to include probe conditioning before each measurement to capture short lived effects. But it was inappropriate to randomly distribute the probe conditioning runs through the experiment, since long lived effects would then also affect subsequent treatment combinations intended to be run without probe conditioning. Therefore, the probe conditioning runs were blocked into the last half of the experiment so that the first half would be influenced by the full force of any effect that might arise from lack of probe conditioning. To clarify, the first eight treatment combinations were run without probe conditioning, while each of the second eight was preceded by a probe conditioning procedure, so that the probes were conditioned a total of eight times during the course of the experiment.

The experimental design is shown in Table 2.3. It was set up as an Addleman design[2] because of the three level factor. Note that operator Curly has twice the work of the other two operators. Indeed, "Curly" is the primary operator of the tool in the day-to-day activity of the laboratory. Note also that the run order (given in the first column of Table 2.3) has been randomized, except for the probe conditioning factor, whose run order has been blocked for the reason explained above. See the Appendix for a more detailed perspective regarding the choice of this design.

To determine whether replications were needed, a sample size calculation was performed.[3] The null hypothesis was that the factor would cause no effect and the alternative hypothesis was that the factor would cause an effect. The desired probability of falsely finding an effect was set at $\alpha = 0.10$, while the desired probability of falsely finding no effect was set at $\beta = 0.05$. The size of the effect judged to be important from an engineering perspective was set at $\delta = 3\sigma$. The calculation yielded a sample size of $2^4 = 16$. Since the "hidden replication" in the design provides eight values at each level of each two-level factor, and a minimum of four values at each level of the three-level factor, no additional design replication was required. In other words, it was adequate to use one wafer at each treatment combination.

The design shown in Table 2.3 was run over a period of five days to capture, in addition to the sources of variation explicitly listed, any other sources that might appear in the course of several workday cycles. This was a mistake in strategy, since a source of variation in workday cycles could have overwhelmed the sources of variation explicitly included as factors. The right design and analysis should have allowed it to be separated out. A better strategy would have been to run the design in as short a period as possible and then to look for variation over time in another phase of the study, or by repeating a few of the runs several days later. Fortunately, the strategic blunder did not result in a failure to meet the

---

[2] See Diamond (1989) for a treatment of Addleman designs. In this case, the Addleman design resulted from modifying an experimental design in 16 treatment combinations for factors at two levels each (a $2^{(6-2)}$ design of resolution IV). Two columns (and their interaction) were combined to accommodate the three-level factor.

[3] See Diamond (1989), pp. 27 ff.

objective, which was to determine whether the measurement system was capable of monitoring the EIP goals (see the Analysis and Interpretation of Results section).

Table 2.3. *Experimental Design. The three-level factor Operator is proportionately balanced. The design is shown in standard order, but the run order (first column) has been randomized, except that the Probe Conditioning run order has been blocked.*

Run	Operator	Probe Condition-ing	Cass Type	Slot Loading	Prior Film	C2C Station
9	Larry	Yes	A	Bottom	TiN	Front
1	Larry	No	A	Top	W	Back
16	Larry	Yes	B	Top	TiN	Front
8	Larry	No	B	Bottom	W	Back
10	Curly	Yes	A	Top	W	Front
2	Curly	No	A	Bottom	W	Front
12	Curly	Yes	B	Bottom	W	Front
6	Curly	No	B	Top	W	Front
14	Curly	Yes	A	Top	TiN	Back
3	Curly	No	A	Bottom	TiN	Back
13	Curly	Yes	B	Bottom	TiN	Back
7	Curly	No	B	Top	TiN	Back
11	Moe	Yes	A	Bottom	W	Back
4	Moe	No	A	Top	TiN	Front
15	Moe	Yes	B	Top	W	Back
5	Moe	No	B	Bottom	TiN	Front

A Prometrix standard 49-point measurement site pattern was used at a measurement diameter of 179 mm on the 200 mm wafer. It should be understood that the EIP goals are stated in terms of the process tolerance for the average and the coefficient of variation of the measurements from these 49 sites on the wafer. Therefore, the gauge study was designed to determine the precision with which these two statistics are obtained. It was not an objective of the study to measure the precision of the individual site measurements.

ANALYSIS AND INTERPRETATION OF RESULTS

The raw sheet resistance and sheet resistance uniformity for the 16 runs are shown in Table 2.4. The column of data from the tungsten test wafer is labeled "7 kÅ W." The other columns of data are from the 3 kÅ tungsten wafer and the TiN wafer used as the Prior Films. The last two rows give the average and the standard deviation of the corresponding columns.

Although the design of the experiment was complex, the analysis was simple, especially since the standard deviations had such low values. The objective of the study was to determine whether the tool could adequately assess the project goals shown in Table 2.1.

The standard deviations in the last row of Table 2.4 represent estimates of the total measurement precision under all potential sources of variation. These values were used with the values of Table 2.1 to compute precision-to-tolerance (P/T) ratios as shown in Table 2.5. The ratios are much better than the value of 0.3 normally considered acceptable.

Table 2.4. *Raw Data. The sheet resistance and uniformity data are shown for the 7 kÅ tungsten film under test, and for the 3 kÅ tungsten and TiN films used for prior measurements.*

Run	Sheet Resistance (mΩ)			Uniformity (%)		
	3kÅ W	7kÅ W	TiN	3kÅ W	7kÅ W	TiN
1	279.9	84.86	13550	1.76	1.29	4.59
2	280.0	84.92	13550	1.69	1.25	4.58
3	279.9	84.81	13550	1.78	1.28	4.60
4	279.7	84.80	13540	1.76	1.28	4.59
5	279.9	84.86	13550	1.79	1.28	4.61
6	280.2	84.93	13550	1.77	1.28	4.59
7	279.8	84.80	13540	1.76	1.29	4.60
8	280.1	84.94	13550	1.75	1.28	4.59
9	280.1	84.91	13550	1.78	1.28	4.59
10	279.9	84.86	13550	1.72	1.29	4.59
11	279.9	84.78	13550	1.87	1.31	4.60
12	279.8	84.86	13550	1.73	1.27	4.63
13	280.2	84.96	13560	1.72	1.28	4.59
14	280.0	84.89	13550	1.73	1.28	4.59
15	280.0	84.90	13550	1.72	1.27	4.58
16	279.9	84.90	13550	1.74	1.27	4.59
Average	280.0	84.87	13550	1.75	1.28	4.59
Std dev	0.14	0.055	4	0.041	0.013	0.012

Table 2.5. *Precision-to-Tolerance Ratios for the 7 kÅ Tungsten Film. The ratios are much better than the 0.3 normally considered acceptable.*

Response	Precision	Tolerance	P/T
Sheet resistance	0.055%	2%	0.03
Uniformity	0.013%	2%	0.007

Having established that the total measurement precision is adequate, the objective of the gauge study was met with no further analysis. However, for completeness of this case study, an analysis of variance was performed. The results are summarized in Tables 2.6a and 2.6b for the sheet resistance and uniformity responses, respectively. None of the factors was found to have a significant effect on either response.[4]

Table 2.6a. *ANOVA Table for Sheet Resistance. None of the factors had a significant effect on the sheet resistance response.*

Source of Variation	Degrees of Freedom	Sums of Squares	Mean Squares	F Value	Pr > F
Operator	2	0.00951250	0.00475625	1.50	0.2788
Cassette Type	1	0.00640000	0.00640000	2.02	0.1926
Slot Loading	1	0.00062500	0.00062500	0.20	0.6683
Prior Film	1	0.00090000	0.00090000	0.28	0.6081
C2C Station	1	0.00062500	0.00062500	0.20	0.6683
Probe Cond'g	1	0.00122500	0.00122500	0.39	0.5509
Model	7	0.01928750	0.00275536	0.87	0.5654
Error	8	0.02528750	0.00316094		
Total	15	0.04457500			

Table 2.6b. *ANOVA Table for Uniformity. None of the factors had a significant effect on the uniformity response.*

Source of Variation	Degrees of Freedom	Sums of Squares	Mean Squares	F Value	Pr > F
Operator	2	0.00015000	0.00007500	0.35	0.7130
Cassette Type	1	0.00010000	0.00010000	0.47	0.5121
Slot Loading	1	0.00002500	0.00002500	0.12	0.7404
Prior Film	1	0.00000000	0.00000000	0.00	1.0000
C2C Station	1	0.00040000	0.00040000	1.88	0.2073
Probe Cond'g	1	0.00002500	0.00002500	0.12	0.7404
Model	7	0.00070000	0.00010000	0.47	0.8317
Error	8	0.00170000	0.00021250		
Total	15	0.00240000			

---

[4] The Addleman design has a more complex confounding scheme than ordinary fractional factorials, but since no factors appeared important, confounding was not considered in this application.

Conclusions and Recommendations

The Prometrix RS35e four-point prober is adequate to monitor the progress and completion of the Novellus Concept One-W EIP with precision-to-tolerance ratios much better than 0.3 in both types of uniformity responses (wafer-to-wafer and within-wafer uniformity).

Under normal operation, the response of the instrument is not significantly affected by operator, cassette type, slot loading, prior film, C2C station, or probe conditioning.

References

Diamond, William J. *Practical Experiment Designs for Engineers and Scientists*. 2nd ed. New York, NY: Van Nostrand Reinhold, 1989.

Shah, Raj, ed. *Novellus Concept One-W Characterization Interim Report*. 92051143A-ENG. Austin, TX: SEMATECH. 1992.

Shah, Raj, ed. *Novellus Concept One-W EIP Project Final Report: E36*. 93011464A-ENG. Austin, TX: SEMATECH. 1993.

Appendix

Perspective on the Choice of The Addleman Design

For the one three-level factor and five two-level factors identified by the project team, a full factorial design would have required $3 \times 2^5 = 96$ treatment combinations. This number of treatment combinations was unacceptably high for the team.

Since the measurement instrument under study is highly automated, the team felt it was highly unlikely that more than one or two of the six factors would prove to be important. Hence it was decided to forgo resolving interactions and to look for main effects only. Possible interactions could be resolved at a later time if needed. A resolution III design in the five two-level factors could be obtained, and the number of treatment combinations reduced, by using a one-quarter fractional factorial in the two-level factors. The number of treatment combinations would now be reduced to

$$3 \times 2^{5-2} = 24.$$

In this design, each of the three operators would have been running an independent replicate of a one-quarter fractional factorial in the five two-level factors. However, replicates were not needed to estimate the effects arising from the two-level factors (see the sample size discussion in the Data Collection Plan), nor were all eight values (from the 2^{5-2} fractional factorial hidden replicates) needed for estimation of operator effects.

The Addleman design in 16 treatment combinations has eight treatment combinations less than the previous experiment.

The Addleman design was derived from a Resolution IV 2^{6-2} matrix by assigning the three level factor to the 234 and 235 interactions, and by assigning the factor 6 to the 2345 interaction. The resulting design was found to be Resolution III by the method described in Diamond (1989).

BIOGRAPHIES

James Buckner has been a technologist with Texas Instruments since 1987, where he has been responsible for metal sputtering, CVD-tungsten deposition, and improvement of CVD-tungsten deposition equipment at SEMATECH. He is currently responsible for the Equipment Migration program in Texas Instruments' Manufacturing Science and Technology Center. James has a Ph.D. in physical chemistry from the University of North Carolina at Chapel Hill.

Barry L. Chin received his B.S. degree from Cornell University and the M.S. and Ph.D. degrees from the University of California at Berkeley. Since 1980 he has been involved in thin film deposition, plasma etching, and process integration programs at several semiconductor manufacturing companies. He joined Novellus Systems in 1987 where he has worked on the product development of Plasma TEOS and CVD-Tungsten deposition systems.

Jon Henri is an applications engineer at Novellus Systems. He works on CVD-Tungsten deposition systems.

CHAPTER 3

CALIBRATION OF AN FTIR SPECTROMETER FOR MEASURING CARBON

Peter C. Pankratz

The following case study shows how design of experiments, statistical process control, and ordinary least squares regression each play a role in the calibration of a Digilab QS-200 FTIR apparatus. Detection limits for the apparatus are calculated. Confidence limits for the unknown measured quantity are also determined.

EXECUTIVE SUMMARY

Problem

Installing a new measurement apparatus and calibrating it so that it produces verifiable results can be a daunting procedure that requires a large number of steps. This case study shows how design of experiments (DOE), statistical process control (SPC), and ordinary least squares regression each play a role in the calibration process. How to use the information generated during the calibration process to establish detection limits for the measurement process will also be shown. Confidence limits for the unknown measured quantity are also determined.

The instrument described here is a Digilab QS-200 Fourier Transform Infrared (FTIR) spectrometer which was acquired by our applications laboratory then modified in order to measure carbon contamination in silicon wafers. The statistical methods that are described here could easily apply to any number of measurement devices.

The goals of the calibration study were to show that the Digilab QS-200 was operating properly and that sufficient controls were in place to provide reasonable assurance that the machine's performance did not vary during its use. We also needed to provide confidence limits for what the true value of carbon would be for a measurement made using the FTIR. Additionally, three kinds of statistical limits needed for the routine operation of the apparatus were computed: calibration limits, detection limits, and SPC limits. Calibration limits were provided to allow specification of a confidence interval for any measurement. These limits were provided as a function of the number of times a specimen was measured in order to allow the laboratory to exploit the fact that precision increases with additional measurements on the same specimen. In a similar fashion, the detection limit for carbon was also calculated as a quantity that improves with repeated measurements. As before, this gives the laboratory some flexibility, especially when working with specimens containing only trace amounts of carbon. Finally, SPC limits for standard specimens were calculated from the data. Two standards were measured routinely each day in order to detect changes in the way carbon measurements were reported by the FTIR. Using the data from the calibration experiment allows a direct link between the SPC procedure and the original calibration.

Solution Strategy

The strategy to achieve the goals of the study used the following steps:
1. examination of basic assumptions
2. design of a calibration experiment
3. execution of the calibration experiment
4. estimation of the calibration curve and its corresponding calibration limits
5. calculation of detection limits
6. reporting of measurements
7. using routine SPC on daily measurements to assure consistent performance capability.

Conclusions

The goals of the study were achieved. When the FTIR repeatedly measured a set of specimens, it was able to report measurements that were predictable. That is, with one exception which will be discussed later, repeat measurements on a set of standard specimens could be shown to have a constant mean and constant variance. This result was established for five specimens over ten days of testing with 50 measurements per specimen. Using x^* (LCL, UCL) as a way to report measurements below the detection limit of the gauge so that all relevant information was retained, revealed the full quality of the reported measurements, including data that would have been censored by other methods.

PROCESS

Calibration Defined

We need to carefully define what we mean by calibration since the word is routinely used to describe a variety of different procedures. For purposes of this case study, calibration is the estimation of the relationship between measurements of specimens and their known properties. In this case, we had a number of silicon specimens for which we knew the carbon contained in each. These were measured on our FTIR multiple times over several days. Our objective was to determine an equation that we could use to adjust any future measurements so that any discrepancy we found between the measured values and the known values would be eliminated.

The Michelson Interferometer

The Digilab QS-200 FTIR is built using the principle of the Michelson Interferometer (see Figure 3.1) as is common practice for most infrared spectrometers (see Strobel, 1989). The Michelson Interferometer works by directing a beam of infrared light covering a broad frequency range at a thin semi-reflective mirror called a beamsplitter. Approximately half of the original light beam is reflected by the beamsplitter and sent to a fixed mirror. The remaining light is allowed to pass through the beamsplitter where it strikes a mirror that can be moved. Both halves of the beam are then reflected back to the beamsplitter where they are recombined.

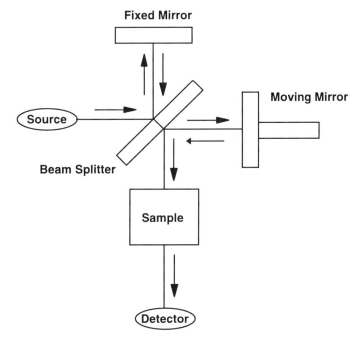

Figure 3.1. *Schematic Diagram of a Michelson Interferometer.*

As the mirror moves, the intensity of the recombined beam at each optical frequency changes. By continuously adjusting the distance between the beamsplitter and the movable mirror, it is possible to establish interference patterns in the recombined beam. The frequency of the change depends upon the distance between the movable mirror and the beamsplitter as well as on the speed at which the mirror is moving. Fourier transforms are used to convert the intensity-versus-mirror position information into a plot of absorptance-versus-wave number, the reciprocal of frequency. This is called a single beam spectrum. At some frequencies, the energy is absorbed by the sample. By comparing the single beam spectrum that is produced when the energy beam is directed through a sample with a reference single beam spectrum, the amount of a chemical species of interest (such as carbon) present in the sample can be determined. An absorptance spectrum is obtained in this manner. The logarithm (base 10) of the ratio of the reference absorptance spectrum divided by the sample absorptance spectrum is calculated at each wave number. The height of peaks on the absorptance spectrum are associated with the amount of chemical species present in the sample.

Statistical Considerations

The preceding description of the Digilab QS-200 FTIR brings up some considerations which affect the statistical model we use. They are examined in the following few paragraphs.

The response from an FTIR is not a single value. An FTIR measures absorptance as a function of wave number. The response is therefore multivariate in nature. However, for this case study, we were only interested in the response at a single wave number that corresponds to carbon content. This simplified the problem to one involving a univariate response.

As with any measurement apparatus, there is some amount of variability intrinsic to FTIR measurements. Because of this, when a sample contains sufficiently small amounts of carbon, it is quite common for the apparatus to report negative values. The reference used for this study was taken from float zone silicon containing only a small amount of carbon. Our best estimate of the carbon contained in the float zone reference was 0.062 parts per million atomic (ppma). Standard specimens, each with an assigned carbon content, were available to this study from the company quality assurance (QA) department. These were used to calibrate other FTIRs in the company, as well as the one described here. The response of interest (carbon) was not assumed to be constant at all points on the specimen. All measurements were made, as closely as possible, at the center of each specimen. The engineer in charge of making the measurements fabricated a device to hold the specimen so that the error associated with locating its center would not be a factor. One measurement on the Digilab QS-200 is actually the average of some set number of scans, where each scan represents a reading of carbon content without moving the specimen. One measurement at n scans is really n measurements taken with no variation contributed from our inability to locate precisely the same spot on the specimen each time.

Before a calibration experiment could be designed, it was essential to examine the various assumptions we made when we developed a statistical model of the measurement process.

The Calibration Model

We seemed to have an example of the standard calibration model (alternatively referred to as an inverse regression model) that appears in the statistical literature (see Strobel,

1989). We had an independent variable (x) that was assumed to be measured subject to little or no measurement variability: the QA standards. We had a dependent variable (y), the carbon reported by the FTIR, which could be assumed to have variability associated with it. The objective was to determine an equation that would allow us to convert the carbon content reported by the FTIR into a value that is calibrated to our reference specimens.

Linear Model, Normal Errors, and Constant Variance

We were working with measured carbon versus known carbon, and so a straight line seemed to be the most reasonable way to expect the measured y to be related to the known x. The statistical model was $y_i = \beta_o + \beta_i x_i + \varepsilon_i; \; i = 1, 2, \ldots, n$. The normal probability distribution was used as a reference distribution for the ε_i's and for the y_i's. We initially assumed that the ε_i's variance would be constant regardless of x. These assumptions were checked after the data were collected.

Independent Errors

In a calibration model, the ε_i's are assumed to be independent of each other. Each realization of ε_i cannot be affected by any other. For the FTIR, this restriction was not one we could easily assume. Time of day could have affected the measurements in the way that room temperature and humidity change with differing levels of activity in the room, the weather outside, etc.

The approach we used was not to build a more sophisticated statistical model to incorporate temperature and humidity. Instead, we chose to try to understand the extent of the effect due to temperature and humidity and then to alter the laboratory environment so that the temperature and humidity changes were too small to affect the FTIR's operation. The work to control temperature and humidity was performed as a separate study.

Representative Data and Stability of the Process

Implicit in all this was the assumption that the measurements we eventually generated were a random sample taken from the hypothetical population of all future measurements made on the equipment, representative of all conditions under which the apparatus would be expected to run. Another assumption was that the measurement process was stable, now and in the future. Without perfect stability, the calibration intervals (which are the final product of the measurement process) are too narrow since the statistical model represents only a portion of the total process variability. This aspect was addressed by implementing SPC on the measurement process once the FTIR was calibrated.

DATA COLLECTION PLAN

The design of a calibration experiment is an especially crucial step. It defines the range of measurement values for which the device is to be used as well as the precision with which each measurement can be reported. Initially, the plan was to select five double side polished standard specimens (2 mm slugs) with carbon values evenly spaced over the range of interest, from as large as 2.2 ppma down to as small as the detection limits of the FTIR.

Unfortunately, the standards that were available did not entirely conform to the desired pattern and some compromises were made. Eventually two low carbon standards were selected, and three others were chosen so that they were equally spaced over the desired range. This was done recognizing that the additional low carbon specimen would tend to narrow the width of the calibration limits at the low end of the carbon scale. This was considered possibly fortuitous because once the FTIR was operational, the majority of its required carbon measurements would be from specimens with only trace amounts of carbon.

Table 3.1. *Carbon Specimens Selected for the Calibration Experiment.*

Sample ID	Assigned Carbon, ppma
A	0.01579
B	0.02994
C	0.37096
D	1.14472
E	2.24033

The assigned carbon content for each of the five standard specimens is listed in Table 3.1 and shown graphically in Figure 3.2. For the duration of the study, these values were treated as though they were known values subject to no variability.

After choosing the levels for the standard specimens, the remainder of the experimental design was extremely simple as can be seen from Appendix A: five measurements per standard were taken every day for ten days. This statement is by no means an apology since simplicity was an important consideration in selecting an appropriate design for the gauge study. This was to assure that all tests would be conducted as specified, with no mistakes introduced by the operator during the tedious repetition of measurements.

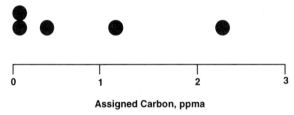

Figure 3.2. *Schematic Representation of the Carbon Levels Used in the Calibration Experiment. The experiment was designed to give more weight to low carbon levels.*

The only additional complication to the experimental design was that the operator had to measure (only one time) each specimen randomly selected from the group of five standards, before making a measurement from another standard. That way, all five standards were measured once before any of them was measured the second time, and so on. The motivation for measuring this way rather than measuring specimen A five times followed by measuring specimen B five times, etc., was to include the additional variability that may be added from moving the specimen into position. Additionally, there was a concern as to whether the measurements would somehow systematically change with time. Randomizing the order in which the specimens were tested was intended to provide some defense against that possibility.

DATA ANALYSIS AND INTERPRETATION OF RESULTS

SPC and Other Diagnostics Used During Execution of the Calibration Experiment: a Reexamination of the Basic Assumptions

It is extremely rare to conduct an experiment without having unexpected revelations about the process or problems in execution.

Various diagnostics should be used during data collection and later during their analysis to learn what went wrong with the experiment under consideration. The appropriate posture to take is to assume things have gone wrong, find the problems, and devise a plan for recovery. It is best to discover mistakes as quickly as possible.

The use of SPC during the calibration experiment helps to assure that the data will be useful at the end of the experiment. It does this by alerting the person running the experiment to any unexpected changes in the behavior of the measurement device while the supposedly representative data are being collected. As stated before, the calibration procedure is based on the assumption that data being collected are representative of measurements that will be made on a daily basis. Therefore, SPC is crucial during the data collection step. No other tool provides the insight into the stability of the apparatus while the calibration data are being generated.

But if we know little or nothing about the apparatus being calibrated, how could we establish SPC limits during the short duration of the experiment? The answer is simple, any apparatus with completely unknown measurement capabilities would not be ready for calibration. So, the scientist or engineer in charge of its set up would not allow the device to be calibrated until he or she has tried it out and gained some experience with its operation. The information collected during this trying-it-out stage can be used to set the limits.

For the calibration experiment used in this case study, the work was done by a graduate student-intern who was working for us for a six-week period. SPC was one of the many tasks the student was asked to perform during the calibration of the FTIR. After frantically struggling to get all of the measurements made before his time was up, the student came to me the night before his presentation of his six-week project and said that he was now ready to do the SPC portion of the study. My response to him was to point out that he had a lot of other things to do in the few remaining hours, and so the SPC should be dropped. This was a mistake as will be demonstrated.

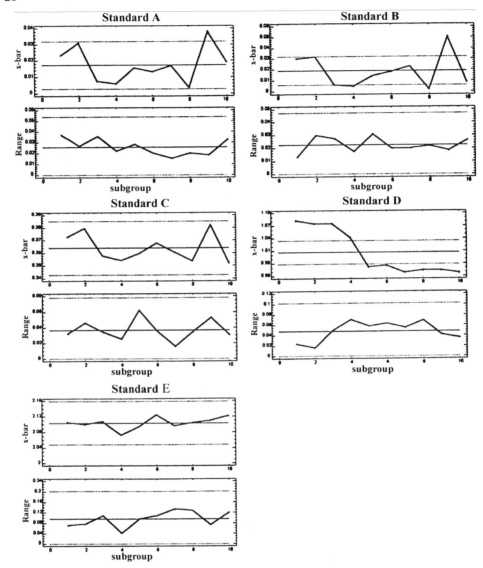

Figure 3.3. *SPC Charts for the 5 Carbon Specimens Where the Observations Were Summarized Daily Over the Course of the Calibration Experiment. Notice the abrupt shift in carbon measurements for specimen D after the third day of testing. Also notice the unusual peaks on day 9 for specimens A, B, and C.*

Figure 3.3 shows SPC charts (x-bar and range) for the five standards. Each point on the charts represents one day's five measurements. A listing showing the complete data set appears in Appendix A. The control limits for the SPC charts are based on all 10 days worth of data. Had this step taken place during the calibration of the QS-200, we would have noticed the abrupt shift in average carbon measurements from Sample D as is evident in Figure 3.3. Notice that none of the ten averages is between the control limits for the sample. (Actually the larger than typical variability for Sample D did get noticed when residual plots were viewed when the calibration curve was being estimated.) Also, it might have been noticed that samples A, B, and C each shows a suspicious looking peak for the average measurements made on day 9 of the experiment. (These were not noticed.)

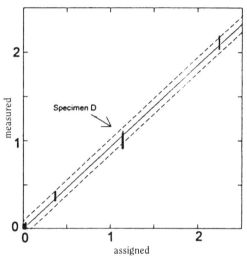

Figure 3.4. *Plot of the Data Collected as Part of the Calibration Experiment. The solid line is the calibration curve ($y_i = -0.006 + 0.931\ x_i$). The broken lines are the upper and lower 95% calibration limits for $m = 1$. The equation for the calibration limits appears in equation 3 of Appendix B. The plotted points seem to indicate a relationship between the assigned values and measurements on the FTIR that is not a straight line. By referring back to Figure 3.3, we conclude that the unusually low measurements from specimen D are the source of the apparent non linearity, and so the straight line model seems appropriate after all.*

The problem with the change in specimen D during the calibration experiment was investigated, but no special cause was identified. It will be addressed by the daily SPC measurements on two standard specimens. If a similar shift occurs again, we expect the SPC to catch it. Then we will be in a better position to understand the nature of the problem.

The results reported in this paper include the influence of Specimen D. Had the SPC charts been prepared as planned, we would have probably stopped the work until the cause for the shift was known. At the very least, Specimen D would have been dropped from the analysis. Since the problem was discovered quite late in the analysis, it was decided to calculate the desired limits recognizing that they would be somewhat wider than if Specimen D had been dropped. The daily use of SPC will alert us in the event of a similar shift in the future.

Estimation of the Calibration Curve and its Corresponding Calibration Limits

Figure 3.4 shows a plot of the data along with the calibration curve and its calibration limits for the case where one measurement is made. Details of how the calibration limits were derived appear in Appendix B. Had we not seen the SPC charts in Figure 3.3, the way that the points from specimen D appear to be outside the lower calibration limit might have seemed to indicate a need for a calibration model that is not a straight line. But we see from Figure 3.3 that specimen D produced a number of low measurements, and so a straight line is probably a good representation of the data.

Calculating Detection Limits

The *detection limit* is the smallest amount of carbon that can be reported as having a non-negative lower calibration limit. This is pictured in Figure 3.5. The detection limit x_d is the value in units of assigned (i.e., actual) carbon that corresponds to the value y_d, the level of measured carbon at the point where the lower calibration limit intersects the y axis.

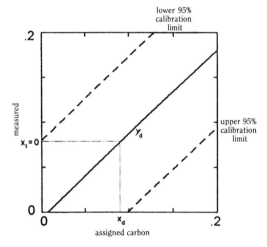

Figure 3.5. *The calibration limits shown in Figure 3.4 are magnified to illustrate how the detection limit is being defined. The detection limit for carbon, here denoted as x_d, is found by finding the assigned carbon y_d that has 0 as its lower calibration limit, x_l.*

The value of x_d for a given number of measurements (m) can be calculated by setting the expression in Equation 3.2 in Appendix B to zero and solving for $\bar{y}_{i_m}$. The regression equation is then used to solve for x given $\bar{y}_{i_m}$ (i.e., y_d). The result is:

$$x_d = -0.0062 + \frac{0.0070\sqrt{149.28+m}}{\sqrt{m}}.$$

In the case where $m = 1$, $x_d = 0.092$. Detection limits for values of m ranging from 1 to 10 are tabulated in Table 3.2 and plotted in Figure 3.6.

Table 3.2. *Tabulated Values used in Figure 3.6.*

Number of Measurements	Detection Limit
1	0.092
2	0.065
3	0.054
4	0.047
5	0.042
6	0.038
7	0.035
8	0.033
9	0.031
10	0.030

As can be seen in Figure 3.6 and Table 3.2, increasing m the number of times a specimen is measured, results in a reduction of the detection limit for the FTIR. This gives the laboratory performing the carbon measurements some flexibility. When trying to measure specimens containing only trace amounts of carbon, multiple measurements can be made from the specimen in order to increase the precision of the FTIR. For more routine samples, the laboratory can use fewer measurements to increase the number of samples it measures during a day.

CONCLUSIONS AND RECOMMENDATIONS

Reporting of Measurements

Scientists and engineers are confused regarding how to report measurements, especially when a measurement is less than the detection limit for the apparatus. Two common but inappropriate methods for reporting measurements appearing to be below the detection limit are to censor the value (indicating only that the measurement was below detection), or replace the measurement with the value of the detection limit. Both methods should be avoided because they throw away possibly useful information and make statistical analysis of the data extremely difficult. The format x^* (LCL, UCL) is proposed as a way to report measurements so that the relevant information is retained where

- x is the measured value, regardless of whether it is less than the detection limit (or even whether it is less than zero).
- An asterisk (*) indicates that the measurement is less than the detection limit. It would not appear for measurements greater than the detection limit.
- LCL is the lower 95% calibration limit for x.
- UCL is the upper 95% calibration limit for x.

This allows the user of the measurement service to understand the quality of the measurements being reported to him. Should the single measurement be part of a larger study, then the user is in a position to work with the data without having to worry about how to handle censored data. If the measurement really is in the noise of the measurement apparatus, that too will be apparent to the user.

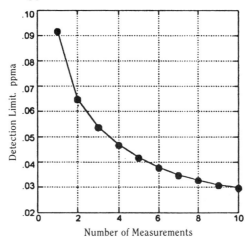

Figure 3.6. *Relationship Between the Number of Measurements at 32 Scans and the Detection Limit for Carbon. This figure indicates that it is possible to decrease the detection limit for the apparatus and increase its precision by measuring a specimen multiple times.*

Routine SPC on Daily Measurements to Assure Consistent Performance Capability

One of the concerns that was addressed in the section where the basic assumptions were discussed was whether the data collected during the calibration experiment could honestly be considered to be representative of the day-to-day operation of the apparatus. Statistical process control is useful to assure that no shifts in the process or changes in the variability of the process occur in the routine operation of the apparatus.

The calibration experiment provides a very good estimate of the variability of the process that can be applied directly to the SPC procedures. Because 250 measurements were used in the calibration of the Digilab, what we get from the calibration experiment is a much better estimate of process variability than one would get using the usual 20 or 30 observations in the typical setup. In addition, by using the estimates from the calibration experiment, we are asking whether the measurement process has deviated significantly from its condition when the calibration was performed. This also implies that if we see a shift in the measurement process and cannot remove the cause for the shift, then the calibration of the apparatus needs to be redone or at least adjusted.

Summary

Did we accomplish the goals we established in the Calibration Defined section? Yes. Except for some concern about what happened while measuring specimen D, the Digilab FTIR was shown to provide measurements that had constant mean and variance.

Acknowledgments

Claus Renner, of the University of Bochum, was the student-intern who took an ill-defined project and transformed it into a quality piece of work in a very brief six weeks. Brenda Schulte and Bill Meder were the laboratory technicians who helped on innumerable occasions and, in general, tolerated our invasion into their lab. Thanks to all.

References

3200 Data Station, Getting Started Manual. Digilab Division, Bio-Rad Laboratories, Inc. Cambridge, MA 02139.

Draper, Norman and Harry Smith. *Applied Regression Analysis*. 2nd ed. New York: John Wiley & Sons, 1981.

RS/1. Domain Solutions Corporation. 150 Cambridge Park Drive, Cambridge, MA 02140.

Statgraphics. STSC, Inc. 2115 East Jefferson Street, Rockville, MD 20852.

Strobel, Howard A. and William R. Heineman. *Chemical Instrumentation: A Systematic Approach*. 3rd ed. New York: John Wiley & Sons, 1989.

Appendix A

Carbon Metrology Data

Room Temperature Carbon Data Collected Over a Period of Ten Days Using the Digilab QS-200 FTIR with 32 Scans. Five measurements were made each day on each of five double side polished specimens.

	\multicolumn{5}{c}{Specimen ID}
	A	B	C	D	E
	\multicolumn{5}{c}{Assigned Carbon ppma (x)}
	0.0158	0.0299	0.3710	1.1447	2.2403
Day	\multicolumn{5}{c}{Carbon Measurements, ppma (Y)}
1	0.0246	0.0275	0.3575	1.0730	2.1321
1	0.0032	0.0377	0.3895	1.0825	2.1385
1	0.0401	0.0246	0.3799	1.0711	2.0915
1	0.0248	0.0311	0.3704	1.0878	2.0941
1	0.0233	0.0289	0.3664	1.0949	2.0680
2	0.0151	0.0315	0.3725	1.0778	2.0853
2	0.0416	0.0285	0.3767	1.0664	2.1476
2	0.0229	0.0222	0.4130	1.0824	2.0733
2	0.0372	0.0519	0.3671	1.0704	2.1194
2	0.0382	0.0242	0.3679	1.0770	2.0717
3	0.0234	0.0089	0.3363	1.0599	2.1428
3	0.0180	0.0073	0.3601	1.0754	2.1446
3	0.0062	-0.0108	0.3697	1.0573	2.0371
3	-0.0121	0.0088	0.3704	1.0721	2.0892
3	0.0010	0.0162	0.3516	1.1073	2.1226
4	0.0025	0.0158	0.3683	1.0047	2.1027
4	0.0149	0.0056	0.3590	1.0752	2.0613
4	0.0183	0.0055	0.3510	1.0389	2.0608
4	-0.0037	-0.0015	0.3494	1.0639	2.0692
4	-0.0028	-0.0006	0.3431	1.0267	2.0684
5	-0.0041	0.0298	0.3860	0.9996	2.0351
5	0.0211	0.0037	0.3461	0.9928	2.0766
5	0.0241	0.0322	0.3707	0.9514	2.1144
5	0.0124	0.0013	0.3238	0.9692	2.1127
5	0.0240	0.0039	0.3703	0.9417	2.1312
6	0.0092	0.0285	0.3556	0.9539	2.1688
6	0.0049	0.0091	0.3757	0.9977	2.1471
6	0.0249	0.0088	0.3739	1.0022	2.1689
6	0.0087	0.0248	0.3846	0.9822	2.0608
6	0.0179	0.0186	0.3492	0.9390	2.0762

APPENDIX A (CONTINUED)

Room Temperature Carbon Data Collected Over a Period of Ten Days Using the Digilab QS-200 FTIR with 32 Scans. Five measurements were made each day on each of five double side polished specimens.

	\multicolumn{5}{c}{Specimen ID}				
	A	B	C	D	E
	\multicolumn{5}{c}{Assigned Carbon ppma (x)}				
	0.0158	0.0299	0.3710	1.1447	2.2403
Day	\multicolumn{5}{c}{Carbon Measurements, ppma (Y)}				
7	0.0112	0.0249	0.3559	0.9446	2.1079
7	0.0119	0.0122	0.3628	0.9316	2.0946
7	0.0266	0.0261	0.3630	0.9861	2.0593
7	0.0193	0.0198	0.3682	0.9851	2.0429
7	0.0151	0.0321	0.3524	0.9449	2.1760
8	0.0123	-0.0049	0.3569	0.9507	2.1666
8	-0.0078	-0.0052	0.3757	0.9818	2.1217
8	0.0086	0.0166	0.3485	0.9848	2.0391
8	0.0043	-0.0017	0.3455	0.9850	2.1018
8	0.0002	0.0067	0.3412	0.9157	2.0903
9	0.0303	0.0416	0.4028	0.9555	2.0953
9	0.0404	0.0459	0.3655	0.9692	2.1205
9	0.0486	0.0555	0.3857	0.9915	2.1504
9	0.0303	0.0492	0.3524	0.9500	2.1049
9	0.0400	0.0598	0.4055	0.9513	2.0766
10	0.0283	0.0237	0.3503	0.9393	2.1695
10	0.0106	-0.0024	0.3445	0.9628	2.1546
10	0.0048	-0.0005	0.3414	0.9614	2.0903
10	0.0138	0.0165	0.3515	0.9494	2.0498
10	0.0373	0.0059	0.3720	0.9746	2.1427

APPENDIX B

Estimation of the Calibration Curve and Its Corresponding Calibration Limits

As stated earlier, the straight line calibration equation is of the form

$$y_i = \beta_o + b_1 x_i + \varepsilon_i \quad i = 1, 2, \ldots, n,$$

where x is the known carbon value of the standard specimen and y is the measurement made on the FTIR.

The equation and its 95% calibration limits are calculated using inverse regression as described by Draper and Smith (1981). The technique is to determine b_0 and b_1, the estimates of β_o and β_1, using ordinary least squares regression. Replacing the y_i with y_{i_m}, the average of the m measured responses[5], the function is then inverted to give the calibration curve.

$$x_i = \frac{\bar{y}_{i_m} - b_0}{b_1} \quad i = 1, 2, \ldots, n.$$

Confidence limits are calculated for the x_i's, as illustrated in Figure 3.7.

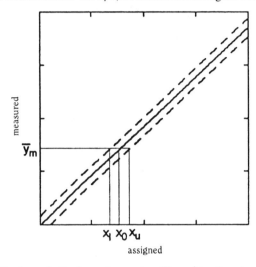

Figure 3.7. Actual Carbon Amounts. The objective is to find the actual amount of carbon x_0 given a measurement y_0. The calibration limits x_l and x_u are 95% confidence limits on the estimated value x_0.

The width of the calibration interval is a function of n, the number of measurements made during the calibration of the equipment, and m, the number of times the specimen with an unknown carbon content is sampled. For the case study, $n = 250$ and the value of m is determined each time a specimen is measured for carbon. The 95% calibration limits for the average of m measurements of a single specimen with unknown carbon content is presented below as a function of $\bar{y}_m$:

$$95\%CI = \bar{x} + \frac{b_1(\bar{y}_m - \bar{y})}{k} \pm \frac{ts}{k}\sqrt{\frac{(\bar{y}_m - \bar{y})^2}{\sum_{i=1}^{n}(x_i - \bar{x})^2} + \left[\frac{1}{n} + \frac{1}{m}\right]k} \quad .$$

(3.1)

---

[5] Switching the notation from y_i to $\bar{y}_{i_m}$ is intended to allow for m measurements of the same specimen ($m = 1, 2, \ldots$) in order to increase the precision or in order to decrease the detection limit of the apparatus.

where:
- $\bar{x}$ = average of the n assigned values
- $\bar{y}$ = average of the n values measured during calibration
- $\bar{y}_m$ = average of the m measurements measured on the specimen of interest
- b_1 = estimated slope
- t = appropriate value from a table of the t-distribution for $\alpha = 0.025$ with $n - 2$ degrees of freedom
- s = standard deviation about the regression line

$$k = b_1^2 - \frac{t^2 s^2}{\sum_{i=1}^{n}(x_i - \bar{x})^2} = b_1^2 - t^2 s_{b_1}^2.$$

s_{b1} = the standard deviation of b_1, the regression coefficient.

Commercially available statistics packages, such as *RS/1*[6] or *Statgraphics*[7], do not calculate calibration limits directly. However, they can be used to generate the information needed to calculate the calibration limits given in Equation 3.1. All but one of the values used in Equation 3.1 are found in the standard output. The only value missing is the sum

$$\sum_{i=1}^{n}(x_i - \bar{x})^2.$$

But this can be found easily by remembering that:

$$s_{b1}^2 = \frac{s^2}{\sum(x_i - \bar{x})^2}.$$

The trick is then to find s and s_{b1} on the computer output and calculate the ratio of their squares:

$$\sum_{i=1}^{n}\left(x_i - \bar{x}\right)^2 = \frac{s^2}{s_{b1}^2}.$$

We will now calculate the calibration limits for the FTIR. From the data in Appendix A, we calculate

$$\bar{x} = \frac{1}{5}(0.0158 + 0.0299 + 0.3710 + 1.1417 + 2.2403)$$
$$= 0.7604$$

and

$$\bar{y} = \frac{1}{250}(0.0246 + 0.0275 + \cdots + 2.1427)$$

---

[6] Available from Domain Solutions Corporation, 150 Cambridge Park Drive, Cambridge, MA 02140
[7] Available from STSC, Inc., 2115 East Jefferson Street, Rockville, MD 20852.

$$= 0.7015.$$

Table 3.3. *Regression Analysis Results from Statgraphics. The results needed for calculation of the calibration limits are underlined.*

Regression Analysis—Linear Model: $Y = b_0 + b_1 x$

Dependent variable: measured carbon   Independent variable: assigned carbon

Parameter	Estimate	Standard Error	T Value	Prob Level
Intercept	−0.00606082	0.00372119	1.62873	0.10464
Slope	<u>0.931145</u>	<u>0.0032717</u>	284.606	0.00000

Correlation Coefficient = 0.998473   R-squared = 99.69%
Stnd. Error of Est. = <u>0.0437574</u>

The data in Appendix A were analyzed using *Statgraphics'* Simple Regression procedure. They were prepared by matching each y carbon measurement with its assigned x value in a computer file containing two columns and 250 rows. Table 3.3 shows the *Statgraphics* output. Incidentally, there was nothing special about using *Statgraphics* here. Just about any statistics software would have worked equally well.

From Table 3.3, we find $b_1 = 0.9311$, $s = 0.0438$, and $s_{b1} = 0.00327$. The calibration curve is

$$y_{i_m} = -0.006 + 0.931 x_i \quad i = 1, 2, \ldots, n$$

or

$$x_i = \frac{y_{i_m} + 0.006}{0.931} \quad i = 1, 2, \ldots, n$$

This is the equation that allows us to convert the carbon reported by the FTIR, y_{i_m}, to measurements that are consistent with the assigned values for the standards. To get the calibration limits as in Equation 3.1, we then calculate the values for

$$\sum_{i=1}^{n} (x_i - \overline{x})^2$$

and for k, where

$$\sum_{i=1}^{n} (x_i - \overline{x})^2 = \frac{s^2}{s_{b_1}^2}$$

$$= \frac{(0.0438)^2}{(0.00327)^2}$$

$$= 178.594$$

and

$$k = b_1^2 - \frac{t^2 s^2}{\sum_{i=1}^{n}(x_i - \bar{x})^2}$$
$$= (0.9311)^2 - (1.96)^2 (0.00327)^2$$
$$= 0.8669.$$

It is now possible to calculate the 95% calibration interval for the Digilab FTIR carbon measurements. They are given by

$$95\% CI_{(asym>0)} = \bar{x} + \frac{b_1(\bar{y}_m - \bar{y})}{k} \pm \frac{ts}{k}\sqrt{\frac{(\bar{y}_m - \bar{y})^2}{\sum_{i=1}^{n}(x_0 - \bar{x})^2} + \left(\frac{1}{n} + \frac{1}{m}\right)k}$$

$$= 0.7604 + \frac{0.931(\bar{y}_m - 0.7019)}{0.8669} \pm \frac{(1.96)(0.0438)}{0.8669}\sqrt{\frac{(\bar{y}_m - 0.7019)^2}{178.594} + \left(\frac{1}{250} + \frac{1}{m}\right)0.8669}$$

$$= 0.7604 + \frac{0.931(\bar{y}_m - 0.7019)}{0.8669} \pm 0.099\sqrt{\frac{(\bar{y}_m - 0.7019)^2}{178.594} + 0.003 + \frac{0.8669}{m}} \quad (3.2)$$

When making only one measurement (that is when $m = 1$). This reduces to

$$95\% CL_{(m=1)} = 0.7604 + \frac{0.931(\bar{y}_m - 0.7019)}{0.8669} \pm 0.099\sqrt{\frac{(\bar{y}_m - 0.7019)^2}{178.594} + 0.8669}$$

(3.3)

BIOGRAPHY

Peter C. Pankratz is senior statistician in the Technology Department of MEMC Electronic Materials, Inc., St. Peters, Missouri. He has been with MEMC since 1989. Pete received his Master of Science degree from Northern Illinois University in applied statistics in 1981.

CHAPTER 4

REVELATION OF A MICROBALANCE WARM-UP EFFECT

James Buckner, Barry L. Chin, Todd P. Green, and Jon Henri

A Mettler microbalance was to be employed to monitor the wafer-to-wafer uniformity of a 3500 Å tungsten film deposition to a tolerance goal of 2%. A gauge study revealed that the measurements were influenced by an instrument warm-up effect, which was eliminated by leaving the tool powered on continuously. Without the warm-up effect, the measurement system was then found to be marginally incapable, with a precision-to-tolerance ratio of 0.41. However, the measurement may still be used if the results are corrected for the variation arising from the measurement system.

EXECUTIVE SUMMARY

Problem

One common method of monitoring the chemical vapor deposition (CVD) of tungsten is to measure the mass gain experienced by the wafer after exposure to the deposition process. In particular, the plan for a SEMATECH equipment improvement project (EIP) called for using mass gain measurements to monitor the wafer-to-wafer uniformity of a 3500 Å tungsten film deposited in the Novellus Concept One-W tool to a tolerance goal of 2%. The generic SEMATECH Qualification Plan calls for a gauge study of all metrology tools at the beginning of a project (see page xxi), therefore, a gauge study was performed on the Mettler microbalance to be used for this measurement.

Solution Strategy

The data collection plan was typical of gauge studies. Three operators were to weigh ten sample wafers twice each. Instrument repeatability and measurement system reproducibility would be estimated from the data. A precision-to-tolerance ratio would be computed using the experimentally determined total precision of the measurement system, and a judgment would be rendered on the capability of the measurement system.

Conclusions

Unfortunately, a mistake in data collection precluded separation of the repeatability and reproducibility components of the total precision. However, it was possible to calculate the total precision itself, and it was found to be inadequate for the purposes of the EIP.

A more detailed examination of the existing data revealed the possibility that the measurement system was influenced by an instrument warm-up effect. Follow-up with the operators confirmed that a warm-up effect was contributing to the variation in the data. The microbalance was subsequently left powered on continuously to eliminate the effect.

After removing the data collected during the instrument warm-up period, the measurement system was found to be marginally incapable of monitoring the mass gain to the required precision. However, the measurement system is still usable, provided that the measurement system variation is accounted for in the interpretation of the data.

PROCESS

It is rather uncommon to use a microbalance in semiconductor manufacturing, but there is one important application where it has become common, and that is in monitoring the chemical vapor deposition (CVD) of tungsten. Because of the relatively high density of tungsten (19.3 g/cm³ for the bulk material), microbalances can be used to monitor the amount of tungsten deposited on a wafer by the deposition process, even for thin films less than 1 µm thick which are required for semiconductor manufacturing. For example, the fifth digit found on many microbalances measures in units of 10 µg, which would potentially make the balance sensitive to tungsten layer thickness variations of as little as 16 picometers (pm) on 200 mm diameter wafers. It is not important to measure tungsten film thickness to the tiny fraction of an atomic radius indicated by this simple calculation, but it is important to know the precision with which actual measurements can be made using this measurement system.

The present microbalance gauge study was motivated by the need to monitor nominal 3500 Å tungsten films to be deposited by the Novellus Concept One-W CVD tungsten deposition tool during an equipment improvement project (EIP) sponsored by SEMATECH (see Shah, 1992 and 1993). The study was prescribed as one of the first steps in the SEMATECH Qualification Plan, which is the generic procedure upon which all SEMATECH EIP plans are modeled (see the SEMATECH Qualification Plan, page xxi). The microbalance would be used in assessing the wafer-to-wafer uniformity goal of 2% for a 3500 Å film, so that the measurement tolerance requirement was 70 Å. Using a weight-to-thickness conversion factor of 21909 Å/g based on cross-sectional scanning electron microscopy (XSEM) measurements of thickness, the tolerance requirement in mass units is 0.0032 g.

DATA COLLECTION PLAN

The data collection plan had the three principal operators weigh ten different samples twice each over a period of several days. The ten samples were chosen to span the range of measurements over which the microbalance would normally be applied. The data are shown in Table 4.1. This plan would provide 30 degrees of freedom in the short term repeatability estimate from the 30 dual measurements, and 20 degrees of freedom in the reproducibility estimate from the (3 – 1) operator degrees of freedom over the ten samples.

Table 4.1. *Raw Data. There are two weight measurements by three operators on each sample.*

Wafer	Operator 1		Operator 2		Operator 3	
	First	Second	First	Second	First	Second
1	54.0083	54.0084	54.0098	54.0041	54.0143	54.0096
2	53.7703	53.7693	53.7708	53.7601	53.7670	53.7685
3	54.2083	54.2084	54.2094	54.1975	54.2079	54.2073
4	54.1110	54.1099	54.1111	54.1007	54.1110	54.1083
5	54.0347	54.0341	54.0350	54.0240	54.0360	54.0333
6	54.1152	54.1130	54.1132	54.1017	54.1141	54.1117
7	54.2192	54.2194	54.2195	54.2081	54.2176	54.2189
8	54.1555	54.1561	54.1561	54.1462	54.1516	54.1550
9	54.4405	54.4399	54.4405	54.4293	54.4381	54.4391
10	54.2291	54.2284	54.2290	54.2179	54.2252	54.2278

ANALYSIS AND INTERPRETATION OF RESULTS

The standard analysis of a raw data set such as that listed in Table 4.1 would be to obtain an estimate of repeatability from the variance of the first and second measurements of each sample wafer, and then to obtain an estimate of reproducibility from the variance of the operator means for each sample wafer. However, a mistake in data collection precluded this standard analysis. The first and second measurements of the sample wafer by each operator were taken at least half a day apart, rather than being taken one after the other as intended. Therefore, the variance of the first and second measurements contains not only the repeatability, but also any component of the reproducibility not related to the operator.

Even though this mistake prevented resolution of repeatability and reproducibility, it did not impact the ability to obtain an estimate of the total precision. The total precision was estimated by pooling the sample wafer variances, where the sample wafer variances were computed from the six individual measurements.[8] The sample wafers thus provided five degrees of freedom each, so that the 10 sample wafers resulted in an estimate of total precision with 50 degrees of freedom.

---

[8] The procedure is as follows. The variance of sample 1 was computed from its six weighings. The variance of sample 2 was computed from its six weighings. Likewise, the variance for each of the ten samples was computed, resulting in ten independent estimates of the variance of the total measurement system with five degrees of freedom each. These ten independent estimates were then pooled (averaged) to obtain a single estimate of the variance of the measurement system with $5 \times 10 = 50$ degrees of freedom.

The total precision computed from the data in Table 4.1 was found to be 0.0042 g (1σ). Note that this estimate is for the precision of a single measurement. The tolerance of 0.0032 g would be obtained as the difference from a tare.[9] The total precision, therefore, would actually be computed from the sum of the variances of the tare and gross weighings, which for this case would be 0.0059 g (1σ). The precision-to-tolerance ratio is then 1.84, indicating an incapable measurement system.

An effort was begun to find the sources of variation that were making the measurement system incapable. Unfortunately, the data collection mistake did not allow direct analysis of the variance into components of instrument repeatability and operator reproducibility. However, an independent estimate of repeatability was available from a prior activity. The estimate was obtained by weighing two sample wafers 30 times each, and pooling the resulting variances (data not shown). The result, after doubling the variance to account for tare and gross weighings was 0.00057 g (1σ), yielding a pseudo-precision-to-tolerance ratio of 0.18. The instrument repeatability, therefore, accounted for only about 1% of the total variance (this number was obtained by squaring the precision-to-tolerance ratio). The larger sources of variation had to be in the environment around the instrument.

At this point, the data were displayed graphically. A box plot of the deviations of each sample wafer measurement from the sample wafer mean is shown in Figure 4.1. The data are plotted in chronological order. The fourth measurement period shows both a bias toward lower weight and an increase in variance, while the fourh measurement period shows recovery from the bias with a lingering variance increase. Examination of the data collection forms revealed that the fourth measurement period was noted by the operator as being in the morning, suggesting a possible warm-up effect. The operator who performed the fourth period measurement confirmed that the measurement had been the first use of the tool that morning. It was the practice in this laboratory to power down the microbalance overnight and to power it back on every morning. No other measurements were performed soon after a power on.

Laboratory practice was immediately changed to leave the microbalance powered on continuously. Using the data from Table 4.1 that was not subject to the warm-up effect, and accounting for tare and gross weighings, the total precision is estimated to be 0.0013 g (1σ). The resulting precision-to-tolerance ratio of 0.41, which is larger than 0.3, would generally be interpreted as meaning that the measurement system is marginally incapable for this application, even without the warm-up effect.

All of the outliers that appear in the box plot of Figure 4.1 are from the first wafer of the ten samples (identified by the last digit in the numbers next to the outliers). Since the procedure used by the operators was to measure all ten samples at one measurement period,

---

[9] A tare is normally defined as the weight of the vessel in which a material of interest is contained. Thus the net weight of the material is the difference between the gross weight and the tare:

$$\text{net weight} = \text{gross weight} - \text{tare}$$

In this case, the tare is the weight of the silicon wafer prior to deposition of the tungsten film. The net weight of the film is then the difference between the weight of the wafer before and after deposition.

these outliers represent evidence of a first wafer effect. This observation suggests that the precision of the measurement may be improved by measuring a dummy wafer at the beginning of every contiguous measurement period after the tool has been idle. Unfortunately, this problem was not discovered until after the completion of the EIP, and so the idea of using a dummy wafer was not tested. However, elimination of all first wafer data from the present data set does not alter the precision-to-tolerance ratio (to two decimal places), suggesting that such a procedure would have had only a marginal effect.

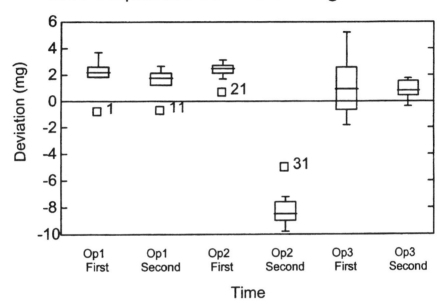

Figure 4.1. *Deviations of the Sample Wafer Measurement from the Sample Wafer Mean. The data are in chronological order. The irregularities in the fourth and fifth measurements are caused by a warm-up effect.*

CONCLUSIONS AND RECOMMENDATIONS

A bias and a variance is associated with the Mettler microbalance warm-up period. The recommended action is to leave the balance powered on to eliminate the warm-up period. This recommendation was implemented.

Even without a warm-up effect, the measurement system is marginally incapable of performing to the required tolerance of 0.0032 g, with a precision-to-tolerance ratio of 0.41. The microbalance may still be used, provided that the relatively large variation of the measurement system is accounted for. This might be done either by using the average of multiple measurements when an accurate value is needed for a single sample wafer[10] or by correcting the variance in control chart data when longer term performance is of interest.[11] It may be possible to further improve the precision of the measurement by using a dummy wafer after idle periods to circumvent a first wafer effect.

REFERENCES

Shah, Raj, Ed. *Novellus Concept One-W Characterization Interim Report*. 92051143A-ENG. Austin, TX: SEMATECH. 1992.

Shah, Raj, Ed. *Novellus Concept One-W EIP Project Final Report: E36*. 93011464A-ENG. Austin, TX: SEMATECH. 1993.

BIOGRAPHIES

James Buckner has been a technologist with Texas Instruments since 1987, where he has been responsible for metal sputtering, CVD-tungsten deposition, and improvement of CVD-tungsten deposition equipment at SEMATECH. He is currently responsible for the Equipment Migration program in Texas Instruments' Manufacturing Science and Technology Center. James has a Ph.D. in physical chemistry from the University of North Carolina at Chapel Hill.

Barry L. Chin received his B.S. degree from Cornell University and the M.S. and Ph.D. degrees from the University of California at Berkeley. Since 1980 he has been involved in thin film deposition, plasma etching, and process integration programs at several semiconductor manufacturing companies. He joined Novellus Systems in 1987 where he has worked on the product development of Plasma TEOS and CVD-Tungsten deposition systems.

---

[10] The precision to which a measured value is known may be improved to any arbitrary level by averaging multiple measurements. If the standard deviation of a single measurement is σ, then the standard deviation of the average of n such measurements is $\sigma/\sqrt{n}$. This is true so long as the *n* measurements are independent and identically distributed. See any introductory statistics text for more information.

[11] Assuming that the variation due to measurement error, σ_m, is independent of variation originating in the process being charted, σ_p, then the total observed variation in the control chart is given by the sum of the variances: $\sigma_t^2 = \sigma_p^2 + \sigma_m^2$. Then the variation arising from the process alone can be obtained as the difference, $\sigma_p^2 = \sigma_t^2 - \sigma_m^2$.

Todd P. Green received a B.S. degree in physics from San Jose State University. He is a process development engineer at Novellus Systems supporting advanced tungsten development.

Jon Henri is an applications engineer at Novellus Systems. He works on CVD-Tungsten deposition systems.

CHAPTER 5

GRR METHODOLOGY FOR DESTRUCTIVE TESTING AND QUANTITATIVE ASSESSMENT OF GAUGE CAPABILITY FOR ONE-SIDE SPECIFICATIONS

Teresa Mitchell, Victor Hegemann, and K.C. Liu

Gauge repeatability cannot be estimated by traditional methods when the measurement process is destructive in that the product is destroyed in the process of being measured. A methodology is presented for estimating gauge repeatability for the destructive bond pull strength tester used in semiconductor manufacturing. In addition, the traditional measure-ment system capability index Cp cannot be used to evaluate the impact of measurement variability on product quality for one-sided specifications such as those on pull strength, where bigger pull strength is better. A new definition of measurement system capability for the case of one-sided specifications is presented, with examples that demonstrate its usefulness.

EXECUTIVE SUMMARY

Problem

Variability in the measurement system impacts the values of the quality characteristics of a process. To ensure meaningful analysis of the process, the measurement instrument must be accurate and repeatable. Variability in a measurement system may mask important information on a manufacturing process, resulting in loss of yield when good product is treated as defective product, and in customer reliability problems when defective product is treated as good product.

Gauge repeatability is the variation in a quality characteristic obtained from repeated sample measurements taken back-to-back under the same measurement conditions. For a destructive measurement process where the sample is destroyed in the process of being measured, gauge repeatability cannot be accurately assessed by standard methods.

The results of a gauge study are used to determine the capability of the gauge to make the required measurements. The variability of the measurement system, given by the standard deviation of the measurement system $\sigma_{measurement}$, can be estimated by $s_{measurement}$, obtained from a gauge repeatability and reproducibility study (GRR). The capability of the measurement process is given by the precision-to-tolerance ratio (P/T)

$$P/T = \frac{6s_{measurement}}{USL - LSL},$$

where $s_{measurement}$ is the estimate of the measurement system standard deviation, USL is the upper specification limit for the process, and LSL is the lower specification limit for the process. P/T is the fraction of the tolerance USL-LSL consumed by measurement system variation. The capability of the measurement process can also be estimated by the capability index Cp,

$$Cp_{measurement} = \frac{USL - LSL}{6s_{measurement}},$$

which is the inverse of the P/T ratio.

In many instances, the quality characteristic of interest has a one-sided specification. Examples include bond pull strength (lower specification limit only), device delay time (upper specification limit only), and device speed (upper specification limit only). For such quality characteristics, the capability of the measurement system cannot be expressed as a fraction of the tolerance that is consumed by measurement system variation.

In the integrated circuit (IC) chip bonding process, the strength of the bond between the wire and the chip pad is an important quality characteristic. The repeatability of the bond strength tester cannot be assessed by standard GRR methods due to the destructive nature of the test. In addition, the impact of this variation cannot be evaluated using the precision-to-tolerance ratio because the process only has a lower specification limit.

Solution Strategy

For a non-destructive measurement process, a single nested design experiment can be used to estimate measurement tool capability. For a destructive measurement process, a method is presented to collect the information in two steps, each consisting of a nested design experiment. A new definition of measurement system capability for one-sided specifications is presented.

Conclusion

A method using two nested experimental designs is used to calculate repeatability for a destructive measurement system. The impact of this repeatability on yield loss and/or customer reliability problems is assessed. A new definition of measurement system capability is given for processes with one-sided specifications. The repeatability and the capability are illustrated with data from a bond pull strength tester. The $6\sigma_{measurement}$ estimate for pull bond tester is 3.79 grams. The yield loss due to measurement error was estimated to be less than 7 PPB (parts per billion).

PROCESS DESCRIPTION

Gold or aluminum wire is used to connect a die (integrated circuit, or IC) to the lead frame on which it is mounted. Sonic or ultrasonic stress is used to bond the wire. Figure 5.1 gives a cross-section of the bonded integrated circuit on the lead frame. Quality suffers if the die and the lead frame are not well connected. A disconnection will result if the ball on the die is lifted from the die, if the neck of the ball on the die is broken, if the wire is broken, if a stitch bond is lifted from the lead frame, or if the wire neck on the lead frame is broken. The bond pull test uses a pull strength tool to pull on a wire to determine the strength required to disconnect the die and the lead frame. The test destroys the sample in the process of measuring the pull strength.

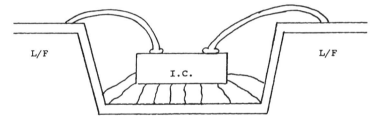

Figure 5.1. *Cross Section of an Integrated Circuit Bonded to a Lead Frame.*

DATA COLLECTION

Non-Destructive Tool

The first part of this section describes a standard nested data collection plan used in a non-destructive measurement tool to quantify repeatability and reproducibility. Table 5.1 shows a typical data collection plan for a GRR study. It has k readings per sample (denoted by xxx), n samples, and m operators, with operators nested in samples.

In this case, three sources of variation for an analysis of variance (ANOVA) can be quantified:

Variance (Sample) = σ_s^2
Variance (Operator) = $\sigma^2_{Operator} = \sigma_R^2$ (Reproducibility)
Variance (Repeated) = $\sigma^2_{Repeated} = \sigma_r^2$ (Repeatability)

Table 5.1. *Example of Data Collection for Standard GRR Study. The xxx's denote multiple readings for each one of n samples, with m operators for each sample.*

Operator	\multicolumn{7}{c}{Sample}

Operator	1	2	3	4	5	...	n
1	xxx	xxx	xxx	xxx	xxx	...	xxx
2	xxx	xxx	xxx	xxx	xxx	...	xxx
3	xxx	xxx	xxx	xxx	xxx	...	xxx
.							
.							
.							
m	xxx	xxx	xxx	xxx	xxx	...	xxx

Repeatability measures the variability between readings for each sample by each operator (the xxx), reproducibility measures the variability between operators, and there is variability between samples due to processing conditions. The corresponding nested design model (see Montgomery, *Design and Analysis of Experiments*, 1997) is

$$Y_{ijl} = u + S_i + O_j(S_i) + e_{l(ij)}$$

where
$i = 1, 2, 3,..., n =$ number of samples,
$j = 1, 2, 3,..., m =$ number of operators,
$l = 1, 2, 3,..., k =$ number of repeated readings.

Expected Mean Squares and Variance Components

Expected mean squares play an important role in the analysis of the data for the non-destructive tool. The expected mean squares associated with the model are used to develop statistics to estimate the different sources of variation and their significance (see Box, 1978). The expected mean squares for the nested model are expressed in terms of the variances as follows:

$$E(MS_{\text{Sample}}) = \sigma_r^2 + k\sigma_R^2 + km\sigma_s^2,$$

$$E(MS_{\text{Operator}}) = \sigma_r^2 + k\sigma_R^2,$$

$$E(MS_{\text{Repeated}}) = \sigma_r^2.$$

Solving for the variances gives

$$\sigma_r^2 = E(MS_{\text{Repeated}}),$$

$$\sigma_R^2 = \{E(MS_{\text{Operator}}) - E(MS_{\text{Repeated}})\}/k,$$

$$\sigma_s^2 = \{E(MS_{\text{Sample}}) - E(MS_{\text{Operator}})\}/mk.$$

Estimating the Variance Components[12]

From the last equations, the population variances σ_r^2, σ_R^2, and σ_s^2 for the non-destructive tool can be estimated by replacing the expected mean squares by the mean squares obtained from the sample

$$s_r^2 = MS_{\text{Repeated}},$$

$$s_R^2 = \{MS_{\text{Operator}} - MS_{\text{Repeated}}\}/k,$$

$$s_s^2 = \{MS_{\text{Sample}} - MS_{\text{Operator}}\}/mk,$$

where

$$MS_{\text{Repeated}} = \frac{\sum_{i=1}^{n}\sum_{j=1}^{m}\sum_{l=1}^{k}(Y_{ijl} - \overline{Y}_{ij})^2}{nxmx(k-1)},$$

$$MS_{\text{Operator}} = \frac{k\sum_{i=1}^{n}\sum_{j=1}^{m}(\overline{Y}_{ij} - \overline{Y}_i)^2}{nx(m-1)}$$

$$MS_{\text{Sample}} = \frac{km\sum_{i=1}^{n}(\overline{Y}_i - \overline{Y})^2}{(n-1)}$$

with

$$\overline{Y} = \frac{\sum_{i=1}^{n}\sum_{j=1}^{m}\sum_{l=1}^{k}Y_{ijl}}{nxmxk} \quad \overline{Y}_{ij} = \frac{\sum_{l=1}^{k}Y_{ijl}}{k} \quad \overline{Y}_i = \frac{\sum_{j=1}^{m}\sum_{l=1}^{k}Y_{ijl}}{mxk}.$$

These statistics are routinely evaluated for nested designs using standard software.

Destructive Tool

Data was collected in two steps to evaluate repeatability and reproducibility for the destructive bond pull measurement tool:

1. Ten units bonded under the same operating conditions were selected at random. Five wires were selected at random for testing on each unit, and the same five wire locations were used on each unit. One operator was used for this step. Data is summarized in Table 5.2.

---

[12] For more information on this, see Montgomery, *Design and Analysis of Experiments* (1997).

Table 5.2. *Pull Strength (in Grams) for 10 Units with Five Wires Each. The five wire locations are the same for all units.*

	Wires				
Unit	1	2	3	4	5
1	11.6	11.3	10.3	11.7	10.3
2	11.3	11.0	10.2	11.1	11.4
3	10.1	10.2	10.8	8.6	10.8
4	10.9	10.8	10.3	11.7	10.6
5	9.7	11.0	10.3	10.8	9.3
6	9.8	10.3	9.2	9.1	10.2
7	10.7	11.2	9.9	9.7	9.7
8	10.8	10.6	9.2	10.6	10.5
9	9.9	9.4	11.3	10.8	11.3
10	9.7	11.2	9.8	11.0	9.6

2. Another 30 units bonded under the same operating conditions as the units in Step 1 were randomly selected. Three operators were chosen. Each operator pulled one wire on each of the ten units; each wire was from the same location on each unit. Data is summarized in Table 5.3.

Table 5.3. *Pull Strength (in Grams) for One Wire on Each Unit for Every Operator. The same wire position was used on each unit.*

	Unit									
Operator	1	2	3	4	5	6	7	8	9	10
A	10.3	10.9	11.0	11.7	9.8	10.5	10.5	10.8	10.8	10.3
B	9.3	11.4	10.1	9.1	10.0	9.9	11.1	10.9	10.3	9.9
C	11.1	10.4	11.5	11.1	11.6	9.7	10.9	9.8	11.4	10.4

ANALYSIS AND INTERPRETATION OF RESULTS

Figure 5.2 gives a schematic representation of the nested design for the data in Table 5.2 collected during Step 1.

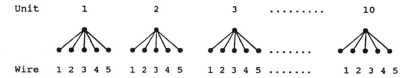

Figure 5.2. *One-Stage Nested Design Model for Wires Nested In Units.*

The expected mean squares equations for Step 1 are

$$E(MS_{\text{Unit}}) = \sigma_r^2 + \sigma^2_{\text{Wire}} = 5\,\sigma^2_{\text{Unit}},$$

$$E(MS_{\text{Wire/Repeated}}) = \sigma_r^2 + \sigma^2_{\text{Wire}}.$$

Note that, since a given wire can only be tested once, the reading, y_{ij}, has the variation due to wire position confounded with the variation for tool repeatability. Solving for σ^2 gives

$$\sigma^2_{\text{Unit}} = \frac{E(MS_{\text{Unit}}) - E(MS_{\text{Wire/Repeated}})}{5},$$

and σ^2_{Unit} can be estimated using the sample statistic

$$s^2_{\text{Unit}} = \frac{MS_{\text{Unit}} - MS_{\text{Wire/Repeated}}}{5}.$$

For the Step 1 data set given in Table 5.2, the following values were calculated:

$$MS_{\text{Unit}} = 0.90,$$

$$MS_{\text{Wire/Repeated}} = 0.48,$$

$$s^2_{\text{Unit}} = \frac{0.90 - 0.48}{5} = 0.084.$$

Figure 5.3 gives a schematic representation of the nested design for the data in Table 5.3 collected during Step 2.

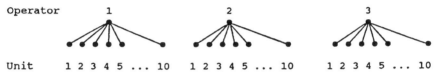

Figure 5.3. *One-Stage Nested Design Model for Units Nested in Operators.*

For Step 2, the expected mean squares equations are

$$E(MS_{\text{Operator}}) = \sigma_r^2 + \sigma^2_{\text{Unit}} + 10\sigma^2_{\text{Operator}},$$

$$E(MS_{\text{Unit/Repeated}}) = \sigma_r^2 + \sigma^2_{\text{Unit}}.$$

In this step, there is confounding between tool repeatability and unit (die). It is very important that the units tested in Steps 1 and 2 be processed under similar conditions, so that the repeatability is the same. Solving for the variabilities yields

$$\sigma^2_{\text{Operator}} = \frac{E(MS_{\text{Operator}}) - E(MS_{\text{Unit/Repeated}})}{10}$$

$$\sigma_r^2 = E(MS_{\text{Unit/Repeated}}) - \sigma^2_{\text{Unit}}.$$

so that $\sigma^2_{\text{Operator}}$ and σ_r^2 can be estimated by the following statistics:

$$s^2_{\text{Operator}} = \frac{MS_{\text{Operator}} - MS_{\text{Unit/Repeated}}}{10},$$

$$s_r^2 = MS_{\text{Unit/Repeated}} - s^2_{\text{Unit}}.$$

From the data in Table 5.3, the following values can be calculated

$$MS_{\text{Operator}} = 0.96,$$

$$MS_{\text{Unit/Repeated}} = 0.43.$$

so

$$s^2_{\text{Operator}} = \frac{0.96 - 0.43}{10} = 0.053.$$

From Step 1, s^2_{Unit} is given by 0.084. Substituting this value in the previous equation, repeatability is estimated as

$$s_r^2 = 0.430 - 0.084 = 0.346.$$

Finally, the total measurement system variation can be calculated as

$$s_{\text{measurement}} = \sqrt{S_r^2 + S^2_{\text{operator}}}.$$

For the bond pull strength measurement tool,

$$s_{\text{measurement}} = \sqrt{0.346 + 0.053} = 0.632$$

and

$$6s_{\text{measurement}} = 3.792.$$

Measurement System Variation Versus Yield Loss[13]

The numerical values of the variances for the bond pull measurement tool just obtained will be used to determine gauge acceptability. This section presents a method of quantifying the acceptability of a tool with only a lower specification limit, such as the pull bond strength tool.

Each recorded reading y_{obs} is the sum of the contribution from the process y_{process} and the contribution from the measurement system $y_{\text{measurement}}$ that is,

$$y_{\text{obs}} = y_{\text{process}} + y_{\text{measurement}}.$$

---

[13] The material in this section was first presented by Hegemann (1991).

The two values, $y_{process}$ and $y_{measurement}$, are assumed to be statistically independent so that

$$\sigma^2_{obs} = \sigma^2_{process} + \sigma^2_{measurement}.$$

Figure 5.4 shows graphically that the variation in the distribution of readings y_{obs} is greater than the variation in the theoretical distribution $y_{process}$ due to measurement error. The yield loss due to the variation resulting from the measurement system is the shaded area between the tails of the distributions.

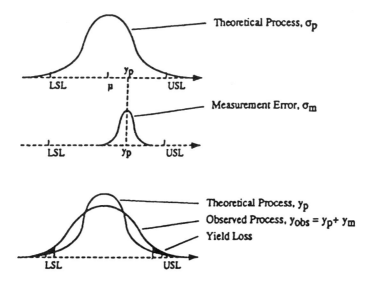

Figure 5.4. *Yield Loss due to Measurement System Error.*

Calculation of Yield Loss Due to Measurement Variation

The estimate of loss (in parts per million, PPM) due to measurement tool variability is derived in the appendix. Operations routinely calculate the average $\bar{y}$ and the standard deviation s from production data, so that the standard normal random variable Z_{obs}

$$Z_{obs} = \frac{\bar{y}_{obs} - \text{LSL}}{\frac{S_{obs}}{\sqrt{n}}}$$

can be computed. A normal probability distribution table can be used to estimate the probability associated with Z_{obs}. PPM_{obs} is then determined by multiplying the estimated probability by 1,000,000.

The standardized normal random variable

$$Z_{process} = Z_{obs} \frac{1}{\sqrt{1 - (\frac{S_{measurement}}{S_{obs}})^2}}$$

allows estimation of the yield loss due to variation in the process, excluding measurement error. The portion of PPM_{obs} loss due to the measurement tool, can be estimated by

$$PPM_{measurement} = PPM_{obs} - PPM_{process}.$$

Table 5.4 gives the estimated yield loss for a wide range of values of Cpk_{obs} (where $Cpk_{obs} = Z_{obs}/3$) and the ratio $S_{measurement}/S_{obs}$. Clearly, the impact of measurement tool variation depends on both $S_{measurement}$ and Cpk_{obs}. The results are tabulated in this manner because many operations routinely track Cpk_{obs}. The estimate $S_{measurement}$ can be combined with S_{obs} obtained using the GRR methodology presented to estimate the yield loss due to measurement tool variation.

Table 5.4. *Yield Loss Due to Measurement System Error—One Sided Specification.*

		\multicolumn{9}{c}{s_m/s_{obs}}
		0.1	0.2	0.3	0.4	0.5	0.6	0.7	0.8	0.9
	1.00	86	243	505	812	1080	1261	1321	1349	1350
	1.10	34	117	214	325	414	466	482	484	484
	1.20	12	38	77	116	142	156	159	159	159
	1.30	4	13	26	38	45	47	48	48	48
	1.33	2	8	17	25	30	32	32	32	32
Observed	1.40	1	4	8	11	12	13	13	13	13
Process	1.50	0.4	1.2	2.3	3	3.4	3.5	3.5	3.5	3.5
Capability	1.60	0.1	0.3	0.5	0.7	0.8	0.8	0.8	0.8	0.8
	1.67	0.04	0.12	0.21	0.26	0.29	0.29	0.29	0.29	0.29
Cp_{obs}	1.70	0.03	0.08	0.13	0.16	0.18	0.18	0.18	0.18	0.18
	1.80	0.01	0.02	0.03	0.03	0.03	0.03	0.03	0.03	0.03
	1.90	0.001	0.003	0.006	0.007	0.007	0.007	0.007	0.007	0.007
	2.00	0.001	0.001	0.001	0.001	0.001	0.001	0.001	0.001	0.001

For the bonding process data

$$LSL = 7.5 \text{ grams,}$$
$$y_{obs} = 11.5 \text{ grams,}$$
$$s_{obs} = 0.7 \text{ grams,}$$
$$s_{measurement} = 0.632,$$

so

$$Cpk_{obs} = (11.5 - 7.5)/(3*0.7) = 1.90$$

and

$$s_{\text{measurement}}/s_{\text{obs}} = 0.90.$$

From Table 5.4, the associated yield loss due to measurement error is estimated to be less than 7 parts per billion (PPB).

When process capability is not good, knowledge of the measurement instrument's variation can be used to guardband product characteristics. A common practice is to use USL $- 3*s_{\text{measurement}}$ and LSL $+ 3*s_{\text{measurement}}$ as the guardbands to prevent bad products from being shipped to customers. The use of guardbands reduces customer reliability problems, but at the same time it increases loss of yield. Management policy needs to strike a balance between loss of reliability and loss of yield.

CONCLUSION AND RECOMMENDATION

The impact of measurement system error on the variation that is seen in process or electrical output parameters can be substantial. In many instances, industries at the leading edge of technology have many measurement systems with poor capability. Even with established technologies, it is necessary to quantify the effectiveness of the measurement systems. There is no standard method to assess measurement system repeatability for destructive measurement tests, and there is no standard method to assess the measurement tool capability when a parameter satisfies only a one-sided specification.

A method using a nested experimental design is presented to assess measurement system repeatability for destructive measurement tests. A method using yield loss due to measurement error is presented to estimate measurement system capability for one-sided specifications. Both methods are applied to a bond pull test measurement system. The methodologies can be easily adapted to similar situations (e.g., die shear), but data collection methods need to be planned carefully to satisfy the assumptions underlying the nested design model. An effective assessment of a measurement tool can also reduce a customer's loss of reliability due to treating bad product as good product.

ACKNOWLEDGMENT

We want to thank Tim Hogan, process engineer, Texas Instruments, for conducting the bond pull experiment and collecting the data for this paper.

REFERENCES

Hegemann, Victor. "Qualitative Assessment of a Measurement System." Doc. No. 11-91-01, Dallas TX: Texas Instruments, 1991.

Montgomery, Douglas C. *Design and Analysis of Experiments*, 4th ed., New York, NY: John Wiley and Sons, Inc., 1997.

Montgomery, Douglas C. *Introduction to Statistical Quality Control*, 2nd ed., New York, NY: John Wiley and Sons, Inc., 1991.

APPENDIX

Estimation of Loss (in ppm) Due to the Measurement Tool Variability

The Cpk capability index of a process is defined by (see Montgomery, *Introduction to Statistical Quality Control*, 1991)

$$\text{Cpk}_{\text{obs}} = \min\left(\frac{\bar{y}_{\text{obs}} - \text{LSL}}{3s_{\text{obs}}}, \frac{\text{USL} - \bar{y}_{\text{obs}}}{3s_{\text{obs}}}\right)$$

and

$$\text{Cpk}_{\text{obs}} = \frac{\bar{y}_{\text{obs}} - \text{LSL}}{3s_{\text{obs}}} \quad \text{(when only LSL is required),}$$

$$\text{Cpk}_{\text{obs}} = \frac{\text{USL} - \bar{y}_{\text{obs}}}{3s_{\text{obs}}} \quad \text{(when only USL is required).}$$

The following results are based on the assumption that only a LSL is required on the process. Similar results can be derived when only an USL is required or both specification limits are required.

$$\text{Cpk}_{\text{obs}} = \frac{\bar{y}_{\text{obs}} - \text{LSL}}{3S_{\text{obs}}} = \frac{1}{3}\left(\frac{\bar{y}_{\text{obs}} - \text{LSL}}{S_{\text{obs}}}\right) = \frac{1}{3}Z_{\text{obs}}.$$

Z_{obs} is a standardized normal random variable which can be used to predict the PPM_{obs} level of the observed characteristic.

In order to determine the portion of PPM_{obs} loss due to the measurement tool, consider the following equations:

$$\text{Cpk}_{\text{obs}} = \frac{1}{3}\left(\frac{\bar{y}_{\text{obs}} - \text{LSL}}{S_{\text{obs}}}\right) = \frac{1}{3}\left(\frac{\bar{y}_{\text{obs}} - \text{LSL}}{S_{\text{measurement}}}\right)\frac{S_{\text{measurement}}}{S_{\text{obs}}}$$

$$= \frac{1}{3}\frac{S_{\text{measurement}}}{S_{\text{obs}}}Z_{\text{measurement}},$$

(5.1)

$$\text{Cpk}_{\text{process}} = \frac{1}{3}\left(\frac{\bar{y}_{\text{obs}} - \text{LSL}}{S_{\text{process}}}\right) = \frac{1}{3}\left(\frac{\bar{y}_{\text{obs}} - \text{LSL}}{\sqrt{s^2_{\text{obs}} - s^2_{\text{measurement}}}}\right)$$

$$= \frac{1}{3}(\frac{\overline{y}_{obs} - LSL}{S_{measurement}}) \frac{S_{measurement}}{\sqrt{S^2_{obs} - S^2_{measurement}}}$$

$$= \frac{1}{3} Z_{measurement} \frac{S_{measurement}}{\sqrt{S^2_{obs} - S^2_{measurement}}}.$$

(5.2)

By combining Equations (5.1) and (5.2),

$$Cpk_{process} = Cpk_{obs} \frac{1}{\sqrt{1 - (\frac{S_{measurement}}{S_{obs}})^2}}$$

Finally,

$$Cpk_{process} = \frac{1}{3} Z_{process},$$

so that

$$Z_{process} = 3 Cpk_{obs} \frac{1}{\sqrt{1 - (\frac{S_{measurement}}{S_{obs}})^2}}.$$

(5.3)

From Equation (5.3), the estimated theoretical PPM can be determined from the normal distribution. Then

$$PPM_{measurement} = PPM_{obs} - PPM_{process}.$$

BIOGRAPHIES

Teresa Mitchell has a Ph.D. in statistics. She is a quality engineer and is responsible for world-wide SPC implementation for Texas Instruments in Dallas, Texas.

Victor Hegemann has a Ph.D. in statistics and works in statistical support for half micron logic production for Texas Instruments in Dallas, Texas.

K. C. Liu is corporate quality, reliability, and assurance director of Hon Hai Group and is responsible for directing the group's quality improvement worldwide.

PART 2

PASSIVE DATA COLLECTION

CHAPTER 6

INTRODUCTION TO PASSIVE DATA COLLECTION

Veronica Czitrom

INTRODUCTION

It is important to understand the current behavior of a process or equipment to be able to improve it. A *passive data collection (PDC)*, or *observational study*, is a systematic approach to collect and analyze data to establish the current process or equipment performance. The term "passive" refers to the fact that the data is collected without making any engineering adjustments (i.e., tweaking) to the process beyond those adjustments that are part of a normal production operation. A passive data collection can be used to:
- establish relationships between process variables
- obtain an initial estimate of stability over time
- obtain an initial estimate of the capability of the equipment to meet process specifications
- understand the sources of variability
- reduce sources of variability
- establish an efficient sampling plan
- indicate opportunities for improvement
- determine the direction for optimization and improvement activities using statistically designed experiments.

A passive data collection can be used for different purposes at different stages of the SEMATECH Qualification Plan. When performed before a statistically designed experiment (Part 3), a passive data collection can help define the objectives of the experiment by identifying areas that require improvement and the factors that can help achieve it. When performed after a designed experiment, a passive data collection can help confirm the experimental results. A passive data collection can be used to establish the baseline process performance and the initial control limits for control charts for statistical process control (Part 4), and to gather initial information for the improvement of equipment reliability (Part 5). In fact, reproducibility studies (Part 1), control charts (Part 4), and marathon runs (Part 5) are types of passive data collections that are used to study sources of variation in the measurement process, establish and maintain process stability, and improve equipment reliability, respectively. While control charts are used primarily to monitor process variability on-line, a passive data collection can be used off-line to reduce process variability by identifying large sources of variability and reducing them. Opportunities for process improvement identified during statistical process control can be addressed using a passive data collection or a designed experiment. A gauge study should be performed before the passive data collection to make sure the measurements are adequate.

The passive data collections in this part of the book are used to:
- identify the largest contributor to variability in a dry etch process and target it for reduction
- establish the feasibility of using cheaper recycled wafers instead of virgin wafers in a new vertical furnace qualification for an oxide process, saving $100,000
- incorporate a predictable bulls-eye pattern into the analysis of a wafer planarization process to give physical understanding of the statistical results and a potential "knob" for further variance reduction
- identify factors that can make a significant difference in the design of bunny suits for clean room use
- reduce the sampling plan of photolithographic registration measurements in a manufacturing plan, decreasing throughput time and personnel
- reduce the resources required and the impact on production of an acceptance sampling plan to determine whether variability in clean room air filters conforms to contract specifications
- predict completion time and track progress of lots through a device development laboratory, which can help managers assess and monitor the health of the laboratory.

The next two sections of this introduction to passive data collection give key concepts and definitions for data collection and data analysis. The following section gives a brief overview of the case studies in the remainder of the part, and the last section gives some suggestions for further reading.

DATA COLLECTION PLAN

A passive data collection consists of a series of independent replications of a process under constant processing conditions and performed over a relatively short period of time without "tweaking". At least thirty independent replications are recommended to obtain reasonably good estimates of variances and means.

Planning is the first and critical step of a passive data collection. The sampling plan should contain a representative sample of the factors of interest and should capture all relevant sources of variability. For example, in a furnace qualification, factors that are likely to affect oxide thickness are furnace run, wafer location in the furnace, and site on wafer. During a passive data collection, it is important to keep track of uncontrolled environmental factors, upstream process parameters, unusual events, and other relevant information. The data must be collected as planned so as not to complicate the analysis. Raw data should be available for analysis in addition to summary statistics, to provide a richer analysis and avoid loss of information. For example, a box plot of the data by site on a wafer might reveal a pattern on the wafer, which cannot be detected using only wafer averages or standard deviations. Intentional over-sampling may be beneficial during the first passive data collection in a qualification. Over-sampling permits informed choices about reducing sampling efforts on subsequent data collection activities, for instance to determine whether it is sufficient to sample 9 sites on a wafer instead of 49, or three times a week instead of seven.

DATA ANALYSIS

After the data has been collected it can be analyzed. Four main areas of data analysis are graphical exploratory data analysis, stability analysis, capability analysis, and analysis of variance. Graphical tools, analysis of variance, and selected statistical terms are defined in the glossary at the end of the book.

Graphical data analysis is the most powerful form of analysis for communication and interpretation. Much of the analysis of the data can be accomplished using simple graphical tools. Graphical analysis should always precede numerical analyses. Kahn and Baczkowski, as well as Czitrom and Reece, analyzed their data graphically before using the analysis of variance for split-plot designs. Exploring data using different graphical tools can help reveal relationships, interesting features, underlying causes of process variability, anomalies such as outliers and non-normality, and directions for subsequent numerical analysis.

Some useful graphs for the analysis of a passive data collection are box plots, bar graphs, Pareto charts, histograms, dot plots, trend and control charts, normal probability plots, multiple variable plots, and scatter plots. Czitrom and Reece (Chapter 8, Figures 8.3 to 8.6) used four *box plots* to display oxide thickness measurements as a function of furnace run, furnace section, wafer type, and wafer site respectively. The box plot by wafer site, for example, shows that observations in oxide thickness vary between sites and within each site. Lynch and Markle (Chapter 7, Figure 7.8) use a *bargraph* to illustrate the relative importance of the run-to-run, wafer-to-wafer, and within-wafer components of the total variance, both before and after a change in the process. Hurwitz and Spagon (Chapter 9, Figure 9.5) give a *Pareto chart* that indicates the relative importance of several contributors to site-to-site variability. Lynch and Markle (Chapter 7, Figures 7.4 and 7.5) give *control charts* for etch rate that point to major changes in the process. They also give *histograms* (Figures 7.6 and 7.7) that indicate major reductions in variability after the major changes in the process. Hall and Carpenter (Chapter 12, Figure 12.1) show that air flow velocities appear to follow a normal distribution, since the observations fall approximately along a straight line on a *normal probability plot*. Kahn and Baczkowski (Chapter 10, Figure 10.3) use a *multiple variable graph* to simultaneously visualize the effect of four factors on particle contamination. Lambert et al. (Chapter 13, Figure 13.5) define a *snapshot* plot to track lot completion time in a device development laboratory. Two curves representing optimistic and pessimistic time estimates allows the snapshot to be thought of as a "control chart" on lots.

A *stable* process or equipment has consistent performance over time. It is important for a process to be stable during manufacture, process optimization, and transfer from development to production. Time trend charts plot data in chronological order, and control charts are time trend charts that include control limits within which a stable process should fall. Trend and control charts can be used to judge stability over time, and to uncover warm-up effects, drifts, and outliers. A passive data collection can provide initial control limits for control charts. Part 4 of the book considers statistical process control and control charts in greater detail.

A *capability analysis* determines the ability of a process on a given piece of equipment to manufacture product that meets specifications. Before process capability can be determined, the process must be stable. To determine *process capability*, the natural spread of the stable process is compared to the process specification limits. The more capable the process, the higher the probability that the measurements will fall inside the specification limits. A histogram of individual measurements that includes the process specifications (capability graph) will provide a graphical assessment of process capability. A more detailed description of process capability is given in the introduction to statistical process control on pages 309–311.

A commonly used numerical measure of process capability is the C_{pk} *capability index* given by

$$C_{pk} = \min\left[\frac{USL - \bar{x}}{3s}, \frac{\bar{x} - LSL}{3s}\right],$$

where $\bar{x}$ is the average and s is the standard deviation of the observations, *LSL* and *USL* are the *lower* and *upper specification limits* of the process, respectively, and min indicates the smallest of the two terms inside the brackets. The larger the value of C_{pk}, the less likely the process is of exceeding specifications. A value of C_{pk} of 1.3 or 1.5 or larger is often considered desirable for observations that come from a normal distribution. Since the estimate of C_{pk} is highly variable, it should be reported with a confidence interval, and it should be computed from one hundred or more observations for greater precision.

Analysis of variance (ANOVA) is a statistical technique that can be used to partition the total variability in the data into *components of variance*. This establishes the sources of variability, or non-uniformity, and their relative contributions to the total observed variability. The sampling plan used to collect the data determines the components that the variability can be partitioned into. Lynch and Markle partitioned etch-rate variability into run-to-run, wafer-to-wafer, and site-to-site (or within wafer) components of variance. For a wafer planarization process, Hurwitz and Spagon incorporated predictable response patterns on wafers into the analysis by further partitioning the within wafer thickness variability into radial and cant components. The variance components can be given as variances or as standard deviations, or as percentages of the total variation, either in a table or in a bar graph. The glossary gives a brief description of analysis of variance.

An important reason for doing a passive data collection is to find the variance components, and in fact all the case studies except the last one do so. The variance components can be used for different purposes. Lynch and Markle as well as Czitrom and Reece identified the large within-wafer component of variance as the greatest opportunity for improvement in etch and oxidation processes, respectively. They planned designed experiments to reduce within-wafer variability. The initial estimates of variability obtained from the passive data collection can be used to compute sample sizes for subsequent designed experiments for process improvement. Components of variance can also be used to optimize resource allocation in sampling plans by sampling more in areas with large variability and less in areas with small variability. Pepper identified lot-to-lot variability as a relatively small component of variance, and developed a reduced acceptance sampling plan that did not need to measure all lots at lithographic layer registration. Kahn and Baczkowski used variance components to determine which factors will need to be included in future experiments in clean room bunny suit designs. Hall and Carpenter compared a variance component to product specifications during acceptance sampling. Czitrom and Reece used variance components as responses to compare the performance of virgin and recycled wafers. To track wafer lot performance in a device development laboratory, Lambert et al. estimate the mean and variability of each process step in lot production, which is analogous to finding the variability of each component of variance in the first five case studies.

Introduction to Case Studies

This section gives a brief overview of the passive data collection case studies presented in the remainder of this part of the book.

Chapter 7. Understanding the Nature of Variability in a Dry Etch Process. *Richard O. Lynch and Richard J. Markle.*

This case study is a straightforward application of the basic principles of a passive data collection. It considers the characterization of a dry etch process that etches silicon dioxide off of silicon wafers in a batch process. Study of etch rate as a function of time was used to identify events after which there was a reduction in the variability of the etch rate and a stabilization of the mean etch rate. Analysis of variance was used to quantify three sources of variability: run-to-run, wafer-to-wafer (between wafers), and site-to-site (within wafer). Within wafer variability was identified as the largest contributor to variability, so its reduction was identified as the greatest opportunity for improvement.

Chapter 8. Virgin Versus Recycled Wafers for Furnace Qualification: Is the Expense Justified? *Veronica Czitrom and Jack E. Reece.*

Semiconductor process or equipment qualification often requires the use of a large number of wafers. This case study presents a strategy for determining whether it is possible to use cheaper recycled wafers instead of the more expensive virgin wafers. Virgin and recycled 200 mm wafers were compared for the qualification of a new three-zone vertical furnace to be used for an oxidation process. The variance components of oxide thickness for each type of wafer were furnace run, location within the furnace, wafer, and site on wafer. The variance components were found to be comparable for the different types of wafer. This led to the decision to use the cheaper recycled wafers in the remainder of the qualification of four furnaces, saving over $100,000.

Chapter 9. Identifying Sources of Variation in a Wafer Planarization Process. *Arnon M. Hurwitz and Patrick D. Spagon.*

A common occurrence in the semiconductor industry is the presence of significant, predictable patterns on a wafer that are consistent across wafers and across lots. This case study considers the strong radial, or bulls-eye, pattern of wafer thickness following wafer planarization. This bulls-eye effect, as well as a cant effect from one side of the wafer to the other, were explicitly accounted for in the analysis of variance by using orthogonal contrasts to partition the site-to-site variability into components corresponding to the effects.

Chapter 10. Factors Which Affect the Number of Aerosol Particles Released by Clean Room Operators. *William Kahn and Carole Baczkowski.*

Particulate contamination is a major issue in semiconductor fabrication. People can be the largest source of particulate contamination in the clean rooms where integrated circuits are manufactured on silicon wafers. This case study describes an experiment to evaluate factors which affect particulate contamination released by "bunny suits," the garments used by people in clean rooms. The results of the experiment are first analyzed graphically, and then with an analysis of variance. The experiment is a full factorial because every level of every factor appears with every other level of every other factor, but restrictions on randomization lead to a split-split-plot structure.

Chapter 11. A Skip-Lot Sampling Plan Based on Variance Components for Photolithographic Registration Measurements. *Dwayne Pepper.*

In a manufacturing environment, measurements are non-value added steps in a process. This case study describes how the number of measurements of lithographic layer registration were reduced by using a sampling plan where some lots are not measured. The reduced sampling plan was developed by considering that lot-to-lot variability is not large in comparison to the other components of variance as revealed by an analysis of variance of historical data, that time and labor are saved by skipping lots at measurement, and that the risk of not measuring lots that are out of specification and that are rejected once they have completed all their expensive processing is acceptably small. The shop operators came up with a procedure to implement the skip-lot sampling plan in practice. The skip-lot sampling plan reduced throughput time by 19 to 30 hours (depending on the product), and by one operator per shift.

Chapter 12. Sampling to Meet a Variance Specification: Clean Room Qualification. *Kathryn Hall and Steven Carpenter.*

It is desirable to minimize the disruption to a production line and the resources required to check whether a new product is in conformance with contract specifications. This was achieved here by using a two-stage sampling plan to determine whether the variability in air velocity through clean room air filters is as low as specified in the purchasing contract. Analysis of variance is used to estimate the variability, and also to remove measurement error. The study was completed after the first stage of sampling when the variability was found to be larger than specified.

Chapter 13. Snapshot: A Plot Showing Progress Through a Device Development Laboratory. *Diane Lambert, James M. Landwehr, and Ming-Jen Shyu.*

Managers need to assess the health of a device development laboratory, and engineers need to predict completion time and track the progress of their lots. This case study presents a new plot of process step completion, called a snapshot, that addresses both these needs. The snapshot can be used when different codes, processes, and mixes of products are present in a DDL, and in spite of the fact that these change constantly. The snapshot can be used for a new code for which a lot has never been run, and it assesses the health of the DDL for short periods and monitors its health over time. Step completion in a snapshot is found by combining information from existing data bases on queue time spent waiting to begin a step, and run time spent in the processing step. The total time needed to process a lot of a particular code is the sum of the step times (excluding engineering holds). The snapshot also shows two curves that represent optimistic and pessimistic time estimates, so that the snapshot can be considered as a "control chart" on lots.

The objective of a passive data collection is to establish current process or equipment performance. Since a snapshot is used to track wafer lot performance in a DDL, the "process" is the lot production process. Lambert et al. estimate the variability and mean of each step of the lot production, which is analogous to finding the variability of each component of variance in the other passive data collection case studies.

SUGGESTED READING

All SEMATECH statistical publications that are available to the general public can be accessed at the following Internet URL: http://www.sematech.org/

Additional information on passive data collection is given in the following article and book:

Hamada, M., R.J. Mackay, and J.B. Whitney. "Continuous Process Improvement with Observational Studies". *Journal of Quality Technology*. Vol. 25, No. 2, pp. 77-84, 1993.

Cochran, W.G. *Planning and Analysis of Observational Studies*. New York: John Wiley and Sons, 1983.

A book that covers several topics of interest in a passive data collection is:

Ryan, Thomas P. *Statistical Methods for Quality Improvement*. John Wiley and Sons, Inc., 1989. Chapter 11: graphical methods, chapters 4 to 10: control charts, chapter 7: process capability.

A book on graphical data analysis is:

Nelson, W. *How to Analyze Data with Simple Plots*. Milwaukee, WI: American Society for Quality Control, 1990.

Books that contain information on variance components are:

Box, George E.P., Hunter, William G., and Hunter, Stuart J. *Statistics for Experimenters*. New York: John Wiley and Sons, 1978. Pages 571 to 582.

Graybill, Richard K., and Graybill, Franklin A. *Confidence Intervals on Variance Components*. New York: Marcel Dekker, Inc., 1992.

Leitnaker, Mary G., Sanders, Richard D., and Hild, Cheryl. *The Power of Statistical Thinking— Improving Industrial Processes*. Massachusetts: Addison–Wesley Publishing Co., 1996.

Montgomery, Douglas C. *Design and Analysis of Experiments*, fourth edition. New York: John Wiley and Sons, Inc., 1997.

CHAPTER 7

UNDERSTANDING THE NATURE OF VARIABILITY IN A DRY ETCH PROCESS

Richard O. Lynch and Richard J. Markle

This paper describes a passive data collection (PDC) carried out for a dry etch process that etches SiO_2 (oxide) off of silicon wafers in a batch process. By studying the etch rate data obtained from the PDC, we saw that the process was statistically out-of-control. Some likely assignable causes were uncovered for the out of control behavior of the process. Analysis of variance (ANOVA) was used to quantify the sources of variability in the oxide etch rate.

EXECUTIVE SUMMARY

Problem

During the start-up of the SEMATECH wafer fab, several etch processes needed to be characterized in order to begin the phase 1 process demonstration. One such process involved an oxide etch using a commercially available multi-wafer etch tool. The goal of the work described in this case study was to develop an understanding of the rate and nonuniformity of the oxide etch process.

Solution Strategy

A passive data collection strategy was used to study the oxide etch rate and nonuniformity. This involved 27 (three less than generally recommended) independent runs of the process. Data collected from the process was analyzed to evaluate process stability and capability as well as to identify opportunities for process improvement.

72

Conclusion

The process was judged to be out of control. Likely causes for the condition were identified. Reduction of the within-wafer nonuniformity was identified as the greatest opportunity for improvement.

PROCESS

As part of the fabrication of integrated circuits, an oxide layer must be etched away in a controlled, uniform, and repeatable manner from selected sites on silicon wafers. One approach to this etching process involves plasma etching a batch of 18 wafers simultaneously. In Figure 7.1, the wafer loading pattern within the etcher is depicted.

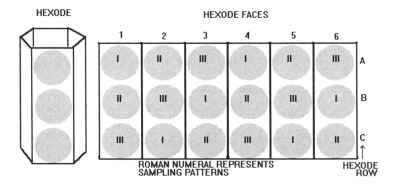

Figure 7.1. *Hexode Loading Pattern for Plasma Etching of an Oxide Layer from a Batch of* 18 *Wafers.*

The etch rate can be measured at a specific site on a wafer by determining the amount of oxide etched at that site, then dividing by the process time. The amount is determined by making oxide thickness measurements at the site before and after processing (see Figure 7.2). The etch rates discussed will be in units of angstroms per minute.

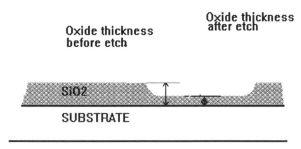

Figure 7.2. *Cross Section of Etched Wafer Indicating Oxide Thickness before and after Etching.*

DATA COLLECTION PLAN

A passive data collection (PDC) strategy was used to study the oxide etch rate and nonuniformity. This data was analyzed after 27 independent runs of the process were obtained. This is three runs short of the recommended number of runs to be used in a PDC. Data collected from the process was analyzed to evaluate process stability and capability as well as to identify opportunities for process improvement. Oxide etch rates were sampled from six wafers in each run, at nine sites within each wafer.

To reduce the amount of work required to collect and analyze the data, only six of the eighteen wafers in each run were measured. The six wafers that were systematically sampled in any particular run are indicated by roman numerals in Figure 7.1. For runs 1, 4, 7, 10, 13, 16, 19, 22, and 25, wafers were sampled from the locations denoted by "I". For runs 2, 5, 8, 11, 14, 17, 20, 23, and 26, wafers were sampled from the locations denoted by "II". Finally, for runs 3, 6, 9, 12, 15, 21, and 27, wafers were sampled from the locations denoted by "III".

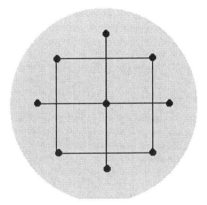

Figure 7.3. *Box and Star Pattern Sampling Plan on a Wafer.*

The nine sites sampled within each wafer were selected systematically, using a "box and star" configuration, as displayed in Figure 7.3.

Passive Data Collection

Because this was to be a passive data collection, there was not supposed to be any process adjustment during the study. The data that resulted from this PDC was intended to exhibit the natural variation that is characteristic of the current process. In effect, we would allow the process to speak to us about its natural capability. The information that we planned to obtain from the PDC included:
- stability of the process
- capability of the process
- contribution of different sources to total process variability.

DATA ANALYSIS

The results of the experiment are given in Appendix A. Highlights of the data analysis follow.

Process Stability

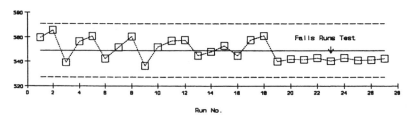

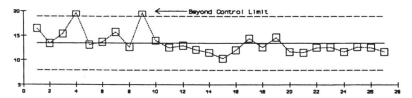

Figure 7.4. *Shewhart Control Charts for the Average Etch Rates and the Standard Deviation of the Etch Rates for the 27 Runs.*

The stability (or instability) of a process must be determined before process improvement efforts are undertaken. The operational definition of stability for this work was "in control based on a Shewhart control chart." Figure 7.4 gives the Shewhart control charts for the within-run averages and within-run standard deviations. The control limits for the control charts were obtained using the rolling range method. The control charts indicate that the process was not stable during the PDC, since there are more than eight points on one side of the center line in the first chart, and there are two points beyond control limits in the second chart.

Assignable Causes

In reviewing the events that occurred in the etch area during the time that the PDC was performed, several possible causes for the process shifts were identified. We will describe two events that may have been the cause for the process changes that were observed in the control charts. These events may not be the true causes for the instability, but they are likely causes that can be tested with additional experimentation.

After run 9, the equipment began to be exercised repeatedly between the PDC runs. This could explain the reduced within-run variability that is observed starting with run 10.

After run 18, there was a recalibration of the CHF3 Mass Flow Controller. This could explain the reduced variability in within-run averages that is observed after run 18.

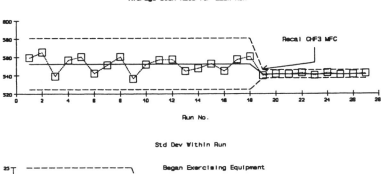

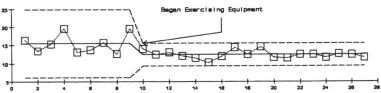

Figure 7.5. *Control Charts with Revised Control Limits. Two likely special causes were identified that changed the mean and the standard deviation of the etch rates: the recalibration of a mass flow controller and the initiation of equipment exercising.*

In Figure 7.5, the Shewhart control charts have been modified to show limits that are recomputed at the times of the process changes. From these charts, there seem to be three stable time frames: runs 1 through 9, runs 10 through 18, and runs 19 through 27. We focused on runs 1 through 9 as a BEFORE case and runs 19 through 27 as an AFTER case for studying the process capability and the analysis of variance.

Process Capability

The capability of a process is generally evaluated in light of some engineering specifications. It is therefore crucial that the engineering specifications be properly determined. Ideally, engineering specifications should reflect the range of process performance that is considered acceptable according to fit and function. At the time that this study was performed, there were no engineering specifications that could be justified as reflecting the true requirements for the manufacturing process to successfully meet fit and function requirements. Therefore, the capability summary is limited to estimating the natural tolerance, given by 6 × (total standard deviation).

Figure 7.6 gives the histogram for the BEFORE case (first nine runs). A normal distribution is superimposed on the histogram. If we assume that the true distribution is the normal distribution, then this superimposed normal distribution can be used to estimate the proportion of process output that would be nonconforming with respect to engineering specifications. The mean and the natural tolerance (six times the standard deviation) for the BEFORE process are estimated to be mean = 552.5 and NT = 113.4 (Å per minute). A 95% confidence interval for the mean is 552.5 ± 8.2, or 544.3 to 560.7, and for the natural tolerance is 97.2 to 135.9. The formulas that were used to estimate the confidence intervals are provided in Appendix B.

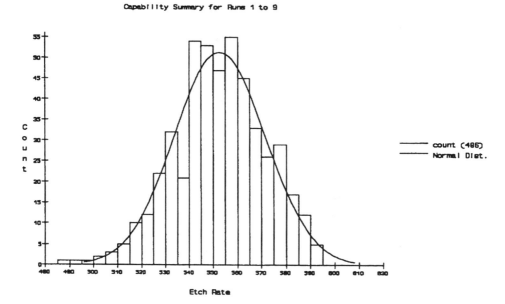

Figure 7.6. *Histogram and Normal Curve for the BEFORE Case (Runs 1 to 9).*

Figure 7.7 shows the histogram and the normal curve for the AFTER case. The estimated mean is 542.7, and the natural tolerance (NT) is 80 (Å per minute). As expected, the natural tolerance (which is a measure of variability) is smaller than in the BEFORE case. The 95% confidence interval for the mean is 542.7 ± 1.4, or 541.3 to 544.1, and for the natural tolerance it is 73.8 to 87.1.

A few words of warning are called for. A capability analysis is only meaningful for processes that are stable. Stability of the process must be established by studying the Shewhart Control Charts. Finally, capability indices are only as meaningful as the Engineering Specifications that are used in the calculation. See Mullenix (1990) for a more complete list of cautions.

Partitioning Process Variability

The data from the PDC were analyzed statistically to gain insight into the sources of variability. Analysis of variance (ANOVA) was used to partition the total observed variability into three sources: Run-to-Run, Wafer-to-Wafer, and Within-Wafer. A description of the calculations involved in this type of ANOVA is given in Box, Hunter, and Hunter (1978, pp. 571 to 582).

In Table 7.1, the ANOVA results are summarized for the BEFORE and AFTER cases. The column labeled "Variance Component" provides the estimates of the contribution to the total variance from each one of the sources listed in the column denoted "Source of Variation." These sources estimate the parameters in a nested model of the process. The nested model may be written as

$$Y_{ijk} = \text{MEAN} + R_i + W_{ij} + D_{ijk}, \tag{7.1}$$

where Y_{ijk} is the observable etch rate at the kth position within the jth wafer of the ith lot. Further,

- R_i is a random variable, with mean = 0 and variance = σ^2_{LOT}.
- W_{ij} is a random variable, with mean = 0 and variance = σ^2_{WAFER}.
- D_{ijk} is a random variable, with mean = 0 and variance = $\sigma^2_{\text{Within Wafer}}$.
- All of the random variables are assumed to be independent.

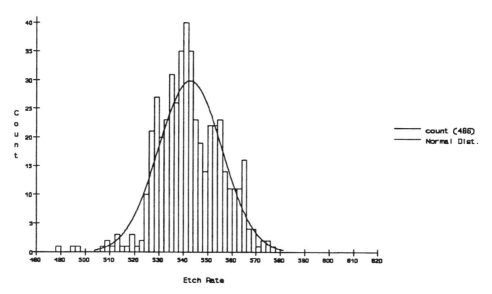

Figure 7.7. *Histogram and Normal Curve for the AFTER Case (Runs 19 to 27)*.

The variance components in the ANOVA table (Table 7.1) estimate the variances of the random variables listed above. The estimate of the total variance of Y_{ijk} is the sum of the variance components. The estimated process standard deviations that were shown in Figures 7.6 and 7.7 are the square roots of the sums of the variance components.

Table 7.1. *ANOVA results from the BEFORE and AFTER cases.*

ANOVA Results for BEFORE (Runs 1 to 9)				
Source of Variation	Degrees of Freedom	Mean Square	Variance Component	Percent of Total
Run to Run	8	6106.5	94.2	26
Wafer to Wafer	45	1019.7	94.6	26
Within Wafer	432	168.3	168.3	47
Total			357.1	100
Overall Mean			552.5	
Overall STD			18.9	
ANOVA Results for AFTER (Runs 19 to 27)				
Source of Variation	Degrees of Freedom	Mean Square	Variance Component	Percent of Total
Run to Run	8	178.0	0.0*	00
Wafer to Wafer	45	623.1	55.7	31
Within Wafer	432	121.7	121.7	69
Total			177.4	100
Overall Mean			542.7	
Overall STD			13.3	

*Negative, set to 0.

The variance components can be used to estimate the total variance and to estimate the relative contribution of each source of variability to the total variability. The last columns of both sections of Table 7.1, Percent of Total, indicate that both BEFORE and AFTER, within-wafer variability was the largest contributor to the total variability with 47% and 69% of the total variability, respectively. The greatest opportunity for improvement will then be the reduction in variability across the wafer.

The estimate of the Run-to-Run variance component for the AFTER case turned out to be less than 0 (actually −8.2), and was therefore set to 0, since variances by definition are not negative.

Notice the difference in variance components in the BEFORE and AFTER cases. The run-to-run variability was dramatically reduced from 94 to 0. The within wafer and wafer-to-wafer variabilities were also reduced. In Figure 7.8, a bar graph is used to highlight the reduction in the variance components.

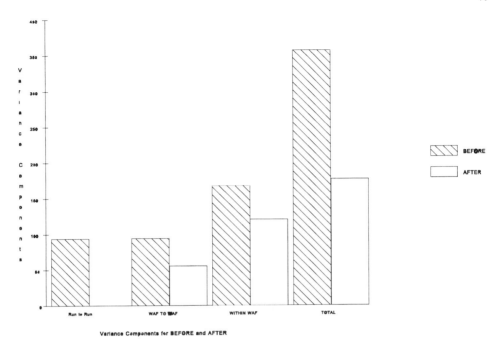

Figure 7.8. Bar Graph Showing the Reduction in Variance Components from the BEFORE Case to the AFTER Case.

CONCLUSIONS AND RECOMMENDATIONS

Passive Data Collection

Using a passive data collection, the performance of a process can be evaluated in a systematic manner. The data resulting from a PDC can be used to evaluate the stability and capability of a process. Furthermore, the components of variability can be estimated using ANOVA. A study of variance components helps identify the greatest opportunities for process variability reduction. The comparison of components of variability allows the success of process improvement efforts to be quantified.

Oxide Etch Process

The process was judged to be out of control. Likely causes of instability were identified to be the increased usage of the etch tool (which reduced within-run variability) and recalibration of the Mass Flow Controller (which reduced run to run variability). The natural tolerance of the process was estimated to be 80 Å per minute based on the last nine runs. Within-wafer nonuniformity was identified as the greatest opportunity for improvement.

Future Plans

Future equipment improvements will be identified that should allow further reductions in wafer-to-wafer variability and within-wafer variability. The PDC will be repeated to verify resulting performance improvements.

REFERENCES

Box, G.E., W.G. Hunter, and J.S. Hunter. *Statistics for Experimenters*. New York, NY: John Wiley and Sons, Inc. 1978.

Lynch, R.O. and R.J. Markle. "Statistical Methodology for the Evaluation Characterization and Optimization of Dry Etch Modules." *Dry Processing for Submicrometer Lithography*. James Bondur, Alan R. Reinberg, Eds. Proc. SPIE 1185, pp. 191 to 197, 1989.

Mullenix, P. "The Capability of Capability Indices with an Application to Guardbanding in a Test Environment." *International Test Conference Proceeding*. IEEE. pp. 907-915, 1990.

Satterthwaite, F.E. "An Approximate distribution of Estimates of Variance Components." *Biometrics Bulletin*. Vol. 2, 1946.

APPENDIX A

Raw Data for PDC

Oxide etch rates for 27 runs, 6 wafers/run, 9 sites/wafer. Sampling is indicated in Figures 7.1 and 7.3, and S1 denotes Site 1, S2 denotes Site 2, etc.

0 RUN	1 WAF	2 S1	3 S2	4 S3	5 S4	6 S5	7 S6	8 S7	9 S8	10 S9
1	1	1	581.14	568.41	571.89	577.14	581.63	545.45	543.04	575.72	540.14
2	1	2	560.81	550.83	554.86	553.39	555.16	537.36	533.59	537.71	526.40
3	1	3	577.13	568.28	569.96	588.71	577.45	556.18	553.25	572.21	541.76
4	1	4	577.58	559.26	561.00	593.21	582.58	528.55	557.79	583.45	540.25
5	1	5	557.95	552.19	553.53	549.39	553.35	552.47	563.81	577.58	552.99
6	1	6	570.51	559.47	563.59	564.91	569.36	547.76	534.69	573.29	532.84
7	2	1	578.21	564.94	569.29	563.66	573.84	593.09	557.05	562.78	548.15
8	2	2	591.81	577.81	576.86	576.21	591.81	580.14	557.74	567.09	561.59
9	2	3	563.41	556.29	554.58	547.24	556.35	587.33	548.91	550.90	547.61
10	2	4	570.28	560.22	558.24	552.59	568.28	587.39	546.28	552.99	549.34
11	2	5	587.89	566.33	574.26	575.65	586.58	570.53	558.09	566.96	547.36
12	2	6	569.55	558.26	560.15	554.09	561.35	587.31	555.05	557.69	550.54
13	3	1	574.74	561.49	558.80	559.69	570.71	562.00	534.11	544.95	534.46
14	3	2	544.61	536.46	529.47	528.68	541.99	561.85	523.70	540.61	530.49
15	3	3	555.03	542.81	544.95	542.47	551.80	571.34	529.33	544.65	530.11
16	3	4	555.43	532.51	533.31	548.78	560.65	523.19	522.44	543.35	522.38
17	3	5	529.06	514.18	509.75	517.85	531.65	539.39	512.58	532.25	517.39
18	3	6	545.72	529.18	527.30	535.08	548.91	551.66	516.04	541.09	524.59
19	4	1	530.35	509.90	503.69	524.99	542.34	531.25	516.09	548.30	521.63
20	4	2	576.24	563.24	560.25	562.38	577.15	586.31	544.72	564.22	552.53
21	4	3	587.47	575.09	575.44	571.11	584.72	579.58	551.59	556.33	551.94
22	4	4	559.94	552.03	549.45	542.47	552.47	579.22	543.35	546.18	543.05
23	4	5	572.19	558.89	560.22	558.51	567.29	588.13	548.86	557.15	543.76
24	4	6	581.64	568.28	565.79	567.89	581.11	568.16	545.66	560.28	547.71
25	5	1	571.26	557.83	556.20	559.44	570.51	579.66	545.24	563.78	545.79
26	5	2	590.49	575.41	575.22	575.72	589.89	575.16	554.72	561.29	554.78
27	5	3	564.04	555.01	554.24	548.94	558.76	580.74	545.26	551.93	545.91
28	5	4	568.85	558.16	558.38	553.30	566.15	584.93	541.33	549.71	544.10
29	5	5	576.40	552.03	554.69	568.34	581.59	545.25	540.88	563.15	539.78
30	5	6	567.09	555.01	557.47	553.58	559.49	580.22	555.10	554.03	547.30
31	6	1	574.56	562.25	563.36	560.21	570.71	565.59	543.59	546.05	535.35
32	6	2	549.19	539.91	538.83	533.83	544.69	560.84	531.30	535.31	531.61
33	6	3	556.39	543.97	546.74	544.39	553.35	565.84	532.51	544.68	532.19
34	6	4	553.66	532.34	529.24	544.84	561.21	518.35	515.80	543.54	523.78
35	6	5	543.85	529.11	529.96	531.14	540.59	549.41	526.70	534.43	524.15
36	6	6	548.69	533.45	534.26	538.70	549.40	549.66	524.83	541.35	527.08
37	7	1	540.39	517.60	519.93	538.19	544.46	540.09	532.58	549.97	523.47
38	7	2	572.38	559.59	558.70	560.39	572.61	581.70	546.63	561.70	549.88
39	7	3	577.72	555.79	555.09	567.40	582.74	547.24	536.76	559.30	542.14
40	7	4	557.08	549.15	548.65	541.91	549.90	575.64	543.34	542.18	540.93
41	7	5	569.85	555.45	559.13	558.14	566.22	581.83	546.06	552.54	539.46
42	7	6	560.88	538.40	538.35	555.80	567.80	528.91	531.08	557.13	531.91
43	8	1	572.46	559.76	561.80	560.63	569.89	584.80	547.20	559.25	543.94
44	8	2	588.21	574.00	573.84	573.60	587.88	572.09	555.69	561.08	554.83
45	8	3	563.38	553.11	553.58	547.84	556.53	579.46	547.58	549.70	545.59
46	8	4	569.59	559.10	558.88	552.94	567.16	583.84	543.86	550.46	545.33
47	8	5	573.84	551.08	553.20	565.70	579.01	545.06	543.63	561.28	538.43
48	8	6	567.97	555.50	557.51	563.84	560.94	578.71	549.75	552.71	545.53
49	9	1	575.00	563.59	562.85	560.26	571.21	567.09	537.41	544.09	534.60
50	9	2	545.10	537.78	531.40	529.38	542.70	560.15	526.35	533.95	529.76
51	9	3	554.51	541.89	544.04	541.71	551.85	565.68	530.58	540.90	529.96
52	9	4	528.94	504.94	495.86	523.59	547.64	489.74	491.30	537.44	510.04
53	9	5	528.09	515.49	508.61	514.75	530.93	536.10	512.88	526.90	518.25
54	9	6	553.31	540.29	542.79	541.04	551.41	565.88	530.75	540.26	529.15
55	10	1	540.71	526.44	522.59	532.34	544.89	544.89	532.69	545.46	529.59
56	10	2	572.54	559.56	557.66	559.22	572.89	583.06	552.85	560.51	546.09
57	10	3	575.65	559.84	554.26	562.41	578.61	555.81	546.94	556.03	542.31
58	10	4	556.05	548.35	547.46	541.95	549.35	576.38	548.61	543.81	537.96
59	10	5	551.06	537.54	535.33	541.84	555.45	562.41	535.22	549.54	536.30

0 RUN	1 WAF	2 S1	3 S2	4 S3	5 S4	6 S5	7 S6	8 S7	9 S8	10 S9
60	10	6	574.35	561.29	552.33	558.96	576.79	553.84	541.89	555.85	540.56
61	11	1	566.38	555.08	555.08	553.45	562.63	581.05	542.41	554.29	537.80
62	11	2	585.41	572.09	572.53	570.24	584.88	573.19	549.80	558.53	551.55
63	11	3	557.86	549.91	546.96	542.74	552.50	575.44	539.89	550.53	540.39
64	11	4	563.74	553.26	552.80	547.26	559.44	579.06	537.99	546.59	539.91
65	11	5	576.41	554.66	560.24	564.46	576.44	555.10	542.61	557.36	534.55
66	11	6	560.96	550.65	552.33	547.79	552.38	576.24	554.29	557.16	539.10
67	12	1	573.86	561.76	564.08	558.59	568.35	569.51	540.85	548.14	539.47
68	12	2	546.96	540.18	536.00	532.43	543.85	562.64	531.14	538.38	536.22
69	12	3	554.63	543.85	545.59	542.90	550.99	568.96	533.05	542.26	534.81
70	12	4	553.22	536.71	533.54	545.38	559.88	530.75	522.70	545.26	530.60
71	12	5	543.49	532.29	531.55	531.66	539.84	553.81	531.10	537.08	528.97
72	12	6	551.20	540.22	541.71	540.55	548.99	562.26	531.97	540.70	531.94
73	13	1	547.14	534.19	537.60	539.76	543.21	555.43	540.21	540.74	531.59
74	13	2	564.59	555.29	555.28	552.81	565.01	572.46	539.09	552.49	544.26
75	13	3	572.21	559.84	558.65	557.85	570.05	561.45	541.11	547.19	542.56
76	13	4	548.06	540.43	540.41	533.09	541.78	563.09	536.74	534.06	534.28
77	13	5	556.35	544.43	547.53	545.16	551.90	568.36	539.14	545.10	533.75
78	13	6	563.65	551.96	553.30	552.46	561.15	553.49	543.30	547.26	535.49
79	14	1	563.21	551.21	553.89	551.84	563.40	572.72	543.80	552.33	543.63
80	14	2	575.59	562.50	562.69	562.34	576.35	560.90	546.55	552.51	551.96
81	14	3	554.29	545.58	544.90	539.35	551.29	566.60	540.56	541.30	542.40
82	14	4	559.29	549.96	549.41	543.61	558.56	570.70	540.29	544.29	542.36
83	14	5	570.41	551.18	558.22	559.97	570.59	552.79	546.10	553.34	539.40
84	14	6	555.24	544.04	546.66	542.90	549.59	564.56	543.46	543.59	540.35
85	15	1	572.41	560.44	559.24	558.03	568.39	565.79	538.04	548.21	539.84
86	15	2	549.31	541.35	540.31	535.71	545.19	562.08	534.26	539.22	535.10
87	15	3	554.61	544.06	545.25	542.50	552.25	564.84	532.47	543.85	535.08
88	15	4	549.83	536.97	526.10	538.97	560.44	525.01	513.05	545.53	532.04
89	15	5	546.74	535.69	537.97	534.46	540.40	556.85	533.58	536.99	527.33
90	15	6	552.84	541.26	543.35	541.76	551.40	559.29	531.84	544.54	532.21
91	16	1	547.56	539.36	551.41	532.63	529.79	548.36	546.18	529.90	534.94
92	16	2	572.76	556.14	573.21	562.94	562.86	559.11	587.97	561.64	556.30
93	16	3	586.13	567.29	583.29	574.05	575.49	545.35	575.46	564.38	564.58
94	16	4	555.56	546.06	542.26	548.53	547.99	533.31	559.01	539.97	578.88
95	16	5	567.14	553.46	563.78	556.43	558.41	550.28	561.91	553.41	570.78
96	16	6	572.56	559.72	567.90	565.75	560.46	535.35	534.30	541.40	560.91
97	17	1	572.69	558.94	562.66	559.80	569.19	582.78	547.09	554.54	546.11
98	17	2	587.76	571.85	571.97	571.69	587.41	573.13	549.00	558.40	555.09
99	17	3	561.28	552.05	551.84	545.84	556.39	574.05	543.83	545.88	546.97
100	17	4	568.25	557.84	558.30	551.75	565.34	582.49	542.65	547.68	546.11
101	17	5	587.47	565.00	569.53	573.97	588.14	563.28	546.99	560.26	547.70
102	17	6	564.90	552.78	554.28	551.70	559.83	572.88	547.36	550.94	547.28
103	18	1	569.78	547.04	550.94	564.51	573.59	548.13	542.90	561.78	540.50
104	18	2	545.95	539.33	537.38	530.10	540.84	560.22	529.21	529.53	534.16
105	18	3	552.35	541.79	542.90	538.91	550.79	559.86	528.03	537.68	532.15
106	18	4	543.46	535.28	514.44	526.72	556.78	517.68	495.75	533.68	531.31
107	18	5	539.63	528.97	525.34	526.00	536.80	546.16	521.01	526.22	522.94
108	18	6	552.76	543.25	543.59	540.91	551.59	563.05	530.63	538.63	532.15
109	19	1	539.36	525.26	525.72	531.99	541.16	552.45	531.81	551.65	533.26
110	19	2	556.54	547.71	546.34	543.84	557.39	568.58	530.89	543.56	540.65
111	19	3	564.47	553.24	553.50	551.06	563.20	554.41	531.31	539.13	534.83
112	19	4	543.16	536.38	535.15	527.58	537.30	557.70	526.54	526.25	528.38
113	19	5	553.15	541.46	545.25	542.78	549.10	563.26	532.84	536.29	528.64
114	19	6	558.31	546.90	548.28	548.06	556.93	545.90	535.24	540.15	528.51
115	20	1	554.72	543.00	541.96	543.76	554.74	569.66	532.68	550.34	538.72
116	20	2	563.90	550.66	551.71	550.70	562.64	550.28	533.59	539.09	539.13
117	20	3	542.05	536.04	530.10	525.71	537.79	555.72	523.34	528.28	529.39
118	20	4	551.72	543.99	542.04	535.75	548.21	564.94	527.84	532.50	531.80

0 RUN	1 WAF	2 S1	3 S2	4 S3	5 S4	6 S5	7 S6	8 S7	9 S8	10 S9
119	20	5	558.71	541.09	542.47	548.28	559.65	539.00	526.59	540.65	526.74
120	20	6	548.34	538.65	540.72	535.59	540.39	559.59	539.21	533.03	529.83
121	21	1	552.79	542.43	541.89	540.68	551.08	562.59	531.45	540.39	532.31
122	21	2	547.50	537.59	539.40	534.36	540.97	559.04	534.16	532.85	528.53
123	21	3	552.65	539.00	529.18	540.43	560.40	527.96	513.33	540.20	529.90
124	21	4	556.56	546.35	546.63	542.84	553.84	564.51	534.40	541.11	536.46
125	21	5	547.38	539.89	537.44	532.45	543.78	559.47	532.10	532.53	537.39
126	21	6	575.79	555.38	560.91	567.61	576.41	561.31	549.44	563.75	543.61
127	22	1	531.54	515.90	513.78	526.29	536.84	542.56	524.80	550.90	529.55
128	22	2	555.95	546.66	544.75	542.97	556.45	564.65	528.40	542.65	538.80
129	22	3	565.78	554.89	554.81	552.60	564.14	555.84	531.28	539.68	534.46
130	22	4	543.97	536.76	535.83	528.19	538.84	555.19	520.86	527.79	534.50
131	22	5	550.74	540.47	541.66	538.40	548.63	561.68	530.19	536.19	528.69
132	22	6	552.60	545.16	532.10	536.49	557.84	536.28	506.55	536.78	534.95
133	23	1	556.83	546.28	546.66	546.03	556.74	573.68	533.94	551.06	539.66
134	23	2	564.66	553.74	552.01	551.95	565.49	553.25	531.21	541.46	541.09
135	23	3	547.05	540.19	538.10	531.99	542.21	562.56	529.46	533.00	532.71
136	23	4	552.16	544.49	541.69	537.08	550.21	565.95	526.10	535.21	532.97
137	23	5	562.28	545.03	546.01	552.33	564.26	541.78	526.65	543.63	529.54
138	23	6	550.29	540.71	542.09	537.97	544.50	561.33	539.21	537.09	532.59
139	24	1	529.10	511.90	509.68	525.69	535.59	538.91	522.18	549.76	526.84
140	24	2	555.26	545.91	545.09	545.40	555.03	568.88	535.16	544.68	541.28
141	24	3	567.26	556.20	555.76	554.39	564.21	560.49	538.34	543.11	538.26
142	24	4	540.94	535.55	534.21	528.22	535.97	559.90	531.71	527.75	530.56
143	24	5	550.40	540.39	539.34	538.72	548.34	560.59	526.94	538.71	528.44
144	24	6	552.64	543.68	534.38	539.70	555.95	537.14	518.58	538.75	529.09
145	25	1	534.49	519.53	516.68	528.10	539.56	544.35	526.26	551.28	530.93
146	25	2	556.53	547.26	545.59	544.39	555.88	567.85	529.65	543.71	538.15
147	25	3	565.85	554.95	555.78	553.41	563.79	557.54	535.28	540.41	556.90
148	25	4	538.79	528.88	529.45	526.46	535.21	550.71	525.34	526.34	524.18
149	25	5	551.69	541.22	542.79	540.91	548.56	563.43	530.96	537.21	528.54
150	25	6	555.89	545.83	541.50	545.00	557.09	541.70	525.16	541.41	529.81
151	26	1	553.58	543.90	546.03	543.96	551.84	574.55	535.06	548.81	536.16
152	26	2	564.78	554.06	553.51	552.63	564.15	555.70	534.18	542.49	541.65
153	26	3	544.54	538.59	535.45	529.69	539.53	561.14	527.75	530.90	531.28
154	26	4	550.40	543.66	540.78	535.58	548.55	565.13	524.99	534.66	532.44
155	26	5	559.90	543.90	545.47	550.36	561.06	542.54	525.96	543.20	527.53
156	26	6	549.46	540.03	541.33	537.95	543.49	559.54	538.50	536.65	531.79
157	27	1	573.43	560.45	562.86	561.99	571.04	568.89	546.97	553.25	542.84
158	27	2	544.94	538.74	536.01	530.31	541.09	559.78	529.21	532.49	534.19
159	27	3	555.80	545.65	546.60	543.33	552.90	566.56	532.18	540.97	534.29
160	27	4	527.24	509.20	496.24	519.58	545.79	494.26	488.69	536.74	513.93
161	27	5	546.55	537.33	539.76	534.29	539.96	559.16	533.69	532.13	527.16
162	27	6	555.94	545.70	546.89	544.16	553.61	564.05	534.35	541.59	533.09

APPENDIX B

Formulas For Natural Tolerance and Confidence Intervals

In order to provide confidence intervals, we assume that the random variables in the model given by Equation 7.1 are all normally distributed. Formulas provided below use information in the ANOVA tables (See Table 7.1).

Natural tolerance:

$$NT = 6 \times \text{(total standard deviation)}.$$

95% Confidence interval for the process mean:

$$\text{Sample Mean} \pm t(.025, df_R) \times \sqrt{(MSE_R/(n_R \times n_W \times n_D))}.$$

95% Confidence interval for the total variance:

[df´ × (Sum of Variance Components)/$\chi^2_{.975}$, df´ × (Sum of Variance Components)/$\chi^2_{.025}$],

where the approximate degrees of freedom df´ are defined below.

95% Confidence interval for natural tolerance: Take six times the square root of the lower limit for total variance as the lower limit and six times the square root of the upper limit for total variance as the upper limit.

Approximate Degrees of Freedom for Total Variance (see Satterthwaite, 1946)

The Satterthwaite approximation to the number of degrees of freedom for the total variance is given by

$$df' = \frac{(a \times MS_R + b \times MS_W + c \times MS_D)^2}{\frac{(a \times MS_R)^2}{df_R} + \frac{(b \times MS_W)^2}{df_W} + \frac{(c \times MS_D)^2}{df_D}},$$

where

$$a = 1/(n_W \times n_D), \quad b = (n_W - 1) \times a, \quad c = 1 - a - b,$$

and where

df_R = Degrees of freedom for run-to-run in the ANOVA table
df_W = Degrees of freedom for wafer-to-wafer in the ANOVA table
df_D = Degrees of freedom for within wafer in the ANOVA table
MS_R = Mean square for run-to-run
MS_W = Means square for wafer-to-wafer
MS_D = Mean square for within wafer
n_R = Number of runs
n_W = Number of wafers per run
n_D = Number of positions within wafer
df = Approximate degrees of freedom of the estimated total variance [t(.025,df_R) is the 2.5th percentile of the t distribution indexed by degrees of freedom (df_R)]
$\chi^2_{.975}$ = 97.5th percentile of the chi-square distribution with df´.

Calculations for the Before Case

Natural tolerance:

$$NT = 6 \times 18.9 = 113.4.$$

95% Confidence interval for the mean:

$$552.5 \pm t(.025, 8) \times \text{sqrt}(6106.5/486) \cdots \pm 8.2.$$

Approximate degrees of freedom for total variance:

$$a = 1/54, \quad b = 5/54, \quad c = 48/54,$$

$$df' = \frac{357.1^2}{\frac{(6106.5/54)^2}{8} + \frac{(5 \times 1019.7/54)^2}{45} + \frac{(48 \times 168.3/54)^2}{432}} = 69.$$

95% Confidence interval for total variance:

For 69 degrees of freedom, $\chi^2_{.975} \cong 94$, $\chi^2_{.025} \cong 48$.

Interval: $[69 \times 357.1/94, 69 \times 357.1/48] = [262.2, 513.3]$.

95% Confidence interval for natural tolerance:

Interval: $[6 \times \text{sqrt}(262.20), 6 \times \text{sqrt}(513.3)] = [97.2, 135.9]$.

BIOGRAPHIES

Richard O. Lynch, Ph.D. is a statistician at Harris Semiconductor. He served as an assignee to SEMATECH From 1988 to 1990. During his assignment, Richard coined the name "Passive Data Collection" and helped define the elements of what is now called the SEMATECH Qualification Plan.

Richard J. Markle is a senior process engineering member of the technical staff at Advanced Micro Devices in Austin, Texas. He is responsible for the development and demonstration of contamination sensors for increased productivity within the manufacturing fabs. In addition to his sensor responsibilities, he leads several wafer-based defect detection and reduction programs, wafer defect detection capacity modeling, and developing and maintaining infrastructure systems for sustainable manufacturing yield management. Prior to joining AMD, Rick managed SEMATECH's in situ sensor program and was instrumental in its efforts applying in situ sensors for substantial cost of ownership reductions on a variety of process equipment. Rick was also SEMATECH's first assignee to one of its member companies. Prior to managing SEMATECH's contamination prevention and reduction programs, Rick was in etch process engineering. He holds a bachelor's degree in chemistry from Miami University and a master's degree in inorganic and organometallic chemistry from The University of Texas at Austin. He has authored over 60 SEMATECH Technology Transfer Reports and over 16 technical publications.

CHAPTER 8

VIRGIN VERSUS RECYCLED WAFERS FOR FURNACE QUALIFICATION: IS THE EXPENSE JUSTIFIED?

Veronica Czitrom and Jack E. Reece

Conducting an equipment or process qualification involves the use of a substantial number of wafers. In this study, we present a strategy for determining whether or not it is possible to use cheaper recycled wafers instead of more expensive virgin wafers. We illustrate the strategy in the qualification of an oxidation process in a new three-zone vertical furnace. Using recycled 200 mm wafers saved over $25,000 for the remaining steps of the furnace qualification. Partial application of the results of this study to other furnace qualifications provided an opportunity to save over $100,000.

EXECUTIVE SUMMARY

Problem

Full qualification of semiconductor equipment for production use requires a relatively large number of wafers. Because virgin (new) wafers are expensive, substituting recycled (i.e., reclaimed) wafers for virgin wafers in the qualification is an attractive option. To protect the integrity of decisions made, one must establish the equivalence, or interchangeability, of virgin and recycled wafers in terms of performance measures of interest. This is further complicated when, in some circumstances, several pieces of equipment of the same type require qualification.

SEMATECH received four new three-temperature-zone vertical thermal reactors (furnaces), as reported in SEMATECH Technical Transfer Document #92031008A-ENG. One of these furnaces was intended for the growth onto 200 mm (8 inch) silicon wafers of a critical thin (90 Å) silicon dioxide layer that was part of the gate of the transistors in the integrated circuit.

Solution Strategy

A reasonable way to proceed when a study involves several pieces of similar equipment is to first establish the wafer equivalence for one piece of equipment using suitable performance measures. If wafer equivalence is established, then recycled wafers can be used for that piece of equipment as well as for the others in later steps of the qualification process. As a control, virgin wafers are included at critical steps.

This study established wafer equivalence for one piece of equipment, the furnace used for oxide layer growth. It compared three types of 200 mm silicon wafers: virgin wafers, internally recycled wafers, and wafers recycled by an external supplier. Four performance measures were used to establish the equivalence of the three wafer types: oxide thickness, wafer-to-wafer oxide thickness variability at different positions along the furnace, across-wafer (or within-wafer) oxide thickness variability, and number of particles added to the wafer during oxidation. Establishing wafer equivalence in the furnace used for oxidation allowed internally recycled wafers to be used in the screening and response surface experimental designs for the next two steps of the qualification process. Confirmation runs for these experiments used both internally recycled and virgin wafers. In addition, the demonstrated utility of recycled wafers in this process encouraged their use in other processes in other furnaces.

A passive data collection (PDC) experiment was used to compare the three wafer types. The PDC in this case study included ten replicated runs under operating conditions prescribed by the tool supplier. The sampling plan placed a wafer of each type at four different locations in the furnace tube. The operators measured the number of particles on each wafer before and after oxidation, and oxide thickness at nine sites on each wafer. Analysis of variance techniques on the oxide thickness measurements estimated the variation between furnace runs, variation between wafers in a furnace run, and variation across wafers.

Conclusions

The results of the study indicated there was wafer type equivalence. In other words, substituting recycled wafers for virgin wafers in the qualification of the oxide furnace did not reduce the effectiveness of the qualification. The study provided important clues for the next step of the qualification process: active process improvement to center the process and to reduce variability using designed experiments. While differences existed in oxide thickness from wafer to wafer at different locations along the furnace, and from site to site across a wafer, wafer type played essentially no role in these differences. Particles did not present a problem in this process, since the oxidation process effectively removed many of them. Particles present were apparently burned off in the high temperature oxygen environment in the furnace. Use of recycled wafers offered a considerable cost advantage without noticeable effect on the results. It saved about $25,000 during the qualification of the oxidation process, and over $100,000 for the qualification of all four furnaces.

Process

Oxide Layer Growth

The oxidation of silicon in a well-controlled and repeatable manner is an important step in the fabrication of modern integrated circuits. The process under consideration grows a 90 Å (nine nanometer) layer of silicon dioxide on 200 mm silicon wafers in a vertical furnace. The high-quality thin layer of oxide is part of the gate of each transistor in the integrated circuit. The oxide thickness and its variability are critical parameters in the manufacture of each transistor.

A quartz boat in the vertical furnace has slots for 160 wafers (see Figure 8.1), but each furnace run in this study included only 120 wafers. The boat supports each wafer at six points around its periphery. The top ten wafers were dummy wafers, the next 100 were product wafers, the next ten were dummy wafers, and the slots below the last dummy wafer were empty. The boat rests on the door of the furnace with all the wafers facing up. An elevator mechanism raises the wafers into the preheated vertical furnace. When the furnace is sealed, a program ramps the temperature to 900° slowly (to prevent the wafers from warping). Introduction of oxygen gas slowly oxidizes the silicon on the surface of the wafer, creating a layer of silicon dioxide. Once the oxidized silicon layer is 90 Å thick, the flow of oxygen stops and the process maintains an elevated temperature to anneal the wafer surface to relieve stress. Finally, the furnace temperature ramps down to allow removal of the wafers. The entire process requires approximately five hours.

Wafer Recycling

The wafer recycling process requires cleaning the oxidized wafers in hydrofluoric acid to remove the silicon dioxide layer followed by conditioning of the wafer surface to make it suitable for further process use. The quality of the recycling process plays a critical role in whether or not recycled wafers are reliable for further use.

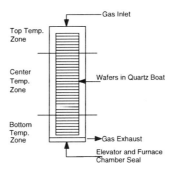

Figure 8.1. *Three-Temperature Zone Vertical Furnace with Quartz Boat Holding Wafers. Oxygen enters at the top of the furnace and leaves at the bottom of the furnace.*

DATA COLLECTION PLAN

A passive data collection (PDC) was used to establish the equivalence of three wafer types for a furnace used for oxide layer growth. The three types of 200 mm wafers considered were:

V = virgin wafers;
E = wafers externally recycled at a vendor site;
I = wafers internally recycled at SEMATECH.

The responses or performance measures used to establish wafer equivalence were
- **Thickness of the Oxide Layer.** The target value was 90 Å.
- **Wafer-to-Wafer Variability of Oxide Thickness Along the Furnace.** The study placed wafers in four general locations within the furnace, with one wafer of each wafer type at each general location. Location 1, at the top of the furnace, consisted of wafer slots 11, 12, and 13; location 2 consisted of wafer slots 43, 44, and 45; location 3 consisted of wafer slots 76, 77, and 78; and location 4, near the bottom of the furnace, consisted of wafer slots 108, 109, and 110. To simulate normal processing conditions such as thermal mass and flow patterns, dummy wafers occupied the remaining wafer slots except for the 40 empty positions at the bottom.

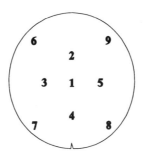

Figure 8.2. *Oxide Thickness Sampling Plan of Nine Sites Per Wafer. The location of the sites on the wafer is intended to maximize coverage of the wafer surface.*

- **Across-Wafer (Within-Wafer) Variability in the Oxide Thickness**. Operators measured oxide thickness at nine sites on each wafer. Figure 8.2 illustrates the sampling pattern on each wafer.
- **Difference in the Number of Particles on the Wafer Before and After Oxidation**. In most furnace runs the process *removed* particles from the wafers, probably because they burned off. This was especially true for virgin wafers, so later designed experiments to minimize sources of variability and to center the process using recycled wafers should produce even better particle counts with virgin wafers in actual production. For this reason, we will not analyze the particle count data.

Ten batches of wafers were processed over a three-day period without changes in processing conditions between runs or within a run. Table 8.1 illustrates the data collection plan that was used to minimize a possible effect due to the relative ordering of the different wafer types inside the furnace. The order of the wafers is relatively "balanced"[14] in the following senses:
- The six possible orderings of the wafer types (VEI, VIE, EVI, EIV, IEV, and IVE) appear almost the same number of times (6 or 7 times each).
- Each ordering of the wafers appears one or two times in each furnace position or location.

Table 8.1. *Ordering of Wafers by Type at Each of Four Locations Within the Furnace for Each of Ten Furnace Runs.*

RUN	LOCATION 1	LOCATION 2	LOCATION 3	LOCATION 4
1	VEI	EVI	EIV	VIE
2	VIE	IVE	EVI	EIV
3	IEV	EIV	IVE	EVI
4	EIV	VEI	IVE	IEV
5	VEI	VIE	EVI	IVE
6	EIV	IVE	VEI	EVI
7	EVI	EIV	IEV	VIE
8	IEV	VIE	EIV	VEI
9	IVE	VEI	VIE	IEV
10	EVI	IEV	VEI	IVE

Figure 8.8 gives a graphical representation of some of the runs of the experimental design of the PDC for the externally recycled wafers.

DATA ANALYSIS AND INTERPRETATION OF RESULTS

The first part of this section considers the graphical analysis of the oxide thickness data and interpretation of results. The second part quantifies variability in oxide thickness using analysis of variance (ANOVA) techniques.

Graphical Data Analysis

The oxide thickness data is given in the Appendix. The total number of observations (1080) corresponds to nine observations per wafer, twelve wafers per furnace run (four wafers from each of the three types of wafer), and ten furnace runs.

Figure 8.3 shows box plots of oxide thickness as a function of furnace run. The figure is also a short-term time trend chart which suggests no apparent time trend in oxide thickness from batch to batch through run eight. The high outliers in run 9 are probably due to a film residue on a wafer, and there was an assignable cause for the low values in run 10. It was decided to exclude runs nine and ten, as well as the eight low outliers (five from externally recycled wafers and three from internally recycled wafers) in the first eight runs, from the

---

[14] The design consists of 10 of the 15 runs in a balanced incomplete block design (BIBD), where only four of the six orderings of the wafer types are considered in each one of the ten runs (blocks). See Montgomery (1991). The ten runs were selected using the D-optimal criterion to maximize the "balance" in the two senses described.

remainder of the study. Including nine runs (minus the nine observations on the dirty wafer) gives similar results. The fact that the average thickness is approximately 92 Å instead of the target 90 Å is not important, since adjusting the oxidation time will adjust the average thickness.

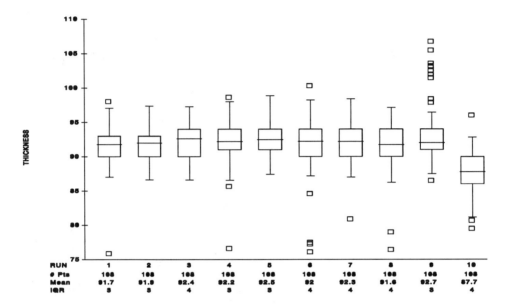

Figure 8.3. *Box Plots of Oxide Thickness as a Function of Furnace Run. The low mean thickness in the last run had an assignable cause. The nine thickness readings above 100 Å in run nine all came from one internally recycled wafer and were probably due to a film residue on the wafer. The eight low readings in runs one through eight were traced to the gauge which sporadically gave unreasonably low readings. The analysis that follows includes only runs one through eight and excluded the eight low outliers in those runs.*

Figure 8.4 shows a box plot of oxide thickness as a function of location in the furnace with the observations in each furnace section plotted by wafer type. Oxide thickness increases slightly from top to bottom in the furnace.

Figure 8.5 shows box plots of oxide thickness for the three types of wafers. The mean thickness of the virgin wafers appears slightly lower than the mean thickness of the recycled wafers. An analysis of variance indicated that the mean oxide thickness is significantly different for the three types of wafers, but this difference has no practical significance since adjusting the oxidation time will adjust the thickness to the target value of 90 Å.

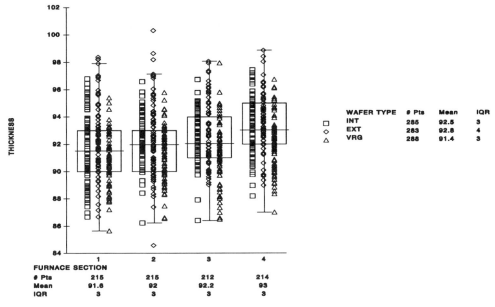

Figure 8.4. *Box Plots of Oxide Thickness as a Function of Furnace Location and Wafer Type. The oxide thickness increases towards the lower part of the furnace.*

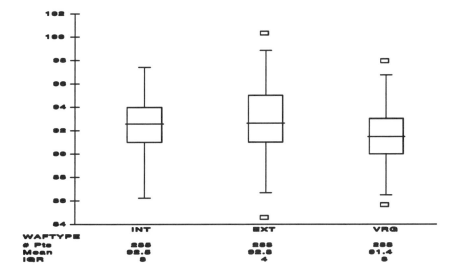

Figure 8.5. *Box Plots of Oxide Thickness as a Function of Wafer Type. Both the mean oxide thickness and its variability appear slightly smaller for the virgin wafers.*

Figure 8.6 shows box plots of oxide thickness at the nine sites on a wafer. The spread of the oxide thickness observations is similar at the nine wafer sites. However, the mean (average) oxide thickness seems to vary from site to site. In spite of this, Figure 8.7 does not indicate a clear pattern in oxide thickness as a function of wafer site. Figure 8.7 combines the mean oxide thickness information given in Figure 8.6, and the location of the wafer sites given in Figure 8.2.

Analysis of Variance

Figures 8.3, 8.4, 8.6 and 8.7 illustrate the differences in the oxide thickness from furnace run to furnace run, from location to location in the furnace, and from site to site on a wafer. The variability associated with these three sources can be quantified using analysis of variance (ANOVA) techniques. The discussion that follows describes only the analysis for externally recycled wafers in some detail. It then presents the results for the other two types of wafer and compares the sources of variability for each of the three wafer types.

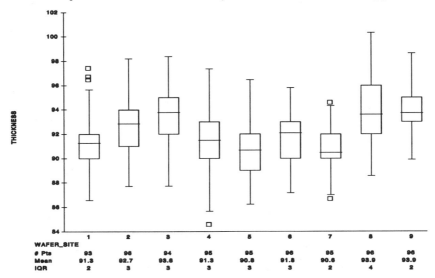

Figure 8.6. *Box Plots of Oxide Thickness as a Function of Wafer Site. The average thickness changes from site to site, but the spread of the thicknesses is similar at the nine sites.*

Consider the 32 externally recycled wafers only. The experimental design of the PDC in the externally recycled wafers is a split-plot design[15] with eight runs. This design is illustrated graphically in Figure 8.8. Each furnace run includes one externally recycled wafer at each one of the four furnace locations. Each furnace run forms a homogeneous unit (block) in which to compare the four locations along the furnace, without including run-to-

---

[15] See Mason, Gunst, and Hess (1989), Box and Jones (1992), Milliken and Johnson (1984), or Cox (1958), for presentations on split-plot designs.

run variability. Each wafer forms a homogeneous unit (whole plot) in which to compare the nine wafer sites (split plots) without including wafer-to-wafer variability. Excluding the five outliers mentioned in Figure 8.3 gives $8 \times 4 \times 9 - 5 = 283$ thickness measurements.

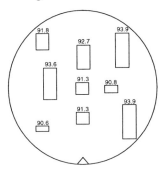

Figure 8.7. *Oxide Thickness by Wafer Site. The height of the bars indicates oxide thickness ABOVE the target value of 90Å at each wafer site. The sites have different thicknesses, but no clear pattern, such as a radial effect or an effect from left to right on the wafer, is apparent.*

Analysis of the data as a split-plot design takes into account that wafer-to-wafer variability does not contribute to differences in oxide thickness at different sites on a wafer (for each wafer, the analysis compares the nine wafer sites). Analysis also takes into account that wafer-to-wafer variability does contribute to differences in oxide thickness at different furnace locations (a different wafer is at each furnace location). The run-to-run variability does not contribute to the differences in furnace locations (the four furnace locations are compared within each furnace run).

The split-plot design is a full factorial design with restrictions on randomization. Nine sites are measured on each wafer (nine wafers with one site measured on each would be used in the completely randomized design). Each run contains four furnace locations (a completely randomized design would include four runs with one wafer in each furnace location). One could consider the design as nested with wafer sites nested in wafers at the different furnace locations, and wafers at the different furnace locations nested in furnace runs. In this case realization that the furnace locations are the same for all runs and that the wafer sites are the same for all wafers leads to the crossed or full factorial design, for which the restrictions on randomization lead to the split-plot design.

To determine whether systematic differences exist between furnace sections and between sites on a wafer, consider furnace Run (R) as a random factor, and Furnace section (F) and Site on wafer (S) as fixed factors. Table 8.2 gives the analysis of variance (ANOVA) for the split-plot design[16] for the externally recycled wafers.

---

[16] The degrees of freedom (DF) and the sums of squares in the ANOVA for the split-plot design can be found from the ANOVA for the full factorial design in run, furnace section, and wafer site, by pooling (adding) the Run x Site and Run x Furnace x Site sums of squares and degrees of freedom in the factorial design, to obtain the residual sums of squares and degrees of freedom in the split-plot design. The fact that the design is unbalanced (5 of the

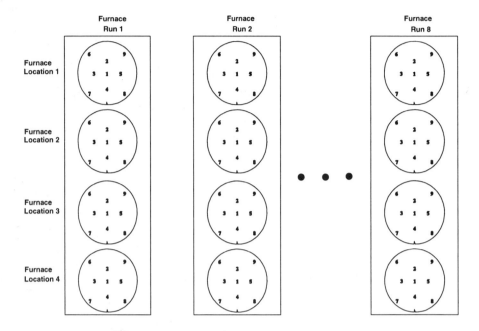

Figure 8.8. *Graphical Representation of the PDC for the Externally Recycled Wafers. Oxide thickness is measured at nine sites on each wafer, one externally recycled wafer is located at each one of the four locations in the furnace, and eight furnace runs are considered. The four furnace locations can be compared in each furnace run, and the nine sites on a wafer can be compared for each wafer.*

Dividing one mean square value by another based on information in the expected mean square column produces the F statistics in Table 8.2. For example, dividing the mean square for site (60.44) by the mean square for residual (4.95) produces 12.20, the F statistic for site. These F statistics, or equivalently the p-values, indicate that

- there is strong evidence of systematic differences in oxide thickness at different sites on a wafer, although no pattern is obvious from Figure 8.7. The screening experiment in the next step of the Qualification Plan should include factors (such as oxygen flow) that may help reduce the thickness variability across a wafer.
- there is strong evidence of systematic differences in oxide thickness at different furnace locations. In fact, Figure 8.4 indicates that oxide thickness increases towards the lower part of the furnace. The screening experiment should include

---

288 observations are missing) implies that the sums of squares in the ANOVA have slightly different numerical values depending on the order in which the factors are specified and the terms that are considered in the model. The small value of $[(1904.69 - 1084.54)/1904.69] = 0.43$ for R^2 in the ANOVA reflects the fact that the residual error is inflated by the large measurement error.

factors expected to affect differences in thickness along the furnace, such as the values of the three temperatures along the furnace.
- furnace location does not affect site-to-site differences in oxide thickness (i.e., no F x S interaction).

Table 8.2. *ANOVA Table of Oxide Thickness for Externally Recycled Wafers, with Location and Site as Fixed Factors. The two small p-values indicate a difference in thickness at different sites on a wafer, and a difference in thickness for wafers in different locations along the furnace. The F x S interaction is not significant.*

Source of Variation	Degrees of Freedom	Sum of Squares	Mean Squares	Expected Mean Squares	F-Statistic (p-value)
Run	7	21.96	3.32	$\sigma^2 + 9\sigma_w^2 + 36\sigma_r^2$	
Furnace Location	3	65.76	21.92	$\sigma^2 + 9\sigma_w^2 + 24\Sigma F^2$	3.48 (0.034)
R x F	21	132.41	6.31	$\sigma^2 + 9\sigma_w^2$	
Site	8	483.48	60.44	$\sigma^2 + 4\Sigma S^2$	12.20 (< 0.0001)
F x S	24	116.54	4.86	$\sigma^2 + \Sigma \frac{(F \times S)^2}{3}$	0.98 (0.49)
Residual	219	1084.54	4.95	σ^2	
Total	282	1904.69			

A common practice in the semiconductor industry produces numerical measures of the components of variability (run to run, furnace location to furnace location, and site to site) by considering furnace run, furnace location, and site on wafer as random factors with variances σ_r^2, σ_f^2, and σ_s^2, respectively.[17] Let σ_w^2 represent the variance from wafer to wafer, and let σ^2 represent the variance of the remaining sources of variability, such as the variability of the measurements introduced by the gauge used to measure the oxide thicknesses. Since the FxS interaction was not significantly different from noise in the ANOVA in Table 8.2, pool it with the residual sum of squares to give the ANOVA in Table 8.3.

Equating the mean squares to the corresponding expected mean squares in Table 8.3 and solving for the variances produces the following estimates of the variances: $\hat{\sigma}^2 = 4.94$, $\hat{\sigma}_s^2 =$

---

[17] Strictly speaking, the factors site on wafer and furnace location are fixed factors. Site on wafer is a fixed factor since the same sites given in Figure 8.1 are measured for each wafer. The measurement procedure is automatic, and the gauge takes the measurements at the prescribed sites for all wafers. Furnace location is also a fixed factor since each run used the same wafer slots (11, 12, 13, 43, 44, 45, 76, 77, 78, 108, 109, and 110). Since the sampling of sites on a wafer and locations along the furnace were not random, the "estimates" of the variances obtained by considering them random are not truly estimates of population variances; but they ARE measures of dispersion. This is the same type of measure of dispersion that Taguchi obtains by calculating a sample variance over an outer array (see Kackar and Phadke, 1989). For example, the nine sites chosen to calculate the variance in oxide thickness across a wafer systematically span the possible sites on a wafer.

1.74, $\hat{\sigma}_w^2 = 0.15$, $\hat{\sigma}_f^2 = 0.22$, and $\hat{\sigma}_r^2 = -0.08 = 0$. The negative value of the variance component for run-to-run variability is set to zero, which indicates short-term stability. A prior gauge study gave an estimate of 4.09 for the variance in the oxide thickness measurements due to measurement tool error, which indicates that most of the residual or error variance appears to come from the measurement tool.[18] In fact the measurement tool error is very large compared to the other variance components, which later in the furnace qualification led to a change of gauge from the interferometer to an ellipsometer. Taking the square roots of the variance components produces the standard deviations given in Table 8.4.

Figure 8.9 graphically illustrates the components of variability given in Table 8.4 (excluding $\hat{\sigma}$) for externally recycled wafers, as well as the components of variability for internally recycled wafers and virgin wafers. For each wafer type, variability from site to site on a wafer is larger than variability from section to section of the furnace; variability from section to section of the furnace is larger than variability from run to run; and most of the residual variability comes from measurement error (not shown). In addition, the actual numerical values of the standard deviations are very small compared to the 92Å average thickness. These results indicate that the three wafer types are equivalent, so that there is no practical difference between the wafer types. For this reason, the cheaper internally recycled wafers were used in the remaining steps of the furnace qualifications. If the designed experiment performed as the next step in the furnace qualification gives processing conditions that reduce the variability in oxide thickness for recycled wafers, one should expect the same conditions to reduce the variability for virgin wafers. To validate this, confirmatory furnace runs under the improved processing conditions should include both virgin and recycled wafers.

CONCLUSIONS AND RECOMMENDATIONS

This case study illustrates a successful example of a cost-saving strategy that allows replacement of virgin wafers with recycled wafers in the qualification of semiconductor equipment. It shows that virgin and recycled wafers are equivalent in the furnace used for oxide layer growth in terms of the performance measures of interest.

---

[18] The design is sometimes treated as a nested design. The values of the variance components considering the design as a nested design are $\hat{\sigma}_s^2 = 6.35$, $\hat{\sigma}_f^2 = 0.23$, and $\hat{\sigma}_r^2 = 0$. The corrected variance component for site is $\hat{\sigma}_s^2 = 2.26$, obtained by subtracting the previous estimate of 4.09 for the variance of the measurement tool. The numerical values of the variance components in this example are similar for the nested and split-plot designs. Analysis of the data as a split-plot design allows extraction of two more variance components from the nested design ($\hat{\sigma}_w^2$ from $\hat{\sigma}_f^2$, and $\hat{\sigma}^2$ from $\hat{\sigma}_s^2$), and does not require a previous estimate of measurement error to find the variance component for site on wafer.

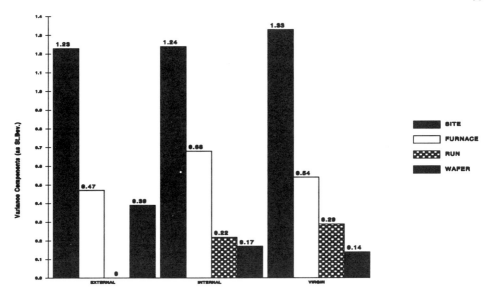

Figure 8.9. *Variance Components Expressed as Standard Deviations for the Three Wafer Types. The different components of variability are almost equal in magnitude for the three wafer types and are all very small compared to the 92Å thickness. This establishes equivalence between the three wafer types. The remainder of the furnace qualification used the cheaper internally recycled wafers instead of virgin wafers.*

Table 8.3. *ANOVA Table of Oxide Thickness for Externally Recycled Wafers, with Location and Site as Random Factors. To estimate the variance components for run, furnace location, and wafer site, equate the mean squares to the expected mean squares expressions and solve for the unknown variances.*

Source of Variation	Degrees of Freedom	Sum of Squares	Mean Squares	Expected Mean Squares
Run	7	21.97	3.32	$\sigma^2 + 9\sigma_w^2 + 36\sigma_r^2$
Furnace Location	3	65.76	21.92	$\sigma^2 + 9\sigma_w^2 + 72\sigma_f^2$
R x F	21	132.41	6.31	$\sigma^2 + 9\sigma_w^2$
Site	8	483.48	60.44	$\sigma^2 + 32\sigma_s^2$
Residual	243	1201.07	4.94	σ^2
Total	282	1904.69		

For the three types of wafers (virgin, externally recycled, and internally recycled), the relative importance of the different sources of oxide growth variability is practically the same, and the magnitudes are very similar and very small compared to the target thickness of 90Å. The three types of wafers produce a small statistically significant difference in mean thickness, but it is of no practical significance since adjusting the oxidation time will adjust the oxide thickness. Particle counts are not a concern in this study since the process tends to remove rather than to add particles.

Table 8.4. *Standard Deviations (in Angstroms) of the Variance Components for the Externally Recycled Wafers*

Term	Symbol	Value
Residual	$\hat{\sigma}$	2.22
Wafer Site	$\hat{\sigma}_s$	1.32
Wafer Type	$\hat{\sigma}_w$	0.39
Furnace Loc	$\hat{\sigma}_f$	0.47
Run	$\hat{\sigma}_r$	0.00

The results of the PDC experiment led to the extensive use of recycled wafers in the designed experiments required for the next two steps of the qualification process. Using 200 mm recycled rather than virgin wafers saved over $25,000 in completing the qualification of the oxide process. Selective use of recycled wafers for qualifying the three other new furnaces (and eventually other equipment) increased the savings to more than $100,000. More widespread use of recycled instead of virgin wafers will lead to even greater savings in semiconductor process and equipment qualifications involving 200 mm wafers.

ACKNOWLEDGMENTS

The authors acknowledge Bill O'Toole and the manufacturing technologists in the SEMATECH Fab for running this experiment and Peg Rumaine for reviewing the manuscript. They also acknowledge Bill Kahn, of W.L. Gore & Associates, Inc., Elkton, Maryland, for helping set up the data collection plan of Table 8.1.

REFERENCES

Box, George, and Stephen Jones. "Split-plot Designs for Robust Product Experimentation." *Journal of Applied Statistics*, Vol. 19, No. 1, pp. 3-26, 1992.

Cox, D.R. *Planning of Experiments*. New York, NY: John Wiley and Sons, Inc.,1958.

Kackar, R.N. "Off-Line Quality Control, Parameter Design, and the Taguchi Method." *Journal of Quality Technology*. Vol. 17, No. 4, pp. 176-209, 1985.

Mason, Robert, Richard Gunst, and James Hess. *Statistical Design and Analysis of Experiments*. New York, NY: John Wiley and Sons, 1989.

Milliken, George, and Dallas Johnson. *Analysis of Messy Data, Volume 1: Designed Experiments*. New York, NY: Van Nostrand Reinhold, 1984.

Montgomery, Douglas. *Design and Analysis of Experiments*, 4th ed. New York, NY: John Wiley and Sons, Inc., 1997.

Phadke, G. *Quality Engineering Using Robust Design*. New Jersey: Prentice Hall, 1989.

SEMATECH. *SVG Thermco Systems Series 7000 PLUS Vertical Thermal Reactor 200mm Oxide and Silicon Nitride Qualifications Report.* 92031008A-ENG. Austin, TX: SEMATECH, June 29, 1992.

APPENDIX

Data for Passive Data Collection

Oxide Thickness Measurements for 10 Runs, Four Furnace Sections (FSCT), Three Wafer Types (WTYP), and Nine Wafer Sites (ST1, ST2,..., ST9).

RUN	FSCT	WTYP	ST1	ST2	ST3	ST4	ST5	ST6	ST7	ST8	ST9
1	1	VRG	90.1	90.1	92.8	87.8	88.2	88.2	90.4	92.1	91.8
1	1	EXT	92.6	90.8	95.6	90.7	89.3	87.2	91.7	90	93
1	1	INT	90.4	91.3	87.7	90.3	90.4	88.5	91	91.7	91.1
1	2	EXT	88.3	90.2	91.8	92.8	89.9	91.8	89.6	96	91.9
1	2	VRG	91.9	93.3	94.1	89.1	91.4	92.2	87.5	91.2	91.1
1	2	INT	90.3	94.5	90.5	90.1	92.4	92.6	89.9	91.8	93.3
1	3	EXT	91.5	93.4	92	91.2	94.4	95.1	75.9	95.9	98.1
1	3	INT	91	96.7	93.2	90.4	91.3	91.9	92.9	91.5	92.5
1	3	VRG	88.1	90.8	91.5	88.2	90.5	92.3	87.4	92.6	92.4
1	4	VRG	90	93.1	92.7	91.6	89.2	92.6	87	93.2	95.2
1	4	INT	89	94.7	94.9	93.5	89.9	91.7	94	93.8	93.6
1	4	EXT	95.6	94.8	97.1	92	96.5	93.1	94.6	93.3	96.5
2	1	VRG	90.7	90.8	90.3	92.7	88.4	89	89.1	92.6	92.8
2	1	INT	88.8	92.1	96.1	88.2	89.3	88.6	90.8	93.1	95.4
2	1	EXT	90	92.7	94.1	92.3	91.7	90.2	90.9	88.6	93.8
2	2	INT	91.9	92.9	93.2	91.7	94.1	91.4	91.3	94.8	94.7
2	2	VRG	88.6	89.1	91.5	89.5	86.6	93.4	89.9	91.8	92.3
2	2	EXT	88.2	92.3	97	90.8	88.9	93.8	89.7	92.5	91.7
2	3	EXT	95	93.5	93	97.4	91	94.2	90.1	96.1	93.1
2	3	VRG	90.2	90.4	90.9	94.7	91.3	91.3	90	91.6	92
2	3	INT	90.4	93.8	92.4	93.5	90.3	92.8	90.4	94.5	92.6
2	4	EXT	89.8	89.6	94.5	94.7	92.5	92.2	94.2	95.7	92.3
2	4	INT	91.3	91	94.3	93.3	92	93.5	90.4	95.6	94.5
2	4	VRG	90.8	92.6	92.6	88.4	92.4	89.9	89.9	91.9	94.1
3	1	INT	90.6	92.6	95.2	87	90.9	89.2	90.3	95.8	94.7
3	1	EXT	93.3	95	91.6	93.3	92.3	92.9	92.4	96.7	94.9
3	1	VRG	89.4	90	93	90.4	90.4	89.9	91.6	92.6	93
3	2	EXT	89.4	92.6	89.3	88.7	90.1	95.7	92.6	93.5	92.4
3	2	INT	91.5	92.8	92.3	92.7	89	94.4	91.2	93.6	94.7
3	2	VRG	89.7	90.1	92.1	88.6	90	92.6	89.2	92.5	93
3	3	INT	90.2	91	94.6	93.2	93.2	93.9	94.3	95.1	95.8
3	3	VRG	86.6	94.9	91	89	90.9	92.3	90.5	93.6	93.6
3	3	EXT	89.8	93.4	96.2	95	89.2	93.5	92.8	97.3	89.9
3	4	EXT	95.3	93.2	96.2	89	93.5	95.4	91.3	95.7	95.2
3	4	VRG	93.2	93.9	91.7	90.3	90.5	93	89.7	92.5	94.6
3	4	INT	92.3	93.1	96.9	94.7	93.2	91.6	91.9	93	97.2
4	1	EXT	92.2	93.5	91.6	91.2	92.7	91.5	86.7	96.9	94.5
4	1	INT	93.2	93.2	95.3	89.7	92.2	88.7	91.3	95.7	94.8

APPENDIX (CONTINUED)

Oxide Thickness Measurements (Continued)

RUN	FSCT	WTYP	ST1	ST2	ST3	ST4	ST5	ST6	ST7	ST8	ST9
4	1	VRG	87.8	93.2	91.7	85.6	90.3	87.9	89.1	93.2	90.9
4	2	VRG	86.6	92.4	90.9	90.9	91.4	90.4	89.7	92.6	92.7
4	2	EXT	92.9	94.2	94.8	94.1	91.8	91.6	90.5	90.7	98.6
4	2	INT	90.9	93.5	94.7	91.6	88.4	94.8	92.1	92.4	94.3
4	3	INT	76.6	92.9	93.9	92.1	90	93.9	91.6	92.6	94.9
4	3	VRG	91.9	93.5	97.9	90.1	87.7	92.1	89	92	93.4
4	3	EXT	92.6	94.7	93.6	91.5	91.1	89.9	91.5	92.2	91.1
4	4	INT	93.4	94.6	94.9	93.5	88.2	92.1	90.6	94.8	95.6
4	4	EXT	89.6	93.8	94.6	93.1	89.7	94.8	93	94.7	98
4	4	VRG	89.1	92.1	94.6	92	89.6	92.4	92.9	96.2	96.1
5	1	VRG	91.8	90.4	91.7	91.8	89	90	88.9	93.8	92.3
5	1	EXT	91.5	95.7	98.4	88.9	91	87.6	90.3	90	95.8
5	1	INT	93.7	92.6	93	88.9	87.9	88.7	87.4	92.9	96.8
5	2	VRG	89.3	94.5	94.6	95.8	93	91.7	89.2	93.3	95.2
5	2	INT	93.8	94.8	94	92	92.4	92.5	91.2	93.2	92.7
5	2	EXT	89.9	93.6	97	91.6	91	95.4	89.7	93.8	94.2
5	3	EXT	91.5	91.7	92.7	89.9	91.4	92.5	91.4	94.2	96.4
5	3	VRG	90	92	95	92.7	88.5	91.3	90	92.1	93.9
5	3	INT	92.2	92.1	94.7	90.3	93.9	95	89.7	93.6	95
5	4	INT	97.4	97	96.4	94.9	92.7	91.4	91.2	94.6	94.3
5	4	VRG	90.2	90.4	93.4	92.4	88.8	91.7	89.4	96.7	92.5
5	4	EXT	94.2	96.2	95.3	93.1	90.6	94.5	90.1	98.8	92.6
6	1	EXT	90	92.2	94.9	92.7	91.6	88.2	92	98.2	96
6	1	INT	91.8	94.5	93.9	77.3	92	89.9	87.9	92.8	93.3
6	1	VRG	90.3	91.1	93.3	93.5	87.2	88.1	90.1	91.9	94.5
6	2	INT	92.6	90.3	92.8	91.6	92.7	91.7	89.3	95.5	93.6
6	2	VRG	91.1	89.8	91.5	91.5	90.6	93.1	88.9	92.5	92.4
6	2	EXT	76.1	90.2	96.8	84.6	93.3	95.7	90.9	1003	95.2
6	3	VRG	92.4	91.7	91.6	91.1	88	92.4	88.7	92.9	92.6
6	3	EXT	91.3	90.1	95.4	89.6	90.7	95.8	91.7	97.9	95.7
6	3	INT	96.7	93.7	93.9	87.9	90.4	92	90.5	95.2	94.3
6	4	EXT	92	94.6	93.7	94	89.3	90.1	91.3	92.7	94.5
6	4	VRG	94.1	91.5	95.3	92.8	93.4	92.2	89.4	94.5	95.4
6	4	INT	91.7	97.4	95.1	96.7	77.5	91.4	90.5	95.2	93.1
7	1	EXT	92.2	98.2	92.1	92.8	93.1	89.6	89.3	97	97.9
7	1	VRG	90.3	91.2	93	89.7	88.1	91	89.7	95	95.4
7	1	INT	93.8	87.7	90.2	90.2	90.1	93.1	91.3	96.4	95.4
7	2	EXT	89.4	94.1	92.2	92.6	94.4	87.4	92	98.2	95.6
7	2	INT	92.8	90.3	94.8	91.5	91.8	92.4	91.4	96.6	94.4
7	2	VRG	92.7	89.3	90.9	90.2	88.8	92.5	89.9	94.2	93.6
7	3	INT	91.5	92.2	94	90.8	91.2	92.2	90.5	94.4	92.6
7	3	EXT	89.6	96.6	94.7	89	90	93.1	89.4	96.6	92.2
7	3	VRG	87	94	95.8	91.7	89.7	88.7	90.7	94.9	91.4
7	4	VRG	91.8	91.8	91.6	94.7	92.7	92.5	90.1	94.9	92.8
7	4	INT	90.9	91.8	94.8	89.8	90.4	90.9	91.9	95.9	92.4

APPENDIX (CONTINUED)

Oxide Thickness Measurements (Continued)

RUN	FSCT	WTYP	ST1	ST2	ST3	ST4	ST5	ST6	ST7	ST8	ST9
7	4	EXT	96.5	93.9	80.9	95.3	93.8	91.2	91.7	98.4	94.4
8	1	INT	90	94.7	93.5	86.7	90.9	88	90.3	94.6	93.5
8	1	EXT	91.8	97	94	89.2	90.4	91.8	90.7	90.5	94.8
8	1	VRG	89	89.8	89	90.5	90.1	88.6	90.5	91.3	93.3
8	2	VRG	89.9	90.6	90.4	91.8	88.3	93.1	88.4	92.1	93.1
8	2	INT	95.1	91.8	94.8	94.8	86.2	90.8	92.1	95.8	91.2
8	4	EXT	91.7	93.6	92.9	92.5	92.7	95	92.2	92	93.8
8	4	INT	95.7	94.2	96.5	94.6	93.7	92.4	94.1	95.1	91
9	1	INT	93	89.9	93.6	89	93.6	90.9	89.8	92.4	93
9	1	VRG	91.4	90.6	92.2	91.9	92.4	87.6	88.9	90.9	91.6
9	1	EXT	91.9	91.8	92.8	96.4	93.8	86.5	92.7	90.9	92.8
9	2	VRG	90.6	91.3	94.9	88.3	87.9	92.2	90.7	91.3	93.6
9	2	EXT	93.1	91.8	94.6	88.9	90	97.9	92.1	91.6	98.4
9	2	INT	90.8	91.5	91.5	91.5	94	91	92.1	91.8	94
9	3	VRG	88	91.8	90.5	90.4	90.3	91.5	89.4	93.2	93.9
9	3	INT	88.3	96	92.8	93.7	89.6	89.6	90.2	95.3	93
9	3	EXT	94.2	92.2	95.8	92.5	91	91.4	92.8	93.6	91
9	4	INT	101.5	103.1	103.2	103.5	96.1	102.5	102	106.7	105.4
9	4	EXT	92.8	90.8	92.2	91.7	89	88.5	87.5	93.8	91.4
9	4	VRG	92.1	93.4	94	94.7	90.8	92.1	91.2	92.3	91.1
10	1	EXT	88.3	91.7	92.8	88.3	89.4	81.2	88.7	90.6	89.5
10	1	VRG	85.6	89	89.8	90	89	87	84.8	88.6	90.2
10	1	INT	91.7	83.8	92.7	96	87.8	85.5	90.8	90.7	89.3
10	2	INT	80.7	88.9	91.1	87	84.6	88	86.9	87	90
10	2	EXT	85.8	85.4	89.2	88.1	90.1	90.7	87.3	90.5	89.5
10	2	VRG	85.7	87.2	87.5	84.4	86.9	86.4	87	86.4	87.6
10	3	VRG	88.5	86.4	89.7	84.3	90	87.9	85.4	85.5	89.3
10	3	EXT	91.5	88.3	89.6	84.3	87.8	91.9	84.1	90.1	87.2
10	3	INT	85.1	89.6	90.2	86.8	87	89.3	83.8	90.9	89.1
10	4	INT	79.5	90.1	88.8	86.3	83	88.5	85.1	86.9	89.1
10	4	VRG	82.8	85.3	88.2	85	83.6	87.1	84.5	87.2	87.6
10	4	EXT	86.7	87	90.8	87.5	87.8	85.5	86.4	88.7	88.9
10	4	EXT	86.7	87	90.8	87.5	87.8	85.5	86.4	88.7	88.9

BIOGRAPHIES

Veronica Czitrom is a Distinguished Member of Technical Staff at Bell Labs, Lucent Technologies. She leads consulting and training efforts for the application of statistical methods, primarily design of experiments, to semiconductor processing. Dr. Czitrom was on a two year assignment at SEMATECH. She has published research papers on design of experiments and mixture experiments, and four engineering textbooks. She received a B.A. in Physics and an M.S. in Engineering from the University of California at Berkeley, and a Ph.D. in Mathematics with concentration in Statistics from the University of Texas at

Austin. She was a faculty member at The University of Texas at San Antonio, and at the Universidad Nacional Autónoma de México.

Jack E. Reece, Ph.D., retired as a Fellow of the SEMATECH statistical methods group in June, 1996. In his capacity as a statistician at SEMATECH, he generated training materials to help engineers understand statistical methods, particularly design of experiments, and provided consulting and training in the use of applied statistical methods across all semiconductor fab operations. Dr. Reece has more than 25 years' experience as a process engineer/statistician in industries such as petrochemicals, photographic products, coating technology, and semiconductors. Currently Dr. Reece is a private consultant based in Lake George, CO. He may be reached on the Internet at jreece@pcisys.net.

CHAPTER 9

IDENTIFYING SOURCES OF VARIATION IN A WAFER PLANARIZATION PROCESS

Arnon M. Hurwitz and Patrick D. Spagon

At any given semiconductor manufacturing process step, the wafer surface will never be perfectly flat. Measurements taken at different sites on a wafer surface often reflect large, predictable site-to-site differences that are consistent across wafers and lots. These consistencies are evidence of a nonrandom structure. This nonrandom structure can be taken into account and quantified for analysis by using contrasts to break down certain sums of squares in the analysis of variance (ANOVA). This approach (using contrasts) extracts more information from the data than when site-to-site differences are treated as random effects or as nested random effects.

EXECUTIVE SUMMARY

Problem

SEMATECH engineers (Sivaram, 1992) were engaged in making performance improvements to a planarization device. The device, essentially a rotating abrasive table, was designed to polish silicon wafers to a high degree of smooth planarity. Much data had already been gathered, and this study examines a part of that data base—nine lots of twenty-three wafers each.

The thickness of each polished wafer was measured at nine site locations on its surface, and the basic question addressed was whether the observed variation in thickness could be attributed to lot-to-lot, wafer-to-wafer, or site-to-site differences. As part of the project requirements, maximum specifications had been set for these respective variables.

In order to set up the correct ANOVA table to complete the data analysis, it was necessary to know how to write a correct model. SAS (the SAS Institute) was the chosen statistical software because of its ability to deal with variance component analysis with unbalanced data. Data is said to be unbalanced if there are unequal numbers of observations in the classification cells of an experiment, for example, if some lots had less than 23 wafers. Part of the analysis involved determining which SAS options were appropriate for dealing with such unbalance.

Solution Strategy

A model was formulated which, in some respects, is generic to semiconductor applications. SAS GLM Type III sums of squares were used to give the best available estimates for the ANOVA, and this will work even for unbalanced data.

In addition, site sums of squares could further be broken down by the use of contrast commands. This partition could be based on a geometric interpretation of the planarized wafer surface. Contrasts can be used for fixed factors.

Conclusion

Site-to-site variability was the largest contributor to thickness variability, contributing about 73% of the observed variability. In addition, a single contrast related to a dishing, or bullseye, planarization phenomenon explained about 73% of the site-to-site variability. Incorporation of the bullseye phenomenon into the analysis was found to be very useful. As a recommendation for future work, engineers should track this bullseye phenomenon as a measure of their progress towards ideal planarity.

PROCESS

Planarization is emerging as an important and viable technology to achieve ultra-flat wafer surfaces (Monnig, 1991). The technology appears at first sight crude: A wafer, held by a robotic arm, is pressed against a rotating circular table; the tabletop has an abrasive pad attached, and this pad is kept covered by a chemical slurry, hence the term chemical-mechanical planarization (CMP). Various temperatures, pressures, and speeds are controlled to achieve a high material removal rate as well as a uniformly flat wafer. See Iscoff (1993) for an article on the nature, benefits, and issues for planarization by chemical-mechanical polishing.

Project engineers analyzed recent data to determine the major remaining sources of thickness non-uniformity of post-planarization wafers. The sources of variation could be categorized by lot, wafer, and site.

DATA COLLECTION

Nine lots of twenty-three wafers per lot were selected and planarized. Each wafer's thickness was measured by a Prometrix® tool at nine fixed site locations. The position of the nine site locations is shown in Figure 9.1. Some of the data is included in the Appendix, and the entire data set is given on the disk.

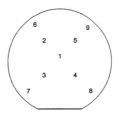

Figure 9.1. *Nine Sites for Prometrix Thickness Measurements. Each wafer was measured at the identical nine locations, making site a fixed factor in the analysis of variance model.*

DATA ANALYSIS AND INTERPRETATION OF RESULTS

Form of the Model—Analysis of the Model

From a statistical point of view, a likely model has to be constructed to test for potential sources of variation and to find out where most of the variation is coming from. To this end, a "mixed" model was constructed (see Appendix). The full model specified potential sources of variability as arising either from site-to-site (SITE), wafer-to-wafer (WAFER), lot-to-lot (LOT), or the interaction between lot and site (LOT*SITE).

The results of running the SAS GLM and SAS VARCOMP analysis of variance procedures on the data are given in Tables 9.1 and 9.2.

Table 9.1. *Type I and III ANOVA Sums of Squares (SS) are identical, indicating balanced data. The overall F = 32.72 and the R^2 = 85.16% indicate an adequate model fit.*

General Linear Models Procedure
Dependent Variable: THICK

Source	DF	Sum of Squares	Mean Square	F Value	Pr>F
Model	278	99301064.624	357198.074	32.72	0.0001
Error	1584	17293464.166	10917.591		
Corrected Total	1862	116594528.790			
	R-Square	C.V.	Root MSE		THICK Mean
	0.851679	3.032131	104.48728		3446.0016
Source	DF	Type I SS	Mean Square	F Value	Pr>F
SITE	8	85399007.519	10674875.940	977.77	0.0001
LOT	8	1281523.528	160190.441	14.67	0.0001
LOT*SITE	64	10341997.498	161593.711	14.80	0.0001
WAFER(LOT)	198	2278536.079	11507.758	1.05	0.2999
Source	DF	Type III SS	Mean Square	F Value	Pr>F
SITE	8	85399007.519	10674875.940	977.77	0.0001
LOT	8	1281523.528	160190.441	14.67	0.0001
LOT*SITE	64	10341997.498	161593.711	14.80	0.0001
WAFER(LOT)	198	2278536.079	11507.758	1.05	0.2999

Table 9.2. *Output from SAS VARCOMP procedure showing how total variability is split between the fixed factor SITE, and the random factors (variance components) LOT, LOT*SITE, WAFER (Nested in LOT), and "Error."*

Variance Components Estimation Procedure
Dependent Variable: THICK

Source	DF	Type I SS	Type I MS
SITE	8	85399007.519338	10674875.939917
LOT	8	1281523.527841	160190.440980
LOT*SIZE	64	10341997.497565	161593.710899
WAFER (LOT)	198	2278536.079405	11507.757977
Error	1584	17293464.165698	10917.591014
Corrected Total	1862	116594528.789846	

Source Expected Mean Square
SITE Var(Error) + 23 Var(LOT*SITE) + Q(SITE)
LOT Var(Error) + 9 Var(WAFER(LOT)) + 23 Var(LOT*SITE) + 207 Var(LOT)
LOT*SITE Var(Error) + 23 Var(LOT*SITE)
WAFER(LOT) Var(Error) + 9 Var (WAFER(LOT))
Error Var (Error)

Variance Component	Estimate
Var(LOT)	-9.63012987
Var(LOT*SITE)	6551.13564721
Var(WAFER(LOT))	65.57410701
Var(Error)	10917.59101370

The SAS "Varcomp" procedure can be used to obtain variance components. The output of the procedure for the data we used is given in Table 9.2. The fixed effect SITE explains about 73% of the total variation, or sum-of-squares (SS) since SS(SITE)/SS(Total) = 0.73. All other model factors (LOT, SITE*LOT, WAFER, and Error) are random,[19] and they explain 27% of the total variability. The variance component part of Table 9.2 breaks down their relative contributions into 0% for LOT[20], 37% for LOT*SITE, 0.3% for WAFER, and 62% for the pure Error term. These percentages add to 100% within rounding errors.

Contrast Analysis

The discovery that site-to-site differences are the major source of variability was of great importance for the work of process improvement. If lot-to-lot had turned out to be a major source, then ways of reducing the variation in incoming lots would have been a primary focus.

---

[19] Although each lot is made up of a different set of wafers (wafers nested in lots), the position of the wafers in the lot and the order that those positions are processed are *fixed* and do not change from lot to lot. Therefore, in some process steps wafers can be a "hybrid" factor that includes *both* a fixed effects component due to the position and order of processing (crossed with lots) and a random effects wafer-to-wafer component (nested in lots). In this chapter, the systematic effects due to position and process order are assumed to be a negligible portion of the wafer-to-wafer differences. In cases where this assumption is valid, wafers can be treated as nested (in lots) random effects.

[20] In this approach, a negative variance component is always set to zero, as a negative variance is considered to be produced by sampling error.

At this stage it was of interest to take a closer look at the site-to-site differences, and try to find a further partition of the sum of squares, preferably one that would correspond to the mechanics of the planarizing operation.

By examining Figure 9.2, a three-dimensional plot of a typical wafer's thickness normalized about zero, we note certain geometric characteristics that are of engineering interest. Two of these are the rather obvious bowl-shape or bullseye curvature and the suggestion of a linear tilt or cant of the plane of the wafer. By observing that sites 6 to 9 form an outer ring (see Figure 9.1) and that sites 2 to 5 form an inner ring, the obvious bowl-shape can be visualized.

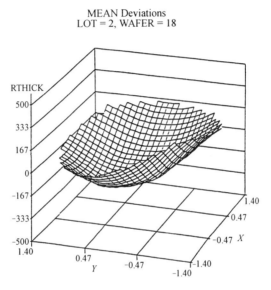

Figure 9.2. *A Three-Dimensional Wafer Thickness Map of a Typical Post-Planarized Wafer. Readings are in Angstroms with the average normalized to zero.*

Contrast Analysis Using SAS

If s_i is the mean thickness at site I, we might ask if $s_1 - s_2$ is a significant difference. If the answer is yes, then we look for an engineering explanation as to the persistence of this difference. The difference, $s_1 - s_2$, is termed a contrast since the weights (or coefficients) attached to the s_i, namely 1 and −1, add up to zero. A contrast is estimable and has its own sum of squares. In general, we want to work with estimable functions (see Milliken and Johnson, 1984, p. 104) of the site parameters, such as $s_1 - s_2$, or in the general form $\Sigma c_i s_i$. If $\Sigma c_i = 0$, then $\Sigma c_i s_i$ is called a contrast in the s-effects. Figure 9.3 shows four different contrasts in the site means.

The SAS GLM program was augmented by adding the four contracts listed in Figure 9.3, and the results—discussed below—are shown in Table 9.3.

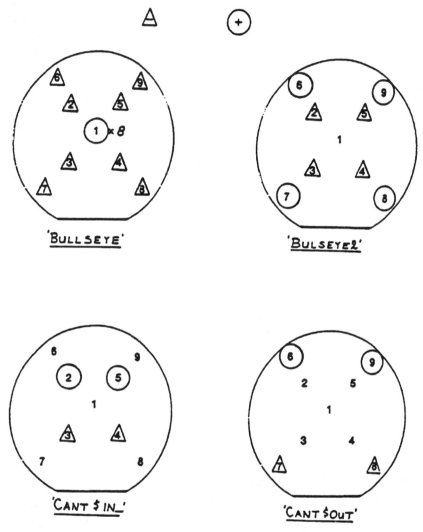

Figure 9.3. *Four Examples of Site Contrasts. A site contrast is a weighted linear combination of site observations, where the weights must sum to zero. For the BULLSEYE contrast, readings at 2 through 9 are subtracted from eight times the reading at site 1; for the BULSEYE2, the readings at sites 2 through 5 are subtracted from the readings at sites 6 through 9, and so on.*

Interpretation of Results

Our analysis, now including the four contrasts, is summarized in an ANOVA table, Table 9.3. This shows how the total variation may be split into that due to model plus that due to error. The model variation is further split into the SITE contribution and the LOT*SITE contribution. Splitting the variation of SITE further, the SITE contribution may be partitioned by the contrasts BULLSEYE, BULSEYE2, CANT$IN, and CANT$OUT. The fixed factor SITE explains 73% of total variability, with the BULSEYE2 contrast explaining 73.5% of that.

Table 9.3. *ANOVA for the Final Model, Including Contrasts. Four of the eight SITE degrees of freedom (df) are given to the four contrasts, each of which absorbs one df. The four leftover df in SITE represent the site-to-site variation that we were unable to find an explicit explanation for in the way of other meaningful contrasts. Only 2.3% of the site-to-site variation remains unexplained.*

Source of Variation		DF	Sum of Squares	Mean Square	F	Pr>F
Model		80	97022528.55	1212782	110.42	0.0001
Site		8	85399007.52	10674876	66.06	0.0001
	BULLSEYE	1	13528445.98	13528446	1231.74	0.0001
	BULSEYE2	1	62762394.61	62762395	5714.42	0.0001
	CANT$IN	1	2009352.841	2009353	182.95	0.0001
	CANT$OUT	1	6000095.214	6000095	546.3	0.0001
LOT*SITE		72	11623521.03	161437.8	14.7	0.0001
Error		1782	19572000.25	10983.17		
Total		1862	116594528.8			

CONCLUSIONS AND RECOMMENDATIONS

In future work involving planarization, engineers should track the bullseye phenomenon as described above as a measure of progress toward ideal planarity. In particular, the bulseye2 metric is most informative.

In general, engineers should construct sensible contrasts and use them as performance metrics for their own particular process. Different processes will require different contrasts. The methods given in this paper enable one to choose those contrasts most applicable to any particular process.

ACKNOWLEDGMENTS

Thanks to Hu Bath (AMD) for providing data and some first contrast structures, to the SEMATECH Statistical Methods Group, and in particular to Professor Peter John (UT) for commenting on an initial draft of the paper, and to Kyle Merrill (now of Hewlett-Packard) for a SAS program to generate 3-D wafer maps. A number of anonymous referees are also thanked for their valuable comments.

REFERENCES

Milliken, G. and D. Johnson. *Analysis of Messy Data*, Vol.1. New York, NY: Van Nostrand Rheinhold, 1984.
Iscoff, Ron. "CMP Takes a Global View," *Semiconductor International*, pp. 72-78, May, 1993.
Monnig, K., R. Tolles, A. Maury, R. Leggett, and S. Sivaram. "Overview of Planarization by Mechanical Polishing," *Extended Abstracts*. Vol. 91-1, no. 385. The Electrochemical Society; Spring Meeting, Washington, DC. May, 1991.
Rosenthal, R. and R. L. Rosnow. *Contrast Analysis*. Cambridge University Press, 1985.
SAS. The SAS Institute, Inc., SAS Campus Drive, Cary, North Carolina 27513.
Satterthwaite, F.E. "An Approximate Distribution of Estimates of Variance Components," *Biometrics Bulletin*, Vol. 2, 1946.
Sivaram, S., H. Bath, E. Lee, R. Leggett, and R. Tolles, "Measurement and Modelling of Pattern Sensitivity During Chemical Mechanical Polishing of Interlevel Dielectrics," *Proc. Advanced Metallization for USLI Applications*, Pittsburgh, PA: Materials Research Society, Oct. 1991.

APPENDIX A

Mixed Models in General

It is useful to have some standard modeling definitions. This is helpful when specifying the code for model analysis. The following definitions were used in this paper.

A factor is *random* if its levels (i.e., observed instances) are a random sample of levels from a population of possible levels. In this case study, the nine lots were selected at random from a larger group of lots.

A factor is *fixed* if its levels are selected by a nonrandom process or if its levels consist of the entire population of possible levels. In this case study, SITE is a fixed factor.[21]

A model[22] is called a *fixed* model or a *fixed-effects* model if all of the factors in the model are fixed.

A model is called a *random* model or a *random-effects* model if all of the factors in the model are random effects (except for the mean μ).

A model is called a *mixed* model or a *mixed-effects* model if it contains *both* fixed and random factors.

A first factor is *nested in* a second factor if the levels of the first factor are *different* within each of the levels of the second. For example, in this case study, WAFER is nested inside LOT. This is written, in SAS, as WAFER(LOT).

A first factor is said to be *crossed with* a second factor if the levels of the first factor are the *same* for each of the levels of the second factor. For example, the same level of SITE appears with every LOT. This is written, in SAS, as SITE*LOT.

---

[21] Note that a factor being either fixed or random is determined by how its levels were selected for the experiment (see Milliken, 1984).
[22] By model, we mean a linear model in the usual sense, as is used in regression or ANOVA.

APPENDIX B

Data from Passive Data Collection

Wafer Thickness (in Angstroms) for Nine Sites per Wafer, 23 Wafers per Lot, and Nine Lots. (The remaining numbers in this table are given in the data sets on the disk.)

OBS	LOT	WAF	SITE1	SITE2	SITE3	SITE4	SITE5	SITE6	SITE7	SITE8	SITE9
1	1	1	3238.8	3092.7	3320.1	3487.8	3181.7	3290.6	3370.9	3855.9	3243.5
2	1	2	3201.4	3212.6	3365.6	3291.3	3193.6	3494.4	3663.3	3511.5	3214.7
3	1	3	3223.7	3320.8	3406.2	3336.8	3217.4	3566.8	3606.2	3529	3209.7
4	1	4	3213.1	3300	3281.6	3312.5	3229	3604.6	3497.9	3555.7	3378.8
5	1	5	3247.5	3277.1	3289.1	3369.6	3313	3392.4	3584.4	3488.2	3534
...	...	...	...	...	...	...	...	...	...	...	...
203	9	19	3138.3	3217.9	3288.7	3213.8	3188.4	3915.8	4028.6	3752.6	3672.1
204	9	20	3154.6	3261	3296.1	3174.9	3154.7	3954.2	3767.4	3813.5	3725
205	9	21	3098.7	3201.2	3354.3	3209.8	3086.6	3817.7	4095.6	3708.8	3741.4
206	9	22	3015.3	3187.7	3248	3130.7	2933.5	3808.5	4206.1	3828.7	3685.2
207	9	23	3093.6	3169.9	3295.1	3249.7	3164.4	3883.9	4083.2	4010.5	3670.5

BIOGRAPHIES

Arnon Hurwitz is a founder and Director of Qualtech Productivity Solutions, a corporation registered in South Africa and doing productivity consulting internationally. He is also a Vice President of MiTeX Inc., a Michigan-based corporation that specializes in run-to-run (batch) control of industrial manufacturing. Dr. Hurwitz was a Senior Statistician in the Statistical Methods Group of SEMATECH, a semiconductor research consortium based in Austin Texas. He also served for three years as a Senior Process Control Project Manager there, developing advanced process control methods and software. Before joining SEMATECH in 1991, he worked for six years in the petrochemical and optic-fiber manufacturing industries in the United States. Arnon holds B.S., B.S. (honors), and Ph.D. degrees in Mathematics and Statistics from the University of Cape Town, South Africa, and an M.S. in Applied Statistics from Oxford University. He is a member of the American Society for Quality Control. Currently resident in South Africa, his e-mail is arnon@global.co.za.

Pat Spagon currently works at Motorola University, Southwest where he develops applied statistics courses for engineers and teaches and consults in statistical methods. Prior to joining Motorola, Dr. Spagon was a Senior Member of the Technical Staff in the Statistical Methods Group at SEMATECH where he did training and consulting, primarily in experimental design and analysis. Pat has held a variety of positions in industry and academia. His experience includes quality assurance and systems analysis in engineering and manufacturing environments. He has been a senior consultant in statistical methods for BBN Software Products Corporation and a consultant for statistical methods for FMC Corporation. He has been a quality assurance manager for Hewlett-Packard and a member of the technical staff at Bell Telephone Research Laboratories. Pat has been a member of the faculty at the Business School at San Francisco State University and The Technological

Institute at Northwestern University. Dr. Spagon holds a Ph.D. in Industrial Engineering from Stanford University, an MS in Electrical Engineering from UC Berkeley, and a BS in Electrical Engineering from the University of Arizona. He is a member of the American Statistical Association, the American Society for Quality Control, and the Institute for Operations Research and Management Science.

CHAPTER 10

FACTORS WHICH AFFECT THE NUMBER OF AEROSOL PARTICLES RELEASED BY CLEAN ROOM OPERATORS

William Kahn and Carole Baczkowski

Semiconductor processing requires a clean manufacturing environment. One of the major sources of wafer contamination are micron sized airborne particles called aerosol particles. People can be the largest source of particulate contamination in the clean room. This experiment studied the effect of eight factors on the number of aerosol particles released by clean room operators. We found that expanded-polytetrafluoroethylene (EPTFE) suits release five times fewer particles than polyester suits. We also found that there is significant person-to-person, day-to-day, and residual variability. Garment size and individual garment factors were negligible effects. Different activity protocols generate different amounts of particulation at different heights in the clean room. This work shows that future evaluation of bunny suits must compare suit designs on the same person during the same day, for several persons, and measure the particulation in multiple ways. The experimental design used in this passive data collection can be usefully considered as a full factorial design, but actually has the more complex split-plot structure.

EXECUTIVE SUMMARY

Problem

Particle contamination is a major issue in semiconductor manufacturing. By working in increasingly cleaner environments, line widths can be reduced, thereby allowing yields, chip speed, reliability, and capability all to improve. Human presence in manufacturing clean rooms is a major source of contamination (see Lange, 1983 and Joselyn, 1988). Even with bunny suit protection, up to 20% of particulate contamination can be assigned to humans (see Burnett, 1986). While efforts are being made to limit human occupancy in manufacturing environments, it is not reasonable to expect that fully automated production will be available in the foreseeable future. Thus, we seek to minimize the contamination caused by human presence in clean rooms.

There are many activities which contribute to the maintenance of a "clean" clean room. One of these activities is the wearing of clean room garments, sometimes called bunny suits. These suits act as whole body filters, allowing air to pass, so as to maintain comfort, but not allowing contamination to pass.

Studies of filtration materials used in bunny suits have characterized filtration efficiencies (see Swick, 1985), material linting (see Swick, 1985 and Helmke, 1982), and breathability (see Swick, 1985). Recent work (see Liu, 1993) has started to establish the "in-use" effectiveness of the entire bunny suit.

This chapter describes a careful experiment undertaken to evaluate factors which affect particulate contamination released by bunny suits. The source of this contamination may either be the suit itself, breathing gaps from the closures, or less-than-100% filtration efficiency.

Solution Strategy

Previous work in this area has not achieved adequately reproducible results. This is because of several previously unquantified, but large, variance components. This experiment was designed to estimate person-to-person variability, day-to-day within-person variability, garment-to-garment variability, and within-day within-person variability. It also looked at four fixed effects: material type used in the bunny suit (expanded-polytetrafluoroethylene and polyester), the size of the suit (small, medium, large), activity protocol (three levels), and monitoring location (workstation, pedestal).

The experiment was run as a full factorial. Every person wore every suit of the appropriate size and was tested twice on both the morning and afternoon of two days. During each test the particulation was measured during three different activities at two different monitoring locations.

Conclusions

The EPTFE suits produced one-fifth the particulation of the Polyester suits. There was large person-to-person variability, day-to-day variability, and residual error variability. The garment-to-garment variability was small. This means that to compare bunny suit designs researchers must compare the designs by using them on the same person during the same day. Researchers must also replicate the comparisons on multiple people.

PROCESS

People are a major source of aerosol contamination in clean rooms. When they must enter one, it is important not to let particles emitted from skin, hair, cosmetics, or clothing escape to the clean room environment. By enclosing people in whole-body filters, called bunny suits, we can capture the particles to prevent them from entering the clean room. A certain percentage of uncaptured particles land on wafers and decrease the yield.

We want to learn how to design bunny suits which simultaneously maximize comfort and filtration efficiency. If the bunny suits are not comfortable, people will overheat, decreasing their productivity. Further, by moving in particular ways they will learn to create small, temporary, breathing gaps in the bunny suit—giving them a puff of cool air, but also releasing a puff of contamination into the clean room.

People have very different particulation levels. Further, a single person may particulate very differently on one day than on another. Recent smoking, exposure to barbecue, or even cosmetics can dramatically change particulation. In this experiment we sought to quantify these variance sources.

DATA COLLECTION PLAN

In this experiment, we wanted to examine the effect of eight factors on the measured particulation. The factors of interest and their levels were

1. Garment Type: Tightly woven polyester, and expanded polytetrafluoroethylene laminate
2. Garment Size: Small, medium, large
3. Person: Two people of each size (six people total)
4. Day: Each person tested on two days
5. Time: Each person tested at two times on each day tested
6. Garment: Two garments for each type of each size (12 garments total)
7. Location: Sampled at workstation level (near hand level) and pedestal level (near foot level)
8. Protocol: Three activities (deep knee bends, marching and slapping, and reaching).

We studied every possible appropriate combination of the selected levels of the eight factors. This is called a full factorial design. Table 10.1 gives the design.

Each of the six people wore all four garments appropriate to his or her size. Each of the 12 garments were worn by both people of the appropriate size. Each of these 24 pairings of person and garment was run in both the morning and afternoon of two days. This gives a total of 96 tests. During each test, the person performed each of the three activity protocols. During each protocol we measured particulation at both workstation level and pedestal level.

Testing was performed in a Class 1 clean room. This clean room was 125 square feet and was completely void of equipment so as not to disrupt the vertical laminar flow or add particles. The clean room has temperature and humidity control, a change room, and an entry through a change shower. Background counts in this room were less than one particle per cubic foot greater than .3 microns.

Table 10.1. *The 96 Tests of the Design. During each of these 96 tests, the subject performed all three activity protocols. During each protocol, we measured the subject at both levels: workstation and pedestal.*

PERSON SIZE	PERSON	DAY	TIME	GARMENT SIZE: Small				GARMENT SIZE: Medium				GARMENT SIZE: Large			
			GARMENT TYPE	E		P		E		P		E		P	
			GARMENT	1	2	3	4	5	6	7	8	9	10	11	12
Small	A	1	AM	x	x	x	x								
			PM	x	x	x	x								
		2	AM	x	x	x	x								
			PM	x	x	x	x								
	B	3	AM	x	x	x	x								
			PM	x	x	x	x								
		4	AM	x	x	x	x								
			PM	x	x	x	x								
Medium	C	5	AM					x	x	x	x				
			PM					x	x	x	x				
		6	AM					x	x	x	x				
			PM					x	x	x	x				
	D	7	AM					x	x	x	x				
			PM					x	x	x	x				
		8	AM					x	x	x	x				
			PM					x	x	x	x				
Large	E	9	AM									x	x	x	x
			PM									x	x	x	x
		10	AM									x	x	x	x
			PM									x	x	x	x
	F	11	AM									x	x	x	x
			PM									x	x	x	x
		12	AM									x	x	x	x
			PM									x	x	x	x

During each of the three protocols, particle counts were measured simultaneously by two particle counters. Each one-cubic-foot-per-minute laser counter was connected to a GORPLER® aerosol sampler. This probe has 80 ports which sample an area of 30 cm x 38.5 cm, providing a representative sample of submicron particles over the whole probe area. One probe was at workstation level, and the other at pedestal level.

The three activity protocols are exactly defined physical movements well explained by their names. The motions are performed to a metronome. These three protocols are designed to span the range of typical human activities in a clean room. Protocol "Reaching" simulates gentle activity, protocol "March & Slap" simulates hurried and active motion, and protocol "Deep Knee Bends" is vertical movement associated with activities such as equipment movement or repair.

Protocol Reaching: A gentle protocol where the subject reaches across the probe changing arms at a rate of 30 reaches per minute. Aerosol particle counts were measured for five minutes and normalized to a number per cubic foot (see Figure 10.1).

Protocol March & Slap: A more vigorous protocol where the subject marches in place while patting his or her chest, thighs, and seat at the rate of one pat per second. Again, aerosol particle counts were measured for five minutes and normalized to a number per cubic foot.

Protocol Deep Knee Bends: Three deep knee bends were performed over a one minute period. Aerosol particles were counted for this minute.

All clean room garments were worn over street clothes and included a disposable mask, goggles, and two pairs of vinyl gloves.

DATA ANALYSIS AND INTERPRETATION

Transformation

The 96 tests with three activity protocols per test and two measurement locations during each protocol produced 576 particle measurements. These readings ranged over more than a factor of 100. The data is given in the Appendix. We took logs (base 10) of all the values before any analyses. Counts of 0 were replaced with counts of 1 for the purpose of the transformation. This gave us reasonable homoscedasticity—the variability of the high values was now similar to the variability of the low values. Further, this transformation allows us to easily talk about the proportionate increase and decrease of different effects.

Graphical Analysis

The first analysis of data is nearly always graphical. We have produced two graphs of our data. Figure 10.2 averages over the 96 tests and shows the effect of Activity Protocol and Location of Sampling on the Log10 of particle counts. From the graph we learn that the march & slap protocol produces the most particulation at both locations. For the workstation level we see that reaching particulates slightly more than deep knee bends. For the pedestal level we see that reaching hardly particulates at all—counts stay at the background level of about 1 count per cubic foot.

Two additional analysis for these two factors may be of interest. We may want the statistical significance of the observed effects. For this, we need to do an analysis of variance as described in the next section. We may also be concerned about whether the effect of activity protocol and sampling location on measured particulation is the same for all garments on all people on all days. This is a question regarding the interaction between the protocol or location with the other six factors. The advanced analysis given in the next section shows that indeed these interactions exist, but have small magnitude in comparison to the effects we have seen in Figure 10.2.

To produce the second graph of interest we averaged together the six measurements made on each of the 96 tests. Figure 10.3 shows these 96 measurements plotted versus person-day. We immediately see the large effect of garment type. The polyester particulates about .7 log10 units more than the EPTFE. The antilog of .7 is about 5. Thus we learn that EPTFE particulates about five times less than Polyester.

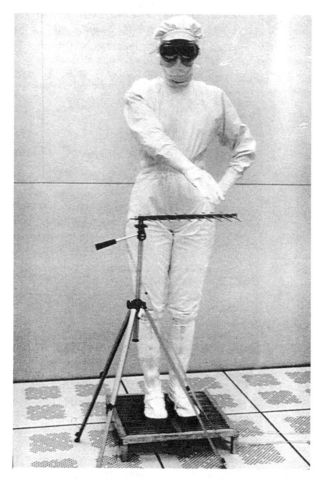

Figure 10.1. *Subject In the Clean Room Executing the Reaching Protocol While Particulation is Measured at the Workstation Level. Particles larger than .3 microns are detected. This activity generated, on average, nine particles per cubic foot.*

We also see in Figure 10.3 the large person-to-person variability, large day-to-day variability within person, and the large residual variability. We see that the difference between the three sizes of people is small given the large person-to-person variability within size.

We may still be interested in quantified estimates of the variance associated with these different sources. While the graphical analysis shows us what is large, the numerical estimates come from the more formal statistical analysis discussed in the next section.

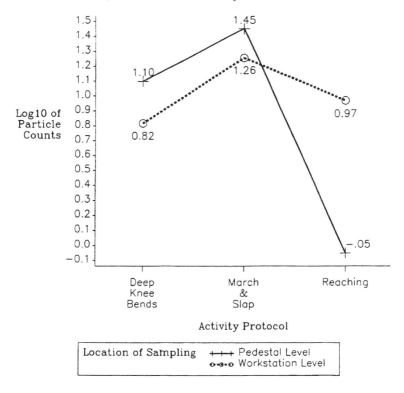

Figure 10.2. *Effects of Activity Protocol and Location of Sampling on the* Log10 *of Particle Counts. March and slap produces the most particulation at both sampling locations.*

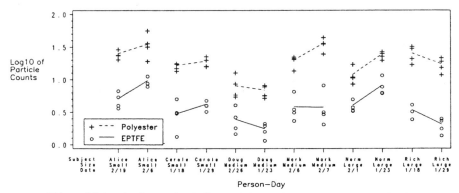

This graph helps visualize two fixed effects and three variance components. Note that the particulation is on a log scale. Thus a shift of 1 unit corresponds to a factor of 10 change. We see that the polyester bunny suits particulated about 5 times more than the EPTFE suits. We also see that there is high within-person within-day variability, day-day variability within person, and person-person variability within size. The size effect is small given the person-person variability within size. We have averaged over three physical activities, called protocols, which include Reaching, March & Slap, and Deep Knee Bends. We have also averaged over Pedestal and Workstation levels. We conclude that to effectively evaluate particulation reduction due to bunny suit design, we must test multiple suits on a single person on a single day, and then repeat on multiple days with multiple people.

Figure 10.3. *Six Measurements Plotted Versus Person-Day. This shows effect of garment type.*

Advanced Analysis[23]

Because this is an eight factor full factorial design, it may seem easy enough to run an analysis of variance. In this design, however, it is not quite straightforward. The six factors which define the 96 tests will have a fairly straightforward analysis, but the two factors, activity protocol and sampling location, which were studied on each of the 96 tests must be treated specially. These two factors are called split-plot factors (see Snedecor, 1980). The reason for this special consideration is that all six measurements were made during a single test. We did not put a person into a bunny suit and measure just reaching at workstation level and then put another person into a bunny suit and measure just march & slap at the pedestal level.

Whenever you have this kind of grouping of measurements, you have what is called a blocked or a split-plot experiment. Each of the 96 tests is called a block or a main-plot.

---

[23] The reader who is not comfortable with analysis of variance may wish to go directly to the Conclusions and Recommendations Section. The important ideas are demonstrated by the graphical analysis.

Within each of these main-plots we now perform a small experiment. The effects studied in this small experiment are called split-plot factors. Such designs are often very efficient at estimating the effect of the split-plot factors because each combination of split-plot factors are compared during the same test. Not only do these designs produce excellent estimates of the split-plot effects, but also they are often much cheaper to run than fully randomized designs. It would be very expensive to run 96x6=576 separate tests because each test requires a new suit-up—a somewhat difficult procedure.

In the present study we have an additional complexity. During each test, the subject did not perform six protocols, twice through the set of three, once for workstation level measurement and once for pedestal level measurement. The subject ran each of the three protocols once, and during each protocol, we measured at both locations simultaneously. During each protocol (called a split-plot), we measured at two locations. Naturally enough, we call location a split-split-plot factor.

There are several possible ways to analyze a split-plot design. We have chosen a simple procedure which requires two analysis of variance runs. The first run tells us about the split-plot factors and the split-plot-by-main-plot interactions. The second run tells us about the main-plot factors.

Split-Plot Analysis

We model the two split-plot factors with a three way analysis of variance. The third factor is a 96 level categorical variable called test. It has a different value for each of the 96 tests. We can also add interactions among the split-plot factors and selected interactions between the split-plot and main-plot factors.

We used SAS® for the analysis (see SAS Institute, 1990). Our PROC GLM code is in Table 10.2.

Table 10.2. *The SAS GLM code for the Split-Plot Analysis.*

```
proc glm;
   class test protocol location type size person;
   model logcount =test protocol location protocol*location
                                type*protocol size*protocol size*type*protocol
                                person*protocol(size) type*location
                                size*location size*type*location
                                person*location(size);
```

The effects and interactions of interest are specified in the model statement. The notation effect1(effect2) is used to specify nested effects. This says that effect1 is nested within effect2. Nesting is a way to combine an effect with its interaction with another effect and treat them as a single term. Note that all the interactions specified in this model involve either location or protocol, because this is a split-plot analysis and, aside from the factor test, only effects which involve the split-plot factors are modeled.

The output produced is shown in Table 10.3.

Table 10.3. *The SAS GLM Output for the Split-Plot Analysis.*

Source	DF	Sum of Squares	Mean Square	F Value	Pr > F
Model	124	270.	2.18	24.68	0.0001
Error	451	39.	0.088		
Corrected Total	575	310.			

R-Square	C.V.	Root MSE	LCOUNT Mean
0.871538	32.19	0.297	0.923

Source	DF	Type I SS	Mean Square	F Value	Pr > F	
TEST	95	100.95	1.062	12.02	0.0001	
PROTOCOL	2	77.00	38.501	435.52	0.0001	←Big
LOCATION	1	4.75	4.754	53.78	0.0001	
PROTOCOL*LOCATION	2	50.82	25.414	287.48	0.0001	←Big
TYPE*PROTOCOL	2	4.94	2.472	27.97	0.0001	
SIZE*PROTOCOL	4	9.69	2.424	27.43	0.0001	
SIZE*TYPE*PROTOCOL	4	0.09	0.024	0.27	0.8941	
PERSON*PROTOCOL(SIZE)	6	4.10	0.683	7.74	0.0001	
TYPE*LOCATION	1	5.29	5.190	58.71	0.0001	
SIZE*LOCATION	2	0.31	0.156	1.77	0.1722	
SIZE*TYPE*LOCATION	2	2.38	1.190	13.46	0.0001	
PERSON*LOCATION(SIZE)	3	10.22	3.409	38.57	0.0001	

The F Value column is the square of a signal-to-noise estimate. We see that the two effects we already noted, protocol and location, are large effects. There is a small interaction effect type*location. This suggests that the EPTFE improvement over the Polyester is different at the two sampling locations. Because all terms have the same noise estimate, the ratio of F Values is the square of the ratio of effect sizes. The protocol*ratio interaction effect is 2.2 times larger than the type*location effect because sqrt(287.48/58.71)=2.2. Because this effect is physically small with respect to the other effects it was not studied further.

Note that the column labeled "Pr > F" gives a p-value. This is the chance of seeing an effect this large just by chance (if there really was no effect). Most of our p-values are small. It is nice to note that the effects we saw in Figure 10.2 are indeed statistically significant—they are not just artifacts of the natural variability of the system.

In a different kind of study, we might be interested in any effect which was statistically significant, i.e., had a p-value under .05. In the current study, however, we were only interested in identifying and quantifying the physically big effects. The existence of real, but physically small effects neither surprises nor interests us.

We have now analyzed the split-plot structure of our experiment in a particularly easy way. We fit the full main-plot structure with the categorical variable TEST, along with the interactions of interest between the split-plot factors and the main-plot factors. All the p-values and effect estimates are correct. Note that we are not testing the main-plot effects here, only blocking them out of the split-plot analysis.

The careful reader of this split-plot discussion will note a small complication we have not yet discussed. During each of the 96 tests we did not perform each of the three activity protocols twice, once for the workstation level measurement and once for the pedestal level measurement. We performed each protocol once and measured each of them twice: at the two levels. This constitutes another level of the analysis and is called the split-split-plot level. In analyzing a split-split-plot design, the averages, and thus the graphs, do not depend on how we model the error term. Only our judgement of the statistical significance of the effects depends on our modeling of the error term. Thus the protocol effect is actually being compared to the wrong error term. However, the location and protocol*location interaction are using the correct error term in the above analysis. By learning that the protocol*location effect is statistically significant, we know the protocol used is significant and thus have no need to look at the protocol main effect with the correct error term. More advanced treatment of the analysis of split-plot designs are described in the statistical literature (see Heiberger, 1989, and Snedecor, 1980).

Main-Plot Analysis

Having analyzed the split-plot structure, we can now analyze the main-plot structure. To do this, we average the six readings from each of the 96 tests to get a single particulation level for each test. Each of the 96 tests is called a main-plot. We use the average over each main-plot as our response variable. Because we are interested in two different kinds of main-plot effects, fixed and random, the main-plot analysis will require two calculations.

We are interested in a quantitative estimate of the difference between the two types of garments. We also want to know the effect of size: do large people particulate more or less than small people? We also want estimates of the variance associated with garment, person, day, and time. Finally, we want an estimate of the residual variance—whatever cannot be explained by the specified terms.

A typical analysis of variance will estimate effects called fixed effects. If the difference in average response between two or more levels is of interest, we call the factor a fixed effect. Our type and size effects are fixed effects. Our first analysis, using PROC GLM again, will study these two fixed effects.

Sometimes we are not specifically interested in the average of each selected level. Rather, the variability of the averages is what matters. Then the effect is called random. This is the case for our garment, person, day, and time factors. We are not interested in the effect of the specific days selected but in estimating how much day-to-day variability we should expect. Typical analysis of variance programs will not estimate these variance components. We have to use a program specifically designed to estimate random effects. Our second analysis, using the SAS procedure, VARCOMP, will study these variance components.

Note that comments in SAS start with an asterisk (*) and end with a semicolon (;).

Table 10.4 shows the SAS code used to analyze the effect of size and garment, the two fixed effects.

Table 10.4. *The SAS GLM code for the Main-Plot Fixed Effects Analysis.*

```
proc glm;
  class day time size garment person type;
  model mlcount=type
                size
                garment(type size)
                person(size)
                day(person size)
                time(day person size); *fit model;
  lsmeans type size;
  random  garment(type size)
          person(size)
          day(person size)
          time(day person size)/test;  *compute correct
p-values;
```

The output from this code is shown in Table 10.5.

Table 10.5. *Analysis of Variance Table for Fixed Effects Analysis.*

Source	DF	SS	MS
type	1	11.160	11.160
size	2	.856	.428
garment(type size)	8	.098	.012
person(size)	3	1.986	.662
day(person size)	6	.860	.143
time(day person size)	12	.273	.023
Model Total	32	15.233	.476
Error Total	6	1.592	.025
Total	95	16.825	
R-square = .91	Mean = .92	Root(MS error) = .16	

The random statement selects the correct error term and performs the correct test for each effect. This is shown in Table 10.6.

From this we learn that we have reliably detected type, person, and day effects. Size, garment, and time effects are small.

We now can see the value of the nested notation. For example in this model, we have person nested within size, symbolized as person(size). This says that to learn whether or not there is a systematic difference between the three sizes, we need to take into account the variability between people of the same size.

Table 10.6. *The Correct p-value Analysis.*

Effect	Correct Error Term	F-ratio	p-value
type	garment(type size)	907.88	.0001
size	garment(type size) +person(size) -error	.66	.58
garment(type size)	error	.49	.86
person(size)	day(person size)	4.62	.05
day(person size)	time(day person size)	6.31	.0034
time(day person size)	error	.90	.55

Similarly, to learn whether or not the six people are different from each other, we need to compare the day-to-day variability of each person. This happens because of our model terms person(size) and day(person size). The analysis of variance for models with multiple error terms is an advanced topic in statistics. Here is an example of it, but it is just icing on the cake–we already learned the important ideas with the graphical analysis.

The LS means statement gives us the average over each level of type and each level of size and is shown in Table 10.7.

Table 10.7. *The LS means for the Two Fixed Effects: Type and Size.*

TYPE		EPTFE	0.58
		POLY	1.26
SIZE		LARGE	0.94
		MED	0.80
		SMALL	1.03

The difference between POLY and EPTFE types is 1.26 − .58 = .68. The antilog of .68 is 4.8. Thus we learn that Polyester particulates about 5 times more than EPTFE. From Table 10.6 we know that this effect is statistically significant.

We also know that the effect of size is not statistically significant, thus we need look no further at it.

We now go to the second analysis of variance—the estimation of the five variance components: garment, person, day, time, and residual error. Table 10.8 shows the SAS code which estimates the variances.

This code is very similar to the GLM code; the model statement is nearly identical. The /fixed=2 option tells VARCOMP that the first two effects are fixed effects, and while we want to partition their effect out of the variance components estimate, we do not want variance estimates for these effects. We have already estimated these two fixed effects with the LSMEANS statement of the first GLM analysis.

Table 10.8. *SAS VARCOMP Code to Estimate the Variance Components.*

```
proc varcomp method=type1;
class day time size garment person type;
    model mlcount=type
            size
            garment(type size)
            person(size)
            day(person size)
            time(day person size)/fixed=2;
```

This VARCOMP code gives the output shown in Table 10.9.

Table 10.9. *Variance Components Estimates.*

Source	DF	Variance	Standard Deviation	Antilog10 of sd
garment(type size)	8	0	0	1
person(size)	3	.032	.18	1.5
day(person size)	6	.015	.12	1.3
time(day person size)	12	0	0	1
error	63	.025	.16	1.4

This analysis shows us the size of the five variance components. The effects of garment and time are very small. Our best estimate of the variability caused by those two sources is zero. It doesn't seem to matter which garment we pick of a particular size or type. Similarly, the small time effect tells us that people do not change their particulation very much during a single day.

Person, day, and residual error are all about the same size. They add .032, .015, and .025, respectively, to the variance associated with the log10 of the particulate count. These numbers correspond to standard deviations on a multiplicative scale of 1.5, 1.3, and 1.4. If we use plus and minus two standard deviations as an approximate 95% confidence interval, then randomly selected people will be expected to vary by a factor of $1.5 \times 1.5 = 2.25$ high or low—i.e., a range of a factor of 5.

If we select a random person on a random day and give a single test, then the sum of the five variance components is .07. The square root of .07 is .27 and the antilog10 of .27 is 1.85. Thus we expect 95% of these measurements to be within a range of a factor of 12 (1.85**4).

CONCLUSIONS AND RECOMMENDATIONS

This experiment has shown us two classes of results. First, and of immediate interest to the authors, is that bunny suit design can make a significant difference in the particulation released to a clean room. Second, and of greater long-term importance, is that future experiments in bunny suit design must compare multiple designs on a single person within a single day and replicate the work on multiple people on multiple days. This kind of experiment is called a replicated blocked study where the blocking factors are person and day.

The person-to-person and day-to-day variability within person can be attributed to differences in skin type, street clothes, hair type, personal grooming, cosmetics, smoking behavior, and recent environments. By learning of and quantifying these effects, we are now in a much better position to develop and evaluate new bunny suits. We do not have to be concerned with a person's natural particulation systematically changing from morning to afternoon.

A subject has the physical ability to do a full suite of tests up to eight times per day. Thus we can compare up to eight bunny suit designs in a simple blocked experiment. This is more efficient than blocking on only four tests performed per half-day. If we have more than 8 possible designs (which we normally do during our research and development cycle) we need to use a more complicated class of experiments called incomplete block designs. The error variability allows us to determine how many replicates we need to detect differences of commercial importance.

It is also nice to learn that there is small garment-to-garment variability. This means we don't have to test many instances of a particular design. Finally, it is good to learn that size of person is not a very important factor. This makes it easier to recruit subjects because we can do so without size constraints.

The significance of the split-plot factors protocol and location, and their interaction, tells us that we must be concerned with where we sample and what physical activity the subjects execute. The march & slap protocol produced the most particulation and the reaching protocol produced almost no particulation at the pedestal level. They all responded similarly to the main-plot effects, but not identically. There were statistically significant interactions of the two split-plot effects with the main-plot effects. The size of the interactions, in relation to the size of the split-plot or main-plot effects, was small, and so not worth careful analysis here. Because of these interactions, however, we will continue to use the entire suite of six measurements for future tests.

While the analysis reported here is not comprehensive, we believe it to be revealing and not misleading. Other models are likely to bring to light additional features and possibly new ideas for clean room maintenance.

This one experiment estimated four fixed effects and five variance components. We hope that experiments like this one become more popular. We also hope others can use these results as an aid in bunny suit evaluation.

REFERENCES

Burnett, E.S. "Clean Room Garments for Contamination Control," *Microelectronic Manufacturing and Testing*, December, 1986.

Heiberger, R.M. *Computation for the Analysis of Designed Experiments*. New York, NY: John Wiley and Sons, Inc., 1989.

Helmke, G.E. "A Tumble Test for Determining the Level of Particles Associated with Clean Room Garments and Clean Room Wipes," Proc., *Institute of Environmental Sciences*, 1982.

Joselyn, L. "Clothing Protects Product from People," *European Semiconductor*, May, 1988.

Lange, John A. "Sources of Wafer Contamination," *Semiconductor International*, p. 129, April, 1983.

Liu, B.Y.H., D.Y.H. Pui, and Y. Ye. "Development of Advanced Test Methods for Clean Room Garments," 1993 Proc., *Institute of Environmental Sciences*, Vol. 1, 1993.

SAS Institute, Inc., *SAS/STAT User's Guide*, Release 6.03 ed., Chapters 20 and 35, 1990.

Snedecor, G.W. and W.G. Cochran. *Statistical Methods*. 7th ed., Ames, IA: Iowa State University Press, 1980.

Swick, R. and V. Vancho, "New Concepts in Clean Room Environments," *Microcontamination*, February, 1985.

APPENDIX

Printout of Average Particle Count Data for the Six Response Variables.

It shows each one of the three protocols [deep knee bends (DKBS), march and slap (March), and reaching (Reach)] at the two locations [pedestal (P_) and workstation (W_)]. MLCOUNT is the mean of the Log10 of each of the six response variables. Counts of 0 were replaced with counts of 1 when taking the logarithm.

TEST	TYPE	SIZE	PERSON	DAY	TIME	GARMENT	P_DKBS	P_REACH	P_MARCH	W_DKBS	W_REACH	W_MARCH	MLCOUNT
1	EPTFE	LARGE	NORM	FEB-1-91	AM	L1	0	0.20	24.6	8	3.60	29.4	0.60
2	EPTFE	LARGE	NORM	FEB-1-91	AM	L2	1	0.00	15.8	5	4.60	26.6	0.55
3	EPTFE	LARGE	NORM	FEB-1-91	PM	L1	1	0.25	7.2	5	9.60	25.0	0.56
4	EPTFE	LARGE	NORM	FEB-1-91	PM	L2	4	0.00	15.4	21	7.25	12.4	0.73
5	EPTFE	LARGE	NORM	JAN-23-9	AM	L1	32	0.40	45.4	23	11.00	24.6	1.09
6	EPTFE	LARGE	NORM	JAN-23-9	AM	L2	5	0.00	30.6	13	13.00	16.0	0.82
7	EPTFE	LARGE	NORM	JAN-23-9	PM	L1	10	1.20	17.0	6	14.40	18.8	0.92
8	EPTFE	LARGE	NORM	JAN-23-9	PM	L2	4	0.60	38.8	9	7.40	14.6	0.83
9	EPTFE	LARGE	RICH	JAN-18-9	AM	L1	26	0.00	36.6	2	1.60	2.8	0.54
10	EPTFE	LARGE	RICH	JAN-18-9	AM	L2	29	0.20	41.6	2	2.40	6.2	0.64
11	EPTFE	LARGE	RICH	JAN-18-9	PM	L1	4	0.00	4.6	5	4.20	4.2	0.42
12	EPTFE	LARGE	RICH	JAN-18-9	PM	L2	13	0.40	27.4	5	0.80	3.0	0.54
13	EPTFE	LARGE	RICH	JAN-29-9	AM	L1	11	0.00	15.8	3	1.60	2.2	0.43
14	EPTFE	LARGE	RICH	JAN-29-9	AM	L2	2	0.20	4.0	1	0.80	7.8	0.17
15	EPTFE	LARGE	RICH	JAN-29-9	PM	L1	8	0.00	6.0	3	3.20	2.6	0.40
16	EPTFE	LARGE	RICH	JAN-29-9	PM	L2	0	0.00	9.8	0	5.40	4.2	0.27
17	EPTFE	MED	DOUG	FEB-26-9	AM	M1	13	0.00	28.6	1	3.20	2.2	0.45
18	EPTFE	MED	DOUG	FEB-26-9	AM	M2	3	0.00	17.2	3	0.80	0.6	0.20
19	EPTFE	MED	DOUG	FEB-26-9	PM	M1	2	0.80	41.8	4	6.40	4.0	0.64
20	EPTFE	MED	DOUG	FEB-26-9	PM	M2	1	0.40	10.2	1	5.00	2.6	0.29
21	EPTFE	MED	DOUG	JAN-23-9	AM	M1	1	0.20	14.0	0	0.80	1.6	0.09
22	EPTFE	MED	DOUG	JAN-23-9	AM	M2	1	0.40	8.4	2	2.60	1.4	0.23
23	EPTFE	MED	DOUG	JAN-23-9	PM	M1	0	0.80	15.6	0	3.40	2.0	0.32
24	EPTFE	MED	DOUG	JAN-23-9	PM	M2	0	0.40	7.0	2	5.60	4.2	0.35
25	EPTFE	MED	MARK	FEB-6-91	AM	M1	2	2.20	2.6	3	7.60	5.4	0.52
26	EPTFE	MED	MARK	FEB-6-91	AM	M2	5	1.00	7.6	2	7.20	5.0	0.57
27	EPTFE	MED	MARK	FEB-6-91	PM	M1	4	0.20	7.8	1	6.80	5.8	0.40
28	EPTFE	MED	MARK	FEB-6-91	PM	M2	7	2.80	4.2	10	19.60	7.4	0.85
29	EPTFE	MED	MARK	FEB-7-91	AM	M1	6	1.80	13.6	13	11.40	20.6	0.94
30	EPTFE	MED	MARK	FEB-7-91	AM	M2	1	0.60	3.4	6	9.00	9.0	0.50
31	EPTFE	MED	MARK	FEB-7-91	PM	M1	3	0.40	5.6	2	9.40	12.2	0.53
32	EPTFE	MED	MARK	FEB-7-91	PM	M2	1	0.20	7.0	0	10.60	7.8	0.34
33	EPTFE	SMALL	ALICE	FEB-19-9	AM	S1	4	0.60	12.6	8	13.60	11.6	0.76
34	EPTFE	SMALL	ALICE	FEB-19-9	AM	S2	4	1.00	12.2	6	19.00	24.8	0.86
35	EPTFE	SMALL	ALICE	FEB-19-9	PM	S1	1	0.60	1.8	20	19.20	8.4	0.59
36	EPTFE	SMALL	ALICE	FEB-19-9	PM	S2	0	0.20	4.6	13	19.60	27.8	0.64
37	EPTFE	SMALL	ALICE	FEB-6-91	AM	S1	2	3.00	29.8	15	32.80	34.6	1.08
38	EPTFE	SMALL	ALICE	FEB-6-91	AM	S2	2	1.60	15.8	15	28.20	31.8	0.97
39	EPTFE	SMALL	ALICE	FEB-6-91	PM	S1	3	3.20	6.4	20	24.60	23.2	0.97
40	EPTFE	SMALL	ALICE	FEB-6-91	PM	S2	4	2.60	7.4	15	17.00	17.8	0.92
41	EPTFE	SMALL	CAROLE	JAN-18-9	AM	S1	0	1.00	5.4	4	6.80	8.0	0.51
42	EPTFE	SMALL	CAROLE	JAN-18-9	AM	S2	3	0.20	8.0	6	7.20	6.8	0.52
43	EPTFE	SMALL	CAROLE	JAN-18-9	PM	S1	1	0.00	0.2	6	4.80	7.2	0.15
44	EPTFE	SMALL	CAROLE	JAN-18-9	PM	S2	3	1.00	7.2	9	9.20	14.2	0.73
45	EPTFE	SMALL	CAROLE	JAN-29-9	AM	S1	11	2.80	16.2	0	7.40	4.8	0.71
46	EPTFE	SMALL	CAROLE	JAN-29-9	AM	S2	21	0.80	25.8	0	4.00	3.6	0.63
47	EPTFE	SMALL	CAROLE	JAN-29-9	PM	S1	5	1.00	4.8	3	12.20	6.8	0.63
48	EPTFE	SMALL	CAROLE	JAN-29-9	PM	S2	1	0.40	3.6	10	11.80	10.0	0.54

APPENDIX (CONTINUED)

TEST	TYPE	SIZE	PERSON	DAY	TIME	GARMENT	P_DKBS	P_REACH	P_MARCH	W_DKBS	W_REACH	W_MARCH	MLCOUNT
49	POLY	LARGE	NORM	FEB-1-91	AM	L3	20	0.20	110.4	10	9.00	58.4	1.06
50	POLY	LARGE	NORM	FEB-1-91	AM	L4	17	0.00	67.2	8	5.80	50.2	0.95
51	POLY	LARGE	NORM	FEB-1-91	PM	L3	19	0.20	72.8	10	17.20	41.2	1.05
52	POLY	LARGE	NORM	FEB-1-91	PM	L4	18	1.40	67.0	22	17.20	52.8	1.25
53	POLY	LARGE	NORM	JAN-23-9	AM	L3	49	0.40	235.4	32	14.20	91.6	1.38
54	POLY	LARGE	NORM	JAN-23-9	AM	L4	23	1.00	135.0	60	8.20	53.0	1.32
55	POLY	LARGE	NORM	JAN-23-9	PM	L3	25	1.80	119.4	48	23.00	57.6	1.42
56	POLY	LARGE	NORM	JAN-23-9	PM	L4	18	2.80	152.8	27	24.00	95.6	1.45
57	POLY	LARGE	RICH	JAN-18-9	AM	L3	276	2.80	217.2	10	16.20	36.0	1.50
58	POLY	LARGE	RICH	JAN-18-9	AM	L4	241	0.40	335.2	24	6.20	23.4	1.34
59	POLY	LARGE	RICH	JAN-18-9	PM	L3	216	2.00	213.0	22	11.80	61.2	1.53
60	POLY	LARGE	RICH	JAN-18-9	PM	L4	119	0.60	109.2	20	6.20	32.0	1.25
61	POLY	LARGE	RICH	JAN-29-9	AM	L3	221	2.00	92.8	10	4.00	37.6	1.30
62	POLY	LARGE	RICH	JAN-29-9	AM	L4	270	0.60	128.6	7	2.20	52.0	1.20
63	POLY	LARGE	RICH	JAN-29-9	PM	L3	179	1.60	186.4	8	9.80	31.6	1.35
64	POLY	LARGE	RICH	JAN-29-9	PM	L4	167	0.20	177.0	6	3.40	30.0	1.09
65	POLY	MED	DOUG	FEB-26-9	AM	M3	12	0.40	32.8	2	6.20	18.6	0.76
66	POLY	MED	DOUG	FEB-26-9	AM	M4	19	1.20	46.8	1	3.00	17.0	0.79
67	POLY	MED	DOUG	FEB-26-9	PM	M3	23	1.20	78.0	0	9.20	28.6	0.96
68	POLY	MED	DOUG	FEB-26-9	PM	M4	25	2.20	108.6	4	11.40	21.0	1.13
69	POLY	MED	DOUG	JAN-23-9	AM	M3	8	0.40	42.2	0	6.60	30.6	0.74
70	POLY	MED	DOUG	JAN-23-9	AM	M4	11	0.80	47.8	3	8.20	29.4	0.91
71	POLY	MED	DOUG	JAN-23-9	PM	M3	9	1.20	41.8	4	11.40	19.4	0.93
72	POLY	MED	DOUG	JAN-23-9	PM	M4	9	0.60	28.6	3	7.00	14.4	0.78
73	POLY	MED	MARK	FEB-6-91	AM	M3	91	2.40	19.0	12	25.20	78.0	1.33
74	POLY	MED	MARK	FEB-6-91	AM	M4	63	3.00	21.4	25	26.00	46.8	1.35
75	POLY	MED	MARK	FEB-6-91	PM	M3	55	4.00	21.8	19	40.40	44.6	1.37
76	POLY	MED	MARK	FEB-6-91	PM	M4	53	1.40	9.2	15	22.80	36.2	1.15
77	POLY	MED	MARK	FEB-7-91	AM	M3	153	2.20	63.2	31	18.40	210.8	1.57
78	POLY	MED	MARK	FEB-7-91	AM	M4	101	3.40	64.2	57	15.00	157.8	1.58
79	POLY	MED	MARK	FEB-7-91	PM	M3	79	2.80	44.0	24	9.60	137.4	1.41
80	POLY	MED	MARK	FEB-7-91	PM	M4	100	6.40	68.8	57	25.20	161.4	1.67
81	POLY	SMALL	ALICE	FEB-19-9	AM	S3	22	4.20	63.8	25	36.40	45.4	1.40
82	POLY	SMALL	ALICE	FEB-19-9	AM	S4	26	5.00	117.8	27	32.80	59.8	1.48
83	POLY	SMALL	ALICE	FEB-19-9	PM	S3	16	2.60	79.2	15	49.00	38.6	1.33
84	POLY	SMALL	ALICE	FEB-19-9	PM	S4	12	6.20	69.0	14	55.60	55.4	1.39
85	POLY	SMALL	ALICE	FEB-6-91	AM	S3	25	8.00	191.0	11	30.40	111.8	1.53
86	POLY	SMALL	ALICE	FEB-6-91	AM	S4	57	16.20	229.4	18	87.80	128.0	1.77
87	POLY	SMALL	ALICE	FEB-6-91	PM	S3	20	2.80	78.8	8	35.00	54.6	1.30
88	POLY	SMALL	ALICE	FEB-6-91	PM	S4	23	5.20	136.0	38	42.80	111.0	1.58
89	POLY	SMALL	CAROLE	JAN-18-9	AM	S3	68	2.60	108.4	3	9.00	15.8	1.15
90	POLY	SMALL	CAROLE	JAN-18-9	AM	S4	70	3.40	57.6	5	18.00	29.8	1.26
91	POLY	SMALL	CAROLE	JAN-18-9	PM	S3	97	3.00	66.8	9	14.20	17.0	1.27
92	POLY	SMALL	CAROLE	JAN-18-9	PM	S4	52	3.40	43.2	6	14.80	21.6	1.19
93	POLY	SMALL	CAROLE	JAN-29-9	AM	S3	343	1.80	97.8	12	9.40	27.0	1.38
94	POLY	SMALL	CAROLE	JAN-29-9	AM	S4	396	1.60	174.6	1	6.80	27.6	1.22
95	POLY	SMALL	CAROLE	JAN-29-9	PM	S3	210	1.00	102.8	10	17.00	23.2	1.32
96	POLY	SMALL	CAROLE	JAN-29-9	PM	S4	221	2.20	67.6	4	9.00	19.4	1.23

BIOGRAPHIES

William Kahn has been with Mitchell Madison Group, a financial services management consultancy, since 1995. From 1988 until 1995 he was with W. L. Gore and Associates. In both positions he has been a practitioner, researcher, teacher, and cheerleader of applied statistics. Earlier jobs were in scientific programming, biostatistics, and semiconductor manufacturing. His B.A. (physics) and M.A. (statistics) are from the University of California, Berkeley; his Ph.D. (statistics) is from Yale. Bill's main professional interest is in the interaction between hard technological problems and the human teams which wrestle with them.

Carole Baczkowski received a Bachelors degree in biology at St. Mary College of Maryland. Carole is a Research Associate for W.L. Gore & Associates, Inc., where for the last eleven years, she has been measuring the effectiveness of clean room apparel and developing new products. She is a member of The Fine Particle Society and the Institute of

Environmental Sciences. Carole participates regularly in contamination control conferences and has presented her work at Digital Clean Room Conferences as well as The Fine Particle Society.

CHAPTER 11

A SKIP-LOT SAMPLING PLAN BASED ON VARIANCE COMPONENTS FOR PHOTOLITHOGRAPHIC REGISTRATION MEASUREMENTS

Dwayne Pepper

A reduced sampling plan that skips lots was found for the measurements at lithographic layer registration. The skip-lot sampling plan is based on variance components. It captures enough process variation to control the process and minimize loss from failing to disposition material that is not within specifications. Implementation of the reduced sampling plan resulted in a savings of 19-30 hours of throughput time per lot, and one operator less per shift. The cost of reducing sampling was estimated to be 0.003% die yield loss per layer, which is small enough that it has not impacted a continuous yield improvement program.

EXECUTIVE SUMMARY

Problem

In a manufacturing environment, measurements are non-value added steps in a process. For lithographic layer registration measurement, the current procedure was to measure 10 sites on each of five wafers from every lot, at every layer. This requires considerable staffing at registration measurement, and adds up to 40 hours of throughput time to every lot. Engineering intuition suggested that this level of sampling could be reduced for a high volume factory running a mature process.

Sample reduction would only be allowable if a new sampling scheme continued to capture enough natural process variation to control the process. Additionally, any sample reduction must provide adequate protection from failing to disposition material that is not within specifications.

Solution Strategy

Production data was gathered so that a variance components analysis could be performed. The variance components could be used, for example, to determine if a smaller number of sites per wafer would still capture enough natural process variation. That is, if the variation from site to site on a wafer was especially small, in a relative and absolute sense, then sampling fewer sites would only fail to capture a small amount of natural variation. After the variance components analysis was interpreted, a risk-benefit analysis was performed. The risks involved with reducing sampling needed to be quantified and weighed against the benefits realized in reduced throughput time and labor savings.

Conclusions

The variance components analysis suggests that the largest source of variation is from one stepper setup to the next. An insignificant amount of variation is shown from lot to lot within a setup. Likewise, wafers within lots and sites on wafers show little variation. Furthermore, reducing the number of sites per wafer, for example, saves very little time. The biggest time savings are realized by not measuring all lots in a stepper setup. This conclusion also offers an exceptional return from a risk-benefit point of view since there is a 19-30 hour reduction in throughput time and a one operator per shift reduction in labor. The disadvantage is a 0.003% per layer reduction in die yield from failing to disposition lots that are not within specifications for registration.

PROCESS

The skip-lot sampling plan is developed for registration measurement after photolithographic (litho) processing. Therefore, a brief overview of the litho process, registration measurements, and litho manufacturing practice is in order.

Lithography and Registration

In semiconductor wafer fabrication, the litho process prints patterns on silicon wafers. These patterns define, among other things, components of transistors, metal connecting lines, and contacts. Patterns are printed on wafers in a sequence of layers, and layers for a given product are unique. The alignment of each layer to the one underneath is called registration. Registration is clearly a critical component in litho processing. For example, the top and bottom of a contact must be aligned to prevent a contact from being placed in an undesirable position.

A stepper moves a reticle across a wafer and exposes the pattern contained on the reticle onto the wafer. This individual print is called a field and may contain patterns for several die. Each time the stepper stops movement, it must align to the previous layer before it can expose the wafer to the pattern on the reticle. Registration monitoring is necessary to control the stepper's alignment process.

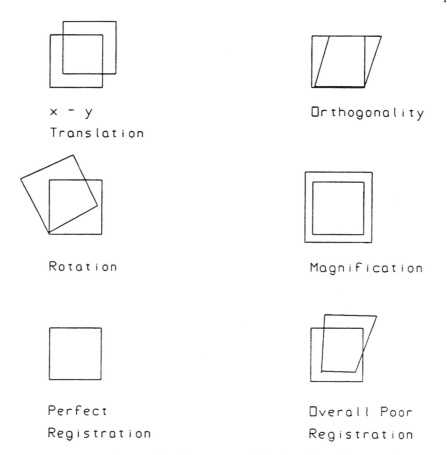

Figure 11.1. *Some of the Components of Registration. These are the components of registration that were monitored for this case study. These components comprise registration in any lithographic process, such as printing money, or in IC fabrication, as considered here.*

In simplistic terms, registration is based on measurements of how well a square on the current layer sits on top of an identical square on the previous layer as illustrated in Figure 11.1.

Four of the basic components of registration are considered for a given field:
- Linear translation measures a linear shift in either the x or y direction.
- Rotation measures the degree to which one box is rotated relative to the other.
- Orthogonality measures the degree to which the stepper preserves right angles.
- Magnification measures how well the stepper preserves dimensions.

In a manufacturing environment, each component of registration contributes to a weighted average, which will be referred to simply as "registration." The registration measurement is typically plotted on a control chart.

Lithography and Manufacturing

Since a stepper must be set up with a reticle that is unique for a given layer of a given product, production operators allow several lots requiring the same reticle to gather at a stepper. The stepper is then set up for that product-layer combination, and the lots are generally processed continuously without breaking the setup. This keeps the fixed time of establishing a setup to a minimum. There is no minimum or maximum number of lots that an operator is required to have before establishing a setup. The number of lots at the time of setup tends to vary with factory loadings and with operator.

DATA COLLECTION

Before the reduced sampling plan was developed, operators measured 10 sites on five (specified) fields across a wafer, for five randomly selected wafers from each lot. At each site, each of the components of registration was measured. These components were processed through the black box weighted average to produce a registration value for each site. Although data from each site was retained, site data was not used for process control or product dispositioning. Site registration values were averaged to give wafer registration values. For each lot, the average of all wafer registration values was used for process control and product dispositioning.

The data used in this study represented all products in the factory over a six month period. For simplicity of calculations, any lot that had a non-standard amount of data gathered for it was not included in the study. For example, certain engineering lots might have registration measured at 3600 sites on all wafers in the lot and were excluded to eliminate the necessity of estimating variance components in an unbalanced situation.

DATA ANALYSIS AND INTERPRETATION OF RESULTS

Registration is recorded for two reasons:
- To check for time trends in the alignment process of the stepper.
- To prevent material whose registration makes it unfit for its intended function from being processed further.

This sets two constraints on sample reduction. First, any sampling plan must be capable of capturing enough natural process variation to successfully capture changes in the stepper's alignment process. Second, any sampling plan which increases the risk of sending on material that is out of specification must have a benefit that justifies the risk. More fundamentally, all decisions must ultimately be cost-effective. Therefore, the cost-benefit balance will be considered at each step in the development of this sampling plan.

Variance Components Analysis

To make sure that any new sampling plan captures enough natural process variation, a variance components analysis was performed on historical registration data from each layer. For a discussion of nested variance components and formulae for their computations, see Montgomery (1991). The variance components analysis separated nested variation by setup, lot within setup, wafer within lot/setup combination, and site within wafer/lot/setup, for each layer. A variance components analysis of the weighted average registration for a typical layer is given in Table 11.1. Variance components are given assuming 100 total units of registration variation (in μm^2). This convention makes the variance component equal to its percentage contribution to total variation.

Table 11.1. *Variance Components for Some Typical Data. The value of the variance components are given as the percentage of total variation.*

Source	Variance Component
Setup	66
Lot	10
Wafer	14
Site	10
Error	0
Total	100

To determine whether the weighted averages de-emphasized problems that might be identifiable from individual components of registration, a variance components analysis was also performed for each component of registration for each layer. In each case, the overall pattern in variance components was the same as that shown in Table 11.1 for the weighted averages.

A brief inspection of the variance components analysis suggests that variation from one setup to the next captures a majority of the natural process variation. Furthermore, it is almost as wise to reduce the number of lots per setup as it is to reduce the number of wafers per lot or the number of sites per wafer. The skip-lot sampling plan that follows is consistent with Cochran's (1977) prescription for stratified random sampling, where the number of units sampled from a stratum is inversely proportional to the variance within that stratum.

Before considering where to reduce the sample, a more thorough consideration of the registration measurement process is in order. Once a lot has been processed by a stepper, it goes into a queue for registration measurement. Typical times from the end of stepper processing to the end of registration measurement are given in Table 11.2.

As can be seen from Table 11.2, very little time is saved by reducing the number of sites per wafer. Also, only modest time savings are realized by reducing the number of wafers measured per lot. However, skipping measurements on a lot completely saves 15 minutes (10 minutes fixed time per lot + 1 minute for each of the five wafers) of the operator's time and decreases lot throughput time by 2 hours (15 minutes measurement time + 1 hour 45 minutes queue time). Depending on factory conditions and business needs for faster throughput, the savings in throughput time can be more financially rewarding than the savings in labor.

Table 11.2. *Time Spent at Registration Check. It is important to note that a lot goes through registration check ten to twenty (or more) times, depending on the product, during its processing.*

In queue waiting for available station	1hr 45min.
Lot time on station + other fixed time activities	10min.
Wafer 1 inspection	1min.
Wafer 2 inspection	1min.
Wafer 3 inspection	1min.
Wafer 4 inspection	1min.
Wafer 5 inspection	1min.
approximately 5 sec. per site	

Skip Limits

The variance components analysis and the time analysis leave little doubt that some sampling scheme that skips lots could be appropriate for registration check. The variance components might suggest some sort of simple skip-lot sampling plan whereby if the first lot in a setup is within specification, then the following lots in the setup are exempted from measurement. Due to the potentially harsh consequences of letting material that is out of specification pass undetected, more rigid requirements needed to be imposed before skipping lots could be accepted. Material that is rejected at registration check can be reworked. Material that is not reworked but fails to meet specifications will be scrapped at wafer sort, when the wafers have completed all of their expensive processing.

The criteria for exempting lots from measurement can be tightened by setting a prescribed distance that the first measurement of a setup must be inside the specification limits before skipping is permitted. The first measurement of a setup is actually recorded as the average of readings from the five wafers from the first lot processed with a new setup. Figure 11.2 shows a trend chart with registration specification and skip limits.

An upper skip limit is calculated as:

$$\text{Skip Limit} = \text{spec. limit} - z_\alpha * [s^2_{\text{Wafer}} + s^2_{\text{Wafer(Lot)}}/w + s^2_{\text{Site(Wafer Lot)}}/(w*s)]^{1/2},$$

where w is the number of wafers per lot and s is the number of sites per wafer. Z_α is the $(1 - \alpha/2)$ percentile of the normal distribution, taken here to be 2.57 for 99% confidence. As shown in Figure 11.3, the skip limit is far enough from the specification limit that all of the lots in the setup are likely to be within specification, whenever the first lot in the setup is inside the skip limits. The significance level (α) is chosen to quantify exactly how likely the successive lots are to be within specification. Without loss of generality, all discussions and illustrations in this paper will be in terms of the upper skip, control and specification limits.

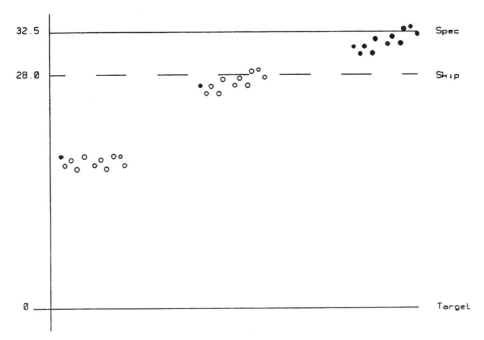

Figure 11.2. *Registration of Lots on a Trend Chart. Solid points represent lots that would be measured. Open points represent lots that would be skipped.*

Assuming that registration errors follow a normal distribution across a wafer, in setups whose first lot does not violate skip limits, each successive lot has at least a 99% probability of being inside the specification limits when only natural variation is present. This is verified by taking the mean for the first lot of a setup as an estimate of the mean of the distribution of the setup. The standard deviation for that distribution is estimated as $s^2_{Wafer} + s^2_{Wafer(Lot)}/w + s^2_{Site(Wafer\ Lot)}/(w*s)$. The probability of at least 99% therefore follows as shown in Figure 11.3.

This means that if the first lot in a setup is within the skip limits, skipping measurements on successive lots is relatively safe. This also assumes that lots failing to meet registration specifications due to natural variation occur randomly throughout a setup.

Risk/Benefit Analysis

This risk/benefit analysis evaluates the risks of failing to catch material that should be reworked and the savings from reduced labor and throughput time for the purpose of evaluating the applicability of the sampling plan. The findings here are a 92% reduction in the resources used for registration measurement while facing a potential 0.003% yield loss from missed registration.

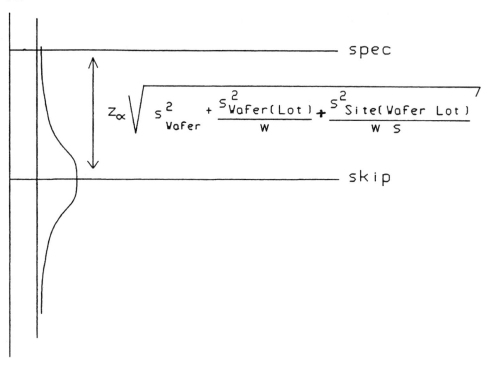

Figure 11.3. *If the first lot of a setup is within the skip limits, no more than 100 α /z % of the rest of the lots in the setup are expected to be out of spec.*

For discussion purposes, the values of certain variables must be assumed. For computational ease, material that should be reworked but is missed will be assumed to fail completely at wafer sort (i.e., 100% yield loss). Likewise, material that is reworked will be assumed to have no yield loss due to the rework process. Assume also that 0.25% of the factory line will need to be reworked because of registration problems. Also, it will be convenient to assume that every stepper setup contains 11 lots. No consideration will be given to the resources required to rework a lot.

Before computing time savings, it will be necessary to estimate the average number of lots that will be measured taking a fixed number of setups and a fixed number of lots per setup. First, however, the proportion of setups that are measured on all lots is denoted by ISetup and is given by:

$$\text{ISetup} = 2*[\Phi((\text{skip} - \text{target})/s_{\text{total}})]$$

where Φ is cumulative normal probability, skip is a skip limit, and σ_{total} is the total process variation. Here, the overall process is assumed to be in control and centered about the target. This is illustrated in Figure 11.4.

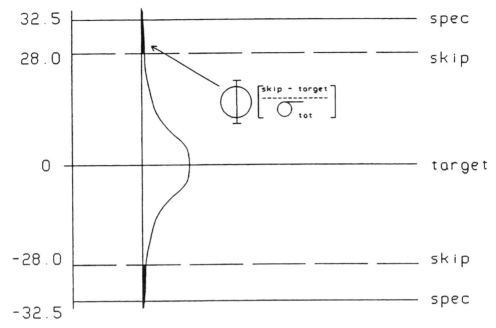

Figure 11.4. *The Expected Number of Lots to be Measured. The value of the variable ISetup is taken as a percentage from the normal distribution.*

Using Figure 11.2 as a specific product and layer, the proportion of setups that are measured completely is:

$$\text{ISetup} = 2 * [\Phi((28 - 0)/10)]$$

$$= 0.0052.$$

The expected number of lots being inspected from a fixed number of setups is given by:

expected number of lots inspected = (11 lots per setup × ISetup × number of setups) + 1 × (number of setups not 100% inspected).

Taking, for example, 2200 lots in 200 setups, the number of lots measured for registration is:

$$(11 \times .52\% \times 200) + 199 = 210.$$

That is, (11 × .52% × 200) = 11 lots will be measured as part of a setup being measured completely and 199 lots will be measured as the first lot of a setup. For this example, the result is a 10% rate of registration measurement. This equates to approximately 90% savings at registration check. Dividing this between operator labor and holding costs while a lot is in queue, a time savings of 13.5 min./lot operator time and 94.5 min./lot queue time is realized. Together, these save 108 min./lot of throughput time per layer. Without quantifying time savings in units of dollars, an analysis of the risks involved is given.

Under the above set of assumptions, of 2200 lots, about 5.5 will need to be reworked because of registration problems. With the skip limits developed thus far, about 99% of those lots will be in setups that get 100% inspected. Hence about $((1\% \times 5.5)/2200)$ 0.003% will be an estimate of missed rework opportunity. Under earlier assumptions, this equates to 0.003% die yield loss per layer. This is 0.003% die yield loss due to natural process variation. Such an estimate is only safe to assume when sufficient guards against special cause variation are present.

In many manufacturing environments, either labor savings or queue time savings easily justify the yield loss. In the factory where this skip-lot sampling plan was developed, labor savings alone, which are the least expensive costs, justify its use.

A very large base of historical data was used to test the actual (versus theoretical) effectiveness of this sampling plan. Testing verified that a predictable proportion of setups with failing lots had their first lot outside of the skip limits. A very fortunate discovery was also made: a vast majority of special causes were seen to have initiated with the first lot of a setup. This adds tremendous robustness to an assumption that some monitor other than registration check will catch special cause variation. Historical data also established a 0.25% rework rate for registration issues. Because some lots that should be reworked will be missed (and therefore not reworked) with skip-lot sampling, this rework rate should drop to about 0.06%. The expected new rework rate is the current rework rate of 0.25%, multiplied by the percent of lots not skipped. Rework due to registration was monitored carefully for several weeks after implementation to assure that the observed rework rate did not drop systematically without reason. A drop in rework rate that could not be explained by process changes might indicate that variance components based skip-lot sampling is failing to catch the predicted amount of reworkable material. The changes in sampling were monitored closely and the rework rate never changed without cause.

Implementation

Realizing the maximum benefit from this sampling plan is entirely dependent on successful implementation. Implementation of this sampling plan in the factory where it was developed was indeed challenging.

Implementation would be trivial if one stepper fed lots to one registration measurement station and registration measurement was a quick first-in/first-out station. If this were the case, the first lot of a setup would reach registration measurement first and would be measured, presumably fast enough to prevent successive lots from that setup from going into queue for registration measurement unnecessarily. In practice, several steppers feed several registration measurement stations, and there is no assurance that the first lot of a setup will get to registration measurement first. Most importantly, there is no assurance that the first lot will be measured fast enough to prevent successive lots from going into queue unnecessarily. When several steppers are running the same layer of the same product, lots with a specific combination can accumulate at registration measurement, and the operators will be left with no indication of which lots came from which stepper in what order.

Listing a few of the possible scenarios where confusion and inefficiency mount helped develop some criteria for an implementation plan. First, implementation of this sampling plan had to save labor. This suggested that operators should know which lots to measure and which to skip with little time invested in the decision-making process. Second, lots that did not need to be measured could not sit in queue waiting for registration measurement or a decision to skip or not. Hence, the logistics could not consume the throughput time savings. With an engineering staff unresolved on some of the implementation issues, a team consisting primarily of operators and a line supervisor was assembled to formulate an implementation plan. Their plan is currently being used. Instead of randomly selecting five wafers from the first lot of a setup, five wafers are randomly selected from the first 15 wafers processed in the first lot of the setup. The stepper operator then takes the wafers for registration measurement. The operator measuring registration returns the five wafers to the stepper operator and does so before the rest of the lot is finished. If the first lot passes, then the stepper operator makes a 'SKIP' notation on the run card for each lot in queue for that stepper. Each lot that has 'SKIP' noted on it will be moved from the stepper to the next operation in the flow, without stopping at registration check. While this system is somewhat manual, its simplicity and effectiveness are providing maximum benefit from skip-lot sampling.

In addition to successful implementation in the factory, a system had to be designed to monitor and maintain the sampling plan. For various reasons, all lots in some setups would need to be measured. Likewise, all wafers in some lots would need to be measured. This, along with data gathered during production, would be a constant source of data to verify that the pattern in variance components did not shift from that outlined in Table 11.1. Engineering resources would be needed periodically for this as long as this skip-lot sampling plan was being used. While no prescribed interval has been set for recalculating skip limits, every time that a variance components analysis is done, skip-limits will be re-evaluated. Establishing an interval for skip limit recalculation is in order.

CONCLUSION AND RECOMMENDATIONS

Generalizations

Although the skip-lot sampling plan was developed in answer to a need for sample reduction at registration measurement, the notion of skip limits may be generalized and implemented whenever some key characteristics are present. An obvious requirement for a reduced sampling plan is the measurement of variables such as registration or layer thickness instead of attribute characteristics such as conforming and nonconforming on the product. The key element of a variance components based skip-lot sampling plan is that lot-to-lot variation must be small both in a relative and in an absolute sense. The presence of stepper setups in the above example gave a convenient subgroup within which lot-to-lot variation is small. Another important element of a variance components based skip-lot sampling plan is that it should be economically unsound to use a simple skip-lot sampling plan, such as skipping all the lots in a setup if the first lot is found to be within specifications. That is, a conventional skip-lot sampling plan must be too lenient; for if it were not, it would likely be preferred to a variance components based plan which has the more stringent requirement that the first lot in the setup be not only within specifications, but also within the skip limit. A sampling plan must also be appropriate for its environment. For example, the skip-lot sampling plan described in this case study is suited for a high volume manufacturing environment, where it is to be implemented.

Summary

After more than 18 months of use, no negative consequences of this sampling plan have been uncovered. A few product-layer combinations that, in engineering judgement, have registration problems have been excluded from this sampling plan. These layers did not benefit from skip-lot sampling. Overall, variance components based skip-lot sampling has saved more than one operator per shift and 19–30 hours of throughput time (110 min per layer) for every lot in the factory while maintaining a predictable rework rate and not hindering factor-wide yield improvement.

ACKNOWLEDGMENTS

This study was initiated when Dan Rosenberg, lithographic registration engineer, approached the author with the need to reduce sampling at registration measurement. Dan is acknowledged for educating the author and many of his colleagues on the science of registration control. Dan works for Intel's Portland Technology Development facility in Portland, Oregon.

REFERENCES

Cochran, W.G. *Sampling Techniques*. New York, NY: John Wiley and Sons, Inc., 1977.

Montgomery, D.C. *Design and Analysis of Experiments*. 4th ed. New York, NY: John Wiley and Sons, Inc., 1997.

BIOGRAPHY

Dwayne Pepper manages the Statistics Group at Intel Corporation's Chandler Assembly Test Site in Chandler, Arizona. Dwayne received an M.S. in statistics from Iowa State University.

CHAPTER 12

SAMPLING TO MEET A VARIANCE SPECIFICATION: CLEAN ROOM QUALIFICATION

Kathryn Hall and Steven Carpenter

This case study examines the sampling plan used to compare the variability in air velocity through clean room air filters to the variability stated in the purchase order. Practical issues of working in a restrictive clean room environment and their influence on the sampling plan are discussed. Use of statistical methods helped in: a) sampling a minimum number of filters, b) decreasing sampling time and cost of analysis, and c) minimizing impact on production in the clean room. Air filter performance was established, and verification of adherence to contract requirements led to modification of future air filter purchase contracts.

EXECUTIVE SUMMARY

Problem

In 1990, Hewlett-Packard constructed a new clean room in Corvallis, Oregon. Some of the items in the new clean room were ultra low particulate air (ULPA) filters, used to maintain uniform air flow over production areas. The contract specifications were written for 100 feet per minute (fpm) filter face air velocities, but the filters were to be used at 75 and 40 fpm. The problem was to determine whether the uniformity in the filter face air velocities at the lower velocities was satisfactory for production in the clean room. The challenge was to minimize the number of filters used for testing and still maintain a high level of confidence in the test results.

Solution Strategy

The strategy was to sample a small number of filters, analyze the data, and determine if additional filters needed to be tested. Real time statistical analysis would be provided to the technicians conducting the test; results would be computed immediately upon completion of testing. If additional filters needed sampling, a second data set would be taken and analyzed. The process would continue until we were assured that enough filters had been sampled to confidently represent the velocity uniformity for the entire clean room.

Conclusions

The variability in filter face air velocities at 75 and 40 fpm did not satisfy the uniformity requirements for 100 fpm specified in the contract. This led to modifications of all ensuing filter purchase contracts.

The sampling methodology used to establish these results saved time and money by
1. sampling a minimum number of filters to determine velocity uniformity
2. decreasing sampling time, therefore reducing analysis cost
3. minimizing impact on production by minimizing interruptions due to sampling.

PROCESS

One major cause of integrated circuit (IC) failure is particles that fall onto the IC while it is being manufactured from the raw wafer. Consequently, ICs are manufactured in clean rooms, places with carefully controlled environments. Extensive efforts are made to eliminate particles from the clean room environment. Additionally, air flow zones in production areas of the clean room are carefully defined. Products are kept in the 75 feet per minute (fpm) air flow zone, and operators work in the 40 fpm air flow zone. The two zones are contiguous, with the 75 fpm air presenting a differential air flow curtain which pushes any contamination from the high pressure product zone into the operator low pressure zone.

One of the many components of clean room construction is selection of air filters that maintain uniform air flow over production areas. The uniformity of the face velocity of the air coming out of those filters is critical to maintaining the different air flow zones. The design intent was to have filters with the range of face velocities (or air flow rates) at various points within a filter to be within 20% of the average velocity for that filter.

The filters are composed of a pleated sheet of fiberglass medium, typically 92% air and 8% medium, surrounded by a 2 x 4 foot frame. The frame is set into the bottom of a metal box. Air is delivered through a collar in the top of the box, and runs over a diffuser plate before it reaches the upper surface of the filter.

Measured flow rates are a function of the filter air flow rate and the distance below the filter that the measurements are taken. Data for 40 fpm filters collected 12 inches below the filter will be discussed throughout the remainder of this case study.

DATA COLLECTION

Before data collection begins, project goals and acceptable risk levels of wrong decisions need to be selected. These are used to determine the minimum sample sizes needed to meet the goals without exceeding the acceptable risks. Additionally, the distribution of air flow rates within a filter needs to be established as well as a cursory estimate of the within-panel variance.

The project goal was to design a sampling scheme to test the velocity variance across the filter faces of two sets of targeted filter velocities (40 and 75 fpm) using the *minimum number* of samples to convince both the filter manufacturer and filter customer that the filters did or did not meet the specifications of the filter purchase contract. The contract specified that for any given filter, all measured filter face velocities would be within ±20% of the average velocity for that filter.

The project goal will be put in statistical terms by using the variance between readings in the same filter panel (σ_{wf}^2) and comparing it to the value specified in the contract. The required variance cannot be calculated directly from observed air flow rates. The observed total within-filter variance is the sum of σ_{wf}^2 and σ_{me}^2, where σ_{me}^2 is the measurement error. The measurement error was estimated, and its effect was removed from the observed total within-filter variance. Clearly, we would like a small measurement error.

More formally, the goal is to test the hypothesis that the within-filter variance is less than the within-filter variance specified in the contract (i.e., test $H_0: \sigma_{wf}^2 \leq \sigma_0^2$ against $H_a: \sigma_{wf}^2 > \sigma_0^2$, where σ_0^2 is specified by contract).

Previously collected filter data showed that individual readings on any given filter followed a normal distribution (see Figure 12.1). Hence, approximately 95.44% of the observations will fall in the interval $\mu - 2\sigma_{wf}$, $\mu + 2\sigma_{wf}$ (see Neter 1974), where μ is the mean and σ_{wf} is the standard deviation of the face velocities on a filter. For purposes of sample size estimation, assume that all observations will fall in this interval. In other words, assume that all the face velocity observations fall within two standard deviations of the mean. With this in mind, the contract specified that 20 percent of the mean of any filter had to be less than two times the within-filter variance for that filter. Hence, the contractual standard deviation, σ_0, can be found by solving $.2\mu = 2\sigma_0$. For the 40 fpm filters,

$$\sigma_0 = \frac{.2 \times 40}{2} = 4.0.$$

Derivation of the appropriate sample size formula is given in Mace (1964, pp. 91–94). This technique for determination of the sample size is discussed in Appendix A.

No one ever wants to make a wrong decision, either saying that the filters meet the contract specification when they do not (Type II or β error), or saying that they don't meet the specification when they do (Type I or α error). The probability of making either type of error depends on the sample size and the actual variance of the observations. After estimating this variance, sample sizes for several α and β levels are calculated. With the resulting table of risk levels and sample sizes, a discussion of data collection costs versus certainty of a correct decision ensued. If the actual within-filter variance was larger than the initial estimate, we would have more confidence that the variance exceeded the contracted value. If the variance were smaller than estimated, the option of collecting data from additional panels was available, allowing attainment of fixed α and β levels.

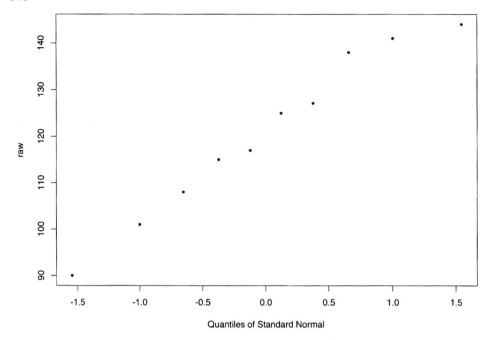

Figure 12.1. *Normal Probability Plot of the Preliminary Within-Filter Air Flow Data. If the data follows an approximate straight line, as it does here, it is considered to follow a normal distribution.*

Because multiple measurements per filter were very simple to make, engineers for both vendor and customer had agreed to take eight measurements per filter. To ensure that results were not biased by the test technician, each site was tested three times by each of two technicians (see Appendix B).

Data was collected in a production clean room during normal production hours. This meant that the persons doing the testing had to become certified to safety and clean room operating specifications and gown-up to work in the clean room. Special care was taken to inform production personnel in advance that the measurements were to take place. The test operators had a maximum of one and a half days available for data collection. To maximize the available time for sampling, it was desirable to minimize the time going in and out of the clean room, an operation that can take up to 30 minutes for one round-trip into the controlled environment. We realize that experienced personnel can enter and exit in less time, but even a technician experienced in working in clean rooms must adjust to the special configurations and requirements for each room in which they work.

Because of this time constraint, it was decided that the sampling plan would call for data collection from 15 filters. The data would then be analyzed to determine if more sampling was needed and the minimum number of samples required to adequately represent the filter face velocity uniformity.

Test equipment consisted of a Velgrid® intake probe connected to a Shortridge Airdata® multimeter. To minimize the measurement error, two multimeters were sent to Shortridge Instruments, Inc. for calibration. The gauges were calibrated on the same day by the same technician and mailed to our site. Two experienced testers from independent companies arrived to conduct the tests.

A probe was fixed onto a tripod and positioned 12 inches beneath the face of a filter. A series of three readings was taken with one multimeter at this location. The multimeter was then disconnected from the stationary probe, the second multimeter connected, and a second set of data taken.

DATA ANALYSIS AND INTERPRETATION OF RESULTS

Despite the decision to sample 15 filters, each technician measured eight sites on each of 11 filters before they brought the first set of data out of the clean room. Each measurement was repeated three times. The data collectors emerged from the clean room with this data set when they took their dinner break. Their cursory examination of the data as they collected it made them feel that the within-panel variance was much larger than the hypothesized value, meaning that this smaller data set could potentially provide enough data to decide if the system met the contracted value and/or if further data was needed.

A graphical display of the flow rates for each panel is given in Figure 12.2. Note that each data box and set of whiskers covers the expected range of observations within any one filter panel. The range of observations for any panel corresponds to roughly

$$6\sqrt{\sigma_{wf}^2 + \sigma_{tech}^2 + \sigma_{me}^2}$$

since, for a normal distribution, six times the total observed standard deviation for any single panel should cover 99.7% of the data range. The technicians guessed that the variability between technicians (σ_{tech}^2) and the variability of the measurement error (σ_{me}^2) would be negligible. Note that $6\sigma_0$ for the hypothesized within-panel standard error is 24. The observed range for most of the panels is much larger than 24, making the technicians suspect that they had collected adequate data.

Computation of each of the variance components was done using SAS, a standard statistical software package. A summary of the results of this analysis is given in Table 12.1.

Since the technician is an integral part of the measurement system, a common practice is pooling (or adding) the last four terms together to represent the variance of the measurement system. Here, that sum is 21.51, so σ_{ms} is 4.64. The technicians component was separately included in the analysis at the request of the air filter vendor.

Appendix B contains the mean of the three observations recorded by a technician at a site on a filter. The raw data is no longer available.

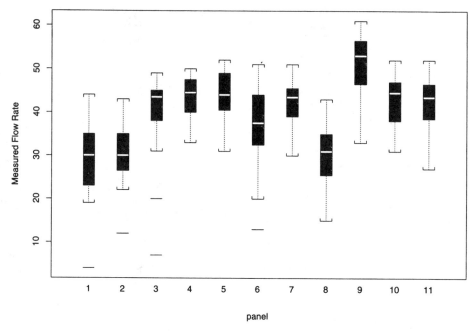

Figure 12.2. *Flow Rate for Each Panel. Each box covers the range of the 25th to 75th quartile of the six measurements taken at each of the eight sites. The whiskers mark at most 1.5 times the length of this box from the 25th and 75th quartile. Individual values outside the box and whiskers are marked with a -.*

Table 12.1. *Analysis of Variance Components.*

Variance Component	Symbol	Variance	Stnd. Error
Total	σ_T^2	113.58	10.66
Filter-to-filter	σ_p^2	41.87	6.47
Within filter	σ_{wf}^2	50.20	7.09
Technician		8.54	
Filter x Tech		2.04	
Tech x site-in-filter		5.61	4.64
Measurement error	σ_{me}^2	5.319	

To test the hypothesis, $H_0: \sigma_{wf}^2 \leq \sigma_0^2$, compute the test statistic $\hat{\chi}^2$ by

$$\hat{\chi}^2 = (n-1)f \frac{S_{wf}^2}{\sigma_0^2} = (8-1)11\frac{50.20}{4^2} = 242.$$

This number should be compared with a $\hat{\chi}^2$ value with $(n-1)f = 77$ degrees of freedom. The difference between the observed and hypothesized values are significant at the 99.99% level if the calculated value exceeds 132. Since 242 is much larger than 132, there is clear evidence that the variability in the filter flow rates is much larger than that specified in the filter purchase contract.

Note that if the calculated value of the $\hat{\chi}^2$ test statistic had been between 87 and 98, the critical χ^2 values at the 80 and 95 percent levels, we would have sampled additional filters. However, that was clearly not necessary.

Also note that there is a large filter to filter variance. This suggests the need to address the problem of accurately setting the air flow rate in any given filter.

CONCLUSIONS AND RECOMMENDATIONS

Statistical methods were very useful to minimize sampling time and clean room disruption. Regardless of the final filter uniformity, we were able to

1. sample a minimum number of filters and confidently determine the status of velocity uniformity
2. decrease the sampling time, and therefore the cost of analysis
3. minimize impact on production.

Sampling of any kind in a production clean room is extremely difficult due to a restrictive environment that requires a great deal of time to decontaminate equipment and one that necessitates working around the needs of the production staff. This obviously increases sampling time and, therefore, the cost of sampling.

Our pause in sampling was planned; this gave us time to determine if additional sampling was required based on statistical principles. In our case, we were fortunate that the minimal number was reached in the first pass. However, if we had not attained the minimum number of samples needed to be statistically confident that we adequately represented the filters in the clean room, we would have continued sampling until we reached a number that satisfied the need.

As it turned out, the 40 fpm filters did not completely meet our expectations for filter uniformity. The variability in the 40 fpm filter flow rates was larger than the variability specified in the purchase contract for 100 fpm filters.

REFERENCES

Mace, Arthur E, *Sample Size Determination*. New York, NY: Reinhold, 1964.

Mood, Alexander M., Franklin A. Graybill, and Duane C. Boes, *Introduction to the Theory of Statistics*. 3rd ed., New York, NY: McGraw-Hill, 1974.

Neter, John and William Wasserman. *Applied Linear Statistical Models*, Homewood, IL: Irwin, 1974.

Snedecor, George W. and William G. Cochran, *Statistical Methods*. 7th ed., Ames, IA: Iowa State University Press, 1980.

APPENDIX A

Sample Size Calculation

According to Mace (1964), for data that is a sample of n observations with a sample variance of S_{wf}^2 from a normal distribution with a variance of σ_0^2,

$$\frac{(n-1)s_{wf}^2}{\sigma_0^2}$$

follows an χ^2 distribution with $n-1$ degrees of freedom, where s_{wf}^2 is the sample estimate of the variance of readings on a filter.

To compute the appropriate sample size, use the fact that if a variable x follows an χ^2 distribution with $n-1$ degrees of freedom, then $\sqrt{x}$ is approximately normally distributed with a mean of $\sqrt{2(n-1)-1}$ and a variance of 1. This means that the appropriate sample size is given by

$$n = \frac{3}{2} + \frac{1}{2}\left(\frac{z_\alpha + \lambda z_\beta}{\lambda - 1}\right)^2$$

Here, z_α is the upper limit of the cumulative standard normal probability integral, where α is the probability of accepting the alternative hypothesis, H_a, when the null hypothesis, H_0, is true. λ and z_β are functions of the probability of detecting shifts of the process standard deviation to some level greater than or equal to D, for $D > \sigma_0$. λ is defined as

$$\frac{D}{\sigma_0}.$$

z_β, like z_α, is the upper limit of the standard normal curve, β being selected so that H_a will be accepted with probability $1-\beta$ when the system is operating with variance of D^2.

The sample size formula is set to determine the number of units that would be needed from a *single* population to estimate and test the variance of that population. In this case study, the concern is the within-filter variance for a population of several filters. The unmodified basic formula is appropriate only if there is only one filter.

To determine the total number of filters needed, note that if the within-filter variance is fairly constant from one filter to the next,

$$(n-1)\sum_{i=0}^{f} \frac{s_i^2}{\sigma_0^2}$$

follows a χ^2 distribution with $f(n-1)$ degrees of freedom, where f is the total number of filters sampled, n observations per filter are taken and s_i^2 is the sample within-filter variance for the ith filter. Analysis of previously collected data showed reasonably constant within filter variance across multiple filters with the same air flow rate.

Again, making use of the normal approximation to the χ^2 distribution, the previous formula is applicable, but the resulting number will be the *total* number of observations needed. Hence, for $\alpha = \beta = .05$, a total of 135 observations (20 filters) are needed for D^2 values that are 150% of the target. This corresponds to a 22% increase in the standard error. 100% increases in the variance could be detected with seven filters.

APPENDIX B

Flow Rate Data

Mean of the Three Flow Rates Recorded by Two Technicians at Eight Sites on Eleven Filters

Filter	Technician	Site 1	Site 2	Site 3	Site 4	Site 5	Site 6	Site 7	Site 8
1	AAF	32	37	44	33	23	38	30	23
1	DEC	26	34	36	23	4	30	28	19
2	AAF	34	43	40	37	31	33	27	25
2	DEC	29	36	34	29	22	27	26	12
3	AAF	45	45	49	36	20	44	45	40
3	DEC	43	45	47	31	7	41	47	42
4	AAF	47	49	49	37	36	45	50	45
4	DEC	45	44	44	33	34	44	48	43
5	AAF	42	49	52	37	43	48	50	40
5	DEC	41	45	51	31	38	48	49	42
6	AAF	20	40	51	44	36	46	39	29
6	DEC	13	38	48	37	37	44	36	22
7	AAF	36	46	51	48	45	45	42	32
7	DEC	30	43	46	45	43	44	43	31
8	AAF	24	32	35	38	43	38	31	27
8	DEC	15	29	31	32	35	28	22	15
9	AAF	61	59	54	43	38	51	55	50
9	DEC	58	61	55	42	33	50	52	54
10	AAF	36	45	49	45	45	50	52	39
10	DEC	31	41	44	38	38	47	47	34
11	AAF	39	48	43	44	52	47	48	30
11	DEC	34	44	46	41	44	41	38	27

BIOGRAPHIES

Kathryn Hall has been a research, teaching, and consulting statistician for the government, universities, and industry and is currently working for Hewlett-Packard's IC business division. A member of the American Statistical Association and the American Society for Quality Control, she enjoys solving real puzzles with sketchy data.

Steve Carpenter is a Micronaut with the Inkjet Supplies Business Unit of Hewlett-Packard in Corvallis, Oregon. Previous to that he was Manufacturing Development Engineer with HP's Integrated Circuits Business Division and Senior Research Associate at Oregon State University. He has published on contamination control for semiconductor manufacturing, microbial ecosystems of old-growth coniferous forests, microbial recovery following the 1980 Mt. St. Helens eruptions, and Ascomycete taxonomy. In 1991 he was awarded Best Technical Paper at the Hewlett-Packard Silicon Technology Conference. Steve received his B.S. at Oregon State University, M.S. at Cornell University, and Ph.D. at the City University of New York. He is a member of the Society of Sigma Xi, Oregon Biotechnology Association, Mycological Society of America and a Senior Member of the I.E.S.

CHAPTER 13

SNAPSHOT: A PLOT SHOWING PROGRESS THROUGH A DEVICE DEVELOPMENT LABORATORY

Diane Lambert, James M. Landwehr, and Ming-Jen Shyu

Codes, processes, and the mix of products at a device development laboratory (DDL) change constantly, and lots may take months to progress through the DDL. Nonetheless, managers need to assess the health of the DDL during short periods and to monitor its health over time. Engineers need to predict completion time and track the progress of their lots, even when there are no previous lots of the same code. This paper presents a new plot of step completion, called a snapshot, that addresses both these needs. A snapshot combines information on run times and queue times from existing data bases to show step completion for a code of interest even if no lots of that code were completed during the time period monitored.

EXECUTIVE SUMMARY

Problem

A device development laboratory (DDL) is a wafer producing facility that develops new processing steps (e.g., bake for a specified time at a specified temperature) and new codes which represent sequences of processing steps. Lots do not move smoothly from start to finish through a DDL, because the objective is to learn about codes and processes rather than to manufacture wafers. Some lots are split, some are held for engineering tests, and others are removed after partial processing. Even lots for codes that are nearly ready to be transferred to production may take weeks to finish. Consequently, the mix of codes at a DDL changes constantly as processes are modified, abandoned, or moved to a factory and as new manufacturing processes and codes are introduced.

Despite all the changes and disruptions, engineers need to predict the time required to finish a lot of a particular code and to track the actual progress of current lots relative to predicted progress, even if the code has never been produced before. Also, despite the lack of standard products and lack of stable conditions, managers need to monitor the DDL's performance over time.

Solution

Ignoring engineering holds, which are delays requested by the design engineer and are not under the control of the DDL, the total time needed to process a lot of a particular code is the sum of the step times. The steps are specified completely, so, for example, bake in a given type of oven at a given temperature for 10 minutes is a different step from bake in the same type of oven at the same temperature for 5 minutes. Each step time is the sum of the queue time spent waiting to begin the step and the run time spent in processing the step. Our goal was to characterize the total time needed to process a code of interest through a particular step.

Our approach was to extract queue times and run times for each step needed by the code from existing databases. Because each step is specified in detail, step run times are the same, apart from random lot-to-lot variability, for all lots, regardless of their code, that pass through the step. Queue times for a particular step depend on the priority of the lot, but are the same for all lots of the same priority, apart from random lot-to-lot variability, regardless of their code. Thus, there can be data relevant to the code of interest, even if no lots of that code were produced from start to finish during the time period being monitored.

From the data on all lots that pass through a given step, we estimated the mean and standard deviation of queue time and the mean and standard deviation of run time for that step. Then we used cumulative sums of the estimated step time means and variances to describe the time needed to process lots from the code of interest from the start through each step.

The cumulative times are displayed in a snapshot plot, which represents a particular code of interest and a specific time period of DDL operation. The snapshot shows the cumulative mean time to complete processing through a step on the vertical axis against step number on the horizontal axis. Thus, for example, the mean time to complete all processing is shown above the last step number in the code. The snapshot also shows two curves that represent optimistic and pessimistic time estimates. The lower curve shows the estimated time required for completion through each step by the fastest 5% of lots of the code of interest, plotted against step number; the upper curve shows the estimated time required by the slowest 5% of lots.

Conclusions and Recommendations

A snapshot of step completion shows the expected progress for a lot of a particular code if the DDL continues to operate as it did during the time period covered by the snapshot. Since the required information on run times and queue times is obtained from all the lots that pass through a step, regardless of their codes, a snapshot estimation can be produced even if no lots of the specified code were completed during the period that the snapshot describes.

Snapshots have two important uses. Snapshots for the same code but different time periods can be used to track the overall operation of the DDL over time. This comparison is valid even if the mix of products in the DDL changes, as long as the steps for the code are not redefined. In addition, a snapshot can be produced for a particular code when a lot enters the DDL, as long as there is information on the run times and queue times for each of its steps. The snapshot will show the mean predicted time and optimistic and pessimistic predictions of the time needed to process the lot; the lot's actual time to complete each step (excluding time on engineering hold) can be added to the plot to monitor its progress.

PROCESS

Yield and standard measures of throughput such as cycle time, are sensible measures of performance for a production line making at most a few different codes. Farrell et al. (1992) discuss measurement and analysis systems to monitor and improve the efficiency of production lines. But overall yield and standard measures of throughput may be inadequate for a DDL that develops new codes and new processes in a rapidly changing environment. For example, one standard throughput measure is the total or average number of steps completed per day in the DDL by all wafers in all lots. The rationale is that steps are easy to count, and the more steps completed, the more productive the facility. In a manufacturing facility with little change from day to day, comparing step counts from day to day makes sense. But when steps differ in complexity and time required to complete them and the environment changes rapidly, the number of steps completed is of limited value for tracking the DDL as a whole or for monitoring the progress of specific lots.

Because the contents of individual steps in a DDL are more stable than codes or sequences of steps, we believe that performance measures for a DDL should consider the information from each step separately rather than information obtained by considering all steps together, without regard to their differences. While information from individual steps, such as step yield and step run times, is important for monitoring DDL activity, information from the separate steps also needs to be combined to assess the DDL's performance as a whole. This last aspect of performance is the one addressed by the snapshot plot in this case study. Namely, the snapshot is a way to extract information from the separate steps and combine it to understand DDL output as a whole without following lots of one particular code.

The key idea is that different codes have different, lengthy sequences of steps, but often steps are shared by several codes. Moreover, all lots of the same priority and size that pass through a particular step, regardless of their code, should require the same queue time and run time, except for the random lot-to-lot variability that pertains to the step. Thus, instead of measuring throughput by tracking lots of a particular code from start to finish of processing, we measure completion time for individual steps and then sum completion times for individual steps.

Data Collection Plan

For each lot entering a step, the following are routinely reported to a database:
- number of wafers at the start of the step
- time of entry into the queue
- start of processing for the step
- end of processing for the step.

From these times, a queue time and run time are calculated for the lot, taking into account holiday and weekend periods when the DDL does not operate. The number of wafers that can be processed simultaneously in each step, called the batch size, is also reported.

Engineers and data base managers at a DDL chose a code, and for each step in the code, they provided the run times and queue times for all lots that used the step in a six week period. There were 148 steps in the code, but some steps, such as cleans, were used more than once. Five steps had been used by only a few lots during the six week period; the engineers replaced these steps with similar steps that had more data. In the end, we had data for a code that had a total of 148 steps, some of which were repeated. There were 120 different steps in all. Ideally, the data would include only lots of the same priority. In our data, however, lot priority could not be extracted from the databases, although the DDL engineers believed that most lots observed during this time period had the highest priority. Queue times depend on the lot's priority, but run times should not.

The observed run time for each lot was converted to a batch or unit run time by dividing the lot run time by the number of batches in the lot. For example, if six wafers could be processed simultaneously in a step and there were 40 wafers in the lot, then the batch size was six and the unit run time was the lot run time divided by the number of batches of a size no more than six (seven in this case).

Queue times and lot run times were calculated from data manually entered into the database, and so were sometimes unreliable. For example, an operator should have time stamped the lot manually when a step began to run, but sometimes did not. If the time that processing ends had to be entered before the lot could move to the next step, the technician may have entered the start and end of processing simultaneously, giving an incorrect run time of zero. Not all zeros were erroneous, however. When only a random sample of the lots was inspected, many of the observations for an inspection step were zero.

There were also unreasonably large unit run times for some steps. For example, one cleaning step had a unit run time of more than eight days, when its median unit run time was 40 minutes. Such comparisons of observed lot run times to nominal values or to reasonable minimum and maximum values can play an important role in process improvement by identifying steps that require more attention or further development. Similar evaluation of queue times can identify bottlenecks. In practice, reasonable minimum, maximum, and typical unit run times should be provided by process engineers. In this test case, however, we identified aberrant run times informally ourselves, rejecting only values that were extremely far from the bulk of the data.

ANALYSIS AND INTERPRETATION OF RESULTS

The snapshot plot requires estimates of the mean and variance of lot run time and queue time for each step. In this section we first show how we estimated the means and variances for each step of a code using the data provided by the DDL. Then we show how to construct a snapshot.

Estimated Means and Variances

First consider unit run time (discussion for queue times is similar). The obvious estimates of the mean and variance of unit run time for a step during a particular period are the sample mean m and the sample variance s^2 of the observed unit run times $x_1,...,x_n$ during the period. The sample mean and variance are attractive because they are widely understood, unbiased, and the best possible estimators if the data are normally distributed.

But if the data are not normally distributed, or if the data are normal but include frequent outliers, then there are better estimators of the true mean and variance than m and s^2. Sometimes it is obvious that a value is an outlier and should be ignored before making any calculations. For example, if only one unit run time for a step exceeds 40 minutes, and that unit run time is eight days, then we would ignore the eight days. But if a unit run time is 35 minutes when all other run times are between 20 and 30 minutes, it is not obvious whether this value is an error that should be ignored. While the data might follow a normal distribution with this value as an outlier, the data might instead follow some non-normal distribution, in which case better estimators could be used. Instead of using m and s^2 as estimators (along with some rules for eliminating possible outliers), we prefer to estimate the mean and variance by using the parameters of a distribution that fits the data. Choosing an appropriate distribution may require the help of a statistician, but once chosen, it is straightforward to update the estimates of its mean and variance.

In our study, we found three standard distributions to be adequate for modeling all the queue times and unit run times: the normal, lognormal, and Weibull (Johnson and Kotz (1970) discuss these distributions). As Figure 13.1 shows, normal observations are symmetric around the mean. However, lognormal and Weibull observations are not symmetric, and large observations are likely. The logs of lognormal data are normal while the logs of Weibull data may be asymmetric and have a right tail that is longer or shorter than that of a normal. Some lognormal and Weibull densities are similar, however.

A lognormal(a,b) distribution has mean $A\sqrt{B}$ and variance $A^2B (B - 1)$, where $A = \exp(a)$ and $B = \exp(b^2)$. A Weibull(a,b) distribution has mean $a\Gamma(1 + b^{-1})$ and variance $a^2\Gamma(1+2b^{-1}) - a^2\Gamma(1+b^{-1})^2$, where Γ is the (complete) gamma function, which is available in many statistical packages and languages such as S (see Becker, Chambers, and Wilks, 1988). Thus, to estimate the mean and variance of unit run time, we need to estimate a and b.

For the normal(a,b), the parameter a is estimated by the sample mean m and the parameter b is estimated by the sample standard deviation s. For the lognormal(a, b), a is estimated by the sample mean of the $\log(X_i)$'s, and b is estimated by the sample standard deviation of the $\log(X_i)$'s. For the Weibull, the maximum likelihood estimators solve

$$a^b = n^{-1} \Sigma X_i^b,$$

$$b^{-1} = \Sigma X_i^b \log(X_i)(\Sigma X_i^b)^{-1} - n^{-1}\Sigma \log(X_i).$$

These equations are easy to solve by iterating for b and then calculating a.

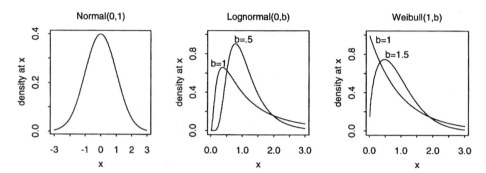

Figure 13.1. *Three Families of Densities Used to Model Run Times for the DDL Data*

Once (a,b) are estimated for the normal, lognormal, and Weibull, we have a fitted normal, fitted lognormal, and fitted Weibull distribution for the data. To choose one of these, we plot the ordered data against the $k/(n+1)$, $k = (1,...,n)$, quantiles of the fitted distribution. These quantiles act like the expected order statistics for a random sample of size n from the fitted distribution (David, 1981). Therefore, if the points in the quantile plot follow a straight line with an intercept at 0 and a slope of 1, the fitted distribution is a good model for the data.

Figure 13.2 shows the fits for an etch step that was used 20 times during the six week period of interest. The solid line in each panel is the reference line with intercept 0 and slope 1. For this step, the lognormal does not fit as well as the normal or Weibull, because it overestimates large observations. That is, the large observations fall below the line, implying that the observation would have been larger if this lognormal were appropriate and that the right tail of the lognormal is too long. There is little to distinguish the normal and Weibull fits, so we chose the normal because it is simpler. We also chose the normal when there was little data. For example, some steps were used only four times during the six week period.

Figure 13.3 shows the normal, lognormal, and Weibull fits to an apply step that was used 110 times during the six week period of interest. All observations but one were below 35 (we omitted a large outlier which was above 100). Here the Weibull provides the best fit. The lognormal seriously overestimates the larger observed run times, and if used it would overestimate the mean and standard deviation of unit run times. The normal fit underestimates the small run times and the large run times, so if used it would underestimate the mean unit run time.

The lognormal and Weibull distributions assume that the data values are all positive and can extend as low as zero (see Figure 13.1), but for some steps very short run times are clearly unreasonable. For these cases we shift the lognormal and Weibull distributions to have a larger threshold that is a sensible value for the step, rather than zero. For example, Figure 13.4 shows 46 observations from a bake step (five observations of 0 and five observations less than 30 were omitted), using a threshold of 180 for the lognormal and Weibull.

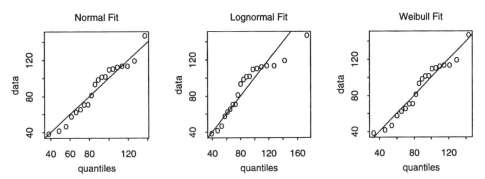

Figure 13.2. *Normal, Lognormal, and Weibull Fits to the Unit Run Times from an Etch Step. The right tail of the lognormal is too long, as it overestimates the largest observation. The normal and Weibull fits are similar. For this step, we chose the normal because it is simpler.*

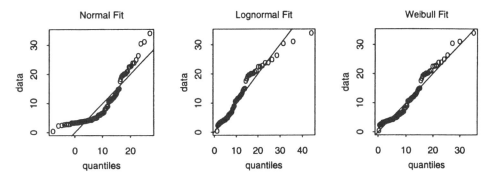

Figure 13.3. *Normal, Lognormal, and Weibull Fits to the Unit Run Times from an Apply Step. The normal underestimates both small run times and long run times. The right tail of the lognormal is too long. The Weibull provides the best fit.*

Out of 120 steps for unit run times, we chose 50 to be Weibull, 50 lognormal, and 20 normal. Of the 100 that are lognormal or Weibull, only six needed a nonzero threshold, and those fits were not too sensitive to the actual choice of threshold.

Sometimes, a step was applied to only a random sample of lots, so many unit run times are zero. For example, for one inspection step, 23 of the 46 unit run times were zero. For such steps, we modeled the distribution of unit run times in two parts. First, with probability p, the lot was inspected, and the unit run time was positive. The positive run times were assumed to be either normal, lognormal, or Weibull. Suppose the true mean and variance of the positive inspection times are M_+ and σ_+^2. Then the mean wait for a random lot that might or might not be inspected is the mean wait for an inspected lot times the fraction of lots inspected. The variance of a random lot is the variance within inspected lots (the variance is zero for uninspected lots) multiplied by the fraction of inspected lots plus the variability between the mean of the inspected group (M_+) and the mean of the uninspected group (zero). That is, the mean M and variance σ^2 of the unit run times for a random lot that might or might not be inspected are given by

$$M = p M_+,$$
$$\begin{aligned}\sigma^2 &= p\sigma_+^2 + (1-p)(0-M)^2 + p(M_+ - M)^2 \\ &= p\sigma_+^2 + (1-p)p^2 M_+^2 + p(1-p)^2 M_+^2 \\ &= p\sigma_+^2 + p(1-p)M_+^2.\end{aligned}$$

See DeGroot (1975) for more discussion of conditional means and variances.

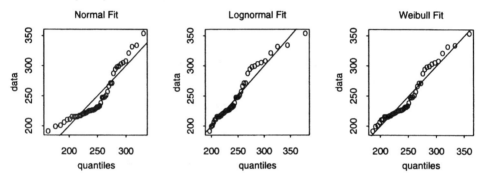

Figure 13.4. *Normal, Lognormal, and Weibull Fits to the Unit Run Times from a Bake Step. For the lognormal and Weibull, a threshold of 180 minutes is assumed. The normal distribution does not fit the observations below the mean; the right tail of the lognormal is too long. The Weibull provides the best fit.*

The probability p of a lot being inspected can be estimated by the fraction of nonzero observations in the data. The mean and variance of unit run time for inspected lots can be estimated by fitting either a normal, lognormal, or Weibull distribution to the run times for inspected lots.

Although the discussion so far has been in terms of unit run times, the same ideas apply to queue times (except that queue times apply only to lots, not to batches within lots). Often, there is a large probability that a lot does not have to wait before being processed, so the queue times must be treated as the run times for randomly inspected lots. That is, lots that do not wait must be treated separately from those that do wait, to find a distribution that fits the data well. Since queue times may be longer for lots with lower priority, the distribution of queue time should be estimated from lots with the same priority.

The Snapshot Plot

A snapshot describes the cumulative mean time needed to process a code through its steps, as well as the optimistic and pessimistic estimates of the cumulative time for processing. The estimated mean time to complete a step is the sum of the estimated mean queue time and the mean run time for the lot. The mean run time for the lot is the product of the number of batches in the lot and the estimated mean unit run time for the step. The mean of a sum of steps is then the sum of the means for each step, even if the steps have different distributions or even if the distributions are statistically dependent (see DeGroot, 1975).

The estimated cumulative time for the fastest 5% of lots depends on the distribution of the sum of step times. Here, the exact distribution of the cumulative time to complete processing through a particular step is the sum of normal, lognormal, and Weibull random variables. The distribution of such a sum is difficult to find. Therefore, we simplified the problem in three ways. First, we assumed that, for any step, the queue time and lot run time are independent. In particular, lots that wait a long time to begin a step of processing cannot be processed faster than lots that are processed immediately. Secondly, we assumed that step times are independent, so if a lot finishes one step quickly it is not more or less likely to finish the next step quickly. Process engineers agreed that both these assumptions were reasonable.

Finally, we simplified our task by using the fact that a sum of independent random variables is approximately normal with mean equal to the sum of the means of the random variables and variance equal to the sum of the variances of the random variables (some weak regularity conditions are needed; see DeGroot (1975) for a detailed statement). Thus, the total time needed to complete processing through a step is approximately normal. For a small number of steps, normal approximation may be poor. But after several steps, normal approximation should be adequate. We used the .05 quantile and .95 quantile of the approximate normal distribution to estimate the completion time for the fastest 5% of lots and the slowest 5% of lots.

Figure 13.5 shows a snapshot for a standard code representing a specific six week period of data, assuming a constant lot size of 40 wafers throughout processing. The middle curve shows that, on average, it takes 14.6 days to complete the first 50 steps of the code and 37.1 days to complete all steps of processing. The fastest 5% of lots complete step 50 in 9.7 days and the entire process in 31.2 days, while the slowest 5% of lots complete step 50 in 19.4 days and the entire process in 43.1 days. Note that the optimistic, lower curve is negative at the first two steps because the normal approximation to the cumulative completion time is poor when there are only a few steps.

Two alternatives to our construction of the snapshot deserve mention. First, instead of using the means and variances from the fitted normal, lognormal, and Weibull distributions, we could just remove the outliers from the data and use the sample means and sample variances. Using the sample means and sample variances would simplify the calculations slightly, but at the risk of introducing more variability into the upper and lower curves in the snapshot. For example, if the true distribution is lognormal, then the variability of the sample variance can be an order of magnitude larger than the variability of the variance of the fitted distribution (see Johnson and Kotz, 1970, for example). In the data discussed here, the total estimated completion times would be about 10% shorter for the fastest 5% of lots and 10% longer for the slowest 5% of lots if the sample means and sample variances were used instead of the means and variances of the fitted distributions. That is, for this data there is only a difference of about four days out of 40 at the last step if sample means and variances are used for all steps and not just those that are normally distributed. A difference of four days may not have any practical importance. In any case, we suspect that, in general, the most important issue is to remove outliers before estimating the means and variances.

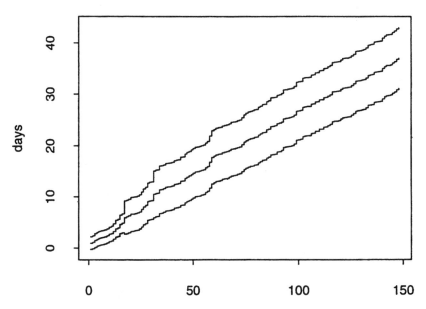

Figure 13.5. *The middle curve shows expected time to complete processing through a step-against-step number for a particular* 148 *step code based on data obtained during a particular six week period at a DDL. The lot size is assumed to be fixed at* 40 *wafers throughout processing. The upper curve shows the estimated time to complete the slowest 5% of lots, and the lower curve shows the estimated time to complete the fastest 5% of lots.*

A second alternative to the snapshot is to use the fitted distributions of run times and queue times to simulate the time-until-completion for a particular code using discrete event simulation (see Pooch and Wall, 1993). Since the snapshot and discrete event simulation could use the same fitted distributions, the major advantage of the discrete event simulation is that it might approximate the expected and estimated completion times for the slowest and fastest 5% of lots better than the normal approximation that the snapshot relies on. If a code contains a reasonable number of steps, say more than 30, and none of the steps dominates the process, then the normal approximation is unlikely to be misleading and, of course, the normal approximation is simpler to obtain than a discrete event simulation. On the other hand, if run times and queue times are dependent and the dependence can be modeled realistically in a discrete event simulation, then the discrete event simulation might give better estimated completion times than the snapshot. The dependence would not affect the average estimated time shown on the middle curve of the snapshot, but it would affect the outer curves that give optimistic and pessimistic estimates of step completion times. In any case, the crucial issue is whether the underlying assumptions about run times and queue times are reasonable; this assumption can be checked only through data analysis, not by simulation.

To track a DDL's performance over time in terms of this standard code, the snapshot can be re-calculated with data from a different, possibly overlapping, time period. The three curves in the new plot can be compared to those from the earlier plot to determine how the estimated time required to produce this code has changed. To make comparisons reflect changes in distributions of queue and run times rather than time changes resulting from differences in yields, it is reasonable to fix both lot sizes at 40 wafers throughout processing.

To track the progress of a specific lot, the snapshot can be calculated when the lot enters the DDL as long as data are available for queue and run times for each of its steps. If the code requires a few new steps or ones that have not been run recently, process engineers may be able to identify similar steps that have been run recently. The lot's actual time to complete each step can be added to the plot as it progresses through the DDL. Of course, the time should exclude engineering holds and periods such as holidays and weekends when the DDL is not operating. If a new lot represents a new process with many new steps for which there is no comparable data available, then a snapshot cannot be calculated to monitor its progress. This is reasonable, because it makes clear that it is impossible to predict processing time for this lot from past data.

When tracking a specific lot, information on wafer loss can be incorporated as it becomes available. If some wafers do not move to the next step, the snapshot can be re-drawn to reflect the reduced lot size in later steps. It is also possible to provide snapshots for lots used in experiments in which some lots are moved to one step, some to another, and at a later step the split lots are combined. The snapshot can show the expected time to finish the slowest lot or the average expected time to finish the split lots.

CONCLUSIONS AND RECOMMENDATIONS

Snapshots can be used for monitoring a DDL's status, for process improvement, and for managing a DDL queue. Because the snapshot does not require data on completed lots, performance on a standard code can be monitored over time even if no lots of that code were produced during the time periods of interest. Engineers can also monitor progress of individual lots with a snapshot. Lots that fall below the pessimistic time estimates would not be considered unduly delayed, and those that fall above would warrant further attention. Although engineers can monitor lot progress by plotting completion time against step number without drawing the snapshot curves, plots on individual lots are more meaningful when standards for performance, such as the outer curves of the snapshot, are shown too. That is, the snapshot can be considered as a control chart on lots, even though it is not derived from data on lots but rather from data on individual steps.

The snapshot can be used not just to monitor performance but also to manage resources. For example, the snapshot could be used in what-if scenarios to predict the reduction in time-to-completion if additional equipment is introduced or additional technicians hired. Finally, the data on distributions of run times and queue times collected for individual steps can play a role in quality improvement programs. Comparing actual distributions to nominal values and past distributions suggests where improvements have been made and where more improvements are needed.

The real difficulty in producing snapshots is not fitting distributions or calculating means and variances but getting the right data at the right time. Some database systems are designed primarily for inventory control. While such systems are simple to program to track where a lot currently is and where it is going, it may be difficult to extract data on processing times and queue times even if the database system routinely collects such data. The data that we have used in this paper were quite difficult to retrieve from the database that was used at that time. Given the wealth of information in the database about processing times and queue times and its importance for quality improvement, easy access to data should be given as much importance in database design as inventory control.

In summary, if the data on individual processing steps can be obtained, snapshots provide a simple but powerful tool for measuring important aspects of the health of a DDL and the progress of lots through it in an ever-changing environment.

Acknowledgments

We thank Jay Reightler and Linda Snyder for their help in obtaining, processing, and understanding the DDL data used in this paper.

References

Becker, R.A., J.M. Chambers, and A.R. Wilks, *The New S Language: A Programming Environment for Data Analysis and Graphics*, Pacific Grove, CA: Wadsworth & Brooks, 1988.

David, H.A. *Order Statistics*, 2nd. ed., New York, NY: John Wiley and Sons, Inc., 1981.

DeGroot, M.H. *Probability and Statistics*, Menlo Park, CA: Addison-Wesley, 1975.

Farrell, B.L., W. Kahan, J.W. Reightler, and A.C. Woodard, "Improving Service Efficiency in Manufacturing Integrated Circuits," *AT&T Technical Journal*, 71, pp. 37-44, 1992.

Johnson, N.L. and S. Kotz, *Continuous Univariate Distributions-1*, Boston, MA: Houghton Mifflin Company, 1970.

Pooch, U.W. and J.A. Wall, *Discrete Event Simulation: A Practical Approach*, Boca Raton, FL: CRC Press, 1993.

BIOGRAPHIES

Diane Lambert holds a B.S. degree from Tufts University and a Ph.D. degree from the University of Rochester. Before joining statistical research at Bell Labs, Lucent Technologies, she was on the faculty of the Department of Statistics at Carnegie Mellon University. Throughout her career, she has focused on developing new methodology for applications with nonstandard data.

James M. Landwehr has worked on a variety of problems in statistical research and applications at Bell Labs, Lucent Technologies since 1973. Several recent projects involve the statistical design and data analysis of industrial experiments to improve quality. He has also pursued research in the areas of cluster analysis, categorical data analysis, and statistical graphics.

Ming-Jen Shyu received B.S. and M.S. degrees in electrical engineering from the Massachusetts Institute of Technology. Before joining Bell Labs, Lucent Technologies, he worked at SRI David Sarnoff Research Center as a MIT graduate intern in the area of high definition television. He is now a member of the technical staff in the statistics research department at Bell Labs, Lucent Technologies, specializing in statistics graphics.

PART 3

DESIGN OF EXPERIMENTS

CHAPTER 14

INTRODUCTION TO DESIGN OF EXPERIMENTS

Veronica Czitrom

INTRODUCTION

Design Of Experiments (DOE) is a structured approach to efficiently characterize, improve, and optimize a process or product by collecting, analyzing, and interpreting data. The methodology is used to
- plan and analyze experiments to study several factors simultaneously.
- reduce experimental cost, time and resources.
- establish cause-and-effect relationships between process inputs and process outputs.
- identify key factors that have the greatest and least impact on product or process performance.
- manufacture products that consistently meet specifications by making them robust to sources of variability.
- establish the process window, or region where the factors may operate, by finding the sensitivity of the responses to the factors.
- improve yield, reliability, and performance.
- assist in process troubleshooting.

Design of experiments does not replace good engineering practice, it enhances it. Engineering knowledge should be used to plan a designed experiment and to interpret the results.

The designed experiments in the case studies in this unit were used to
- characterize a tungsten chemical vapor deposition process to eliminate TiN peeling, improve film uniformity, and experiment efficiently to avoid a project deadline crisis.
- improve within-wafer uniformity in the characterization of the chemical vapor deposition of tungsten by incorporating a bulls-eye uniformity inversion into the model.
- improve uniformity in an etching process by adding a reduced number of observations to the experimental design to resolve confounding.
- improve throughput (removal rate) and yield (removal rate uniformity) in a wafer chemical mechanical polishing process.

- optimize an etching process and make it robust (insensitive) to circuit design characteristics.
- improve a photolithographic wafer stepper alignment system by making it robust (insensitive) to manufacturing steps preceding alignment.
- characterize a polysilicon rapid thermal chemical vapor deposition, and model the deposition rate by fitting an Arrhenius equation.

Designed experiments are also described in case studies in other parts of this book. Lewis et al. (Chapter 28) used a robust design experiment to improve equipment reliability. Buckner et al. (Chapter 2), Pankratz (Chapter 3), and Mitchell et al. (Chapter 5) used designed experiments during gauge studies.

The need for improvement in process or equipment performance identified during a gauge study (Part 1), a passive data collection (Part 2), statistical process control (Part 4), or equipment reliability (Part 5) can be actively pursued by performing a designed experiment. After a designed experiment has been performed to characterize, improve, or optimize a process, a passive data collection can be performed to confirm experimental results, and statistical process control (which can also be seen as long-term confirmation of results) can be used to monitor the process. The experimental plan of a gauge reproducibility study, a passive data collection, and a marathon run in a reliability study, is usually a full factorial designed experiment. Gauge studies should precede a designed experiment to ensure adequate measurement precision for the responses.

Engineers often perform *one-factor-at-a-time experiments* where they change the levels of one factor at a time while holding other factors fixed. For several reasons, statistically designed experiments provide a more effective way to determine the impact of several factors on a response than one-factor-at-a-time experiments. A designed experiment usually requires less resources (experiments, time, etc.) for the amount of information obtained. The estimates of the effects of each factor are usually more precise, often using all observations instead of only two to estimate the effect of each factor. The effect of one factor at different levels of another factor (interaction) can be estimated systematically, while it is usually not estimable in one-factor-at-a-time experiments. Process optimization is usually more efficient because the results are valid over a broader factor region. These advantages of designed experiments will be described more fully in the section on planning the experiment.

The design of experiments methodology can be divided into four stages:
- planning the experiment
- conducting the experiment
- analyzing the data
- reaching conclusions and implementing recommendations.

These four stages will be considered in the next four sections of this introduction to design of experiments. The next-to-the-last section gives a brief overview of the case studies presented in the remainder of this part of the book, and the last section gives some suggestions for further reading.

PLANNING THE EXPERIMENT

We will consider experimental planning in several sections: basic concepts and definitions, models underlying an experimental design, types of experimental designs, advantages of using designed experiments, practical considerations for setting up a designed experiment, and the experimental designs used in the case studies. Planning an experiment involves practical considerations such as determining who is going to perform the experiment and who is going to collect the data, as well as selecting the experimental design and the corresponding underlying model. The different types of experimental designs take up the bulk of the material, and are considered in two sections.

Basic Concepts

Planning an experiment includes formulating the objectives of the study, selecting responses, selecting factors and their settings, selecting the experimental design, and planning the data collection. It is important to use all available information to plan the experiment, including historical data, experimental results, theoretical knowledge, and expert opinion. Experimental planning is an iterative process that takes into account available resources, the desired amount of information, and the acceptable level of risk.

Table 14.1. 2^3 *full factorial experimental design in three factors (tem-perature, pressure, and argon flow) in 8 runs (runs 1 to 8), with one centerpoint (run 9) and with three responses (tungsten deposition rate, tungsten non-uniformity, and stress). Temperature was hard to change, so the runs were grouped by temperature and randomized within each temperature, giving a split-plot design structure. The order in which the runs were performed is 4, 2, 1, 3, 9, 7, 5, 8, and 6. (From Buckner et al., Chapter 15, Appendix B). A two-wafer batch was used for each run.*

Run	Temperature (°C)	Pressure (torr)	Argon flow (sccm)	Tungsten deposition rate (Ång/min)	Tungsten non-uniformity (%)	Stress (Gdyn/cm^2)
1	440	0.8	0	265	4.44	10.73
2	440	0.8	300	329	8.37	10.55
3	440	4.0	0	989	4.48	9.71
4	440	4.0	300	1019	7.89	9.77
5	500	0.8	0	612	6.39	8.35
6	500	0.8	300	757	8.92	7.73
7	500	4.0	0	2236	4.44	7.55
8	500	4.0	300	2389	7.83	7.48
9	470	2.4	150	1048	7.53	8.59

A designed experiment is used to establish a cause-and-effect relationship between *process inputs* (or *input variables*, *factors*, or *predictors*) and *process outputs* (or *output variables*, or *responses*). The designed experiment in Table 14.1 was used to understand the effect of the three factors, temperature, pressure, and argon flow, on three responses, tungsten deposition rate, tungsten non-uniformity, and stress. Factors and responses should be selected in accordance with the objectives of the study using engineering knowledge. The results of the experiment can be used to build a *model* of how changes in the factors affect the responses.

Four types of responses can be continuous numerical variables, discrete numerical variables, discrete categorical variables, and graphs. Data from a *continuous numerical response* takes on values in an interval of numbers. The three responses in Table 14.1, average tungsten deposition rate over 49 sites on a wafer, tungsten non-uniformity over 49 sites on a wafer (standard deviation divided by mean and multiplied by 100), and stress on a wafer, are continuous responses. Data from a *discrete quantitative response* (such as the number of particles on a wafer) can be counted. Data from a discrete *categorical response* (such as the presence or absence of reflow, and device acceptance or rejection at testing) can be categorized. If it is possible to choose between a categorical and a continuous response, the continuous response is preferable because it contains more information. For example, it is better to consider the thickness of a layer than whether the layer is present or absent. A statistic computed from the data, such as the average or the standard deviation of the observations on a wafer, can be used as a response. If possible, use a response that is easy to measure, and consider responses that cover different aspects of the process or product.

The *factors* in an experiment can be *continuous* numerical variables such as time and temperature, or *discrete* categorical variables such as type of equipment and presence or absence of a processing step. A *factor setting* (or *factor level*, or *experimental setting*) is the value or condition of the factor during an experimental run. The designed experiment in Table 14.1 has three continuous factors, temperature, pressure, and argon flow. The factor temperature has three settings, 440, 470, and 500°C. A discrete factor "machine" could have two settings, Machine A and Machine B. Use engineering knowledge to select factors that can help satisfy the objectives of the experiment, and that cover different aspects of the problem. Select factor settings so that there is no basic change in process behavior, such as a change in chemical regime or restrictions in pump capacity. Usually two levels for each factor are sufficient, unless curvature is expected. A *run, treatment combination,* or *(design) point* is the combination of factor settings at which one experimental run is performed. The designed experiment in Table 14.1 has nine runs, the first of which was performed at a temperature of 440°C, a pressure of 0.8 torr, and an argon flow of 0 sccm (no argon flow). An *experimental design* (or simply *design*, or *design matrix*) is a group of treatment combinations (including the factors and factor settings but not including the responses), such as the nine runs given in Table 14.1. The *factor space*, or *experimental region*, is the region defined by the factor settings. For Table 14.1, the factor space is the region where temperature is between 440 and 500°C, pressure is between 0.8 and 4.0 torr, and argon flow is between 0 and 300 sccm. An *experimental unit* is the entity on which a measurement is made. For the experiment in Table 14.1, the experimental unit for each run was a batch of two wafers.

The first eight runs in Table 14.1 are a *full factorial* experimental design, consisting of all possible combinations of two levels of each one of the three factors. An experimental design is *orthogonal* (or *balanced*), and the factors are *orthogonal* to each other, if each level of one factor appears the same number of times with each level of another factor, and this is true for all pairs of factors in the design. For example, the full factorial design in runs 1 to 8 of Table 14.1 is orthogonal. The temperature of 440°C appears twice with a pressure of 0.8 torr and twice with a pressure of 4.0 torr, and the temperature of 500°C appears twice with a pressure of 0.8 torr and twice with a pressure of 4.0 torr. Similarly, the two settings of temperature appear twice with each setting of argon flow, and the two settings of pressure appear twice with the two settings of argon flow.

Some common types of experiments and experimental objectives follow.
- A *screening experiment* in a large number of factors is used to identify a small number of factors that have a large effect on the response. Screening experiments are usually used in the initial exploration of a process. It is usually better to include too many factors than to risk missing an important factor.

- A *second-order design* (or *response surface design*, or *RSM design*) considers curvature to characterize and optimize the response. It is often used after a screening experiment using a reduced number of important factors.
- A *robust design experiment* is used to manufacture products that consistently meet specifications by making the product or process robust, or insensitive, to sources of variability.
- A *sensitivity experiment* is used to explore the factor region, for example by varying the factors ± 10% from the optimal conditions.
- *Confirmation runs* are performed to validate the results of a designed experiment. They can consist of a few runs, a designed experiment, or a passive data collection (Part 2).
- *Exploratory runs* can be used before a larger designed experiment, for example to select factor settings. Factor settings need to be chosen with care if the experiment is close to a change in process behavior, for example near a region of restricted pump capacity, or near a region where there is a change in chemical regime.
- *Sequential experimentation* uses the results from one experiment to plan the next experiment. It is usually more efficient to perform a sequence of small experiments than one large comprehensive experiment designed to answer all possible questions. The sequence of experiments could include some of the following: exploratory runs, then a screening experiment and a second order design, or a robust design experiment, and finally sensitivity experiments and confirmation runs.

It is desirable to perform the runs in an experiment in *random order* to minimize the effect of extraneous sources of variability. For the experiment in Table 14.1, there was a restriction on randomization because it was hard to change the temperature. The runs were ordered by temperature, and then the order of the runs within each temperature was randomized. *Replication* is the repetition of an experimental run or runs. A *replicate* usually refers to a complete repetition of the entire experimental design. Replication allows estimation of the inherent "*error*" or *natural variability* in the response, since repeating a run under what appear to be identical conditions can give slightly different values of the response. The centerpoint in run 9 of Table 14.1 was to be replicated three times to give an estimate of error, but only one replicate was performed due to insufficient wafers. It is important for replicates to incorporate the variability of changing factor settings. If the conditions of the centerpoint are performed once at the beginning of the experiment, once in the middle, and once at the end, the centerpoint is replicated; if three consecutive wafers are processed at the conditions of the centerpoint without making any changes to the factor settings, the centerpoints are not true replicates, and the estimate of natural variability will probably be underestimated because it does not include the variability from changing factor settings. The estimate of error can be compared to the effect of changing the level of a factor to determine whether the observed effect of the factor is *statistically significant* (larger than the background noise, or natural variability). In Table 14.1, the difference in tungsten non-uniformity at the low and high pressures can be compared to the measure of error to determine whether pressure has a significant effect on non-uniformity. Replication can also be used to obtain more precise estimates (estimates with less uncertainty). *Blocking* improves the precision of an experiment by comparing factors in homogeneous groups of runs, or *blocks*. For example, to compare two types of materials, it is better to use only one machine (block) than to use one material on one machine and the other material on another machine. To combine the results of two groups of experiments that were performed at different times, the first group can be considered to be in one block, and the second group in another block. The block effect can be removed from the effects of interest.

The *main effect* of a factor is the average change in the response produced by a change in the level of the factor. For 2^k full factorial and 2^{k-p} fractional experimental designs with factors at two levels, the main effect of a factor has an intuitively appealing interpretation as the difference between the average of the observations at the high level of the factor, minus the average of the observations at the low level of the factor. For Table 14.1, the main effect of argon on non-uniformity is 3.315%, which is the difference between the average non-uniformity of 8.2525% when the argon flow is 300 sccm, and the average non-uniformity of 4.9375% when the argon flow is 0 sccm.

The main effect of argon flow on non-uniformity can also be obtained in another way that illustrates the orthogonality property that is at the heart of why designed experiments work: even though many factors are varied simultaneously in a designed experiment, the individual effects of the factors can be isolated. When temperature is 400°C and pressure is 0.8 torr (runs 1 and 2 in Table 14.1), non-uniformity increases by 3.93% as argon increases from 0 to 300 sccm; when temperature is 400°C and pressure is 4.0 torr (runs 3 and 4), non-uniformity increases by 3.41% as argon increases from 0 to 300 sccm; when temperature is 500°C and pressure is 0.8 torr (runs 5 and 6), non-uniformity increases by 2.53% as argon increases from 0 to 300 sccm; when temperature is 500°C and pressure is 4.0 torr (runs 7 and 8), non-uniformity increases by 3.39% as argon increases from 0 to 300 sccm. The average of the four increases in non-uniformity as argon increases from 0 to 300 sccm is 3.315%, which is the value of the main effect of argon. Each individual comparison of the effect of argon was made at the same temperature and pressure, and when the four comparisons are averaged, the effects of temperature and pressure "cancel out" of the main effect of argon. In addition, there is *hidden replication*, since the effect of argon is estimated four times (once at each temperature and pressure combination) using all eight runs.

Two factors *interact* if the effect of one factor on the response depends on the level of the other factor. For 2^k full factorial and 2^{k-p} fractional factorial experimental designs with factors at two levels, the *interaction effect* AB between factors A and B is the average difference between the effect of factor A at the high level of B, and the effect of factor A at the low level of B. When the interaction between two factors is large, the corresponding main effects have little meaning and cannot be meaningfully interpreted without considering the interaction. Three factors interact if the two-factor interaction between two of the factors depends on the levels of a third factor. Interactions between three or more continuous factors are third order terms in a Taylor series expansion, and are usually assumed to be negligible.

Models

A *model* of the response in terms of the factors underlies every experimental design. The model allows estimation of the effects of the various factors on the response. The model can be used to *predict* process performance over the entire factor region, including points where the experiment was not performed. The predictions from the model can be visualized using contour plots. A *polynomial regression model*, which we will now consider, is a truncated Taylor series expansion.

Consider a response y as a function of two factors A and B. Let the factors A and B be coded numerically in terms of x_1 and x_2 respectively. For example, if factor A (continuous or categorical) has two levels, represent the two levels by the values -1 and $+1$ of the coded variable x_1. The response y can be expressed in terms of the two factors using different polynomial regression models: a *first-order model* in the two factors,

$$y = \beta_0 + \beta_1 x_1 + \beta_2 x_2 + \varepsilon,$$

a first-order model with an interaction term (or a second order model without quadratic terms),

$$y = \beta_0 + \beta_1 x_1 + \beta_2 x_2 + \beta_{12} x_1 x_2 + \varepsilon,$$

or a *complete second-order model* or *quadratic model*

$$y = \beta_0 + \beta_1 x_1 + \beta_2 x_2 + \beta_{12} x_1 x_2 + \beta_{11} x_1^2 + \beta_{22} x_2^2 + \varepsilon.$$

The β coefficients are unknown constants that are estimated using the results of the experiment, β_0 is a constant term, $\beta_1 x_1$ and $\beta_2 x_2$ are directly proportional to the main effects previously described corresponding to the two factors A and B, $\beta_{12} x_1 x_2$ is the AB interaction term, and $\beta_{11} x_1^2$ and $\beta_{22} x_2^2$ are the quadratic terms corresponding to the two factors. *Linear* terms are the first order terms $\beta_1 x_1$ and $\beta_2 x_2$. *Non-linear* terms include the interaction term $\beta_{12} x_1 x_2$, the quadratic terms $\beta_{11} x_1^2$ and $\beta_{22} x_2^2$, and other higher order terms such as $\beta_{112} x_1^2 x_2$ and $\beta_{111} x_1^3$. All three models are said to be *linear models* because they are linear in the β coefficients. The *error term* ε represents the variation in the response y that cannot be explained by the model. The error terms ε corresponding to different observations are assumed to be independently distributed and to come from a normal distribution with mean 0 and variance σ^2. These assumptions imply that the observations are independent, and that they have the same variance at each level of each factor. Error terms include measurement error and random fluctuations in the process. The assumptions on ε are used to determine the statistical significance of the terms in the model. The variance σ^2 can be estimated by replicating one or more experimental runs.

Experimental designs that support the first order model and the first order model with interaction terms require only two levels for each factor, and are often screening experiments. Experimental designs that support the second order model have factors at three or more levels, or by pooling terms that are omitted from the model. A *statistical model* includes an error term.

The results of the experiment can be used to find the *least squares estimates b* of the β coefficients in the model. The model is then used to find the *estimated, predicted, or fitted value* $\hat{y}$ of the response y. The *residual* of an observation is the difference between the observation y and its estimated value $\hat{y}$. The least squares estimates b of the model coefficients β are those that minimize the sum of the squares of the distances between the estimated values and the actual observations, that is, the least square estimates minimize the sum of the squares of the residuals. For the first order model, the estimated value $\hat{y}$ is given by

$$\hat{y} = b_0 + b_1x_1 + b_2x_2.$$

There is no error term ε in this prediction equation because, for given numerical values of b_0, b_1, and b_2 and for specific factor settings (and thus x_1 and x_2), there is a single prediction value ŷ with no error. Analogous models are obtained for the linear model with interactions and for the quadratic model.

For the example in Table 14.1, the model for tungsten non-uniformity is

Tungsten non-uniformity = 6.70 - 0.44$x_{pressure}$ + 1.66x_{argon}.

The term in temperature and all the two-factor interaction terms are missing from the model because they were found not to be statistically significant (their coefficients were very small, close to zero). Since the factor temperature is not in the model, it has a negligible effect on tungsten non-uniformity, for the temperature range (400 to 500°C) considered in the experiment. The constant term 6.70 in the model is the average of all the observations, and 1.66 is one-half of the main effect of 3.315% of argon. The negative coefficient –0.44 indicates that tungsten non-uniformity decreases as pressure increases, and the positive coefficient +1.66 indicates that tungsten non-uniformity increases as argon flow increases. Since 1.66 is larger in absolute value than 0.44, argon has a larger effect than pressure on tungsten non-uniformity, for the factor ranges considered. We can make this comparison "for the factor ranges considered" because both factors are coded as –1 and +1 at their low and high values respectively, so the fact that pressure goes from 0.8 to 4.0 and is measured in torrs and argon flow goes from 0 to 300 and is measured in sccms does not cloud our analysis. At a pressure of 0.08 torr (low pressure, or $x_{pressure} = -1$) and an argon flow of 300 sccm (high argon flow, or $x_{argon} = +1$) for an arbitrary temperature, the predicted tungsten non-uniformity is 6.70 - 0.44 (-1) + 1.66 (+1) = 8.80%.

Experimental Designs

To select an experimental design, it is necessary to keep in mind the engineering objectives of the experiment, the available resources, the number of factors and number of levels for each factor, the model for the response, the need for centerpoints to detect curvature, and the possibility of replication to determine the statistical significance of factors and to construct confidence intervals. The selection of an experimental design is usually an iterative process.

In this section we will consider the most frequently used experimental designs, namely
- full factorial designs, which include all factor level combinations,
- fractional factorial designs, which consist of a fraction of a full factorial design,
- central composite designs to study curvature,
- robust designs to make a process insensitive to sources of variability.

We consider additional experimental designs in the next section.

Full and fractional factorial designs are used in screening experiments to identify a small number of key factors from a larger number of factors. The most frequently used full and fractional factorial designs have factors at two levels each, and the corresponding models have main effect and interaction terms. Central composite designs are used to study curvature and to optimize a response using a reduced number of continuous factors. The factors are at three or more levels each, and the corresponding second order models include quadratic terms in addition to main effect and interaction terms. Robust design experiments are used to make a product or process robust or insensitive to sources of variability. Robust design experiments use the previous designs, or others, to achieve their goal. These designs will now be considered in greater detail.

A *full factorial design* includes all possible combinations of the levels of every factor with the levels of every other factor. The number of experimental runs is the product of the number of levels of each factor. For example, Freeny and Lai (Chapter 18) used a 3×5 full factorial design, which consisted of all fifteen combinations of the levels of a three-level factor and a five-level factor. A full factorial design can accommodate a model with main effects, two-factor interactions, three-factor interactions, etc. up to the k-factor interaction.

Experiments with factors at two levels play a special role, because they are a very efficient form of experimentation. Of special importance is the 2^k *full factorial design* for k factors at two levels each in 2^k runs. The first eight runs in Table 14.1 are a 2^3 full factorial design for 3 factors in $2^3 = 8$ runs. This design is illustrated graphically in Figure 14.1a. When the factor levels of the 2^k full factorial are coded as -1 and 1, then the least squares estimate b_0 of the constant term β_0 in the model is the average of all the observations, the least squares estimate b_i of the coefficient β_i of each linear term in the model is one-half of the value of the corresponding main effect, and the least squares estimate b_{ij} of the coefficient β_{ij} of each interaction terms is one-half of the value of the interaction effect. The reason for the halves is that main effects, for example, are estimated as the change in the response as the factor goes from its low level to its high level (a change of two units in the coded values of -1 to $+1$), while the coefficient b_i in the model represents a slope, which corresponds to the change in the response for a unit change in the factor. The model for tungsten non-uniformity in Table 14.1 is $6.70 - 0.44 x_{\text{pressure}} + 1.66 x_{\text{argon}}$. The average of the nine values of tungsten non-uniformity in Table 14.1 is 6.70, and the main effect 3.315% of argon divided by two gives the coefficient 1.66 in the model. For continuous factors, a *center point* is a run with each factor setting at the middle of its range. Run 9 in Table 14.1 is a centerpoint. A center point can be added to an experiment with continuous factors at two levels each to test for the presence of curvature in the response. If curvature appears to be present, additional experimentation is required to determine which of the factors is the one that has the curvature.

Even for factors at only two levels each, the number of runs in a full factorial can be excessively large. For example, a full factorial in 4 factors at two levels each requires 16 runs, 5 factors require 32 runs, and 6 factors require 64 runs. To reduce the number of runs, it is possible to select a fraction, like one half or one fourth, of the full factorial. A 2^{k-p} *fractional factorial design* in k factors at two levels each in 2^{k-p} runs is a carefully selected $1/2^p$ fraction of the 2^k full factorial experimental design. Fractional factorial designs are among the most widely used experimental designs, especially for screening experiments. Not all terms in the model for the full factorial design are estimable for the fractional factorial design. By assuming three or more factor interactions to be negligible, some or all main effects and two-factor interactions may be estimable from a carefully selected fraction, say half or a quarter, of the full factorial experiment. Figure 14.1b shows a 2^{3-1} fractional factorial design for three factors in $2^{3-1} = 4$ runs. It is a one-half fraction or half-replicate of the full factorial in three factors in 8 runs shown in Figure 14.1a. The remaining vertices of the cube, which comprise the other 2^{3-1} half fraction of the full factorial, could also have been chosen. The 2^{3-1} fractional factorial design indicated by circles in Figure 14.1b corresponds to runs 1, 4, 6, and 7 of Table 14.1, and the other half fraction corresponds to runs 2, 3, 5, and 8.

(a) 2^3 **Full factorial design**	(b) 2^{3-1} **Fractional factorial design**
3 factors at 2 levels, each in 8 runs	3 factors at 2 levels each, in 4 runs
• Design: Includes all combinations of factor levels • Model: main effects, interactions	• Design: Carefully selected fraction of full factorial • Model: main effects, interactions
(c) **Central composite design**	(d) **Box-Behnken design**
3 factors at 5 levels each, in 15 different runs	3 factors in 13 different runs
• Design: Full or fractional factorial, axial points, center points • Model: main effects, interactions, quadratic terms	• Model: Main effects, interactions, quadratic terms

Figure 14.1. *Graphical representation of some experimental designs. In the cubes, the three perpendicular directions (left to right, front to back, bottom to top) represent three factors, A, B, and C. Factor A is at its low level on the left face of the cube, and at its high level on the right face of the cube. The two cubes in (g) correspond to the two levels of a fourth factor D.*

Fractionation of a full factorial leads to confounding, or confusion. Two effects are *confounded*, or *aliased*, if it is not possible to separate their effects. For example, if wafer coating A is always used on machine 1 and wafer coating B is always used on machine 2, it is not possible to separate the effect of the wafer coating from the effect of the machine on the response. Wafer coating and machine are "confounded" with each other. For effects that are confounded, the least squares estimates of the coefficients in the model are identical, so the confounded effects can be written as a sum with one common coefficient. For confounded effects, the estimate of each effect is the sum of the effect in question and the effect(s) that it is confounded with. An *alias chain* gives the set of effects that are confounded with each other in a given experimental design. The alias chain can be derived from the *generators* or from the *defining relations* that specify the confounding pattern of the design. The *resolution* of an experimental design is a measure of the amount of confounding in the design. A design is of *resolution III*, with the main effect of factor A confounded with the two-factor interaction BC, B confounded with AC, and C confounded with AB, if no main effect is confounded with another main effect, but at least one main effect is confounded with a two-factor interaction. A design is of *resolution IV* if no main effect is confounded with a two-factor interaction, but at least one two-factor interaction is confounded with another two-factor interaction. A design is of *resolution V* if no main effect or two-factor interaction is confounded with another main effect or two-factor interaction. The 2^{3-1} fractional factorial design in Figure 14.1b is of resolution III. The resolution of a 2^{k-p} fractional factorial design can be indicated by a subindex, for example 2_{III}^{3-1} or 2_{IV}^{7-3}. For a given number of runs (experimental resources), as the number of factors increases, the amount of confounding increases (the resolution decreases) and the risk of incorrectly identifying an important factor or interaction increases. For a given risk level (amount of confounding, or resolution), as the number of factors increases, the number of runs (experimental resources) increases.

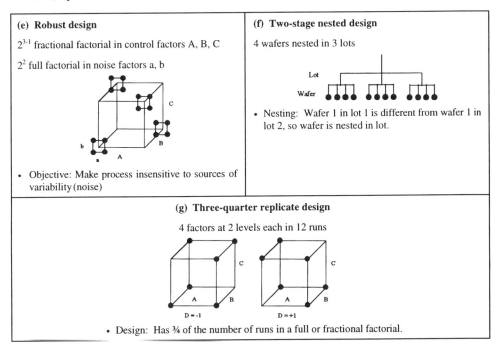

Figure 14.1 (continued).

A *central composite design* in k continuous factors consists of a 2^k full factorial design or a 2^{k-p} fractional factorial design of resolution V, augmented by $2k$ star or axial points, and by center points. A *star* or *axial point* has all except one factor setting at the middle of the factor ranges. The center point (with all factor settings in the middle or their ranges) is usually replicated several times. Figure 14.1c shows a central composite design for three factors in 15 runs. The eight runs at the corners of the cube are the full factorial portion of the design, the six runs "sticking out" of the cube are the axial points, and the run in the middle of the design is the centerpoint. A central composite design can be used to study curvature, and accommodates a second-order model with quadratic terms. A central composite design may be built sequentially. For example, if a 2^k full factorial design has center points that indicate the presence of curvature, the design can be augmented to a central composite design by adding star points to determine the factor(s) which account for the curvature.

Robust design experiments were introduced by the Japanese engineer *Dr. Genichi Taguchi*. The goal of robust design is to make a product or process robust, or insensitive, to sources of variability. A robust manufacturing process will make products that consistently (with reduced variability) meet specifications. A robust design experiment usually has two types of factors, control factors and noise factors. *Control factors* (or *control parameters*, or *design factors*) are factors whose settings are controlled under normal operating conditions. *Noise factors* (or *noise parameters*, or *environmental factors*) correspond to sources of variability that one is unable or unwilling to control under normal operating conditions, but that are manipulated during a robust design experiment. Noise factors can originate in manufacturing (differences in raw materials or equipment), in the environment (particle contamination or customer use), or as a result of deterioration (chemical changes or mechanical wear-out). The goal of a robust design experiment is to find control factor settings that will make the product or process insensitive to sources of noise. The experimental design of a robust design experiment is often a *product array*, or *crossed design*, where an experimental design in the noise factors (*noise array*, or *outer array*) is repeated at every treatment combination of an experimental design in the control factors (*control array*, or *inner array*). Figure 14.1e shows a product array with a 2^2 full factorial in four runs in two noise factors a and b repeated at every treatment combination of a 2^{3-1} fractional factorial in three control factors A, B, and C in four runs. A robust design experiment often considers two quality statistics, the mean and a measure of variability (such as the standard deviation), of all the observations at each treatment combination in the control factors. The measure of variability is used to select the settings of the control factors that reduce variability in the response, and the mean is used to select the settings of the control factors that take the response to its desired value (maximizing it, minimizing it, or taking it to a target value). Taguchi uses *lattices* or *orthogonal arrays* (full factorial and fractional factorial designs presented as tables) to set up the experimental design. The experiments obtained using lattices or orthogonal arrays are (classical) designed experiments.

Additional Experimental Designs

The most frequently used experimental designs are the ones given in the previous section. We will now introduce the following additional topics:
- Response surface methodology for sequential experimentation
- Box-Behnken designs to estimate curvature
- D-Optimal designs for constrained factor spaces
- Nested and crossed designs to define factor relationships
- Foldover designs to resolve confounding
- Three-quarter replicate designs to reduce the number of runs
- Split-plot and strip-plot designs for restrictions on randomization
- Plackett-Burman designs for screening experiments
- Evolutionary Operation (EVOP) for on-line optimization
- Mixture experiments for experiments with mixtures.

Response surface methodology (RSM) is a procedure of sequential experimentation to optimize a response that is a function of several continuous factors. It is a hill-climbing procedure that usually involves a sequence of experiments each of which determines the direction of maximum increase (or decrease) in the response until the local maximum or minimum in the region is reached. The initial designs are usually two-level full or fractional factorials with center points, which are fit with a first-order model. The designs in the general vicinity of the optimum (maximum or minimum) where there may be curvature are usually central composite or Box-Behnken designs, which are used to fit a second order-model with quadratic terms. The models represent *response surfaces* which are often graphed using contour plots, and the experimental designs used to fit them are *response surface designs*. Strictly speaking, a "response surface" can be generated for both first-order and second-order models (see Montgomery 1991, and Box and Draper 1987), although the term "response surface design" is commonly used to refer to "second-order design". A second order design is *rotatable* if a fitted model estimates the response with equal precision at all points in the factor space that are at the same distance from the center of the design.

A *Box-Behnken design* has continuous factors at three levels each and can be used to fit a quadratic model. A Box-Behnken design for three factors in 13 different treatment combinations is illustrated in Figure 14.1d. A Box-Behnken design can be used if there are restrictions on the factor space, for example if the treatment combination at a corner of the cube in Figure 14.1d can't be performed due to restrictions on pump capacity.

The computer-aided *D-optimal designs* are especially useful when there are restrictions on the factor space (irregular shape), or on sample size. For example, a corner of the cube in Figure 14.1a may be unavailable for experimentation because of inability to strike a plasma in an etching experiment. The D in D-optimal stands for Determinant, since a D-optimal design minimizes the value of a determinant that measures variability (variance-covariance matrix) in the least squares estimates of the coefficients in the model. The points in the experimental design are selected from a set of candidate points using a computer algorithm. The disadvantages of using a D-optimal design are that it depends on a proposed *a-priori* model, it depends on the candidate set of points, and it is not always orthogonal so that the analysis may not be as simple and intuitive as in other standard experimental designs.

One factor is *nested* in another factor if the levels of the first factor are similar but not identical at different levels of the second factor. For example, wafers are nested in lots because wafer number 1 in lot A is different from wafer number 1 in lot B. A design with nested factors is a *nested* or *hierarchical design*. Figure 14.1f illustrates a two-stage nested design with four wafers nested in three lots. Two factors are *crossed* if the levels of the first factor are identical at different levels of the second factor. For example, temperature is crossed with pressure because a temperature of 440°C at 0.8 Torr is identical to a temperature of 440°C at 4.0 Torr. All cubes and squares in Figure 14.1 correspond to crossed factors.

A design can be *folded over* to resolve confounding. A design with factors at two levels is folded over by repeating every run with all the levels of all the factors reversed (high levels are set to low levels, and low levels are set to high levels). A *foldover design* consists of the original design and the folded over design. Folding over a design of resolution III (where main effects are confounded with two-factor interactions) gives a foldover design of resolution IV or higher (where main effects are NOT confounded with two-factor interactions). Folding over a design of resolution IV does not necessarily lead to a design of resolution V. When the design of resolution III of Figure 14.1b is folded over, the design in the remaining vertices of the cube is obtained. The foldover design is the full factorial design with no confounding shown in Figure 14.1a. A design is *semifolded* on one factor by reversing the levels of that factor only.

A *three-quarter design* (or *three-quarter fractional factorial design*, or *three-quarter replicate design*), is a carefully selected three-quarter fraction of a full or fractional factorial design. It is convenient to use a three-quarter fractional factorial design when there are insufficient resources to perform all the runs in a full or fractional factorial design, or in order to add a reduced number of runs to a fractional factorial design to resolve confounding issues. The model corresponding to a three-quarter fractional factorial design can accommodate main effects and interactions. Figure 14.1g illustrates a three-quarter fractional factorial design for four factors in twelve runs. The design is of resolution V because it can be used to estimate all main effects and two-factor interactions without confounding with other main effects and two-factor interactions. These effects can also be estimated from the 2^4 full factorial in sixteen runs, but not from the 2_{IV}^{4-1} half fraction in eight runs. Thus the three-quarter fractional factorial design represents an approach to reduce the experimental resources required to estimate the effects.

Split-plot or *split-unit designs*, *split-split-plot* or *split-split-unit designs*, and *strip-plot* or *criss-cross designs* are experimental designs such as full and fractional factorial designs which have restrictions on randomization due to the way in which the experiment was performed. For example, the order of the runs in the experiment in Table 14.1 was not fully randomized because the factor temperature is hard to change. The four runs at the low temperature were performed at the beginning of the experiment, and the four runs at the high temperature were performed at the end of the experiment. The effect of temperature does not include the variability due to changing temperature settings. The data from split-plot and strip-plot designs can be analyzed graphically. A properly done numerical analysis of variance (ANOVA) of a split-plot or strip-plot design takes into account the restrictions on randomization to give estimates of variability that are used to assess the statistical significance of the effects.

Plackett-Burman designs are fractional factorial designs that can be used to study up to $k = N - 1$ factors at two levels each in N runs, where N is a multiple of 4. They are designs of resolution III that are used for screening experiments. Each main effect is confounded with fractions of several two-factor interactions, and conversely, the confounding due to each two-factor interaction is partitioned among several main effects. To remove confounding of main effects with two-factor interactions, a Plackett-Burman design (resolution III) can be folded over to obtain a foldover design of resolution IV. A Plackett-Burman design is a 2^{k-p} fractional factorial design when the number of runs N is a power of 2.

Evolutionary Operation (EVOP) is a method of sequential experimentation that can be used in on-line optimization of a manufacturing process. Usually the changes in the parameters are small enough so that the process continues producing good product, but large enough so that an improvement can be detected.

In a *mixture experiment*, the response depends on the relative proportions (rather than the amounts) of the components that make up a mixture, and the factors are subject to the restriction that the proportions (or percentages) add up to 1 (or 100%). For example, in a chemical bath consisting of two chemicals A and B, two factors can be the proportions, or percentages, of each chemical. The response does not depend on the total amounts of chemicals (a bath of one liter of chemical A and two liters of chemical B has the same chemical properties as a bath of ten liters of chemical A and twenty liters of chemical B) but on their relative proportion (1:2).

Some terms in experimental design refer to specific experimental layouts (such as a 2^{5-1} fractional factorial design), while other terms refer to methodology or objectives. Robust design alludes to the objective of making the response robust, or insensitive, to variability, and different experimental designs can be used for this purpose. Response surface methodology and evolutionary operation are methods of sequential experimentation that can use different experimental designs. Any experimental design can exhibit a split-plot structure because of the way the experiment was performed.

Advantages of Using Designed Experiments

The experimental designs we have described have a number of desirable properties, some of which we now describe.

- *Require reduced resources for the amount of information obtained.* Design of experiments is an efficient experimentation methodology to study the effect of several factors simultaneously on one or more responses.
- *Hidden replication.* Each observation may be used to estimate several effects. For example, all the observations are used to estimate each main effect and each interaction in 2^k full factorial and in 2^{k-p} fractional factorial designs. For the 2^3 full factorial experiment given in the first eight runs of Table 14.1 and represented graphically in Figure 14.1a, the main effect of each factor is the average of four estimates of the effect of that factor, and the average involves all eight observations.
- *Increased precision.* Effects are estimated with increased precision (reduced variability) thanks to hidden replication. If the variance of one observation is σ^2, the variance of the average of n observations is σ^2/n, which is smaller (more precise). In 2^k full factorial and 2^{k-p} fractional factorial designs, the main effects and interaction effects are estimated using averages involving all the observations, so that the estimates have higher precision.
- *Large region of validity.* The results of the experiment are valid over the entire factor space. A model can be used to predict the response at any point in the factor space, even at a point where a run has not been performed.
- *Projection property.* If a factor turns out not to be important (statistically significant), the designed experiment can be projected into another designed experiment in the remaining factors by not considering the unimportant factor. The new design is analyzed to study the remaining factors with higher precision. For example, the single runs in the cube in Figure 14.1a are projected into replicated runs on squares (full factorials in two dimensions) towards the back, the side, and the bottom of the cube. The projection of Figure 14.1b results in one run on each square. In general, a full factorial projects into a full factorial, and a fractional factorial projects into a full or a fractional factorial. The projection of the three-

dimensional central composite design of Figure 14.1c in any one of the three directions results in a two-dimensional central composite design.
- *Analysis at individual factor levels.* In 2^k full factorial and 2^{k-p} fractional factorial designs, the runs at one level of a factor correspond to a designed experiment in the remaining factors. This means that if, for example, all the runs at the high temperature gave poor results, the runs at the low temperature can be analyzed as a designed experiment in the remaining factors.
- *Orthogonality.* Full factorial designs and 2^{k-p} fractional factorial designs are orthogonal with each level of one factor appearing the same number of times with each level of another factor, for all pairs of factors. Orthogonal designs enjoy the following properties:
    * *Simple analysis.* The data can be analyzed using averages, which leads to an analysis that is simple, intuitive, easy to understand, and easy to present to others. For example, the effect of factor A is the difference of the average of all the observations at the high level of A, minus the average of all the observations at the low level of A.
    * *Independent estimates of effects.* The estimate of one effect (which corresponds to a term in the model) does not change if other terms are added or removed from the model. For example, if the AB interaction turns out not to be important (significant), it can be removed from the model without changing the estimates of the effects corresponding to the remaining terms in the model. The estimates of the main effects and of the two-factor interactions will be the same whether the three-factor interaction is included in or excluded from the model.
    * *Effects can be isolated.* Even though many factors are varied simultaneously, the effect of other factors "cancels out" when considering each individual effect. This is the reason the estimates of the effects are independent.

Practical Considerations

Some considerations regarding data collection and sampling are:
- Select a representative sampling plan. Give adequate representation to different areas of a wafer, regions in a furnace, types of films, etc.
- Consider including runs at the current operating conditions.
- Consider replicating the entire experiment or some runs (such as center points), to obtain an estimate of "pure error", or natural variability. The estimate of natural variability can be used to determine the statistical significance of the factors, and to construct confidence intervals.
- Decide on run order. Perform the runs in random order to protect against unknown or unmeasured sources of possible bias. Place center points evenly spaced throughout the experiment to monitor drifts in the process.
- Make plans for collecting the data, and make clear data collection worksheets.
- Perform the experiment under homogeneous conditions. For each experimental condition, use wafers from the same lot or lots, perform the experiments on the same day or week, use the same gauge to measure each response, have the same person perform all the runs. If the experiment can't be performed under homogeneous conditions, use blocking to improve the precision of the estimates of the effects.

If something can go wrong, it often will. Some practical considerations when planning an experiment are:
- *Keep the experiment as simple as possible.* The results will be easier to analyze, and you reduce the risk of complications.
- Check the performance of the measurement equipment using gauge studies (Part 1 of the book) before taking measurements.
- Check that the process is not drifting, for example by performing a passive data collection (Part 2 of the book) or marking some preliminary runs.
- Check that all the runs are feasible. If runs are lost or modified, desirable properties of the design such as orthogonality may be lost, which may complicate the analysis.
- Allow time for unexpected events.
- Obtain buy-in from affected parties, assign duties to personnel, and review results.
- Ensure that operator judgement or unconscious bias doesn't influence the results through treatment, handling, measurement, or otherwise.
- Preserve the raw data. Raw data can be used to identify patterns in the data, and outliers. Outliers can have a strong influence on summary statistics such as the mean and the variance.
- Reset the equipment to its original state after the experiment. This is critical for production equipment.
- Record everything that happens.

Experimental Designs in the Case Studies

The case studies illustrate a wide range of concepts and issues related to planning an experiment, some of which will be mentioned in this section. The following experimental designs were used: full factorial, fractional factorial, robust, central composite, split-plot, foldover, and three-quarter fractional factorial designs.

Buckner et al. (Chapter 15) used response surface methodology to eliminate TiN peeling during tungsten deposition in a Chemical Vapor Deposition (CVD) process. They performed three designed experiments and a confirmation experiment. The first experiment was the 2^3 full factorial in three factors with a center point given in Table 14.1, the second experiment was a central composite design in two factors, the third experiment had one factor at three levels, and the confirmation experiment was a sixteen-run passive data collection. Prior empirical information and process theoretical knowledge of the process were used to select the factors and their ranges, as well as the number of factor levels. The experimental conditions of the first experiment were made as homogeneous as possible by using wafers from the same lot, the same two (of six) wafer positions in the reactor, and by performing all runs in one shift. For practical reasons, the runs were grouped by temperature, giving a split-plot structure to the experimental design.

Buckner et al. (Chapter 16) used a screening experiment to study within-wafer thickness uniformity during tungsten deposition in a Chemical Vapor Deposition (CVD) process. The experiment described in this chapter, the first of four used to improve the process, is a sixteen-run 2_{IV}^{7-3} fractional factorial design with three center points. Five two-factor interactions with one particular factor turned out to be significant, so the design was folded over to estimate them. This gave sixteen additional runs, and three more center points were used to check for possible process drift. Chemical reactions may operate in a "kinetic regime" or in a "feed rate limited regime." Different factors dominate the reaction in each regime, and using a factor region that crosses regimes results in models with spurious interactions. The factor region was selected so the reaction would operate only in the kinetic regime.

Barnett et al. (Chapter 17) used a three-quarter fractional factorial design to improve uniformity in an oxide etch process during the conversion of a vapor phase etch tool to a larger wafer size. Initially, the etch process was characterized using a sixteen-run 2_{IV}^{6-2} fractional factorial experiment, with two center points. Nine interactions appeared in significant alias chains, but it was not possible to single out the significant individual interactions due to confounding. To resolve the confounding, a three-quarter fractional factorial design was used to add eight more runs by semi-folding on one factor. The usual method of adding another sixteen-run quarter fraction to resolve the confounding would have required almost twice the resources, which were unavailable to the engineer.

Freeny and Lai (Chapter 18) used a full factorial experiment to study the removal rate and uniformity of a chemical mechanical wafer planarization process. The 3×5 full factorial design consisted of all fifteen combinations of a three-level factor and a five-level factor. Since both factors had more than two levels each, they were parameterized using *orthogonal polynomials*, which allowed decomposing and quantifying the effect of each factor in linear, quadratic, etc. components. A known tendency of the polisher removal rate to drift lower over time was incorporated into the study by ordering the treatment combinations appropriately (the linear component of time drift was confounded with a high order interaction). This allowed estimation of the effects of the design parameters independently of the time drift, and quantification of the drift in time. To minimize differences between wafers, all wafers came from one lot.

Preuninger et al. (Chapter 19) performed a robust design experiment to optimize a process for etching polycrystalline silicon gates. A process recipe obtained from a previous fractional factorial screening experiment significantly improved line-size uniformity, but unfortunately it tended to undercut. The robust design experiment was performed to characterize the parameter region around the new recipe in order to find conditions that would preserve the good line-size uniformity and avoid the undercut. It was desired to find values of the processing factors (control factors) that would minimize the line-size variability, making it insensitive to circuit design characteristics (noise factors). The robust design experiment is a product array, with a 2^2 full factorial in the noise factors at each treatment combination of a central composite design in three control factors with four center points. The central composite design is illustrated in Figure 14.1c.

Cantell et al. (Chapter 20) performed a robust design experiment to optimize the wafer stepper alignment system used in a photolithographic process. The goal was to maximize the signal strength while making the process robust, or insensitive, to the manufacturing steps used by equipment customers before the photolithographic process. The experimental design is a product array, with a 2^4 full factorial in four noise factors (related to previous manufacturing steps) performed at each one of the eight treatment combinations of a 2^3 full factorial in three control factors for the photolithographic process. The experimental design has a split-plot structure due to restrictions on randomization from the way in which the wafers were processed.

Designed experiments were also used in other parts of the book. Lewis et al. (Chapter 27) used a robust design experiment with a split-plot structure to improve equipment reliability. Lynch and Markle (Chapter 7) used a central composite design as a nine-point sampling plan on a wafer. Mitchell et al. (Chapter 5) used two related nested designs to estimate gauge repeatability in destructive testing. In two passive data collections, Kahn and Baczkowski (Chapter 10) and Czitrom and Reece (Chapter 8) used full factorial experimental designs, and the methods of data collection resulted in split-split-plot and split-plot structures respectively. Both analyzed the data graphically before using analysis of variance.

CONDUCTING THE EXPERIMENT

Care should be taken to perform the experiment as planned in order to avoid losing desirable properties of the design and to avoid complicating the subsequent data analysis. If the data for an entire treatment combination is lost, try to perform it again. Record additional information that might be useful when analyzing the experiment, such as ambient particle count and equipment malfunction. Factors that cannot be controlled during the experiment, but that can be measured, can be included in the analysis as *covariates*.

ANALYZING THE DATA

We will now consider the analysis of data from a statistically designed experiment. The first section presents graphical and numerical techniques for analyzing the data, and the second section describes how these techniques were used in the case studies. The glossary at the end of the book contains additional information on graphical tools, analysis of variance (ANOVA), and selected statistical terms.

Graphical and Numerical Data Analysis

Analyzing data from a designed experiment in different ways gives different insights. The raw data as well as statistics computed from the raw data can be used as responses. Two statistics are often used as responses: the mean and the variance (or standard deviation, or non-uniformity) of all the observations at each treatment combination. The mean gives a measure of the location or central tendency of the observations, and the variance gives a measure of the variability or spread of the observations. Variance components, such as site-to-site variance and wafer-to-wafer variance, can also be considered as responses. The mean and the variance(s) can be evaluated for different outputs such as thickness, sheet resistance, and reflectivity.

Graphical analysis is a powerful, intuitive, and easy way to understand the results of a designed experiment and to present the results to others. The first step in the analysis of a designed experiment should be *exploratory data analysis* of the raw data (before computing averages and variances) to detect outliers and to discover interesting properties that can direct subsequent analysis. The raw data can be examined as a function of run number, factor, wafer site, wafer type, furnace run, location in furnace, chamber, lot, or whatever variables are part of the data. Box plots, wafer maps, and scatter plots are especially useful in this initial analysis.

Graphs of particular importance for full and fractional factorial designs are box plots by treatment combination (run), and main effects and interaction plots. These graphs are intuitive and easy to understand, and can provide enough information for significant process improvement without additional analysis. A *box plot* of the raw data by treatment combination reveals outliers, depicts the variability between treatment combinations and the variability within each treatment combination, can help identify the factor settings that minimize (or maximize) the response, and can help identify the factor settings that minimize (or maximize) the variability in the response. Main effects and interaction plots depict the effect of each factor (magnitude and direction) on the response, and can be used to identify the factors that have the largest and the smallest effects on the response. A *main effects plot* for a given factor gives the average of all the values of the response at each level of the factor, with the averages connected by a straight line. The *interaction plot* between two factors gives the average of the response at each level of one of the factors (connected by straight line(s)), for the different levels of the other factor. To make it easier to compare the effects of different factors on the response, make the scale of the vertical axis (values of the response) the same for all the main effects and interaction plots. The raw data or the values for each run can be added to the main effects and interaction plots. Subsequent numerical analysis can be used to determine the statistical significance of the factors, and to find confidence intervals for the effects of the factors.

We will now describe the main statistical concepts of numerical data analysis, and terms that appear frequently in computer output. We will start with an overview in this paragraph, and will then describe the statistical concepts in greater detail. Numerical data analysis of a designed experiment is used to fit a model to the data, and to predict the response using the model. Model selection is an iterative process which begins by proposing an *a-priori* regression model. The *a-priori* model is fit to the data by estimating the coefficients of the model. The overall fit of the model, and the importance of individual terms in the model, is assessed. Terms can be added to or removed from the model until a satisfactory model is reached. The fit of the model may be improved by transforming the response. Residual analysis is used to check the fit of the model and the assumptions underlying the model. The final model is used to predict the response in order to identify which factors affect the response and how they affect it, to visualize the results using graphs such as contour plots, and to optimize the response. If there are several responses, the best factor settings may require trade-offs between the responses.

Numerical data analysis begins by selecting an *a-priori* regression model, for example, a model with all the terms that can be estimated. The experimental design determines the type of model that can be selected. For example, the model cannot have quadratic terms if all the factors in the experiment are at two levels each. Model selection is an iterative process. The *a-priori* model is assessed, and terms may be added to or removed from the model until a satisfactory model is reached. *Stepwise regression*, or *stepwise variable selection techniques*, are methods to add or remove terms from the model, typically one at a time. It is usually preferable to end up with a relatively simple model.

The model is fit to the data by finding the *least squares estimates* of the coefficients in the model. A *Pareto chart* of the absolute values of the coefficients in the model is useful for identifying the largest coefficients (which correspond to the factors with the largest effects on the response) as well as the smallest coefficients (factors with small effects). The *coefficient of determination* R^2 is a measure of how well the model fits the data, and hence how well the model describes the relationship between the factors and the response. R^2 is the fraction (or the percentage) of the variability in the data that is explained by the regression model. The higher R^2 (i.e., the closer to 1, or 100%), the better the model fits the data. A value of R^2 greater than 0.9 or 0.8 is common in engineering applications. R^2 is usually higher for a model of the mean than for a model of the variance or of the standard deviation, and particle counts often have very low R^2 (particle counts are hard to predict because particles tend to cluster). As the number of terms in the model increases, R^2 gets closer to 1 whether the model is suitable or not. This means it is necessary to strike a balance between including too many terms in the model to obtain a good fit to the data (high value of R^2), and reducing the number of terms in the model to obtain a simpler model. It may be possible to fit a polynomial of degree 19 exactly through 20 points with an R^2 of 1 (with the polynomial oscillating wildly between the points), but a simple line (polynomial of degree one) may give a better model of the trend in the data. If adding a term to the model results in a very small increase in R^2, it may be undesirable to add it. The *adjusted coefficient of determination, R^2 adjusted*, adjusts the value of R^2 to the number of terms already in the model. R^2 adjusted is always less than R^2, and is especially useful (most different from R^2) when the number of terms in the model is almost as large as the number of observations.

A model may have main effects terms, interaction terms, and quadratic terms. A hypothesis test, a t-value, and a p-value are associated with each term in the model. A *hypothesis test* checks whether the coefficient of a term in the model is equal to zero or is different from zero, so that the corresponding term is removed from the model or is kept in the model. If the term is kept in the model, the corresponding factor (if the term is a main effect or a quadratic term) or factors (in an interaction term) affect the response. The *standard error* of the coefficient of a term in the model is the standard deviation of the estimate of the coefficient, which is a measure of the precision with which the coefficient is estimated due to the natural variability of the observations. The *t-value* (or *t-statistic*, or *t-ratio*) of a coefficient is the coefficient divided by its standard error. The t-value can be thought of as a signal-to-noise ratio: if the t-value is much larger than one, the signal is larger than the noise, the coefficient is significantly different from zero (larger than the natural variability, or background noise), so the factor(s) corresponding to the coefficient affect the response. The hypothesis that the coefficient of a term in the model is zero can be tested by comparing the t-value to the values in a table of the t distribution. The *p-value* (or *significance probability*, or *Prob > |t|*) corresponding to a coefficient is the probability that a t-value could be as large or larger than it is just by chance, that is, it is the probability of seeing an effect (coefficient) of that size just by chance if there really is no effect. The p-value and the t-value are equivalent, but using the t-value is more inconvenient because it requires a table of the t-distribution. The p-value is compared to a threshold value such as 0.05 or 0.1. If the p-value of a coefficient is small (less than 0.05, say), the hypothesis that the coefficient is zero is rejected, the coefficient is *statistically significant* (*significantly different from zero*), the term is kept in the model, and the corresponding factor(s) affect the response. If the p-value is large (greater than 0.05, say), the hypothesis that the coefficient is zero is not rejected, the coefficient is not significantly different from zero, and the term is removed from the model. The 0.05 threshold means that there is a 5% chance of making the mistake of keeping a term in the model (saying that the coefficient is different from zero) when in reality it should not be there (the coefficient is zero). Statisticians call this kind of risk the α-risk. Statistical significance is different from *practical significance*. A factor may have a statistically significant effect on the response, but the *magnitude* of the effect on the response may not be of practical importance. Knowledge that a factor does not affect the response can be useful. For example, if changes in temperature do not affect the response, it may be possible to operate at a lower temperature and consume less energy.

Analysis of Variance (ANOVA) is used in the case studies in this part of the book to test the hypothesis that an individual coefficient, or that a group of coefficients, is zero. If a group of coefficients is not zero, it could be because one or more of the coefficients is not zero. The hypothesis test determines whether the corresponding term(s) should be in the model, and hence whether the factor(s) affect the response. Frequently used groups of coefficients correspond to all the terms in the model except the constant term, all the interaction terms, and all the quadratic terms. The hypothesis is tested by comparing the *p-value* in an analysis of variance table corresponding to the coefficient or group of coefficients to a threshold value such as 0.05 or 0.1. If the p-value for a coefficient or group of coefficients is greater than 0.05, the term(s) can be removed from the model. If the p-value for a group of coefficients is less than 0.05, the terms should be kept in the model because at least one of them is statistically significant, and the individual coefficients can be tested to identify the significant one(s). The p-value corresponds to an *F-value*, *F-statistic*, or *F-ratio*. The F-value can also be used to test the hypothesis by comparing it to values in a table of the F-distribution. It is more inconvenient to use the F-value than the p-value to test the hypothesis because it requires an F-table. For a single coefficient with one degree of freedom, the F-statistic is the square of the t-statistic, and the corresponding p-values are identical. If all or most coefficients (or none or almost none of the coefficients) in the model are significant, it is likely that the p-values are too small (or too large) due to an estimate of variability that is too small (or too large). For example, if replicates are taken consecutively without changing the factor settings, the natural variability can be underestimated because it does not include the variability of changing factor settings, and too many terms in the model may be statistically significant. Having too many or too few significant terms in the model for a designed experiment is similar to having too many or too few points out of control in a control chart: both can be due to an incorrect estimate of variability.

A *normal probability plot* of the coefficients in the model can also be used to visually identify significant coefficients as the ones that are large in magnitude and fall far from a straight line passing roughly through the points close to zero. A half-normal probability plot is similar and uses the absolute values of the coefficients in the model. The normal and half-normal probability plots are especially useful when there are at least ten effects. When no effects are likely to be significant, the slope of a line through the group of effects near zero gives an estimate of the standard error. When there are large effects, an estimate of the standard error can be obtained from the slope after the large effects are removed.

Model selection is an iterative process. The fit of the model may be improved by adding terms to the model, or by removing terms from the model. The fit of the model may also be improved by *transforming* the observations, and by using the transformed observations as a new response. Transformations are used for a variety of reasons. One reason is to obtain a simpler model by eliminating the need for interaction or quadratic terms. Another reason is to simplify the model when the largest observation is over 5 or 10 times as large as the smallest observation, in which case a compressive transformation (such as taking the logarithm or the square root of the values of the response) is often appropriate. To avoid a model of the variance (or of the standard deviation) that predicts negative values, one can take the logarithm of the values of the variance. Also, the logarithm of the variance is more nearly normal (one of the model assumptions) than the variance itself. Another reason to transform the response is to obtain a more nearly equal variance σ^2 at all the treatment combinations, which is one of the model assumptions.

Residual analysis uses residuals to assess how well the model fits the data by detecting *outliers* (observations that fall far from the value predicted using the model), assessing the need to transform the response, assessing the need to add terms to the model, and by testing the model assumptions. The *residual* e=y-ŷ corresponding to an observation is the difference between the actual observation y and the estimated value ŷ given by the model. A *studentized residual* (residual divided by its standard deviation) normalizes the residual by giving the distance of the observation to its estimated value in standard deviation units (a studentized residual of 2 means the observation is two standard deviations away from its estimated value). Before accepting the final model, residual analysis is used to check the assumptions underlying the model, namely that the observations are independent, identically distributed with a normal distribution with mean 0 and constant variance σ^2. We will now describe some graphical techniques of residual analysis. The observations are assumed to be normally distributed if the histogram of the residuals is approximately mound-shaped and symmetric; outliers will be far from the bulk of the residuals, and a skewed distribution may be dealt with by transforming the response. Another test for the normality of the observations is to check whether the residuals fall approximately on a straight line on a normal probability plot (outliers will fall far from the line). Unusual trends in the data may be identified by patterns in a plot of the residuals in chronological order. The observations are assumed to be independent of the magnitude of the response if there is no pattern in the plot of the residuals against the predicted values of the response; if there is a pattern, it might be appropriate to add terms to the model or to transform the response. If the observations have a constant variance, a plot of the spread in the residuals is the same at the different levels of each factor, and there is no pattern (for example, no funnel-shape) in the plot of the residuals as a function of time.

Once the final model is selected, it is used to predict the response, to determine which factors affect the response and how they affect it, to visualize the effect of the factors on the response, to optimize the response, and to do trade-offs if there are several responses. Main effects and interaction plots, which are intuitive and easy to understand, are the major graphical analysis tools for full and fractional factorial designs. Contour plots and three-dimensional graphs are the main graphical analysis tools for second-order (RSM) designs. Contour plots are especially useful for making trade-offs among several responses by overlapping all the responses in one contour plot.

Data Analysis of the Case Studies

The case studies illustrate techniques of data analysis that range from graphical displays of many types, models of the responses, analysis of variance to test effects for significance, transformations, residual analysis, and contour plots using the models for the responses. The remainder of this section highlights both straightforward and unusual applications of the analysis methods appearing in the case studies.

Buckner et al. (Chapter 16) used many different graphical techniques: time charts and histograms, Pareto charts to identify important effects and interactions, interaction plots, plots of residuals against fitted values, wafer maps (contour plots of the response over the surface of a wafer), contour plots of the response as a function of two process factors, and a Gantt chart of a project plan.

Freeny and Lai (Chapter 18) analyzed the data graphically with box plots, and then used analysis of variance techniques to quantify the effects of the factors. A plot of residuals against time indicated a slight drift.

Buckner et al. (Chapter 15) found it necessary to make trade-offs between responses in experiment 2. This was done by using overlapping contour plots of the models of two continuous responses and adding a discrete response (presence or absence of a key defect) by circling the treatment combinations of the experiment where it occurred. They give the least squares estimates of the coefficients of the polynomial models of the continuous responses.

The case studies use analysis of variance to give the significance of factors and interactions. Cantell et al. (Chapter 20) use several analyses of variance to analyze the data in different ways.

Barnett et al. (Chapter 17) illustrate the results of performing stepwise regression to add terms to the model. They include contour plots and interaction plots that illustrate the equivalence of the results obtained from both. They also use a logarithmic transformation to reduce the range of the numerical values of a response.

To fully characterize the parameter space, Preuninger et al. (Chapter 19) explicitly considered the effects of two noise factors and their interactions with control factors. Noise factors are usually not considered explicitly in robust design experiments, where the observations at a treatment combination in the control factors are treated as "replicates" even when they arise from different values of the noise factors.

REACHING CONCLUSIONS AND IMPLEMENTING RECOMMENDATIONS

It is important to confirm the results of an experiment before making process changes, especially in a manufacturing line. A *confirmation experiment* will validate the conclusions drawn from a designed experiment. A confirmation experiment can consist of a few runs (for example, at the predicted best factor settings, at the original operating conditions, and at the best treatment combination), a passive data collection, or even another designed experiment varying the factors by $\pm$ 10% around their optimal settings to study sensitivity of the response to changes in the factors. The long-term confirmation that a process meets requirements is that it performs adequately in manufacturing.

The case studies illustrate different types of confirmation experiments. Buckner et al. (Chapter 15) performed a sixteen-run passive data collection to confirm the results of the study before transferring the process to manufacturing. Preuninger et al. (Chapter 19) verified process improvement on the manufacturing line.

The results of the experiment and the conclusions drawn from it should be reported in an effective manner. A graphical presentation of results is highly effective, especially when presenting the results to other people. A technical memorandum will prevent loss of the information. To be effective, recommendations should not only be made, they must also be implemented.

INTRODUCTION TO CASE STUDIES

This section gives a brief overview of the applications of design of experiments to engineering problems provided by the case studies in this part.

Chapter 15. Elimination of TiN Peeling During Exposure to CVD Tungsten Deposition Process Using Designed Experiments. *James Buckner, David J. Cammenga, and Ann Weber.* This case study describes the use of designed experiments to eliminate TiN peeling during a CVD tungsten deposition process. The peeling occurred when a reactor was converted from 150 mm to 200 mm diameter silicon wafer capability. A project deadline crisis was averted by using a sequence of four designed experiments to develop a new process and to confirm that it met requirements, in less than a week and a half. The experiments were a 2^3 full factorial with a center point, a central composite design in two factors, one factor at three levels, and a sixteen-run passive data collection (Part 2) to confirm that the new process for 200 mm wafers was as good as the previous process for 150 mm wafers for all responses except stress, which was intentionally compromised. An excellent introduction to the process describes the use of prior empirical and theoretical process knowledge to select the factors and their levels.

Chapter 16. Modeling a Uniformity Bulls-eye Inversion. *James Buckner, Richard Huang, Kenneth A. Monnig, Eliot K. Broadbent, Barry L. Chin, Jon Henri, Mark Sorell, and Kevin Venor.* One common contributor to loss of yield in integrated circuit fabrication is poor uniformity across a wafer, for example due to a radial or bulls-eye pattern. This case study describes how a designed experiment was used to characterize and improve within-wafer thickness uniformity in the deposition of a thin film of tungsten using Chemical Vapor Deposition (CVD). A very good model of within-wafer uniformity was obtained by recognizing that the bulls-eye pattern of uniformity on the wafers had different directions, with the center of the wafer sometimes having the highest uniformity and sometimes the lowest uniformity. This bulls-eye inversion was accounted for by assigning a positive sign to the uniformity when it was lower in the middle of the wafer than at the edge, and by assigning a negative sign to the uniformity when it was higher in the middle of the wafer than at the edge. The factors and the region for the experiment were found by using previous empirical and theoretical knowledge of the process, to make sure the process operated entirely in the "kinetic regime" and did not cross over to the "feed rate limited regime". The study used a 2_{IV}^{7-3} screening experiment with three center points, which was folded over to estimate five confounded interactions. Hurwitz and Spagon (Chapter 9) use a different approach in terms of contrasts to model patterns across the wafer.

Chapter 17. Using Fewer Wafers to Resolve Confounding in Screening Experiments. *Joel Barnett, Veronica Czitrom, Peter W. M. John, and Ramón V. León.* Rising costs make the possibility of saving resources during experimentation increasingly attractive. This case study describes how resources were saved during the conversion of a vapor phase etch tool for use with 200 mm wafers. The etch process was initially characterized using a 2_{IV}^{6-2} fractional factorial experiment in sixteen runs. Several interactions, that were confounded with other interactions, turned out to be significant. To be able to tell which of the interactions were the important ones, eight additional runs were selected to obtain a three-quarter fractional factorial design. The customary method would have added another quarter fraction of sixteen runs (which in this case would not have been available) instead of the eight that were used in this experiment.

Chapter 18. Planarization by Chemical Mechanical Polishing: A Rate and Uniformity Study. *Anne E. Freeny and Warren Y.-C. Lai.* Wafer planarity is becoming increasingly impor-tant in the semiconductor industry as the technology strives for smaller feature sizes and larger wafers. In this case study, a designed experiment is used to characterize a chemical mechanical polishing process for wafer planarization. The goal is to find the maximum (to maximize throughput) rate of oxide removal which can be used without degrading the uniformity of the removal on the wafer surface. A trade-off between removal rate (throughput) and uniformity was found to be necessary. The experiment was a 3×5 full factorial which incorporated a previously noted tendency of the polisher removal rate to drift lower over time. The effects of the design parameters were estimated independently of the time drift, and the time drift was quantified. The time drift, combined with large wafer-to-wafer variability, results in imperfect process reproducibility, even with automation. This lack of reproducibility is an important factor in judging the cost and timing of introducing chemical mechanical polishing into manufacturing.

Chapter 19. Use of Experimental Design to Optimize a Process for Etching Polycrystalline Silicon Gates. *Fred Preuninger, Joseph Blasko, Steven Meester, and Taeho Kook* One of the most important characteristics of integrated circuits is speed, which is partially determined by the physical size and profile of the gate after etching. A robust design experiment is used to optimize the average etch of a polycrystalline silicon gate etching process. A process that significantly improved line-size uniformity was identified in a previous screening experiment, but unfortunately it tended to undercut. A robust design experiment was performed to characterize the parameter region around the new recipe in order to find conditions which would achieve the line-size control of the new process, without the undercut. The robust design experiment is a product array with a central composite design in three control factors and a full factorial in two noise factors. After identifying trade-offs that gave a practical solution, optimal processing conditions that avoided undercutting were identified. The improvement using the new processing conditions was confirmed by implementation in the manufacturing line.

Chapter 20. Optimization of a Wafer Stepper Alignment System Using Robust Design. *Brenda Cantell, José Ramírez, and William Gadson.* Photolithography is a key process in semiconductor manufacturing that limits the continued increase in wafer size and the reduction of feature size on wafers. Robust design is used in this case study to optimize a wafer stepper alignment system for use after wafer multi-level metal processing. The goal is to make the alignment system robust, or insensitive, to variation in the manufacturing steps preceding the alignment operation. This is accomplished by looking for settings of the controllable alignment system factors that maximize the signal strength while making it robust to manufacturing process variability. The results of the experiment indicated that a trade-off was required between increasing the signal strength and minimizing variation in the response. The experimental design is a product array in a 2^4 full factorial in four noise factors at each level of a 2^3 full factorial in three control factors, with a split-plot structure due to the manner in which the wafers were processed.

SUGGESTED READING

All SEMATECH statistical publications that are available to the general public can be accessed from the following Internet URL: http://www.sematech.org/

Chapters on statistical design of experiments are contained in the following books:

Montgomery, Douglas C. *Statistical Quality Control*, second edition, New York: John Wiley and Sons, Inc., 1991, Chapters 11 and 12.

Ryan, Thomas P. *Statistical Methods for Quality Improvement*, John Wiley and Sons, Inc., 1989, Chapters 13, 14, and 15.

John, Peter W.M. *Statistical Methods in Engineering and Quality Assurance*. New York: John Wiley and Sons, Inc., 1990, Chapters 17, 18, and 19.

Statistical design of experiments is covered in the following books:

Box, George E.P., Hunter, William G., and Hunter, Stuart J. *Statistics for Experimenters*, New York: John Wiley and Sons, 1978.

Montgomery, Douglas C. *Design and Analysis of Experiments*, fourth edition, New York: John Wiley and Sons, Inc., 1997.

Mason, Robert L., Gunst, Richard F., and Hess, James L. *Statistical Design and Analysis of Experiments with Applications to Engineering and Science*, New York: John Wiley and Sons, Inc., 1989.

Milliken, George, and Johnson, Dallas. *Analysis of Messy Data, Volume 1: Designed Experiments*, New York: Van Nostrand Reinhold, 1984.

Case studies illustrating the application of design of experiments to different industries are given in:

Snee, Ronald D., Hare, Lynne B., and Trout, J. Richard. *Experiments in Industry. Design, Analysis and Interpretation of Results*, Milwaukee, WI: American Society for Quality Control, 1985.

For further reading on response surface methodology, see:

Cornell, John A. *How to Apply Response Surface Methodology*, Milwaukee, Wisconsin: American Society for Quality Control, 1990. Volume 8 of the ASQC Basic References in Quality Control: Statistical Techniques.

Box, George E.P. and Draper, Norman R. *Empirical Model Building and Response Surfaces*, New York: John Wiley and Sons, 1987.

Khuri, Andre I. and Cornell, John A. *Response Surfaces, Designs and Analyses*, New York: Marcel Dekker, and Milwaukee, WI: American Society for Quality Control, 1987.

Two articles and a book on robust design are:

Kacker, Raghu N. "Off-Line Quality Control, Parameter Design, and the Taguchi Method," *Journal of Quality Technology*, Vol. 17, pp. 176-209, 1985.

Nair, Vijayan N., editor. Taguchi's Parameter Design: A Panel Discussion, *Technometrics*, Vol. 34 No. 2, pp. 127-161, 1992.

Phadke, Madhav S. *Quality Engineering Using Robust Design*, Englewood Cliffs, NJ: Prentice Hall, 1989.

An article and two books describing split-plot designs are:

Box, George E.P., and Jones, Stephen P. Split-plot Designs for Robust Product Experimentation, *Journal of Applied Statistics*, Vol 19, No. 1, pp. 3-26, 1992.

Mason, Robert L, Gunst, Richard F., and Hess, James L., *Statistical Design and Analysis of Experiments with Applications to Engineering and Science*, New York: John Wiley and Sons, Inc., 1989.

Milliken, George A., and Johnson, Dallas E. *Analysis of Messy Data*, New York, NY: Chapman & Hall, 1992.

A book on mixture experiments is:

Cornell, John A. *Experiments with Mixtures*, New York, NY: Wiley, 1990.

Regression analysis and model fitting is described in the following book:

Draper, Norman and Smith, Harry. *Applied Regression Analysis*, second edition, New York: John Wiley and Sons, 1981.

A classic advanced book on design of experiments is:

John, Peter W. M. *Statistical Design and Analysis of Experiments*, Philadelphia: SIAM, 1998

CHAPTER 15

ELIMINATION OF TiN PEELING DURING EXPOSURE TO CVD TUNGSTEN DEPOSITION PROCESS USING DESIGNED EXPERIMENTS

James Buckner, David J. Cammenga, and Ann Weber

Peeling of the incoming TiN adhesion layer upon exposure to the CVD tungsten reaction environment was eliminated by altering the tungsten deposition process conditions. Tradeoffs among TiN peeling, tungsten film uniformity, and tungsten film stress were identified and characterized. TiN peeling was found to be controlled by the H_2/WF_6 ratio, an unexpected result. A project deadline crisis was averted by using response surface methodology (RSM) to guide in the sequential application of design of experiments (DOE), where factor and range definition were constrained by physical theoretical understanding of tungsten reaction kinetics.

EXECUTIVE SUMMARY

Problem

Conversion of a Genus 8720 CVD tungsten deposition reactor from 150 mm diameter wafers to 200 mm diameter wafers was impacted by decomposition and peeling of the incoming titanium nitride (TiN) adhesion layer upon exposure to the reactive environment surrounding the tungsten deposition (see Cammenga, 1991). The 200 mm deposition process recommended by the supplier produced acceptable film uniformity, but the incoming furnace reacted TiN adhesion layer was chemically incompatible with the reactive environment. The 150 mm process-of-record prevented TiN peeling, but at the expense of unacceptably high film nonuniformity on 200 mm wafers. TiN produced by reactive sputtering was known to be compatible with the tungsten reaction environment, but it was necessary to continue using the problematic furnace reacted TiN to obtain acceptable contact resistance.

Solution Strategy

It was not possible to harden the incoming TiN adhesion layer due to a tight project deadline. Therefore it was decided to search for an interim tungsten deposition process which would eliminate TiN peeling while simultaneously matching film parameters established by the 150 mm process-of-record.

The experimental strategy was to begin at the 150 mm process-of-record, where no TiN peeling occurred, and to sweep out through the process factor space in search of a region where the other 150 mm film responses could be matched. For time and resource efficiency, design of experiments (DOE) and response surface methodology (RSM) were used. Further efficiency was realized by constraining the selection of factors and ranges using prior empirical information and process theoretical knowledge of the influence of physical considerations on factor settings (see Appendix A). Three experiments and a verification run were performed.

Experiment 1 had three factors and eight treatment combinations, with centerpoints. The process space was broad in temperature and backside argon flow, and moderate in pressure. Results from this experiment were used to set the temperature to 480°C and backside argon to zero flow. Pressure would be included in the next experiment and its range would be extended.

Experiment 2 had two factors and nine treatment combinations, with centerpoints. The process space was broad in both pressure and H_2/WF_6 ratio. The response surface obtained from this experiment revealed tradeoffs among TiN peeling, film stress, and film uniformity. Based on the results of this experiment, the project team made a decision to compromise stress in favor of uniformity and TiN peeling.

Experiment 3 was a single factor experiment using three values of the H_2/WF_6 ratio. The experiment was designed to look for an optimum along the H_2/WF_6 ratio axis in the response surface produced by Experiment 2. The optimum uniformity without TiN peeling was found to be at a H_2/WF_6 ratio of 5. These were the conditions selected as the interim deposition process.

Finally, a passive data collection was performed with sixteen runs to verify that the new process would meet requirements.

Conclusions

All process requirements were met with the exception of stress, which was intentionally compromised.

There exists a tradeoff among TiN peeling, uniformity, and stress. Low pressure is required to lower stress, while uniformity requires high pressure. A low H_2/WF_6 ratio is required for uniformity and to optimize stress at high pressures, but elimination of TiN peeling requires a high H_2/WF_6 ratio. Uniformity can be optimized against TiN peeling by setting the H_2/WF_6 ratio just above the peeling threshold, which is between 4 and 5.

A project deadline crisis was averted by using DOE and RSM in combination with prior empirical findings and theoretical process knowledge. The entire sequence of experiments including verification took less than a week and a half!

PROCESS

After the transistors of an integrated circuit have been built on a silicon wafer, it is necessary to connect them electrically to one another and to the outside world. The connections are manufactured by first depositing an insulating material over the transistors, then opening holes through the insulating material to the transistor connection points. Next, a blanket layer of metal is deposited over the entire wafer, contacting the transistors through the holes in the insulator. Finally, portions of the metal layer are selectively removed, leaving behind the metal "wires" that make the electrical connections.

A number of different metals may be used in this "metallization" process. Of late, tungsten deposited by chemical vapor deposition (CVD) has been preferred. This is because the CVD process excels at throwing enough metal down into the contact holes to make good contact. With the continued miniaturization of integrated circuits, the diameter of the contact holes has been diminished to the extent that the more established metallization processes of sputtering and evaporation, which are controlled by a different physical mechanism, have become marginal in their ability to deposit metal into the contacts. The CVD tungsten process, therefore, is an important contributor to the advancement of integrated circuit manufacturing.

CVD tungsten is not without its problems, however. One disadvantage compared to the older, more established, metallization processes, is that CVD tungsten does not adhere to silicon dioxide, the layer of insulating material that isolates the transistors. Since CVD tungsten will adhere to other metals, though, this problem is typically solved by depositing a very thin layer of another metal onto the insulating layer prior to CVD tungsten deposition. The layer functions as a "priming coat," and it has merited several designations, namely, adhesion layer, nucleation layer, glue layer. It will be referred to as the adhesion layer in this study. Among a number of possible metals, titanium nitride (TiN) has emerged as a preferred adhesion layer material with many semiconductor manufacturers.

The problem that is the subject of this study arose when a CVD tungsten deposition reactor was converted from 150 mm to 200 mm diameter wafers to increase the size of the silicon wafers it could process. The larger wafer size required a change in the deposition process in order to obtain a tungsten layer on the 200 mm wafers that would be comparable to that deposited on 150 mm wafers. However, when this process change was implemented, the TiN adhesion layer material was found to be incompatible with the new process conditions in the tungsten reactor. The incompatibility resulted in chemical decomposition of the TiN layer so that its physical integrity was degraded to the extent that it peeled and flaked away from the wafer.

Three strategies for solving the problem were considered:
1. Alter the process used to make the TiN layer, rendering it more resistant to attack by the tungsten process
2. Make the TiN by an entirely different process, which is known to produce TiN film that resists attack
3. Alter the CVD tungsten process to eliminate TiN attack, while simultaneously obtaining a tungsten film on 200 mm diameter wafers that would be comparable with the tungsten film produced on 150 mm diameter wafers.

All three strategies were pursued in parallel. The first was preferred, but it was unsuccessful. The second was next preferred, but it was incompatible with meeting a scheduled project deadline, leaving the third strategy as the only hope for solving the problem before the deadline. This strategy included not only eliminating TiN peeling, but also maintaining the tungsten film parameters that had been established in the historical process-of-record for 150 mm diameter wafers.

DATA COLLECTION PLAN/ANALYSIS AND INTERPRETATION OF RESULTS

Successful completion of this project required three experiments to find a usable process, and a verification period to demonstrate the adequacy of the new process, giving a total of four separate data collection plans. All four plans were roughed out at the beginning of the project, but the later plans could be detailed only after the output of the earlier plans became available because of the sequential nature of the project strategy. Each plan will be described separately in its completed form.

Because the data collections were planned using statistical design principles, the analyses were straightforward. Since the techniques used to analyze statistically designed experiments are well established,[24] the following sections will focus on interpretation, with figures highlighting pivotal items.

Experiment 1

The design of the first experiment relied heavily on prior knowledge of the CVD tungsten process. Just before the conversion of the reactor to 200 mm diameter wafers, an extensive equipment improvement project (EIP) had been completed for 150 mm diameter wafers with thorough characterization of the process.[25] In addition, (unpublished) experimental data were available from the reactor supplier for the 200 mm diameter wafers.

From this body of information, the project team considered three facts to be of overriding importance to the solution of this problem. First was that no TiN peeling was observed in the 150 mm process-of-record. Second was that the 150 mm process-of-record was known to result in unacceptably high nonuniformity of the tungsten layer across a 200 mm wafer. Third was that acceptable uniformity was known to be attainable at process pressures very much higher than the pressure of the process-of-record.

In addition, there were several other considerations relevant to the design of the experiment. Because the TiN peeling was the result of a chemical degradation, those members of the team with a background in chemistry believed that since temperature and absolute reactant concentration (proportional to process pressure) have a powerful effect on most chemical reactions, they could be major controlling factors for the observed peeling. Another observation was that the peeling occurred only in the perimeter region of the wafer where tungsten deposition was inhibited by a mechanical clamp used to hold the wafer down. Since the space between the wafer and the clamp is flushed by argon from the backside, some members of the team believed that increasing the backside argon flow might control TiN peeling by preventing chemical reactants from reaching the susceptible region. Finally, it was known that another important quality of the tungsten layer, namely, the stress in the material, varies with process temperature and process pressure.

---

[24] See, for example, Box, Hunter, Hunter (1978).
[25] Cammenga (1991), reports the characterization method and results fully in Chapter 7.

All of this prior knowledge was incorporated into the experimental design. The experimental process space would include the process-of-record, and sweep out toward lower temperature, higher pressure, and higher backside argon flow, looking for a region of process space where no peeling and acceptable uniformity could be simultaneously achieved. Stress would be among the list of properties to be measured.

The number of factors was held to three in order to meet a very tight schedule. This was justified based on the large body of characterization data available from the recent equipment improvement project (EIP). The design was set up to consider every combination of three factors at two levels each (full factorial in eight treatment combinations) (see Box, 1978) with three center points using the factors and levels shown in Table 15.1. The run order was randomized except that the runs were grouped by temperature, since this is a difficult parameter to change on the reactor. A two-wafer batch was used for each run. This reactor is capable of running up to six wafers in a batch, but only two were used because of the expense of the wafers. The same two wafer positions in the reactor were used for each run. All runs were made within one shift using incoming wafers all from the same TiN lot. The experimental design and the values of the responses are shown in Appendix B.

Table 15.1. *Factors and Levels of Experiment 1.*

Factor	Low	High
Temperature	440° C	500° C
Pressure	0.8 torr	4.0 torr
Backside Argon	0 sccm	300 sccm

Tungsten layer uniformity was measured using four-point-probe sheet resistance measurements. Forty-nine locations were measured on each wafer in the default standard pattern for this measurement reactor (Prometrix Rs50e), which is a de facto industry standard. The uniformity value was defined as the standard deviation of the forty-nine points divided by their average,[26] and expressed as a percentage. A gauge study performed prior to this work showed a total measurement system variability of 0.04% (1σ).

Stress was determined from an optical measurement of wafer bow using Stoney's equation.[27] A prior gauge study showed a total measurement system variability of 0.07 Gdyn/cm² (1σ).

Presence or absence of TiN peeling was determined by visual inspection.

Running this type of experiment had become routine during the recent equipment improvement program, so that no problems were encountered in setting up the treatment combinations or in measuring the responses. However, only one of the three planned center points was performed due to a lack of wafers (see Appendix B for a table with the raw data).

---

[26] The uniformity response thus has the form of a coefficient of variation, which has been shown to be an inefficient statistic when the relationship between the standard deviation and the mean is nonlinear [see Box (1988)]. However, the denominator of the uniformity (the site average sheet resistance) was held approximately constant (within experimental error) for each treatment combination. Any nonlinear functional relationship between the standard deviation and the mean can be modeled by a linear function over a sufficiently narrow range of the mean. The use of uniformity as a response is therefore partially justified in this case study. For a more detailed discussion of uniformity as a response, see the case study in Chapter 16.

[27] Marcus (1983).

The coefficients for the linear regression models for uniformity and for stress are shown in Tables 15.2 and 15.3, respectively. The models were obtained using stepwise regression beginning from the model with three main effects and three two-factor interactions. Uniformity is improved by increasing pressure and decreasing backside argon, while stress is improved (lowered) by increasing both temperature and pressure.

Table 15.2. *Linear Regression Coefficients of Uniformity Response, Experiment 1. An increase in backside argon degrades uniformity, while an increase in pressure improves it.*

Term	Coeff.	Std. Error	T-Value	Signif.
Average	+6.698889	0.229599		
Pressure	−0.435000	0.243526	−1.79	0.1243
Back Argon	+1.657500	0.243526	+6.81	0.0005

No. cases = 9. R^2 = 0.8919. RMS Error = 0.6888
Resid. df = 6. R^2 adj. = 0.8559. Cond. No. = 1.

Table 15.3. *Linear Regression Coefficients of Stress Response, Experiment 1. Increases in temperature and pressure both decrease stress.*

Term	Coeff.	Std. Error	T-Value	Signif.
Average	+8.938333	0.088301		
Temp	−1.205000	0.093658	−12.87	0.0001
Pressure	−0.358500	0.093658	−3.83	0.0087

No. cases = 9. R^2 = 0.9678. RMS Error = 0.2649.
Resid. df = 6. R^2 adj. = 0.9570. Cond. No. = 1.

A representative contour plot of stress and uniformity, plotted as a function of pressure and backside argon at a temperature of 480°C, illustrates these dependencies (see Figure 15.1). This temperature was chosen for convenience, since other processes on the reactor are run at 480°C, and temperature is relatively difficult to change on the reactor. No TiN peeling occurred in any treatment combination, making the entire process space eligible for defining an improved process. The contour plot shows that the high pressure/low backside argon corner is good in terms of both stress and uniformity, but still neither met the stated goals of reaching the process-of-record results for 150 mm wafers with the new 200 mm wafers (see Table 15.9). This corner process was adopted as an improvement, however, in the interim before the results of Experiment 2 became available.

Experiment 2

The results of the first experiment indicated the need for a second experiment with an increased range of pressure. The range chosen extended from the highest pressure in Experiment 1 (4 torr) up to the pressure used in the supplier recommended process (80 torr).

The H_2/WF_6 ratio was chosen as a second factor based on the observation that there was a large discrepancy between the 150 mm process-of-record and the supplier recommended process at 200 mm, which ran at H_2/WF_6 ratios of 23 and 2.3, respectively. The lower limit of the range was chosen to be 2 in order to include the supplier recommended process. The upper limit was arbitrarily chosen to be 10.

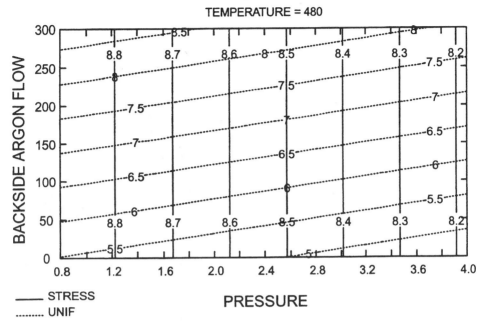

Figure 15.1. *Contour Plot of Stress and Uniformity, Experiment 1. Pressure dominates stress while backside argon dominates uniformity for a fixed temperature.*

It was decided to consider only two factors in this experiment and to use a central composite (inscribed) design, illustrated by asterisks in Figure 15.2. Although the usual empirical approach might have suggested increasing the number of factors and using a linear rather than quadratic design, there were compelling reasons to take the strategy that was chosen. The decision to keep the experiment to only two factors was driven by a theoretical physical argument (see Appendix A) indicating that the H_2/WF_6 ratio should produce curvature in the deposition rate, and that there may be a local maximum in the region. Since a local maximum in the deposition rate implies a local minimum in uniformity, the decision was made to devote the budget of treatment combinations to a search for curvature associated with the H_2/WF_6 ratio rather than to explore other factors and interactions.

The factors and their levels are given in Table 15.4. Sampling and measurement of responses were identical to those described for Experiment 1. The central composite experimental design and the values of the responses are given in Appendix B.

Table 15.4. *Factors and Levels of Experiment 2.*

Factor	Low	High
Pressure	4	80
H_2/WF_6 Ratio	2	10

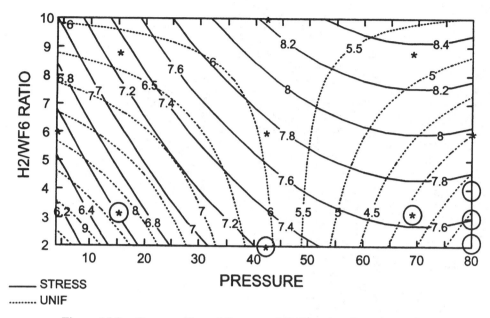

Figure 15.2. *Contour Plot of Stress and Uniformity, Experiment 2. The nine asterisks indicate the central composite experimental design that was used. Circled data indicate all treatment combinations where* TiN *peeling occurred (compiled from both experiments 2 and 3, and the initial supplier recommended process). The contours show tradeoffs between stress, uniformity, and peeling.*

The coefficients of the linear regression models for uniformity and stress are shown in Tables 15.5 and 15.6, respectively. There is an important interaction between pressure and the H_2/WF_6 ratio for uniformity response. The overall model fit is not excellent, at R^2 adj. = 0.814, but this is considered good for uniformity, which is normally a difficult response to model. Stress is also controlled by pressure and the H_2/WF_6 ratio, with significant curvature in pressure. The overall model fit is very good.

Table 15.5. *Linear Regression Coefficients of Uniformity Response, Experiment 2. There is an important interaction between pressure and the H_2/WF_6 ratio.*

Term	Coeff.	Std. Error	T-Value	Signif.
Average	+5.927273	0.185831		
Pressure	−1.912335	0.308142		
H_2/WF_6	−0.224966	0.308142		
Pressure * H_2/WF_6	+1.699487	0.616144	2.76	0.0282

No. cases = 11. R^2 = 0.8695. RMS Error = 0.6163.
Resid. df = 7. R^2 adj. = 0.8136. Cond. No. = 1.

Table 15.6. *Linear Regression Coefficients of Stress Response, Experiment 2. H_2/WF_6 ratio and pressure control stress, with significant curvature in the response to pressure.*

Term	Coeff.	Std. Error	T-Value	Signif.
Average	+7.733551	0.037042		
Pressure	−0.737459	0.044084		
H2/WF6	+0.497877	0.044084	11.29	0.0001
Pressure²	−0.494690	0.070923	−6.97	0.0002

No. cases = 11. R^2 = 0.9849. RMS Error = 0.08817.
Resid. df = 7. R^2 adj. = 0.9784. Cond. No. = 2.364.

TiN peeling was observed in this experiment. Since this was a visual yes-or-no response, a linear regression coefficients table is not shown, but the treatment combinations where peeling occurred are circled in the contour plot of uniformity and stress shown in Figure 15.2. The H_2/WF_6 ratio clearly controls TiN peeling. This was a pivotal finding, and a surprise to the project team, who expected temperature and pressure to be the controlling factors.

The contour plot can be used to do tradeoffs between the three responses. The best uniformity occurs at high pressure, while the best (lowest) stress occurs at low pressure, and both require low H_2/WF_6 ratios. However, TiN peeling occurs at low H_2/WF_6 ratios, rendering this region of process space unusable.

This contour plot served to convince the project team and management that one response would have to be compromised to meet the tight project deadline. Based on this analysis, the decision was made to compromise stress in favor of uniformity.

Experiment 3

The results of Experiment 2 shown in the contour plot of Figure 15.2 indicated that uniformity could be traded off against TiN peeling by setting the pressure at 80 torr and considering several H_2/WF_6 ratios until the peeling threshold was cleared. The third Experiment, therefore, was a single factor experiment in the H_2/WF_6 ratio to find the peeling threshold, holding pressure constant at 80 torr. The treatment combinations and results are summarized in Table 15.7.

The TiN peeling response to the conditions of Experiment 3 is shown in Table 15.7. Treatment combinations where peeling occurred are circled in the right-hand side of the contour plot of Figure 15.2. The TiN peeling threshold appears to be at a H_2/WF_6 ratio somewhere between 4 and 5.

Table 15.7. TiN *Peeling Response, Experiment 3. The TiN peeling threshold appears to be at an H_2/WF_6 ratio somewhere between 4 and 5.*

Pressure (torr)	H_2/WF_6 Ratio	TiN Peeling
80	3	YES
80	4	YES
80	5	NO

Verification

The new process conditions defined by Experiment 3 are given in Table 15.8. It was necessary to verify the results at these new processing conditions prior to its release for production. Verification consisted of a passive data collection of sixteen consecutive runs at the new process conditions.

Table 15.8. *New Process Conditions.*

Parameter	Value
Backside Ar	0 sccm
Temperature	480°C
Pressure	80 torr
H_2/WF_6 ratio	5

The results of the passive data collection used for verification are summarized in Table 15.9. Temperature remained the same. No TiN peeling was observed, and uniformity matched that of the process-of-record. Stress failed to meet the goal due to the decision made after Experiment 2. Resistivity and particle performance (not discussed in this chapter) were better than the objectives, giving lower values than the process-of-record.

Table 15.9. *Goals versus Accomplishments. The verification experiment at the new process conditions shows the goal to match the prior process-of-record was reached for all the responses except stress, which was intentionally compromised. One standard deviation is given in parenthesis.*

Parameter/Response	Prior Process	New Process	Units
Nominal Temp	480	480	°C
TiN Peeling	None	None	(visual)
Resistivity	10.6 (0.6)	9.6 (0.1)	$\mu\Omega$-cm
Stress	6.6 (0.6)	8.4 (0.1)	Gdyn/cm²
Particles	0.3 (0.3)	0.1 (0.1)	cm^{-2}

CONCLUSIONS AND RECOMMENDATIONS

Tungsten film uniformity on 200 mm wafers in the Genus 8720 reactor can match uniformity on 150 mm wafers by increasing pressure from 80 torr to 800 mtorr and by decreasing the H_2/WF_6 ratio from 23 to 5. However, this results in a worse film stress.

If the incoming TiN adhesion layer is susceptible to chemical attack in the tungsten deposition reactor, TiN peeling can be prevented by increasing the H_2/WF_6 ratio. This will compromise uniformity and further compromise stress.

A project deadline crisis was averted by using design of experiments and response surface methodology, applied with the guidance of engineering knowledge. The entire sequence of experiments and verification took only a week and a half!

The result obtained is only a stop-gap measure and the result of a stop-gap strategy. The recommendation is to use the new process as an interim solution until one of the preferred strategies can be implemented. The data suggest that the best strategy would be to produce an incoming TiN adhesion layer that is not susceptible to peeling so that the high pressure/low H_2/WF_6 ratio can be exploited.

References

Box, George E. P., William G. Hunter, and J. Stuart Hunter, *Statistics for Experimenters: An Introduction to Design, Data Analysis, and Model Building*, New York, NY: John Wiley and Sons, Inc., 1978.

Box, George E. P. "Signal-to-Noise Ratios, Performance Criteria, and Transformation." *Technometrics* vol. 30, pp. 1-40, 1988.

Cale, T. S., M.K. Jain, and G.B. Raupp, "Maximizing Step Coverage During Blanket Tungsten Low Pressure Chemical Vapor Deposition," *Thin Solid Films* vol. 193/194, pp. 51-60, 1990.

Cammenga, David, *Genus 8720 Blanket Tungsten Process Equipment Improvement Program Final Report*, 91090700A-ENG, Austin, TX: SEMATECH, 1991.

Cobianu, C. and C. Pavelescu, "A Theoretical Study of the Low-Temperature Chemical Vapor Deposition of SiO_2 Films," *Journal of the Electrochemical Society* vol 130, pp. 1888-1893, 1983.

Marcus, R. B. "Diagnostic Techniques," *VLSI Technology* p. 532. Sze, S.E. Ed. New York, NY: McGraw Hill. 1983.

McConica, C. M. and K. Krishnamani, "The Kinetics of LPCVD Tungsten Deposition in a Single Wafer Reactor," *Journal of the Electrochemical Society* vol. 133, pp. 2542-2548, 1986.

Appendix A

Influence of Physical Considerations on Factor Setting

The tungsten deposition reaction occurs at sites on a solid surface. Because real solids are not continuous but atomic in nature, the number of sites per unit area is finite. The two gaseous reactants in this case study are H_2 and WF_6. Assume that these species (and their reaction intermediates) must be adsorbed onto the surface in order for the reaction to take place, and that they compete for this fixed number of surface sites. Also, for convenience of conceptualization, assume complete coverage of the surface sites (the argument can easily be extended to the case of incomplete coverage, but this is unnecessary for the present purpose).

Define the fraction of sites occupied by species containing tungsten as θ. Then, under the assumption of complete coverage, the fraction of sites occupied by species not containing tungsten is $1 - \theta$. Clearly, when $\theta = 0$, there are no species containing tungsten on the surface and therefore the deposition rate, r, must be zero. Likewise, when $\theta = 1$, there are only species containing tungsten on the surface and therefore the deposition rate, r, must again be zero. Then it follows that there must be a maximum in the deposition rate, r, somewhere between $\theta = 0$ and $\theta = 1$. Intuitively, this maximum would correspond to the fractional coverage where adjacent sites are most likely to contain complementary reactants. For a simple bimolecular reaction with a stoichiometry of 1:1 and random site occupation, the maximum would be at $\theta = \frac{1}{2}$. A simple mathematical form that expresses these properties of r in terms of θ is[28]

$$r \propto \theta \cdot (1 - \theta). \tag{15.1}$$

---

[28] It is possible to derive this form for the deposition rate beginning from the chemical kinetic rate equations and invoking the idea of a "most abundant surface intermediate." See McConica and Krishnamani (1986) for an example of the use of a "most abundant surface intermediate" in the derivation of a deposition rate expression.

Now consider the case where there are dynamic equilibria between the species adsorbed on the surface and those in the gaseous ambient. These equilibria link θ to the gas phase composition, and the gas phase composition may be characterized by, and in fact controlled by, the H_2/WF_6 ratio, ρ. The functional relationship between ρ and θ is unknown, but the limits are obvious. For zero influent hydrogen, $\rho = 0$, the surface sites are all occupied by species containing tungsten, and therefore $\theta = 1$. In the limit as influent tungsten hexafluoride approaches zero, ρ approaches 1, the surface sites are all occupied by species not containing tungsten, and θ approaches zero. Therefore, a convenient functional form to use for the sake of argument is

$$\theta = e^{-\rho/\phi}, \qquad (15.2)$$

where ϕ is a constant whose value depends on the gas-to-surface equilibrium constants. Then substitution of (15.2) into (15.1) gives

$$r \propto e^{-\rho/\phi} \cdot (1 - e^{-\rho/\phi}). \qquad (15.3)$$

The proportionality (15.3) indicates that varying the H_2/WF_6 ratio, ρ, should produce curvature in the deposition rate and that it may be possible to find a local maximum, depending on the values of ρ and ϕ.[29] A local maximum in deposition rate is attractive for improving site-to-site uniformity, since site-to-site variations in delivery of the reactants would then have a diminished effect on site-to-site deposition rate.[30]

APPENDIX B

Experimental Data

Experiment 1: Experimental Worksheet.

Run	Temperature °C	Pressure (torr)	Bkar (sccm)	Peel (boolean)	Uniformity (%)	Stress (Gdyn/cm^2)	Dep. Rate (Å/min.)
1	440	4.00	300	0	7.89	9.77	1019
2	440	0.80	300	0	8.37	10.55	329
3	440	0.80	0	0	4.44	10.73	265
4	440	4.00	0	0	4.48	9.71	989
5	470	2.40	150	0	7.53	8.59	1048
6	500	4.00	0	0	4.44	7.55	2236
7	500	0.80	0	0	6.39	8.35	612
8	500	4.00	300	0	7.83	7.48	2389
9	500	0.80	300	0	8.92	7.73	757

---

[29] Such curvature with maxima has in fact been observed to occur in the chemical vapor deposition of SiO_2, where the ratio is between the reactants O_2 and SiH_4 (see Cobianu and Pavelescu, 1983).

[30] This same basic idea can be found in the analysis by Cale, Jain, and Raupp (1990), who find that the ratio of reactant feed rates may be used to maximize tungsten step coverage (i.e., "uniformity") down a contact hole. The interested reader will find their paper much more rigorous than the qualitative arguments presented in this Appendix.

Experiment 2: Experiment Worksheet.

Run	Pressure (torr)	H_2/WF_6 Ratio (unitless)	Deposition Rate (Å/min.)	Uniformity (%)	Stress (Gdyn/cm^2)	Peeling (boolean)
1	80.00	6.00	12603	4.60	8.04	0
2	42.00	6.00	8979	6.20	7.78	0
3	68.87	3.17	9393	3.40	7.58	1
4	15.13	8.83	5602	6.90	7.27	0
5	4.00	6.00	1984	7.30	6.49	0
6	42.00	6.00	8960	6.40	7.69	0
7	15.13	3.17	3663	8.60	6.66	1
8	42.00	2.00	5007	6.30	7.16	1
9	68.87	8.83	12488	5.10	8.33	0
10	42.00	10.00	10310	5.40	8.19	0
11	42.00	6.00	8979	5.00	7.90	0

BIOGRAPHIES

James Buckner has been a technologist with Texas Instruments since 1987, where he has been responsible for metal sputtering, CVD-tungsten deposition, and improvement of CVD-tungsten deposition equipment at SEMATECH. He is currently responsible for the Equipment Migration program in Texas Instruments' Manufacturing Science & Technology Center. James has a Ph.D. in physical chemistry from the University of North Carolina at Chapel Hill.

David J. Cammenga is currently the senior development engineer specializing in electrochromic devices at Gentex, Inc. in Zeeland, MI. David received a B.S. in chemistry from Calvin College (1982) and a M.S. in chemical engineering from Michigan Technological University (1984). He has worked as a process engineer for Texas Instruments and more recently for IBM, including an assignment to SEMATECH.

Ann Weber received a M.S. in chemistry from the University of California at Santa Cruz. She worked for Genus, Inc. as a field process engineer in Austin, TX during the two year SEMATECH/Genus tungsten equipment improvement program. She is currently working in poly and nitride etch for Lam Research Corp.

CHAPTER 16

MODELING A UNIFORMITY BULLS-EYE INVERSION

James Buckner, Richard Huang, Kenneth A. Monnig,
Eliot K. Broadbent, Barry L. Chin, Jon Henri,
Mark Sorell, and Kevin Venor

A very good model of within-wafer sheet resistance uniformity was obtained in an experiment to characterize the chemical vapor deposition (CVD) of tungsten in a Novellus Concept One-W reactor. The key to obtaining a good model fit for this normally difficult response was to recognize that the uniformity profile passes through a bulls-eye inversion (concave or convex surface of values corresponding to points on the wafer) within the region of examined process space, and that the measured uniformity number alone does not adequately characterize the physical phenomenon. Analysis of the raw uniformity numbers resulted in a poor fit of the model to the data and with the appearance of strong curvature in the response. However, adding a positive or negative sign to the uniformity numbers to indicate the concave or convex direction of the bulls-eye produced a very good fit to the linear design model without the need to resort to higher order terms to accommodate curvature.

EXECUTIVE SUMMARY

Problem

Poor uniformity in the deposition of tungsten thin films has been one contributor to the loss of yield in state-of-the-art integrated circuits manufactured using tungsten Chemical Vapor Deposition (CVD). Early users of the Concept One-W CVD tungsten deposition tool of Novellus Systems, Inc., reported uniformities superior to those produced by competing tools, and members of the SEMATECH consortium requested that SEMATECH evaluate this tool. Therefore, SEMATECH set out to characterize the uniformity response of the tool, along with other parameters.

The uniformity response of many processes used in the manufacture of semiconductors has generally been difficult to explain, and it is common to obtain only poor empirical models whose physical interpretations are elusive. This has often been the case with other CVD tungsten deposition tools.

The challenge was to obtain a good model of the uniformity response on the Novellus tool, while retaining physical interpretation if possible.

Solution Strategy

The strategy was to use a statistically designed experiment, but with the design region constrained within the kinetic regime region that is defined by the well established kinetic rate equation of the CVD tungsten deposition reaction.

The CVD tungsten deposition reaction, like all chemical reactions, may operate in a "kinetic regime" or in a "feed rate limited regime." Different factors dominate the reaction in each regime. Previous studies using unconstrained statistical designs crossed regimes, and the resulting analyses revealed spurious interactions among factors. The present study's goal was to run the entire design in the kinetic regime without crossing over into the feed rate limited regime, hence eliminating spurious interactions (see Appendix A).

At first it appeared that this strategy had not worked because the familiar poor model fit of the sheet resistance uniformity was obtained due to what appeared to be a strong curvature in the response. Furthermore, the analysis indicated the scarcely credible result that the model contained at least five two-factor interactions!

However, further analysis revealed that the raw uniformity values did not fully reflect the nature of the variation in uniformity. In particular, examination of the uniformity contour profiles indicated that there was a bulls-eye pattern inversion in the region of process space studied, with some wafers having a higher sheet resistance at the edge of the wafer than at the center of the wafer and other wafers having a lower sheet resistance at the edge of the wafer than at the center of the wafer.

Prefixing the raw uniformity values with a positive or negative sign (i.e., "signing") to indicate convex or concave direction of the bulls-eye produced quite different results in the analysis. All evidence of strong curvature disappeared, leading to a very good overall fit with the linear design model (adjusted $R^2 = 0.91$). Furthermore, the spurious two-factor interactions vanished, leaving a much more credible model of the response that indicated that uniformity is controlled by individual factors and dominated by one, the partial pressure of hydrogen.

Conclusions

A very good uniformity model was obtained for the Novellus Concept One-W reactor. Uniformity is dominated by the partial pressure of hydrogen, and fine tuning control can be accomplished by adjusting the backside hydrogen/argon ratio. A uniformity bulls-eye inversion can be modeled by signing the raw uniformity values according to the direction of the bulls-eye.

When studying a chemical reaction, the kinetic reaction rate equation may be used to constrain a statistical design to a single reaction regime, eliminating spurious interactions.

Process

One of the enabling technologies in the manufacture of integrated circuits is the ability to connect the individual transistors, capacitors, and other circuit elements, to one another with a conducting metallic material. These connections, which constitute the "wiring" of the circuit, are typically built by first depositing a thin blanket layer of metal over the entire silicon wafer, and then etching away the unneeded portion in a way that leaves behind material in the appropriate wiring pattern.

Recently, tungsten deposited by chemical vapor deposition (CVD) has become the metal of choice for manufacturing certain state-of-the-art integrated circuits. In applying tungsten deposition technology to the manufacture of such circuits, layer thickness uniformity across the face of the silicon wafer has been a common problem. A nonuniform tungsten layer leads to difficulty in the subsequent etch process and to variation in performance of the interconnects, resulting in yield loss.

In order to solve this and other problems with tungsten deposition technology, semiconductor manufacturers embarked on programs to improve the existing commercial tungsten deposition tools, while continuing to seek a better tool. It was in this climate that Novellus Systems, Inc., introduced their Concept One-W reactor. Early users of the tool reported much better layer uniformity, among other things, and members of the SEMATECH consortium requested that SEMATECH evaluate the tool.

As a result of this request, the Novellus tool became the subject of a SEMATECH equipment improvement project (EIP). Some elements of the project plan are shown in Figure 16.1. After gauge studies were completed for the parameters of interest, tool performance was baselined in an initial passive data collection (PDC). Four design-of-experiments (DOE) improvement cycles were then planned to characterize and optimize any shortcomings that would have appeared in the baseline PDC. A final PDC was planned to confirm the improvements that resulted from the DOE's, and the project ended with a marathon run for stability and reliability testing of the improved tool.

The work reported in this chapter corresponds to the DOE1 improvement cycle in the project plan of Figure 16.1. The baseline PDC showed generally good results.[31] Therefore the objective of DOE1 was not to address any particular problem, but rather to characterize the deposition process as broadly as possible around the nominal process parameters recommended by the supplier.

Data Collection Plan

Based on the collective experience of the project team members in working with other tungsten deposition equipment, the seven factors shown in Table 16.1 were chosen for the experiment. Temperature, pressure, partial pressure of hydrogen, tungsten hexafluoride flow, and argon flow are all fundamental quantities taken from the well established kinetic rate equation for the reaction.[32] Backside gas flows are used to prevent deposition of tungsten on the backside of the wafer by excluding the reactive tungsten hexafluoride from this region. The backside H_2/Ar ratio was unique to this tool, since it was the only one among the commercial competitors that used a hydrogen component in the backside flow.

---

[31] See Shah (1992).
[32] See, for example, McConica and Krishnamani (1986), and McInerney, Chin, and Broadbent (1990).

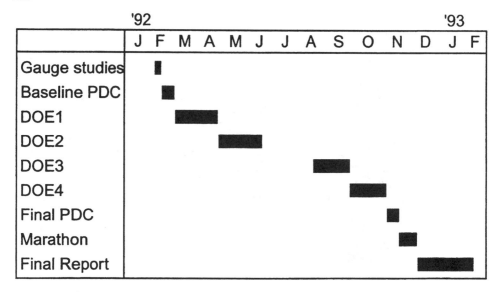

Figure 16.1. *Novellus EIP Project Plan Gantt Chart. The project plan includes gauge studies, baseline and final PDCs, four DOE improvement cycles, and a marathon run to test reliability and stability of the improved tool.*

Factor ranges were chosen that span as much of the process space as possible within some practical limits known from previous empirical and mechanistic studies, while the nominal values recommended by the supplier were used as the centerpoint process. For example, the temperature was not driven lower than the value shown based on previous experience that deposition rates would be unacceptably low in lower temperature ranges, and backside flow limits were based on knowledge of a minimum effective H_2/Ar ratio and the maximum flows accessible with the standard sized mass flow controllers.

Table 16.1. *Factors and Ranges. The first five factors were selected based on the well-established kinetic rate equation for tungsten deposition. Backside gas flows are used to prevent deposition on the backside of the wafer. The backside H_2/Ar Ratio was unique to this reactor, as no competing reactor used hydrogen in the backside flow.*

Factor	Range	Units
Temperature	405 to 425	°C
Pressure	35 to 45	torr
$(P_{H2})^{1/2}$	2.61 to 4.51	$torr^{1/2}$
WF_6 Flow	360 to 440	sccm
Ar Flow	10 to 14	slm
Backside H_2/Ar Ratio	1.5 to 3.5	(unitless)
Backside Total Flow	5 to 9	slm

A primary strategy in selecting factor ranges was to constrain the study to the "kinetic regime" of the reaction. This is in contrast to previous studies,[33] which crossed regimes. In general, a chemical reaction may be operated either in its "kinetic" regime or in its "flow (or feed) rate limited" regime. The kinetic regime is obtained, under most practical circumstances, by flooding the reactor with an excess of reactants so that the inherent nature of the chemical reaction mechanism controls the rate of reaction. The flow rate limited regime, by contrast, is obtained by throttling down the flow of reactants into the reactor so that the reaction cannot proceed at its inherent rate because it is effectively starved of reactants.

Because different physical parameters dominate the reaction rate in the two regimes, the nature of empirical responses to factors in a DOE will also differ in the two regimes. In the analysis of an experiment that crosses regimes, the changes in the nature of the response will appear as spurious interactions among factors. In order to maintain the ability to interpret the experimental results as much as possible in terms of physically meaningful parameters, the project team decided to eliminate the potential for such mathematical phantoms by constraining the experiment to the kinetic regime.

Since temperature, pressure, partial pressure of hydrogen, tungsten hexafluoride flow, and argon flow are all connected through the kinetic rate equation of the reaction, the ranges of all of these factors were subject to the kinetic regime constraint. The details of applying the kinetic regime constraint are outlined in Appendix A. The ranges shown in Table 16.1 span the process space and are constrained simultaneously by the kinetic regime boundary and by a physical limit of the tool, with the nominal supplier-recommended process at the center. The physical limit of the tool was the size of the hydrogen mass flow controller (see Appendix A). If a larger region was needed, it could be obtained by increasing the size of this mass flow controller.

Note that these physical considerations resulted in a choice of factors and ranges based on physical mechanistic ideas, rather than a choice based on the knobs available to adjust the tool.

A total of ten responses were identified for the experiment. Besides uniformity, the topic of this chapter, there was deposition rate, film stress, layer resistivity, tungsten hexafluoride conversion efficiency, film reflectivity at the center and at the edge of the wafer, width of exclusion of film deposition from the edge of the wafer, film adhesion, and backside deposition.[34] Uniformity was measured as the standard deviation of sheet resistance probed at 49 points across the surface of the wafer using the de facto industry standard pattern of the Prometrix RS35e metrology tool, divided by the average of those points and expressed as a percentage. The rationale for this choice of uniformity measure and some concerns regarding it are discussed in Appendix C.

Since many of the responses are functions of the film thickness, it was decided to hold the film thickness as constant as possible at 3500 Å. This was accomplished by performing

---

[33] See, for example, characterizations of the Genus 8720 tool by Joshi, Mehter, Chow, Ishaq, Kang, Geraghty, and McInerney (1990), and by Cammenga (1991); and the characterization of the Applied Materials P5000-W tool by Clark, Chang, and Leung (1991). The author has also examined a (1992) unpublished characterization of the Genus 8720 reactor, and an older (1988) unpublished characterization of the Spectrum 202 reactor, both of which crossed regimes. Riley, Clark, Gleason, and Garver (1990) is an example of a study based on statistical design which appears not to have crossed regimes.

[34] See Shah (1992).

a set-up run for each treatment combination, and adjusting the deposition time to account for the differences in deposition rate before performing the experiment run.

The initial design was set up as a resolution IV screening experiment with seven factors at two levels each in sixteen treatment combinations (2^{7-3}), with three center points. The initial design in actual run order and the corresponding responses are given in the first nineteen rows of data in Appendix B.

Since analysis of the initial design (see the Analysis and Interpretation of Results section below) appeared to show several two-factor interactions involving the square root of the partial pressure of hydrogen, $(P_{H2})^{½}$, a foldover design[35] was generated to resolve the $(P_{H2})^{½}$ interactions. The sixteen new treatment combinations and three new center points (to check process drift) were appended to the initial design (rows 20 to 38 of Appendix B) for analysis, resulting in a total experiment of 32 treatment combinations with six centerpoints. The following analysis will be for the 38 treatment combinations.

DATA ANALYSIS AND INTERPRETATION OF RESULTS

Uniformity is commonly found to be a difficult response to model in semiconductor processes. This was the case with SEMATECH's previous CVD tungsten project on the Genus 8720.[36] The first analysis of the uniformity data from the present experiment yielded similarly poor results with an overall model fit of adjusted $R^2 = 0.40$. Evidence of unusual curvature in the response, however, warranted further investigation. Examination of the tungsten film thickness revealed that the film thickness variation on the wafer had a strong radial effect, so that the contours were like the concentric rings of a target (see Figure 16.6). It was also noted that the centers of the bulls-eyes were thicker than the perimeters for some wafers while thinner than the perimeters for other wafers. This "bulls-eye inversion" became the key to modeling the uniformity. When the raw uniformity numbers were given a positive or a negative sign (depending on whether the bulls-eye was thinner or thicker at the center of the wafer than at its edge), analysis of the signed numbers produced an overall model fit of adjusted $R^2 = 0.91$. In addition, the spurious interactions and evidences of curvature disappeared. In the remainder of this section, the analysis of the unsigned data with its misleading results is presented first, followed by the analysis of the signed data with its straightforward results.

Unsigned Data

The time-ordered plot of the unsigned sheet resistance unformity data (Figure 16.2) shows a fairly stable unimodal distribution. The almost imperceptible downward shift in the last half of the data is attributed to an adjustment in the mechanical wafer positioning in the reaction chamber done after the initial experiment and before the foldover follow-on experiment. The outlier at 12.9% uniformity corresponding to treatment combination 37 was verified to be real with two subsequent runs. It was probably due to the fact that the hydrogen flow was higher than the capacity of the system pressure feedback loop.

---

[35] A foldover design is obtained from an initial design by changing the low values to high values and the high values to low values. When the initial design and its foldover are combined and analyzed as a single data set, the usual result is that some or all of the confounding is resolved. See Box, Hunter, Hunter (1978), pp. 399 ff.

[36] See Cammenga (1991).

Having established that the process was stable during the experiment, factor effects were analyzed. Since the outlier was found to be real, it was included in the analysis. A pareto ranking of the effects shows that two-factor interactions with $(P_{H2})^{1/2}$ dominate, as indicated in Figure 16.3 where $(P_{H2})^{1/2}$ is indicated by "P_".

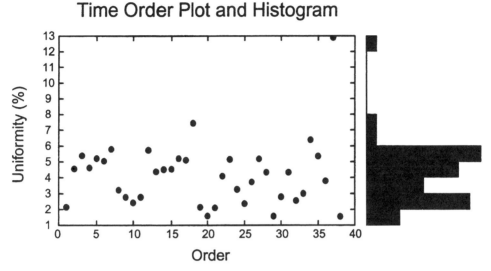

Figure 16.2. *Time-Ordered Plot of Unsigned Sheet Resistance Uniformity Data. The data indicate a unimodal distribution. The outlier is real, and the nearly imperceptible downward shift in the last half of the data is attributed to an adjustment in the mechanical wafer positioning in the reaction chamber.*

Plots of the five dominant interactions with $(P_{H2})^{1/2}$ are given in Figures 16.4a–e. They show evidence of curvature in uniformity, with the centerpoint significantly below the four averages in each interaction plot. These five plots taken together were the first clue that something was wrong with the analysis, because it is unlikely that the model for uniformity would have so many significant interactions.

One reason the model fit was poor was because the design did not account for curvature. Figure 16.5 gives a plot of the residuals, and Table 16.2 gives the least squares coefficients.

Signed Data

To understand the curvature in uniformity, the project team was preparing to plan a third experiment, when one member suggested a bulls-eye inversion might be occurring. Another member immediately realized that this could indeed explain the "curvature" in the response. Subsequent examination of the uniformity contour plots confirmed that an inversion does take place in the process space explored by the experiment. In fact, three patterns were found: a bulls-eye pattern where the tungsten layer is thinner in the center of the wafer than at the edge, a non-bulls-eye pattern, and a bulls-eye pattern where the tungsten layer is thicker in the center of the wafer than at the edge (see Figures 16.6a–16.6c). The non-bulls-eye wafers were produced by the centerpoints at the nominal process. Tungsten layer thickness and tungsten sheet resistance are inversely related, with low sheet resistance for thick layers and high sheet resistance for thin layers.

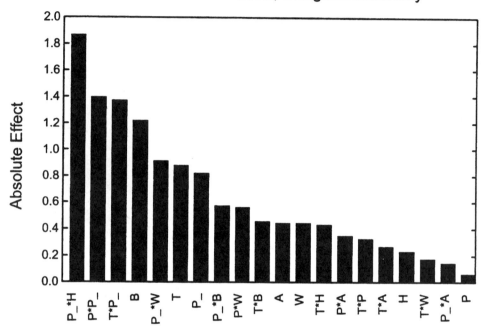

Figure 16.3. *Pareto Chart of Factor Effects for Unsigned Sheet Resistance Uniformity. The chart indicates the importance of at least five two-factor interactions, each involving $(P_{H2})^{1/2}$. $(P_{H2})^{1/2}$ is denoted by "P_" in this computer-generated plot.*

Table 16.2. *Least Squares Coefficients for the Unsigned Sheet Resistance Uniformity Model. The unsigned uniformity model has numerous spurious interactions and a poor model fit due to an appearance of strong curvature in the response.*

Term	Coeff.	Std. Error	T-value	Signif.
Average	+4.184211			
$(P_{H2})^{1/2}$ * Back H$_2$/Ar	−0.932500	0.283070	−3.29	0.0026
$(P_{H2})^{1/2}$ * Press	−0.698750	0.283070	−2.47	0.0197
$(P_{H2})^{1/2}$ * Temp	+0.686875	0.283070	+2.43	0.0217
Back Flow	+0.612500	0.283070	+2.16	0.0389
Temp	+0.436250	0.283070		
$(P_{H2})^{1/2}$	+0.410625	0.283070		
Back H$_2$/Ar	−0.116875	0.283070		
Press	−0.029375	0.283070		

No. cases = 38. R^2 = 0.5260. RMS Error = 1.601.
Resid. df = 29. R^2 adj. = 0.3952. Cond. No. = 1.

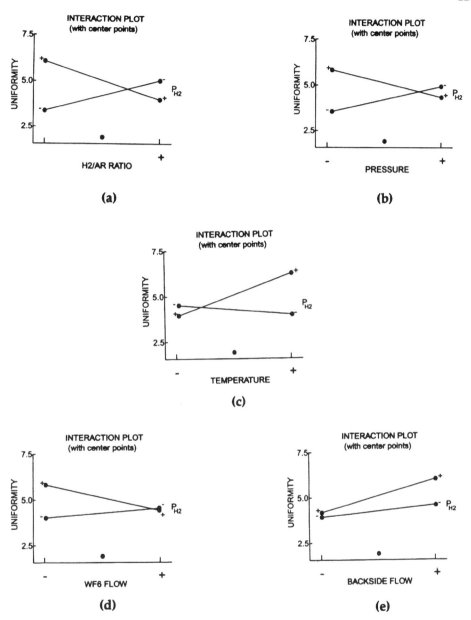

Figure 16.4. $(P_{H2})^{½}$ *Interaction Plots for Unsigned Sheet Resistance Uniformity.* (a) $(P_{H2})^{½}$ *versus Backside H_2/Ar Ratio.* (b) $(P_{H2})^{½}$ *versus Pressure.* (c) $(P_{H2})^{½}$ *versus Temperature.* (d) $(P_{H2})^{½}$ *versus WF_6 Flow.* (e) $(P_{H2})^{½}$ *versus Backside Flow. All interaction plots show evidence of strong curvature, with uniformity for the nominal process (centerpoints) significantly lower than uniformity at the four averages on each interaction plot.*

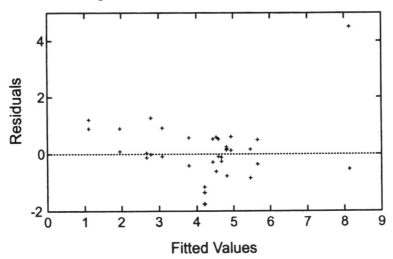

Figure 16.5. *Residuals Plot for Unsigned Sheet Resistance Uniformity. There appears to be a slight downward trend and one outlier.*

To account for the bulls-eye inversion in the analysis, the raw uniformity numbers were signed (that is, a positive or negative sign was added to the numerical value) by the convention illustrated in Figure 16.7. Signing was done by examination of the uniformity contour plots without reference to the treatment combinations so as not to bias the results, although bias was hardly possible because of the unmistakable differences. Signing of the nominal centerpoint process was more difficult. Two analyses were done, one for positive nominals, and the other for negative nominals. In the final analysis, nominal centerpoint processes were signed positive based on the residuals plots, which showed a marked deviation from the model for the negative signed nominals. All signs, including nominals, ended up being the same as the sign of the individual sheet resistance reading at the center of each wafer.

A time-ordered plot of the signed uniformities shows a bimodal distribution (see Figure 16.8). The pareto chart of the factor effects (see Figure 16.9) which shows $(P_{H2})^{1/2}$ to be the dominant factor, is dramatically different from the pareto chart of Figure 16.3. None of the two-factor interactions appears to be significant for the signed model. The largest of the interactions is plotted in Figure 16.10. It shows evidence of a small amount of curvature in the response with the centerpoint lying off center, but the strong evidence of curvature indicated by the unsigned data has vanished.

Stepwise regression was used to remove non-significant terms from the signed non-uniformity model. The model had a very good model fit, within an adjusted R^2 of 0.91. Figure 16.11 gives the residual plot, and Table 16.3 gives the least square regression coefficients.

Appendix C compares the model for uniformity (standard deviation over mean) of the unsigned data to the model obtained from modeling the standard deviation, modeling the mean, and taking the quotient. Both models give approximately the same contour plots, so they are equivalent.

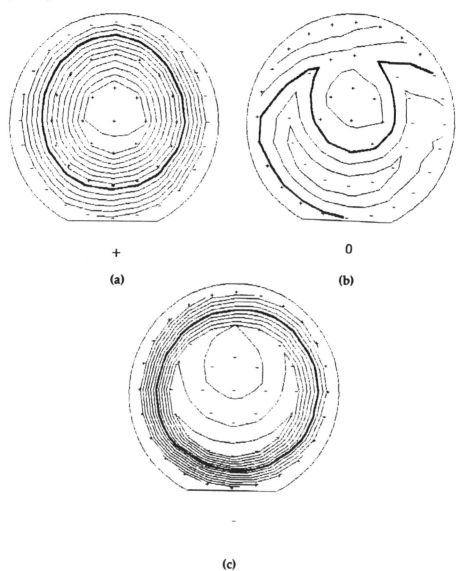

Figure 16.6. *Prometrix Contour Plots of Tungsten Sheet Resistance. The bold contours indicate average values of sheet resistance, the positive values indicate higher than average sheet resistance, and the negative values indicate lower than average sheet resistance. (a) is a positive bulls-eye. (b) is a non-bulls-eye. (c) is a negative bulls-eye. The unsigned raw uniformity numbers do not account for these differences in the uniformity profiles.*

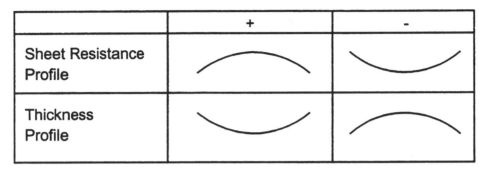

Figure 16.7. *Signing Convention. Uniformity numbers were prefixed with a positive or negative sign, depending on the direction of the bulls-eye uniformity profile. For example, a negative sign was given to the uniformity if the tungsten film was thicker in the middle of the wafer than it was at the edge, in which case the sheet resistance was lower in the middle of the wafer than it was at the edge.*

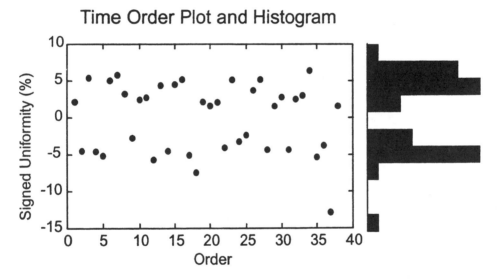

Figure 16.8. *Bimodal Time-Ordered Plot for Signed Sheet Resistance Uniformity. The signed data are bimodal.*

CONCLUSIONS AND RECOMMENDATIONS

Analysis of the signed uniformity data yields more credible results than analysis of the unsigned data. Analysis of the signed uniformity data shows that uniformity is dominated by hydrogen partial pressure with little interaction between the other factors, while analysis of the unsigned data indicates that uniformity is dominated by two-factor interactions.

A contour plot of uniformity for the signed data using the two main factors is shown in Figure 16.12. The nominal process recommended by the supplier lies very close to the zero uniformity contour line, which is the desired value for the signed uniformity.

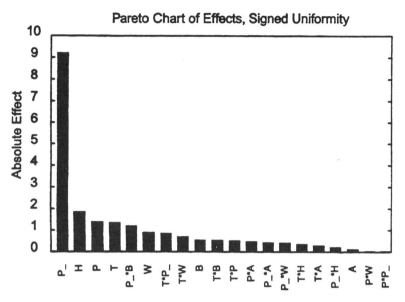

Figure 16.9. *Pareto Chart of Factor Effects for Signed Sheet Resistance Uniformity. This Pareto chart for the signed data is dramatically different from the Pareto chart of Figure 16.3 for unsigned data. The spurious interactions have disappeared and single factor effects are seen to control the uniformity response, with partial pressure of hydrogen dominating.*

Table 16.3. *Least Squares Regression Coefficients for the Signed Sheet Resistance Uniformity Model, Signed Data. The signed data eliminate the spurious interactions and produce a much better model fit than the non-signed data and the response no longer shows evidence of strong curvature.*

Term	Coeff.	Std. Error	T-value	Signif.
Average	−0.044737	0.224733		
Temp	−0.686875	0.244897		
Press	+0.698750	0.244897	+2.85	0.0079
$(P_{H2})^{1/2}$	−4.611250	0.244897		
WF_6 Flow	+0.459375	0.244897	+1.88	0.0708
Back H_2/Ar	+0.932500	0.244897	+3.81	0.0007
Back Flow	−0.284375	0.244897		
Temp*Press	−0.436250	0.244897	−1.78	0.0853
$(P_{H2})^{1/2}$ * Back Flow	−0.612500	0.244897	−2.50	0.0183

No. cases = 38. R^2 = 0.9323. RMS Error = 1.385.
Resid. df = 29. R^2 adj. = 0.9136. Cond. No. = 1.

The model for signed uniformity implies that hydrogen partial pressure should be used for primary control of uniformity, and backside hydrogen/argon ratio should be used for fine tuning. In general, uniformity studies should check for the possibility of bulls-eye inversion and "signing."

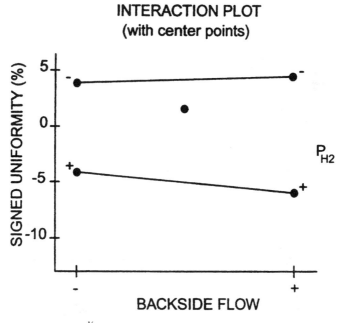

Figure 16.10. $(P_{H2})^{\frac{1}{2}}$ *versus Backside Flow Interaction Plot for Signed Sheet Resistance Uniformity. The only interaction remaining in the signed data analysis shows evidence of some small curvature, but the strong curvature indicated by the unsigned data has vanished.*

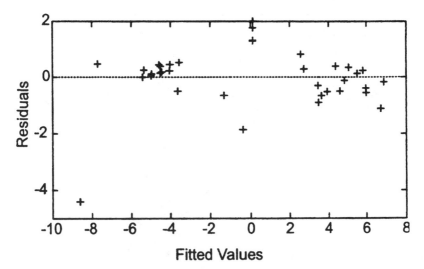

Figure 16.11. *Residuals Plot for Signed Sheet Resistance Uniformity. There is one low outlier.*

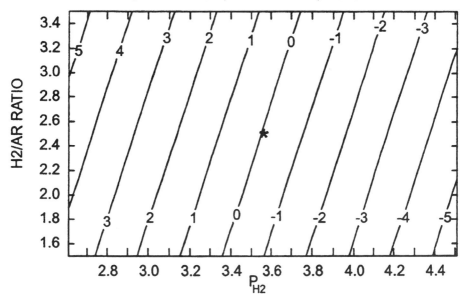

Figure 16.12. *Contour Plot of Signed Sheet Resistance Uniformity Around the Nominal Process.* $(P_{H2})^{\frac{1}{2}}$ *can be used for primary control of uniformity, and the second largest factor, backside* H_2/Ar *ratio, can be used for fine tuning. The nominal process recommended by the supplier is very close to the zero contour line.*

ACKNOWLEDGMENTS

The authors thank Karen Horrell of SEMATECH for help using the RS/1® statistical software package in our time of need, and Todd Green of Novellus Systems, Inc., for providing quick access to uniformity contour plots.

REFERENCES

Box, George E. P., "Signal-To-Noise Ratios, Performance Criteria, and Transformations," *Technometrics*, Vol. 30, pp. 1-40, 1988.

Box, George E. P., William G. Hunter, and J. Stuart Hunter, *Statistics for Experimenters: An Introduction to Design, Data Analysis, and Model Building*, New York, NY: John Wiley and Sons, Inc., 1978.

Cammenga, David, *Genus 8720 Blanket Tungsten Process Equipment Improvement Program Final Report*, 91090700A-ENG, Austin, TX: SEMATECH, 1991.

Clark, Thomas E., Mei Chang, and Cissy Leung, "Response surface modeling of high pressure chemical vapor deposited blanket tungsten," *Journal of Vacuum Science and Technology*, Vol. B 9, No. 3, pp. 1478-1486, 1991.

Joshi, R. V., E. Mehter, M. Chow, M. Ishaq, S. Kang, P. Geraghty, and J. McInerney, "High Growth Rate CVD-W Process for Filling High Aspect Ratio Sub-Micron Contacts/Lines," *Materials Research Society Symposium Proceedings VLSI V*, pp. 157-166, 1990.

McConica, C. M. and K. Krishnamani, "The Kinetics of LPCVD Tungsten Deposition in a Single Wafer Reactor," *Journal of the Electrochemical Society*, Vol. 133, No. 12, pp. 2542-2548, 1986.

McInerney, E. J., B. L. Chin, and E. K. Broadbent, "The Rate Kinetics of High Pressure CVD Tungsten." *Proceedings of the Workshop on Tungsten and Other Advanced Metals for ULSI Applications VII*, October 22-24, 1990. (Dallas, Texas, U.S.A.)

Riley, Paul E., Thomas E. Clark, Edward F. Gleason, and Garver M. Garver, "Implementation of Tungsten Metallization In Multilevel Interconnection Technologies," *IEEE Transactions on Semiconductor Manufacturing*, Vol. 3, No. 4, pp. 150-157, 1990.

Shah, Raj, Ed, *Novellus Concept One-W Characterization Interim Report*, 92051143A-ENG. Austin, TX: SEMATECH, 1992.

APPENDIX A

Constraining the Experiment to the Kinetic Regime

The form of the kinetic rate equation from tungsten deposition resulting from the reduction of WF_6 with H_2 has been independently established by several workers to be[37]

$$R = A \exp(-Q/kT) \cdot P_{H2}^{1/2}, \quad (16.1)$$

where A is the "pre-exponential constant," Q is the activation energy, k is Boltzmann's constant, T is the absolute temperature, and $P_{H2}^{1/2}$ is the partial pressure of hydrogen.

The first step in setting the ranges of the factors was to recognize that temperature has the strongest effect on the rate. The low temperature setting was arrived at by consensus of the experienced engineers on the team, who agreed that temperatures any lower than $405°C$ would result in deposition rates too low to be practical. The high temperature of $425°C$ was then chosen to position the nominal value of $415°C$ in the center of the range.

Next, it was desired to maximize the range of $P_{H2}^{1/2}$. Solving equation (16.1) for $P_{H2}^{1/2}$ yields

$$P_{H2}^{1/2} = (R/A) \exp(Q/kT). \quad (16.2)$$

The settings for $P_{H2}^{1/2}$ can then be determined by evaluating this expression at the extreme values of R and T:

$$P_{H2}^{1/2}{}_{low} = P_{H2}^{1/2}(R_{low}, T_{high}), \quad (16.3a)$$

$$P_{H2}^{1/2}{}_{high} = P_{H2}^{1/2}(R_{high}, T_{low}). \quad (16.3b)$$

The limits on T were set above. Setting the limits on R requires estimating the position of the boundary between the kinetic and feed rate limited chemical reaction regimes. The essence of feed rate limitation is that the inherent reaction rate (determined for a particular set of conditions by the kinetic rate equation) is faster than the feed rate of one of the reactants. The boundary of the feed rate limited regime, therefore, can be estimated by calculating the maximum reaction rate that can be supported by the feed rate of the

---

[37] See, for example, McConica and Krishnamani (1986), and McInerney, Chin, and Broadbent (1990).

reactants, assuming complete consumption of the limiting reactant by the reaction. Since in this example it is desired to set the levels of the H_2 reactant, the other reactant, WF_6, was treated as the limiter.

In order to do this calculation, level settings were required for WF_6 flow. These were set by consensus of the team at ±40 sccm around the nominal. R_{low} and R_{high} were then obtained using the low and high values of WF_6 flow, and then $P_{H2\ low}^{1/2}$ and $P_{H2\ high}^{1/2}$ were calculated using equations 16.3a and 16.3b.

$P_{H2}^{1/2}$ is a physically relevant quantity, but this parameter does not appear as an adjustable knob on the machine. In order to run the experiment, the design parameters had to be converted into machine settings. Rather than partial pressures of the reactants, the machine controls total pressure, P, and gas flows, F_j, for gas j. Ignoring reaction flux effects (that is, considering conditions at the inlet to the reactor only, away from the reaction site), these quantities are related as follows.

$$P = P_{H2} + P_{WF6} + P_{Ar} \qquad (16.4)$$

$$P_j = (F_j / \Sigma F_j) \cdot P. \qquad (16.5)$$

The new parameters required for the machine setup are the total pressure, P, and the argon flow, F_{Ar}. (Argon is an inert "carrier" gas.) These two parameters could have been set at a nominal value, but it was decided to add them to the experimental design since it was felt they may have effects of their own. The levels were set by consensus to those shown in Table 16.1. Finally, the hydrogen flow settings were found by solving equation (16.5) for F_{H2}. The resulting F_{H2} settings (serendipitously) spanned the entire usable range of the hydrogen mass flow controller.

APPENDIX B

Raw Data

Designed Experiment in Seven Factors and Two Blocks in 38 runs with 14 responses.

	temp (°C)	press (torr)	P_H_2 (torr$^{0.5}$)	WF_6 flow (sccm)	argon flow (slm)	H_2–Ar ratio (none)	backside flow (slm)	block	deposition rate (Ang/min)	uniformity (%)
1	415	40	3.56	400	12	2.5	7	2	2071	2.14
2	405	35	4.51	440	14	1.5	5	2	1908	4.55
3	405	45	2.61	440	14	1.5	9	2	1618	5.39
4	405	35	4.51	360	14	3.5	9	2	1830	4.62
5	405	45	4.51	360	10	1.5	9	2	1997	5.20
6	405	35	2.61	440	10	3.5	9	2	1674	5.04
7	405	45	2.61	360	14	3.5	5	2	1610	5.80
8	405	35	2.61	360	10	1.5	5	2	1494	3.22
9	405	45	4.51	440	10	3.5	5	2	2190	2.77
10	415	40	3.56	400	12	2.5	7	2	2111	2.43
11	425	35	2.61	360	14	1.5	9	2	1995	2.76
12	425	45	4.51	360	14	1.5	5	2	2634	5.73
13	425	35	2.61	440	14	3.5	5	2	2140	4.38
14	425	45	4.51	440	14	3.5	9	2	2836	4.51
15	425	45	2.61	440	10	1.5	5	2	2206	4.53
16	425	45	2.61	360	10	3.5	9	2	2405	5.20
17	425	35	4.51	360	10	3.5	5	2	2571	5.11
18	425	35	4.51	440	10	1.5	9	2	2604	7.44

APPENDIX B (CONTINUED)

	temp (°C)	WF$_6$ press (torr)	argon P_H$_2$ (torr$^{0.5}$)	H$_2$–Ar flow (sccm)	backside flow (slm)	ratio (none)	deposition flow (slm)	block	rate (Ang/min)	uniformity (%)
19	415	40	3.56	400	12	2.5	7	2	2113	2.15
20	415	40	3.56	400	12	2.5	7	1	2168	1.59
21	405	35	2.61	440	14	1.5	5	1	1326	2.11
22	405	45	4.51	440	14	1.5	9	1	2149	4.08
23	405	45	2.61	440	10	3.5	5	1	1503	5.14
24	405	35	4.51	440	10	3.5	9	1	2029	3.26
25	405	45	4.51	360	14	3.5	5	1	2034	2.37
26	405	45	2.61	360	10	1.5	9	1	1481	3.72
27	405	35	2.61	360	14	3.5	9	1	1581	5.19
28	405	35	4.51	360	10	1.5	5	1	1935	4.34
29	415	40	3.56	400	12	2.5	7	1	2173	1.58
30	425	45	2.61	360	14	1.5	5	1	2111	2.80
31	425	45	4.51	440	10	1.5	5	1	2891	4.34
32	425	35	2.61	440	10	1.5	9	1	1940	2.55
33	425	35	2.61	360	10	3.5	5	1	2158	3.00
34	425	45	2.61	440	14	3.5	9	1	2196	6.38
35	425	45	4.51	360	10	3.5	9	1	2766	5.35
36	425	35	4.51	440	14	3.5	5	1	2622	3.79
37	425	35	4.51	360	14	1.5	9	1	2669	12.89
38	415	40	3.56	400	12	2.5	7	1	2204	1.55

	stress (Gdyn/cm^2)	resistivity (u-ohm cm)	WF$_6$ efficiency (%)	reflectivity (%)	exclusion width (mm)	adhesion (boolean)	thickness (Angstroms)	edge reflectivity (%)	sheet mean (mohm)
1	13.1	11.8	16.8	85	2.3	1	4176	87	283.70
2	13.1	12.2	14.1	89	2.2	1	4246	90	286.30
3	12.9	11.8	11.9	85	2.7	1	4176	91	281.90
4	12.8	11.8	16.5	88	3.0	1	4347	84	272.20
5	13.6	12.0	18.0	88	3.2	1	4226	94	283.60
6	13.0	11.9	12.3	85	2.3	1	4130	83	288.00
7	12.7	12.0	14.5	85	1.4	0	4187	82	287.40
8	13.0	12.0	13.5	84	1.9	0	4220	81	285.30
9	13.5	12.4	16.1	90	1.3	0	4270	88	290.90
10	13.2	11.8	17.1	85	2.5	1	4345	85	271.20
11	12.4	11.7	18.0	80	3.5	1	4323	85	270.40
12	12.1	11.9	23.7	87	3.3	1	4434	98	268.40
13	12.0	12.1	15.8	81	1.4	1	4351	81	277.10
14	12.4	12.0	20.9	87	3.1	1	4395	97	272.30
15	12.3	11.9	16.3	82	2.2	1	4154	81	286.20
16	12.3	11.8	21.7	85	2.7	1	4108	89	287.70
17	12.1	12.1	23.2	88	2.4	1	4200	95	288.50
18	12.8	12.1	19.2	86	3.4	1	4145	96	291.50
19	13.0	11.7	17.1	84	2.5	1	4261	86	274.30
20	12.5	12.4	17.6	85	3.2	1	4390	84	282.58
21	13.0	12.3	9.8	86	3.0	0	4055	82	304.04
22	13.4	12.5	15.8	90	3.8	1	4531	92	274.92
23	12.8	12.2	11.1	87	2.2	0	4033	80	301.89
24	12.8	12.3	15.0	96	3.3	1	4278	97	287.11
25	12.8	12.1	18.3	93	2.9	1	4084	94	297.44
26	12.8	11.8	13.3	86	3.7	1	4259	82	278.10
27	12.8	11.7	14.2	82	3.1	1	4531	82	257.92
28	12.5	12.3	17.4	92	3.0	1	4273	92	288.93

APPENDIX B (CONTINUED)

	stress (Gdyn/cm^2)	resistivity (u-ohm cm)	WF$_6$ efficiency (%)	reflectivity (%)	exclusion width (mm)	adhesion (boolean)	thickness (Angstroms)	edge reflectivity (%)	sheet mean (mohm)
29	12.5	12.2	17.6	85	3.1	1	4400	85	276.29
30	11.8	12.2	19.0	80	3.1	1	4310	80	283.63
31	12.4	12.2	21.3	89	3.0	1	4337	90	282.30
32	12.3	12.2	14.3	83	3.5	1	4220	81	288.28
33	12.1	12.2	19.4	81	2.6	1	4406	79	276.62
34	11.8	12.2	16.2	87	3.3	1	4191	84	291.85
35	11.9	12.4	24.9	94	4.0	1	3942	98	315.73
36	10.8	12.7	19.3	100	3.0	1	3933	102	322.35
37	10.4	12.6	24.1	100	5.0	1	4182	105	300.20
38	12.8	12.2	17.9	87	3.0	1	4463	86	274.18

	sheet std (mohm)	signed uniformity (%)	signed rs std (mohm)
1	6.07	2.14	6.07118
2	13.03	-4.55	-13.02665
3	15.19	5.39	15.19441
4	12.58	-4.62	-12.57564
5	14.75	-5.20	-14.74720
6	14.52	5.04	14.51520
7	16.67	5.80	16.66920
8	9.19	3.22	9.18666
9	8.06	-2.77	-8.05793
10	6.59	2.43	6.59016
11	7.46	2.76	7.46304
12	15.38	-5.73	-15.37932
13	12.14	4.38	12.13698
14	12.28	-4.51	-12.28073
15	12.96	4.53	12.96486
16	14.96	5.20	14.96040
17	14.74	-5.11	-14.74235
18	21.69	-7.44	-21.68760
19	5.90	2.15	5.89745
20	4.49	1.59	4.48600
21	6.41	2.11	6.41500
22	11.23	-4.08	-11.23000
23	15.51	5.14	15.51000
24	9.35	-3.26	-9.34500
25	7.05	-2.37	-7.05200
26	10.34	3.72	10.34000
27	13.38	5.19	13.38000
28	12.55	-4.34	-12.55000
29	4.36	1.58	4.35600
30	7.94	2.80	7.93800
31	12.26	-4.34	-12.26000
32	7.35	2.55	7.35300
33	8.30	3.00	8.30100
34	18.62	6.38	18.62000
35	16.90	-5.35	-16.90000
36	12.20	-3.79	-12.20000
37	38.60	-12.89	-38.60000
38	4.25	1.55	4.25400

Appendix C

Justification for Using Uniformity as a Response

The uniformity response used in this study is computed as the standard deviation of measurements taken at 49 sites on a wafer divided by the mean of those same 49 measurements and multiplied by 100 to express the standard deviation as a percentage of the mean. Thus it has the form of a coefficient of variation. It has been shown[38] that coefficients of variation can be inefficient statistics under conditions that are not uncommon in experiments, namely, where the standard deviation is not proportional to the mean. "Inefficient" in this context means that the statistic effectively fails to use all the information at hand. Using an inefficient statistic is equivalent to throwing away data, thus creating the need for larger sample sizes.

Uniformity, calculated in the manner described above, was chosen as a response in the present study because it is commonly used in the engineering community. In this study, a relatively good model was obtained for uniformity.

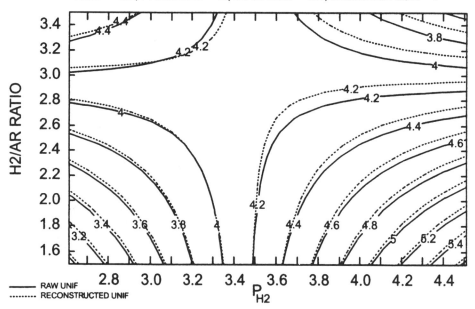

Figure 16.13. *Contour Plot of Unsigned Uniformity. The plot illustrates that there is little difference between the model obtained by direct analysis of unsigned uniformity and the model obtained by reconstructing unsigned uniformity as the quotient of separate models of standard deviation and mean.*

---

[38] See Box (1988).

When the desired response is a coefficient of variation, an approach frequently used by statisticians is to separately model the standard deviation and the mean, and then to construct a new model for the coefficient of variation as the ratio of these two models. The data of this study have been reanalyzed by the author using this approach. Figure 16.13 illustrates the result for the unsigned uniformity data in the form of a contour plot spanning the same region of parameter space as that of Figure 16.12. The surface represented with a solid line was produced from a direct analysis of the uniformity response. The surface represented with a dashed line is a reconstruction of the uniformity obtained as the ratio of the separately modeled standard deviation and mean. Although there are some small differences in the positions of the contours, the essential features of the two analyses agree. In particular, the spurious interactions and curvature in the response still appear.

Figure 16.14 illustrates the results of the two methods of analysis for the signed uniformity data. Signed uniformities were reconstructed by modeling signed standard deviations. Again, the results are essentially all the same.

These figures illustrate that in this case study, the model of the ratio of the standard deviation over the mean, gives approximately the same results as taking the ratio of the model of the standard deviation over the model of the mean and modeling it.

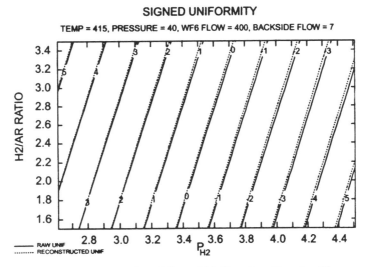

Figure 16.14. *Contour Plot of Signed Uniformity. The plot illustrates that there is little difference between the model obtained by direct analysis of signed uniformity and the model obtained by reconstructing signed uniformity as the quotient of separate models of signed standard deviation and mean.*

BIOGRAPHIES

James Buckner has been a technologist with Texas Instruments since 1987, where he has been responsible for metal sputtering, CVD-tungsten deposition, and improvement of CVD-tungsten deposition equipment at SEMATECH. He is currently responsible for the Equipment Migration program in Texas Instruments' Manufacturing Science & Technology

Center. James has a Ph.D. in physical chemistry from the University of North Carolina at Chapel Hill.

Richard Huang has been at AMD since 1988. He is currently manager of the Advanced Dielectric Development Group. Richard received his M.S. degree from San Jose State University in 1987.

Kenneth A. Monnig received an A.B.A. degree in physics from Thomas More College in Covington, Kentucky, a B.S. degree in engineering from the University of Dayton, and M.S. and Ph.D. degrees in electrical engineering from the University of Virginia. He started his career at Hewlett-Packard, where he designed processes and equipment for tungsten interconnect systems. He then served as a technical director for Genus, Inc. and later joined SEMATECH in 1988, where he has initiated and led projects in chemical-mechanical planarization, tungsten tool, and collimated PVD Ti/TiN tool improvement programs. He is currently program manager for advanced interconnect projects encompassing CVD Ti/TiN deposition Cu interconnects, low epsilon dielectrics, and aluminum planarization. Kenneth was awarded patents for his work at Hewlett-Packard and Genus.

Eliot Broadbent is Chief Scientist for Novellus Systems, where he is responsible for developing advanced thin film deposition systems. From 1979 to 1989, he worked for Philips/Signetics as a research scientist and IC manufacturing engineer, specializing in the areas of CVD metals, silicide formation, diffusion barriers, and aluminum interconnection. The author of over 50 publications, he holds seventeen issued patents. He joined Novellus in 1989.

Barry L. Chin received his B.S. degree from Cornell University and the M.S. and Ph.D. degrees from the University of California at Berkeley. Since 1980, he has been involved in thin film deposition, plasma etching, and process integration programs at several semiconductor manufacturing companies. He joined Novellus Systems in 1987, where he has worked on the product development of Plasma TEOS and CVD-Tungsten deposition systems.

Jon Henri is an applications engineer at Novellus Systems. He works on CVD-Tungsten deposition systems.

Mark Sorell earned a B.S. (1986) and M.S. (1988) in Statistics from Kansas State University. He joined Intel Corporation in 1988 working for the Corporate Technology Manufacturing Group in Santa Clara, CA. Work included supporting central groups with general statistical consulting and providing support for Fab 1, a memory production fab and Fab D2, a memory/microprocessor development fab. In 1991 he joined the SEMATECH Modelling and Analysis Group in Austin, TX., as an Intel assignee. During his time at SEMATECH he provided statistical support to the Multi-Level Metals Group for projects and supporting the 6 to 8 inch wafer conversion. He also focused on statistical metrology issues and provided metrology training. In 1993 he joined Intel's new microprocessor production fab (Fab 11) in Rio Rancho, NM where they started and ramped a new product line. He currently supports the Fab 11 Yield Group doing statistical consulting and training.

Kevin Venor works at Hewlett-Packard in Colorado.

CHAPTER 17

USING FEWER WAFERS TO RESOLVE CONFOUNDING IN SCREENING EXPERIMENTS

Joel Barnett, Veronica Czitrom,
Peter W. M. John, and Ramón V. León

What happens when a designed process characterization experiment is ambiguous about which process factors interact? Usually, resolving this confounding of the interactions involves conducting additional experimental runs. In this case study we present a technique (three quarter fractional factorial designs) for adding fewer runs than commonly done. The technique was used in an oxide etch uniformity experiment where it led to a process that met its uniformity goals.

EXECUTIVE SUMMARY

Problem

The semiconductor industry is increasingly using designed experiments for rigorous process characterization. Designed experiments are used because as semiconductor equipment becomes more complex, more process factors must be investigated in a characterization. Because of the many process factors, the first characterization experiment considered is usually a screening experiment. Screening experiments are used to efficiently decide which process factors are significant and to detect the presence of interactions among these factors. Since screening experiments investigate many factors with relative few runs, their results can be ambiguous. In particular, it is often not clear which process factors interact. Resolving this confounding of the interactions usually involves conducting additional experimental runs.

A situation where additional experimental runs were needed to resolve confounding recently occurred at SEMATECH when a vapor phase etch tool was converted from processing 150 mm wafers to processing 200 mm wafers. To characterize this new vapor phase etching tool for an etching process, a resolution IV screening experiment with six factors and 16 runs plus two center points (18 runs total) was conducted. The experiment analysis of the uniformity of the etch thickness showed that nine interactions in significant alias chains involving all six process factors were confounded. Thus it was impossible to distinguish which of these interactions were significant and which were not. The usual approach to resolving this confounding involves adding 16 more runs. These runs are obtained by adding another quarter fraction of the 2^6 full factorial. But the usual approach would have required a total of 34 experimental runs, essentially doubling the experiment size, which the engineer did not have the time or resources to perform. So the question came up: is it possible to use fewer runs to resolve the confounding among the interactions?

Solution Strategy

Using the technique presented in this paper only eight (instead of 16) additional runs had to be conducted. Thus a total of 26 runs were used rather than the usual 34. The additional eight runs were obtained by semifolding on one factor to form a three-quarter fractional factorial design (see the Appendix for an explanation). With the additional eight runs, seven of the nine confounded interactions of interest could be estimated.

Conclusions

At the completion of the eight new experimental runs, a model was fitted to the data generated by all 26 runs. This model was used to predict optimal settings for the process factors. This prediction was then tested by running the process at the predicted settings. The results of the test showed that the uniformity of the process was as predicted when the amount of oxide etched was 200 Å. When the amount of oxide etched was 50 Å, then the uniformity observed was larger than that predicted by the model. However, starting from the predicted settings, the engineers changed the setting of one factor. After this additional change, the 50 Å target met its uniformity goal.

The results obtained in the etch uniformity experiment show that the confounding among interactions in a screening experiment can be resolved and an appropriate model fitted without the usual doubling of the experiment size. Semifolding on one factor to form a three-quarter fractional factorial design, added only half as many runs while still allowing for the resolution of the confounding between the interactions and the fitting of an appropriate model. This decrease in the number of required runs is not always possible, but it is possible often enough so that semifolding should always be considered *before* automatically doubling the experiment size. For many processes, this decrease in the number of experiment runs could lead to considerable savings in time and other resources.

PROCESS

The objective of the etch uniformity experiment was to certify the vapor phase etching process for 200 mm wafers. This new process was the result of the conversion of the vapor phase etch tool to the 200 mm diameter wafer size.

In the vapor phase etching process, a wafer with thermally grown oxide is mounted on a rotating chuck on a turnable shaft inside a sealed chamber (see Figure 17.1). Nitrogen (N_2) flows over the top of a water tank where it absorbs water vapor, mixes with additional N_2 and enters the chamber. This gas mixture conditions the surface of the rotating wafer.

When the conditions inside the chamber stabilize, a small amount of anhydrous hydrofluoric acid is introduced into the gas mixture to etch the wafer. The amount of anhydrous hydrofluoric acid flow is varied to target the amount of oxide etched.

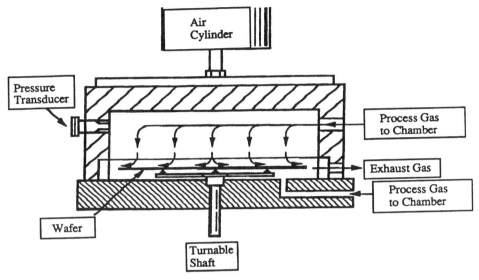

Figure 17.1. *Vapor Phase Etch Tool. The wafer is mounted on a rotating chuck and etched by gas introduced into the chamber.*

The objective of the characterization experiment was to investigate the pre-etch stabilization and etching processes for two oxide etch target thicknesses, 50 Å and 200 Å. In the experiment, the rotating speed of the wafer, the amount of N_2, the amount of water vapor, and the total amount of gas flow (N_2 + N_2 & water vapor) were studied for the two oxide etch targets of 50 Å and 200 Å. (The amount of anhydrous hydrofluoric acid is too small to be considered significant to the total flow calculation.)

DATA COLLECTION PLAN

Table 17.1 shows the factors in the designed experiment. Table 17.2 gives the data collection plan used in the initial eighteen runs of the designed experiment. The actual settings of the factors A, B, C, D, and E (revolutions per minute, pre-etch total flow, etc.) are not given for proprietary reasons. We already know the settings of factor F are 50 Å and 200 Å. In Table 17.2, "–" refers to the low setting of the factor and "+" refers to the high setting of the factor. The eighteen experiment runs were performed in random order.

One thermally grown oxide wafer was used at each one of the eighteen experimental runs. Oxide thickness was measured before and after etching at the nine wafer sites illustrated in Figure 17.2. The difference in oxide thickness before and after etching, which is the amount of oxide etched, was calculated at the nine sites on each wafer. Before each run, test passes were made to ensure that the amount of oxide etched (factor F) was targeted correctly.

Table 17.1. *Six Factors of the Etch Uniformity Experiment.*

A = Revolutions per minute
B = Pre-etch total flow (N_2 + N_2 & water vapor)
C = Pre-etch vapor flow
D = Etch total flow (N_2 + N_2 & water vapor)
E = Etch vapor flow
F = Amount of oxide etched (50 Å or 200 Å)

Three related measures of etch variability were used as responses:
- Standard deviation (S.D.) of the oxide etched at the nine sites of one wafer
- Uniformity of the oxide etched at the nine sites of one wafer. Uniformity is the standard deviation expressed as a percentage of the mean (standard deviation multiplied by 100 and divided by the mean.)
- Natural logarithm of etch uniformity.

Uniformity is a response widely used by engineers, and standard deviation is the response preferred by statisticians. The logarithm of uniformity was considered to improve the statistical analysis, because the values of uniformity in the experiment had an enormous range (the largest uniformity was almost ninety times larger that the smallest uniformity.) All three related responses were considered to see whether they led to the same conclusions, which they did. The mean etch thickness was not considered as a response, since it can be tightly controlled with etching time.

Table 17.2. *Data Collection Plan with Responses for the Initial Eighteen Runs.*

	Factors								
Run	A	B	C	D	E	F	Etch S.D. (Å)	Etch Uniformity	Log_e Etch Uniformity
1	−	−	−	−	−	−	6.49	11.00	2.40
2	−	+	−	−	+	+	19.20	10.10	2.31
3	−	−	+	−	+	+	19.11	8.70	2.16
4	−	+	+	−	−	−	4.05	9.20	2.22
5	−	−	−	+	−	+	7.20	3.20	1.16
6	−	+	−	+	+	−	2.61	4.90	1.59
7	−	−	+	+	+	−	3.10	5.80	1.76
8	−	+	+	+	−	+	6.18	2.90	1.06
9	+	−	−	−	+	−	1.71	3.10	1.13
10	+	+	−	−	−	+	7.00	3.60	1.28
11	+	−	+	−	−	+	7.25	3.60	1.28
12	+	+	+	−	+	−	3.43	7.70	2.04
13	+	−	−	+	+	+	1.60	0.80	−0.22
14	+	+	−	+	−	−	16.58	41.00	3.71
15	+	−	+	+	−	−	38.50	71.00	4.26
16	+	+	+	+	+	+	3.10	1.50	0.41
17	0	0	0	0	0	0	4.83	3.90	1.36
18	0	0	0	0	0	0	6.37	5.20	1.65

The first sixteen runs in Table 17.2 make up a resolution IV quarter fraction of a six factor, 64 run full factorial design. Runs 17 and 18 are center points used to estimate experimental error; they make no contribution to the estimate of the main effects or the interactions and so we shall not consider them in the discussion that follows.

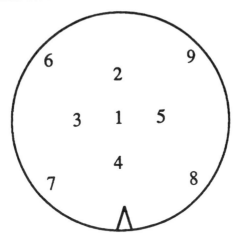

Figure 17.2. *Nine Wafer Sites Where Thickness is Measured for Uniformity Calculations.*

The generators[39] for the first sixteen runs are E = ABC and F = BCD, and the corresponding defining relation is I = ABCE = BCDF = ADEF. The defining relation can be used to derive the alias chains for the two-factor interactions given in the first column of Table 17.3. In deriving these alias chains, we assume that interactions among three or more factors are negligible. Note that each two-factor interaction is aliased with at least one other two-factor interaction.

The four asterisks in Table 17.3 indicate alias chains that were significant for the response $\log_e$ (Uniformity). These chains were also significant for the response standard deviation. Confounding means that, given that the AC+BE alias chain was significant, it was not possible to tell without further experimentation whether this significance was due to the AC interaction, the BE interaction, or both interactions. There was a similar confounding problem with the other three significant alias chains.

To break this confounding, eight additional runs were performed at eight new factor settings. These eight new factor settings were obtained by taking the first eight rows in Table 17.2 (where A was at its low (–) setting) and changing the level of A to its high (+) setting (see Table 17.4). Notice that the levels of the other factors are not changed. This procedure is called *semifolding* on factor A. The resulting design of 16 + 8 = 24 runs (omitting the center points) is a three-quarter fractional factorial design, since the 24 runs are three quarters of the 32 observations in a half replicate in six factors [see the Appendix, John, (1971), or Chapter 10 of Diamond, (1981)].

Semifolding on A makes all interactions involving A or E estimable from the 24 runs, as is shown in the Appendix. Thus the nine interactions that are underlined in Table 17.3 become estimable. The ability to estimate these interactions means that a model with the following nineteen terms can be fit to the data: a constant term, six main effects (A, B, C, D,

---

[39] For an introduction to generators and related topics see Box, Hunter, and Hunter (1978).

E, and F), and twelve interaction terms (AB, CE, AC, BE, AD, EF, AE, [BC or DF], AF, DE, [BD or CF], and [BF or CD]). Semifolding on A did not break the chains involving the letters in the generator BCDF. In the first experiment with sixteen runs, nine interactions in the alias chains were potentially significant, AC, BE, AD, EF, AE, BC, DF, AF, and DE. The eight additional experimental runs obtained by semifolding allow seven of these interactions to be estimated. In addition, the engineer was able to state *a priori* that DF was negligible and so we are able to attribute significance in the chain BC + DF to the BC interaction.

Table 17.3. *Regression Results for the Response Log_e (Uniformity) for the First Sixteen Runs. The table includes the least squares estimates of the coefficients in the regression model and the corresponding standard error, t-ratio, and p-values. The four significant alias chains with confounded interactions are indicated by asterisks (*).*

Term or Alias Chain	Coefficient Estimate	Std. Error	t-Ratio	Prob > |t|
Intercept	1.784375	0.0267	66.83	0.0002
A	−0.048125	0.0267	−1.80	0.2133
B	0.043125	0.0267	1.62	0.2476
C	0.114375	0.0267	4.28	0.0504
D	−0.068125	0.0267	−2.55	0.1254
E	−0.386875	0.0267	−14.49	0.0047
F	−0.604375	0.0267	−22.64	0.0019
*AF+DE	−0.444375	0.0267	−16.64	0.0036
*AD+EF	0.371875	0.0267	13.93	0.0051
*AC+BE	0.146875	0.0267	5.50	0.0315
*AE+BC+DF	−0.509375	0.0267	−19.08	0.0027
BD+CF	−0.066875	0.0267	−2.50	0.1292
AB+CE	0.080625	0.0267	3.02	0.0944
BF+CD	0.041875	0.0267	1.57	0.2573

DATA ANALYSIS AND INTERPRETATION OF RESULTS

We analyzed the data resulting from the complete experiment in twenty-six runs. The experimental design and the values of the etch standard deviation, etch uniformity, and log of etch uniformity for the complete experiment are given in Tables 17.2 and 17.4.

The twenty-six observations of etch uniformity range in value from 0.80 to 71.00. The very wide range of values (71.00 is almost ninety times as large as 0.8) suggests the use of a compressive transformation such as the logarithm or square root. For log_e(uniformity) the range of values is −0.22 to 4.26, which is smaller. The log transformation should allow better modeling of the response, which in turn should allow a more accurate determination of the significance of factors and interactions.

The value 71.00 for etch uniformity in run 15 is flagged as an outlier in the untransformed data but not after the log transformation. If run 15 is removed from the analysis, the regression on uniformity, or on standard deviation, of the remaining twenty-five points yields the same conclusions as the analysis on the logarithm of the uniformity employing all twenty-six points. This result is not surprising, since the logarithmic transformation brings the apparent outlier closer to the remaining data points. In addition, when the response is standard deviation, which like the uniformity also measures variability, the fit of the model is not as good in terms of R² as the fit of the model using log$_e$(uniformity). Thus we prefer to use log$_e$(uniformity) as the response. However, both models lead to the same recommendations for optimal factor settings. For these reasons the remainder of this paper uses log$_e$(uniformity) as the response that measures variability in the amount of oxide etched from a wafer.

Table 17.4. *Data Collection Plan with Responses for the Eight Additional Runs. The additional runs were obtained from the first experiment by semi-folding on A.*

Run	Factors A B C D E F	Etch S.D. (Å)	Etch Uniformity	Log$_e$ Etch Uniformity
19	+ − − − − −	12.96	21.00	3.04
20	+ + − − + +	9.81	4.60	1.53
21	+ − + − + +	6.19	3.05	1.12
22	+ + + − − −	13.67	29.40	3.38
23	+ − − + − +	8.59	4.18	1.43
24	+ + − + + −	6.63	14.05	2.64
25	+ − + + + −	6.60	13.52	2.60
26	+ + + + − +	9.23	4.43	1.49

Since the last eight runs were conducted several weeks after the first 18 runs, we included in the model a blocking term as an additional factor with value −1 for the first eighteen runs and value 1 for the last eight runs. Note that in the regression fits that follow all "−" settings are given the value of −1 and all "+" settings are given the value of 1. Thus a −1 corresponds to the low setting of a factor and a 1 to its high setting.

We performed several regression fits of the log(uniformity) data using a stepwise regression algorithm. Table 17.5 summarizes the results of the algorithm's steps. The Step 1 column gives the estimates of the coefficients and their t-ratios for the regression model that includes only the main effects. The last two rows of the table show the coefficient of determination R² and the standard error s, respectively. The Step 2 column gives the estimates and t-ratios for the model that adds the one two-factor interaction or block factor that gives the highest R², namely the BC interaction. Step 3 adds the next most influential two-factor interaction or block factor, and so on. The model that we select based on this stepwise regression is the one in Step 5 with eleven terms since, after that step, further reductions in the R² are relatively small. This model is the one presented in Tables 17.6 and 17.7.

Table 17.6 indicates that E and F are the only statistically significant main effects for the model of log$_e$(uniformity), and Table 17.7 gives the overall fit of the model. Table 17.8 shows the regression fit for the model of log$_e$(uniformity) that includes only the main effects E and F along with the interactions AF, BC, and EF, and the blocking factor. Table 17.9 shows a summary of fit for this model. This model is the one finally selected as best fitting the data since dropping all main effects other than E and F reduces the R² from .918 to .906 (a negligible amount[40]). Expressed in terms of etch uniformity this final model is given by the following equation:

---

[40] Many computer programs discourage including an interaction in a regression model without also including the main effects of the factors in the interaction (obeying "hierarchy"). However, this recommendation is not

Etch Uniformity

$$= \exp[1.95 - 0.464\,E - 0.532\,F - 0.375\,A\,F - 0.434\,B\,C + 0.285\,E\,F - 0.2\,\text{Block}].$$

Table 17.5. *Estimates and t-Ratios in the Different Steps of the Stepwise Regression for $Log_e(Uniformity)$. The first column indicates the terms that could potentially be in the model. The remaining columns indicate the terms that are in the model at different steps, with values of the coefficients in the regression model and the corresponding t-ratios to test for their significance. The next-to-last row gives the value of R^2 for the corresponding regression. The last row gives the value of the standard error.*

Term	Step 1	Step 2	Step 3	Step 4	Step 5	Step 6	Step 7
Constant	1.856	1.856	1.856	1.856	1.978	1.978	1.978
A	.067	.067	.067	.067	−0.048	−0.05	−0.048
t-ratio	.40	.48	.60	.74	−0.52	−0.57	−0.62
B	.064	.064	.064	.064	.064	.064	.064
t-ratio	.40	.48	.60	.74	.85	.93	1.02
C	.074	.074	.074	.074	.074	.074	.074
t-ratio	.46	.56	.69	.86	.99	1.07	1.18
D	−0.083	−0.083	−.083	−0.013	.013	−0.068	−0.068
t-ratio	−0.52	−0.63	−0.78	−.15	.17	−0.80	−0.88
E	−.318	−0.463	−0.463	−0.463	−0.463	−0.463	−0.387
t-ratio	−1.99	−3.29	−4.08	−5.06	−5.81	−6.31	−5.02
F	−0.657	−0.657	−.533	−0.533	−0.533	−0.533	−0.533
t-ratio	−4.11	−4.95	−4.70	−5.82	−6.68	−7.26	−7.98
BC		−.433	−0.433	−0.433	−0.433	−0.433	−0.358
t-ratio		−3.08	−3.83	−4.74	−5.44	−5.91	−4.64
AF			−0.373	−0.373	−0.373	−0.373	−0.373
t-ratio			−3.29	−4.07	−4.68	−5.08	−5.58
EF				.290	.290	.209	.209
t-ratio				3.17	3.64	2.47	2.71
Block					−0.224	−0.224	−0.224
t-ratio					−2.47	−2.69	−2.95
AD						.163	.163
t-ratio						1.93	2.12
AE							−0.152
t-ratio							−1.97
R^2	.5332	.6941	.8130	.8852	.9185	.9356	.9504
s	.782	.650	.523	.423	.368	.336	.308

Figures 17.3, 17.4, and 17.5 are contour plots of this model for different two dimensional projections of the model. The contour plot in Figure 17.4 revealed an operating point with low uniformity that had not been used before. Figure 17.6 gives the interaction plots for this model. For screening designs in general, the conclusion derived using contour plots and using interaction plots are similar. Hence, because of their simplicity, the interaction plots are preferred by many.

---

universally valid. When there is a significant interaction between two factors, the corresponding main effects have little meaning. In particular for the data in this paper adding these main effects does not improve the predicting power of the model; so for simplicity's sake we only include in the model the main effects E and F.

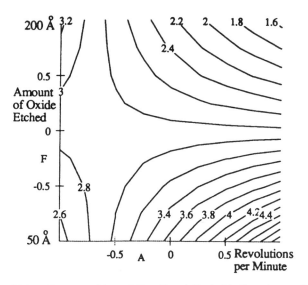

Figure 17.3. *Contour Plot of Predicted Etch Uniformity at B = −1 and E = 1 as a Function of Amount of Oxide Etched (F) and Revolutions per Minute (A). This figure indicates that one should run the 50 Å (low F) process at low A, that is at low revolutions per minute, and the 200 Å (high F) process at high A, that is at high revolutions per minute.*

Table 17.6. *Regression Fit for the Model of $\log_e(Uniformity)$ with All Main Effects, the Interactions AF, BC, and EF, and the Blocking Factor. The estimate of the coefficient, its standard error, the t-ratio, and the significance probability are given for each term in the model.*

Term	Estimate	Std Error	t-Ratio	Prob > |t|
Intercept	1.9776042	0.09069	21.81	0.0000
A	−0.048125	0.09198	−0.52	0.6085
B	0.0641667	0.0751	0.85	0.4063
C	0.0741667	0.0751	0.99	0.3390
D	0.0134375	0.07966	0.17	0.8683
E	−0.462813	0.07966	−5.81	0.0000
F	−0.5325	0.07966	−6.68	0.0000
BC	−0.433438	0.07966	−5.44	0.0001
AF	−0.3725	0.07966	−4.68	0.0003
EF	0.2903125	0.07966	3.64	0.0024
Block	−0.224271	0.09069	−2.47	0.0259

Table 17.7. *Summary of Fit for the Model in Table 17.6. The model includes all main effects, the interactions AF, BC, and EF, and the blocking factor.*

Rsquare	0.918
Rsquare(adj)	0.865
Mean of Response	1.877
Root Mean Square Error	0.368

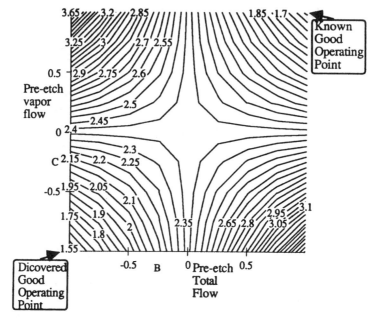

Figure 17.4. *Contour Plot of Predicted Etch Uniformity at A = 1, E = 1, and F = 1 as a Function of Pre-Etched vapor Flow (C) and Pre-Etched Total Flow (B). This figure indicates that (low B, low C) = (low pre-etch total flow, low pre-etch vapor flow) is likely to be a good operating condition as well as the previously known (high B, high C) = (high pre-etch total flow, high pre-etch vapor flow).*

Table 17.8. *Regression Fit for the Model of $\log_e$(Uniformity). The model includes only the main effects (E and F), the interactions (AF, BC, and EF), and the blocking factor.*

Term	Estimate	Std Error	t Ratio	Prob > |t|
Intercept	1.9535417	0.07396	26.41	0.0000
E	−0.462813	0.07537	−6.14	0.0000
F	−0.5325	0.07537	−7.07	0.0000
BC	−0.433438	0.07537	−5.75	0.0000
AF	−0.3725	0.07537	−4.94	0.0001
EF	0.2858333	0.07105	4.02	0.0007
Block[−1-1]	−0.200208	0.07396	−2.71	0.0140

Table 17.9. *Summary of Fit for the Model of $\log_e$(Uniformity). It includes only the main effects E and F, the interactions AF, BC, and EF, and the blocking factor.*

Rsquare	0.906
Rsquare(adj)	0.879
Mean of Response	1.877
Root Mean Square Error	0.348

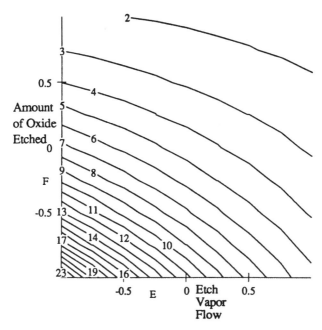

Figure 17.5. *Contour Plot of Predicted Etch Uniformity at B = 1, C = 1, and A = 1 as a function of Amount of Oxide Etched (F) and Etch Vapor Flow (E). This figure indicates that high E (high etch vapor flow) should produce the best uniformity for both the 50 Å (low F) process and the 200 Å (high F) process.*

Confirmation Experiments

Table 17.10 gives the best factor settings as predicted by the model for the 50 Å (low F) process and for the 200 Å (high F) process using the new conditions for factors B and C. A confirmation experiment was performed at these settings. Table 17.10 also compares the predicted uniformity to the observed uniformity at these settings. The result of the confirmation test shows that the uniformity of the process was correctly predicted when the amount of oxide etched is 200 Å, since the observed uniformity lies in the confidence interval. When the amount of oxide etched was 50 Å, the observed uniformity was larger than that predicted by the model. However, starting from the predicted settings, the engineers changed the setting of Factor A to a value close to its center value. With this change, the uniformity for the 50 Å target decreased to 3.7%, which met the process uniformity goal. It would have been desirable to also increase factor E (Etch Vapor Flow) beyond its high value, but this was not possible due to flow limitations.

CONCLUSIONS AND RECOMMENDATIONS

This chapter shows how a screening experiment can be augmented by careful analysis of the data and the experimental design. Semifolding on A allowed the addition of only eight new experimental runs to estimate nearly all the interactions of interest. Had standard experimental practices been followed, sixteen additional runs would have been required. Analysis of the combined data led to a process that surpassed its uniformity goals. Semifolding saved considerable experimental resources as less time, fewer wafers, and fewer experimental set-ups were needed to complete the experiment. The results obtained by analyzing uniformity were the same as those obtained by analyzing the standard deviation.

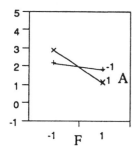 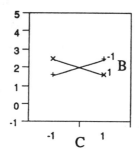 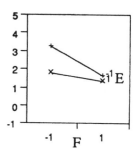

Figure 17.6. *Interaction Plots for Log (Uniformity). Similar to Figure 17.3, Figure 17.6a (AF interaction) indicates that one should run the 50 Å (low F) process at low A (at low revolutions per minute) and the 200 Å (high F) process at high A (at high revolutions per minute) Similar to Figure 17.4, Figure 17.6b (BC interaction) indicates that (low B, low C) = (low pre-etch total flow, low pre-etch vapor flow) is likely to be a good operating condition as well as the previously know (high B, high C) = (high pre-etch total flow, high pre-etch vapor flow). Similar to Figure 17.5, Figure 17.6c (EF interaction) indicates that high E (high etch vapor flow) should produce the best uniformity for both the 50 Å (low F) process and the 200 Å (high F) process.*

Table 17.10. *Predicted Uniformity Versus Observed Uniformity at the Predicted Best Operating Conditions.*

Target	A	B	C	D	E	95% Confidence Interval for Predicted Uniformity	Observed Uniformity
F = 50 Å	low	low	low	either	high	(1.1%, 5.9%)	8.5%
F = 200 Å	high	low	low	either	high	(.7%, 3.5%)	1.5%

Acknowledgments

We would like to acknowledge the comments provided by Bob Witowski and Jack Reece, both formerly of SEMATECH.

References

Box, G. P. E., W. G. Hunter, and J. S. Hunter, *Statistics for Experimenters: An Introduction to Design, Data Analysis, and Model Building*, New York, NY: John Wiley and Sons, Inc., 1978.

Diamond, W. J. *Practical Experiment Designs for Engineers and Scientists*. Belmont, CA: Lifetime Learning Publications, Wadsworth, 1981.

John, P. W. M. *Statistical Design and Analysis of Experiments*, Philadelphia: SIAM, 1998.

APPENDIX

How Semifolding Works

This appendix contains a technical explanation of how factors and interactions are estimated in the three-quarter fractional factorial design. It is not necessary to understand this explanation to analyze this type of design, since the estimates we consider are the least squares estimates that statistical software packages automatically provide. In this discussion, we do not consider the two center points (runs 17 and 18 in Table 17.2) since they do not enter into the calculation of the least squares estimates of the main effects and interactions. The two center points were included in the experiment to obtain an estimate of pure error and to be able to test for lack of fit and curvature. Table 17.11 shows no indication of lack of fit.

Table 17.11. *Lack of Fit Test for Selected Model Given in Table* 17.8.

Source	DF	Sum of Squares	Mean Square	F Ratio	Prob > F
Lack of Fit	18	2.2602021	0.125567	2.9861	0.4300
Pure Error	1	0.0420500	0.042050		
Total Error	19	2.3022521			

The first sixteen runs of the initial experiment in six factors are given in Table 17.2. They constitute a 2^{6-2} fractional factorial design of resolution IV with defining relation

$$I = ABCE = BCDF = ADEF.$$

With this design, the main effects can be estimated clear of two-factor interactions, and each two-factor interaction is confounded with at least one other two factor interaction (see Table 17.3). The eight runs given in Table 17.4 were added by semifolding on A (repeating the eight initial runs at low A and changing only the level of factor A). This gave a three-quarter fractional factorial in twenty-four runs. Ignoring higher order interactions, all interactions involving A or E (that is, nine of the 15 two-factor interactions among the six factors) can be estimated from the twenty-four runs, as we now show.

For convenience, we denote the first group of eight runs (runs 1 to 8) by Group i, the second group of eight runs (runs 9 to 16) by Group ii, and the last group of eight runs (runs 19 to 24) by Group iii. Since the eight runs in Group i are all at the low level of A they constitute the one-eighth fraction generated by

$$I = -A = ABCE = -BCE = BCDF = -ABCDF = ADEF = -DEF. \qquad (i)$$

The eight runs in Group ii form the one-eighth fraction generated by

$$I = A = ABCE = BCE = BCDF = ABCDF = ADEF = DEF. \qquad (ii)$$

The eight runs in Group iii form the one-eighth fraction generated by

$$I = A = -ABCE = -BCE = BCDF = ABCDF = -ADEF = -DEF. \qquad (iii)$$

Note that the defining relation for Group iii (the eight new runs) can be obtained from the defining relations of Group i (the eight old runs that were semifolded on A to give the eight new runs) by replacing A by –A.

We now combine the three-factor groups in pairs to obtain quarter fractions of the full factorial from which, ignoring three and higher order interactions, we can estimate main effects and two-factor interactions. We have already seen that groups i and ii combine to form Quarter Fraction i + ii. These are the 16 initial runs of Table 17.2 with defining relation

$$I = ABCE = BCDF = ADE.$$

The sixteen runs in groups i and iii combine to form Quarter Fraction i + iii with defining relation:

$$I = BCDF = -BCE = -DEF.$$

The alias chains for Quarter Fraction i + iii are

A + ABCDF – ABCE – ADEF	B + CDF – CE – BDEF
C + BDF – BE – CDEF	D + BCF – BCDE – EF
E + BCDEF – BC – DF	F + BCD – BCEF – DE
AB + ACDF – ACE – ABDEF	AC + ABDF – ABE – ACDEF
AD + ABCF – ABCDE – AEF	AE + ABCDEF – ABC – ADF
AF + ABCD – ABCEF – ADE	BD + CF – CDE – BEF
BF + CD – CEF – BDE	

Notice that if we ignore third and higher order interactions, all the two-factor interactions involving A (which are underlined) can be estimated from Quarter Fraction i + iii.

The sixteen runs in groups ii and iii combine to form Quarter Fraction ii + iii with defining relation

$$I = A = BCDF = ABCDF.$$

The alias chains for Quarter Fraction ii + iii are

B + AB + CDF + ACDF	C + AC + BDF + ABDF
D + AD + BCF + ABCF	E + AE + BCDEF + ABCDEF
F + AF + BCD + ABCD	BC + ABC + DF + ADF
BD + ABD + CF + ACF	BE + ABE + CDEF + ACDEF
BF + ABF + CD + ACD	CE + ACE + BDEF + ABDEF
DE + ADE + BCEF + ABCEF	EF + AEF + BCDE + ABCDE
BCE + ABCE + DEF + ADEF	BDE + ABDE + CEF + ACEF
BEF + ABEF + CDE + ACDE	

Notice that if we ignore higher order interactions, all the two-factor interactions involving E (which are underlined) can be estimated from Quarter Fraction ii + iii except for interaction AE. Recall, however, that interaction AE can be estimated from Quarter Fraction i + iii. This ends the proof that the three-quarter fractional factorial design can be used to estimate all the two-factor interactions with factors A and E.

Notice several other facts about the three-quarter fractional factorial design obtained by combining groups i, ii, and iii:

- The main effect A can be estimated from two of the quarters, namely quarters i + ii and i + iii. Its least squares estimate is the average of these two estimates.

- We are not able to break the alias chains BC + DF, BD + CF, and BF + CD because the word BCDF was a generator of all three quarter fractions. We were able to estimate interaction BC in the etch uniformity experiment because experience enabled the engineers to say that interaction DF was negligible.
- The estimates are not orthogonal to one another. An important implication of this lack of orthogonality is that if, for example, we were to decide (after seeing the t-values) that main effect C is negligible and dropped this factor from the model, we should get a different estimate of AC from the one obtained with the factor C included. Lack of orthogonality explains why in the stepwise regression of Table 17.5 some of the coefficients change as more terms are brought into the model.

The main point of this case study is to show how to reduce the work and resources needed to break the alias chains that are significant in the first sixteen runs. The work and resources required was reduced by judiciously choosing only eight more runs. If the engineer in the etch uniformity example had insisted on working only with regular fractions and orthogonal estimates he would have had to take another sixteen runs to get a half replicate of resolution V. Of course he would have then been able to estimate all the fifteen two-factor interactions but at a high cost. In the etch uniformity experiment the engineer, having done his screening experiment, was interested only in a few of the interactions and wanted to get this information in as few runs as possible. As a matter of fact, before he knew of the semifolding technique, the engineer had not finished clarifying the results of some previous experiments because he could not afford the required additional work and resources.

Semifolding on E would have allowed estimation of all interactions involving A and E. Semifolding on B or C would have allowed estimation of all interactions involving B or C, and semifolding on D or F would have allowed estimation of all interactions involving D or F. Therefore, engineers can choose how to semifold based on the results of their initial screening experiments.

BIOGRAPHIES

Joel Barnett is a process development engineer in the surface conditioning division of FSI International. He spent four years as a CVD and ion implant process engineer at LSI Logic in Santa Clara, CA. He spent three years on assignment from LSI to SEMATECH in Austin, TX. At SEMATECH, he worked on the GENUS tungsten deposition equipment improvement program. Joel received a B.S. in chemical engineering from the University of California at Berkeley in 1984.

Veronica Czitrom is a Distinguished Member of Technical Staff at Bell Labs, Lucent Technologies. She leads consulting and training efforts for the application of statistical methods, primarily design of experiments, to semiconductor processing. Dr. Czitrom was on a two year assignment at SEMATECH. She has published research papers on design of experiments and mixture experiments, and four engineering textbooks. She received a B.A. in Physics and an M.S. in Engineering from the University of California at Berkeley, and a Ph.D. in Mathematics with concentration in Statistics from the University of Texas at Austin. She was a faculty member at The University of Texas at San Antonio, and at the Universidad Nacional Autónoma de México.

Peter W. M. John was born in Wales and educated at Oxford University and the University of Oklahoma. He has been a professor of mathematics at the University of California and the University of Texas and a research statistician with Chevron Research Corporation. He has written three books and numerous papers on design of experiments. He is a former chairman of the Gordon Research on Statistics and a fellow of the Institute of Mathematical Statistics, the American Statistical Society, and the Royal Statistical Society.

Ramón V. León is associate professor of statistics at the University of Tennessee at Knoxville. Until June 1991, he was supervisor of the quality engineering research and technology group of the quality process center of AT&T Bell Laboratories. Ramón is coauthor of the AT&T reliability management handbook, Reliability by Design. His papers have appeared in Technometrics, The Annals of Statistics, The Annals of Probability, Mathematics of Operations Research, Statistics Sinica, and The Journal of Mathematical Analysis and Applications. Ramón received Ph.D. and M.S. degrees in probability and statistics from Florida State University and a M.S. degree in mathematics from Tulane University. Before joining AT&T Bell Laboratories in 1981, Ramón was on the faculty of Florida State University and Rutgers University.

CHAPTER 18

PLANARIZATION BY CHEMICAL MECHANICAL POLISHING: A RATE AND UNIFORMITY STUDY

Anne E. Freeny and Warren Y.-C. Lai

Wafer planarization by chemical mechanical polishing (CMP) has gained importance as very large scale integration (VLSI) pushes below the 0.5 μm regime. The global wafer planarity which can be achieved by this technology permits maintaining fine features with increasing metal levels. In this article, we describe a designed experiment used to help identify an optimum processing regime and quantify process reproducibility. Looking at the effects of platen and wafer rotation on rate and uniformity of oxide removal, we discovered a tradeoff between rate (throughput) and uniformity (quality). A process drift over time was quantified. These results contributed to a judgement of cost and timing in introducing a new process and equipment into manufacturing.

EXECUTIVE SUMMARY

Problem

Advanced photolithography in VLSI increasingly demands global planarity across a chip-sized printing field for fine resolution. Planarization by CMP is a simple technique that can achieve this. In a designed experiment we evaluated a polisher which does oxide planarization by CMP for possible use in the wafer fabrication manufacturing process. Our goal was to find the maximum rate of oxide removal which could be used (for throughput reasons) without degrading the uniformity of the removal over the surface of the wafer.

Solution Strategy

In CMP, a wafer is held by a rotating carrier and is polished by pressing the wafer face down onto a polishing pad on a rotating platen. The important parameters for the polishing process are platen and wafer rotation frequencies and polishing pressure. Design of experiments (DOE) is an effective way to investigate the relationship of these parameters to oxide removal rate and uniformity across the wafer, while quantifying the reproducibility of the removal process from wafer to wafer. We designed an experiment to study the effects of platen and wafer rotations for an automatic wafer polishing machine suitable for CMP. All combinations of five wafer and three platen rotation frequencies were used. Three wafers were polished at each combination of wafer and platen rotation frequencies. A tendency of the polisher removal rate to drift lower through time had previously been noticed. Thus an important design issue was to choose the order of the fifteen combinations to estimate the effects of the design parameters independent of a linear drift and to quantify the magnitude of the drift. Graphical methods and analyses of variance allowed detection of the effects of the parameters and estimation of the relationship of uniformity and removal rate to platen and wafer rotation frequencies.

Conclusions

Both the polishing rate and across-the-wafer non-uniformity showed large increases with increasing platen rotation and slight increases with increasing wafer rotation. A simple linear model is appropriate for quantifying these effects. A preferred operating regime for platen and wafer rotation frequencies was derived. The presence of the suspected drift was confirmed as a decrease in polishing rate over time regardless of the rotation frequencies. This drift, combined with large wafer-to-wafer variability, results in imperfect process reproducibility even with automation. This lack of reproducibility is an important factor in judging the cost and timing of introducing CMP into manufacturing.

PROCESS

Chemical-mechanical polishing (CMP) (see Lifshitz, 1988) is a novel approach to planarizing oxide topography. The procedure presses a rotating wafer face-down onto the surface of a rotating padded platen impregnated with a slurry of extremely fine abrasive. The key parameters affecting the removal rate and uniformity across the wafer are the wafer pressure against the polishing pad, and the platen and wafer rotation frequencies.

The polisher used in the experiment was an automatic cassette-to-cassette-loading single-wafer polisher with five stations: input, primary polish, secondary polish, rinse, and output. Wafers are removed singly from the input cassette and transported via the carrier to the primary polishing station. The polishing action consists of three relative motions: the wafer rotation about its center, the platen rotation, and a small oscillation of the arm supporting the wafer carrier. After primary polishing is completed, the carrier transports the wafer to the secondary polishing station, which performs a clean-up step. The polished wafer is unloaded via a water track into the output cassette, and kept submerged until cassette removal.

DATA COLLECTION PLAN

Thickness Measurements and Derived Responses

Figure 18.1 shows the 37 sites from the 49 possible sites where oxide film thickness was measured on experimental wafers. Each wafer was measured before and after polishing so that a comparison could be made on a site-by-site basis. In this study, measurements from sites 27-33 and 36-40 were eliminated, leaving 37 sites: the former because of proximity to the wafer scribe mark; the latter because of proximity to the wafer flat. Both the scribe mark and the flat induce nonuniform oxide removal because of local topography disruption. The amount of oxide removed was calculated as the difference between the thicknesses before and after polishing at each of the 37 sites. The responses of interest were the oxide removal rate and the uniformity of the removed oxide film. The removal rate (R) was determined by dividing the mean of the thicknesses of the removed oxide layer measured at the 37 sites by the polishing time (three minutes in this study). The uniformity (U) was computed as the standard deviation of the thicknesses of the removed oxide layer at the 37 sites.[41]

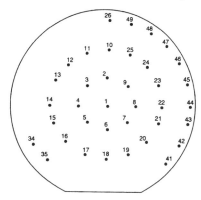

Figure 18.1. *Map of Thickness Measurement Locations. Sites 27–33 were eliminated because of proximity to the wafer scribe mark. Sites 36–40 were eliminated because of proximity to the wafer flat. Both induce nonuniform oxide removal.*

The Design of the Experiment

Previous work (see Lifshitz, 1988) indicated that the polisher may not reach equilibrium immediately after a change in the parameter settings. Therefore we used three wafers in each experimental condition, in case the first wafer polished behaved differently from the other two. To minimize initial wafer-to-wafer variability, all wafers came from a lot of 50 oxide-coated unpatterned wafers, which was generated in two batches of 25 wafers each. Three wafers were randomly selected from the two batches so that each polishing condition would be performed with one wafer from one batch and two wafers from the other.

The two parameters of interest were platen (polishing pad) rotational frequency (P) and wafer (wafer carrier) rotational frequency (W). Previous work (see Lifshitz, 1988) and other recommendations suggested good operating ranges of 10 to 20 rpm for platen frequency and 10 to 55 rpm for wafer frequency. Therefore, we selected three values (11, 15, and 19 rpm) for platen frequencies and five values (varying from 12 to 52 in increments

---

[41] There are two measures of the quality of the polishing process: the final planarity and the removal rate uniformity. For actual topography reduction, the final planarity is the important measure. For process evaluation, because of the number of wafers and measurements involved, planar oxide monitors with no underlying topography are used for the study. The initial uniformity of these wafers is very good, which allows us to infer the higher quality of the final planarity than that of the removal uniformity.

of 10) for wafer frequencies. The 15 polishing conditions combining every platen frequency with every wafer frequency formed a full factorial experiment. Pad pressure was held constant at a single optimum value based on previous work.

Because both P and W are quantitative, they are parameterized using orthogonal polynomials, which allows us to evaluate the shape of the functional relationship of the response to each of the parameters. Thus the effect of each parameter is seen as the sum of a linear, quadratic, etc, component. Consequently, because of the orthogonality, we can quantify the importance of each component. The P by W interaction can also be broken into similar polynomial component combinations.[42]

Since properties of the polishing pad are likely to change with use, we were concerned that evaluation of the effects of the parameters would be affected by a drift (time dependence) of the polisher performance. There is no sequence of runs for a three by five experiment that makes estimates of effects of the parameters completely free of a linear effect of polisher drift. In this experiment, the sequence of runs was chosen to confound the effect of a linear drift with a high order component of the interaction, specifically the P quadratic by W cubic component.[43] This effect is denoted in the figures by PQWC. The Appendix describes how the run sequence was derived. The main effects of P and W, and the low-order components of the P by W interaction are essentially independent of a linear drift in the polisher. The sequence of polishing conditions is given in Table 18.1.

Table 18.1. *Platen and Wafer Frequency Experiment Sequence.*

Run	1	2	3	4	5	6	7	8	9	10	11	12	13	14	15
P	15	19	11	15	11	19	11	15	19	11	19	15	19	11	15
W	22	42	42	52	12	12	32	32	22	52	52	12	32	22	42

It required several hours to polish the 45 wafers. The time at which each of the wafers was polished was not recorded, however the experiment ran smoothly enough so that the assumption of equally-spaced measurements in time needed to estimate the magnitude of the drift is not unreasonable. One polisher problem requiring operator intervention was noted. It occurred when polishing the third wafer at P = 11, W = 32. We discuss an anomalous result for uniformity which may be related to this interruption in Results for Uniformity Section.

ANALYSIS AND INTERPRETATION OF RESULTS

Graphical Presentation

The box plots[44] in Figure 18.2 show the distribution of removal rates over the 37 sites on each wafer. It is organized so that W appears as rows and P as columns. The three

---

[42] A discussion of parameterizing quantitative factors in this way can be found in, e.g., Hicks, 1973.
[43] "Confounding" means that we cannot tell the difference between a linear drift in the polisher and this component of the interaction. However, real high-order interactions of this type are unexpected, so we are willing to assume that the size of this component of the interaction will reflect the linear time dependence of the polisher and not some underlying physical relationship of the response to the factors.
[44] A box is drawn for every wafer summarizing the distribution of the observations over the 37 sites on each wafer. The large dot within the box is plotted at the median, and the box boundaries signify the upper and lower quartiles. "Whiskers" are drawn out to the closest data point not more than 1.5 times the interquartile range from the nearest quartile. Points beyond the whiskers are plotted individually.

boxes within each P and W combination are shown in the order in which the three wafers in that condition were polished.

Information relating to removal rate and uniformity can be found from the box plots by noting that both the mean (removal rate) and median are measures of location, and both the standard deviation (uniformity) and vertical size of the box are measures of spread.

Figure 18.2 indicates that removal rate increases as P increases. Furthermore there appears to be a lot of variability among the three wafers in any experimental condition, both in location and spread. It also seems that the boxes are larger, indicating more variability, in conditions with P = 19 rpm. The third wafer polished with P = 19 rpm and W = 22 rpm appears to be unusual in both location and spread.

Model for Analyses of Variance

To evaluate the effects of the parameters P, W, and the order in which wafers were polished (O), in each experimental condition, we fit the model

$$Y_{ijk} = m + P_i + W_j + (PW)_{ij} + O_{k(ij)}$$

to both Y = R, the removal rate, and Y = log (U), the log uniformity.[45] This model allows us to evaluate the main effects of P and W, and a possible P by W interaction. As previously stated, P and W are decomposed by orthogonal polynomials, and the component of PW, PQWC, which is confounded with a linear drift, is separated from the rest of PW. The parameter $O_{k(ij)}$ specifies that the wafer-to-wafer variability effects will be computed within each of the 15 experimental conditions.

The results of the analyses of variance are initially shown by half-normal quantile-quantile plots (often called half-normal plots)[46] of the single-degree-of-freedom effects (see Daniel, 1959, and Nair, 1986). We discuss the results for removal rate next and uniformity following that.

Results for Removal Rate

Figure 18.3 presents half-normal plots of single-degree-of-freedom effects from analysis of variance of removal rate with and without the largest effect.
- The platen rotation, P, has by far the most important effect on removal rate and its effect is linear (see Figure 18.3a).
- The effects of other factors on the removal rate are less clear because of the indication of non-homogeneous variability (see Figure 18.3b).

---

[45] Taking the logarithm of uniformity is often done when the effects of the design parameters are likely to be multiplicative in the original scale, or because standard deviations are constrained to be non-negative and are right-skewed. The log-transformation addresses both of these problems, so that the assumption of a linear model with additive effects, necessary for analysis of variance, is more likely to be satisfied.

[46] In the context of analysis of variance, half-normal plots show the absolute values of single-degree-of-freedom effects plotted against quantiles of the standard half-normal distribution. Half-normal rather than full normal plots are used when the magnitude of the effects rather than the sign is of interest. If no parameter has any effect, as in a case of only random variability, the configuration of points in the plot should approximate a straight line. The slope of such a line is an estimate of the standard error, which in this case is the wafer-to-wafer variability. Large effects should appear as points on the right side of the plot that are considerably higher than the other points. When large effects do appear, an estimate of the standard error can be obtained by using the slope after these large effects are removed. Nonhomogeneous variability is indicated when the configuration of points appears to be a set of joined line segments with different slopes.

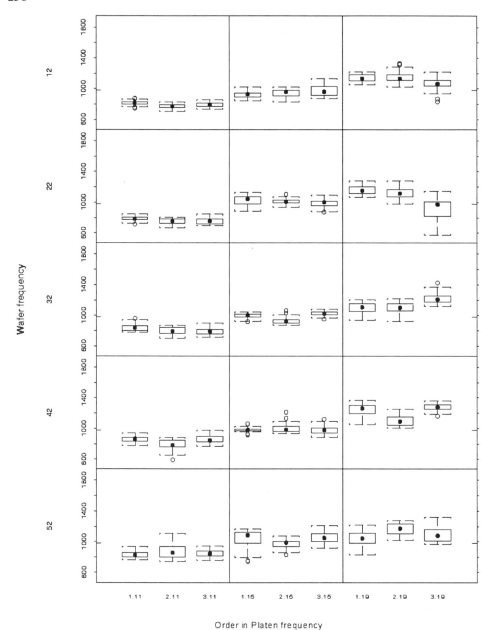

Figure 18.2. *Box plots of Removal Rate in Ångstroms/minute. The figure is organized so that wafer frequency (W) appears as rows and platen frequency (P) as columns, with three boxes for each P and W combination. Each box summarizes the distribution of the 37 removal rates for one wafer.*

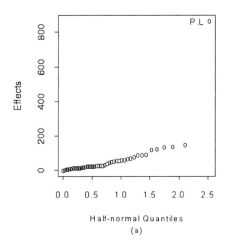
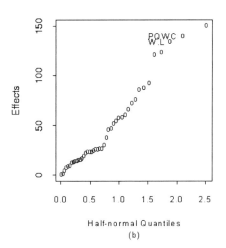

Figure 18.3. *Half-normal Plots of Single-degree-of-freedom Effects From Analysis of Variance of Removal Rate. Figure* 18.3a *shows all effects. P.L., the linear component of P, is much larger than the other effects. Figure* 18.3b *shows all effects except P.L. W.L. is the linear component of W. PQWC is the linear drift of the polisher.*

Figure 18.4 shows the effects in Figure 18.3b separated by type.
- The similar range of Figures 18.4a and 18.4b indicates that the wafer-to-wafer variability is as large as the variability due to all parameter effects, except the linear P effect.
- The effect of the polisher drift is as large as the effect of the wafer rotation, W (Figure 18.4a).
- The wafer-to-wafer variability is not homogeneous (Figure 18.4b).

To find the cause of the inhomogeneous wafer-to-wafer variability, we plotted the variances of the mean removal rates of the three wafers in each P, W combination in Figure 18.5. The wafer-to-wafer variability of wafers polished with platen rotation P = 19 rpm is clearly larger than that for wafers polished at either P = 11 or 15 rpm. One consequence of the unusual observations for wafer 3 in P = 19 rpm, W = 22 rpm, is seen in the extremely large variance at that P and W combination. Even disregarding this condition, this plot indicates that the higher platen frequency of 19 rpm will not give as good reproducibility as will lower platen frequencies.

We can formally test the inhomogeneity of wafer-to-wafer variability for P = 19 rpm versus P = 11 and 15 rpm by computing an F statistic for the respective pooled variances (see Devore, 1982). Table 18.2 shows this calculation, with and without the observations at P = 19 and W = 22. Clearly the wafer-to-wafer variability is significantly larger for wafers polished at a platen frequency of 19 rpm.

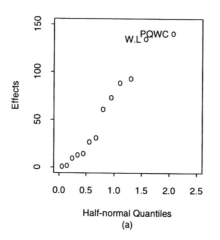
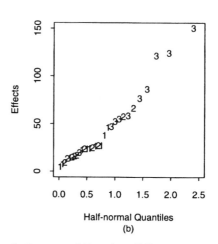

Figure 18.4. *Half-normal Plots of Single-Degree-of-Freedom Effects of Removal Rate. Figure* 18.4a *shows the effects of the parameters when the linear effect of P (PL) is omitted. The linear effect of W (WL) and the linear effect of the polisher drift (PQWC) appear larger than the other effects. Figure 18.4b shows effects of wafer-to-wafer variability. The plotting symbols 1, 2 and 3 correspond to order effects within P = 11, 15, and 19 rpm, respectively. It appears that the slope of a line drawn through points labeled with 3 is different from the slope of a line drawn through the remaining points.*

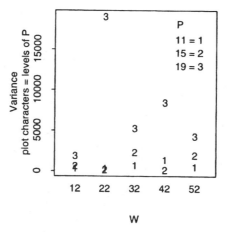

Figure 18.5. *Variances of Mean Removal Rate within Each Polishing Condition. The plotting characters denote the three levels of P. Wafers polished with P = 19 rpm (plotting character 3) have larger wafer-to-wafer variability.*

This result presents a problem for analysis of variance. We would like to compare the effects of the parameters to the wafer-to-wafer variability. But one of the assumptions of analysis of variance is homogeneity of the residual variance, which here is the wafer-to-wafer variability. So to quantify the importance of the parameters at the levels in which we are interested, we omit the data for P = 19, and compute an analysis of variance for the remaining data, first correcting for polisher drift. Table 18.3 presents this analysis of variance. The source (W.QC4) is the higher-order components of wafer frequency, and PW is the interaction between the platen and wafer frequencies, now not affected by the drift.

Table 18.2. *Comparison of Pooled Wafer-to-Wafer Variances.*

Source	DF	Mean Square	F-value	P-value
Variance, P = 19	10	7801.76	8.81	<.0001
Variance, P = 19, omitting W = 22	8	4991.15	5.64	.0008
Variance, P = 11 and 15	20	885.32		

Table 18.3. *Analysis of Variance for Removal Rate Omitting P = 19.*

Source	DF	Sum of Sq	Mean Sq	F-value	P-value
P	1	246724.6	246724.6	279.32	<.0001
WL	1	14058.15	14058.15	15.92	.0007
W.QC4	3	384.12	128.04	0.14	0.94
PW	4	2219.8	555.0	0.63	0.65
Wafer-to-wafer	20	17666.3	883.3		

This analysis shows that for platen frequencies 11 and 15:
- The effect of platen frequency, P, is the most important effect in determining removal rate.
- The linear effect of wafer frequency (WL) is also important.
- There is no non-linear component of wafer frequency.
- There is no interaction between the two rotational frequencies P and W.

There is no quantification of the polisher drift, but we remember from the initial analysis that the effect of drift was as important as the linear effect of wafer frequency.

We want a model which relates removal rate to the parameters at the lower platen frequencies. The linear relationship between the removal rate and the rotational frequencies, omitting data for P = 19 rpm and correcting for the polisher drift is[47]

$$R_{ij} = 264.19 + 45.18\,P_i + 1.53\,W_j.$$

(34.79)   (2.50)   (0.35)

This equation quantifies the dependence of removal rate on P and W without being affected by the polisher drift. Figure 18.6 shows the results of fitting the model.

The model appears to fit well, except for the tendency of one or two groups of points to appear a little higher or a little lower than the fit. This is because the fit is corrected for drift, whereas the data are not. For example, the three points at P = 15, W = 22 are higher than the fitted line. These were the first three wafers polished, so the polisher removal rate

---

[47] Numbers shown in parentheses below the coefficients are the standard errors of the coefficients.

was relatively higher. The normal quantile-quantile plot of residuals[48] in Figure 18.6b has a reasonable linear configuration, indicating that the model is appropriate.

We also evaluated the behavior through time of the residuals from a similar fit including the data for P = 19 rpm. Figure 18.7 shows a plot of these residuals plotted against order number of each wafer in the experiment. The figure displays the behavior of the polisher through time without the effects due to changing the polishing conditions. The tendency of the polisher to drift, that is, to show removal rate reduction over time, is seen.

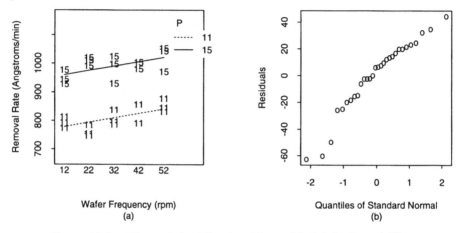

Figure 18.6. *Plots of the Fit of a Linear Model in P and W to Removal Rate. Data for P = 19 rpm is omitted. Figure 18.6a shows the data (uncorrected for drift) and fitted values for P = 11 and 15 rpm. Figure 18.6b shows a normal quantile-quantile plot of the residuals.*

Results for Uniformity

We defined uniformity within a wafer as the standard deviation about the mean of the amount of oxide removed, actually analyzing the log standard deviation as a function of the design parameters. The analysis was similar to that for removal rate; the results are summarized as follows:

- The linear effects of both platen and wafer frequencies on log uniformity are important.
- There is no P, W interaction effect.
- The uniformity is not affected by the polisher drift.
- There is no indication of nonhomogeneity of wafer-to-wafer variability. In terms of uniformity, the higher platen frequency of 19 rpm seems to give as good results as the lower platen frequencies.

---

[48] Under the assumptions for fitting linear models, the residuals should be normally distributed with a mean of 0 and some variance. A plot of the ordered residuals versus quantiles from a standard normal distribution should have a straight line configuration with an intercept of 0 and a slope which estimates the residual standard error.

Figure 18.8a shows the data for uniformity and fitted values computed from the model transformed back to the Angstrom scale, and Figure 18.8b is a normal plot of the residuals on the log Angstrom scale.

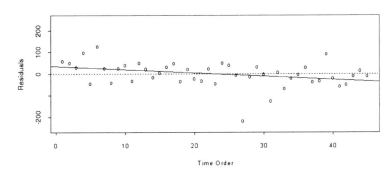

Figure 18.7. *Plot of Residuals from the Fit of Removal Rate to the Linear Effects of P and W. The solid line drawn is the least squares line with slope −1.5 from fitting the plotted residuals to time order. The dashed line has slope 0. The slope of the fitted line is significantly different from 0.*

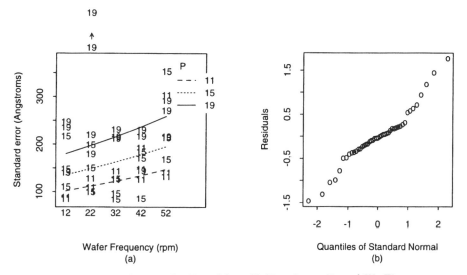

Figure 18.8. *Plots of the Fit of Log Uniformity to P and W. Figure 18.8a shows the data and fitted values on the Ångstrom scale. Figure 18.8b shows a normal quantile-quantile plot of the residuals on the log Ångstrom scale.*

The model appears to be reasonable, although the residual variability is large for some polishing conditions. The normal plot shows good linearity in the central part of the distribution, but has tails that are "too long," indicating that some of the data do not agree with the model as much as we would like. When we investigate the identity of the points in the lower tail of the distribution (better uniformity than expected), we find that all three wafers run at the parameter combination of P = 15 rpm and W = 32 rpm show much better uniformity than is predicted (see Figure 18.8a). This parameter combination immediately followed the wafer on which there was an interruption due to the polisher problem noted. The improvement can be due to optimum conditions or to something that changed during the interruption. Since there is no clear trend indicated by the model, an optimum condition is unlikely. The points that have much worse uniformity than predicted (upper tail of Figure 18.8b) have no consistent values of the experimental parameters.

Conclusions

We have employed a designed experiment to study the effects of platen rotation frequency and wafer rotation frequency on oxide removal rate and uniformity in the chemical-mechanical-polishing process. Some of the important findings are
- Simple linear models are appropriate for relating both removal rate and uniformity to platen and wafer rotation frequencies.
- Wafer to wafer reproducibility of removal rate is worse for a platen frequency of 19 rpm than for one of 15 or lower rpm. This finding aids in narrowing an operating regime. It also points out that rotational frequencies that result in a more uniform polished oxide also give a lower rate of throughput.
- The effect of polisher drift can be separated from the effects of the rotational frequencies. Removal rate is clearly affected by polisher drift in this experiment; however uniformity is not affected.

This experiment demonstrates that statistical design and analysis are powerful tools in quantifying the various effects that experimenters often observe but may not be able to easily characterize. Besides finding the effects of the design parameters, this experiment helped to quantify the polisher drift that had been previously observed but not well understood, and showed that lower wafer-to-wafer variability is achieved at lower platen frequencies.

Wafer-to-wafer variability is of great concern as it relates to the manufacturability of the chemical-mechanical polishing process. The equipment, operating procedures (such as pad dressing in order to rejuvenate the pad), and process control all could be improved to lower the cost in the routine use of this technology.

Acknowledgments

G. Smolinsky participated in the planning and execution of this experiment, and in writing a draft of earlier documentation. We also thank a referee who suggested the simplified explanation of determination of the run order shown in the Appendix.

References

Daniel, C. "Use of Half-normal Plots in Interpreting Factorial Two-level Experiments," *Technometrics*, Vol 1, No. 4, November 1959.

Devore, J.L. *Probability and Statistics for Engineering and the Sciences*, Monterey, CA: Wadsworth, Brooks/Cole, pp. 311-313. 1982.

Hicks, C.R. *Fundamental Concepts in the Design of Experiments*, New York: Holt, Rinehart and Winston, pp. 131-151, 1973.

Lifshitz, N., A. J. Miller, W. Y-C. Lai, and G. Smolinsky. *Planarization of the Interlevel Dielectrics by Polishing: Demonstration of Feasibility*, AT&T Bell Laboratories Internal Memorandum, 1988.

Nair, V.N. and D. Pregibon. "A Data Analysis Strategy for Quality Engineering Experiments," *AT&T Technical Journal*, Vol 65, Issue 3, pp. 73-84, May/June, 1986.

APPENDIX

Details of Run Order

The run order can be determined from the contrast coefficients for the effect to be confounded by ordering the runs in increasing order of these coefficients. In Table 18.4, the levels of W and P are shown along with the values of orthogonal polynomial coefficients for the quadratic effect for three levels and the cubic effect for five levels. The body of the table shows the coefficients for the P quadratic by W cubic effect.

Table 18.4. *Coefficients for the P Quadratic by W Cubic Effect.*

P		W				
		12	22	32	42	52
		−1	2	0	−2	1
11	1	−1	2	0	−2	1
15	−2	2	−4	0	4	−2
19	1	−1	2	0	−2	1

A suggested order of runs is then determined by ordering by the PQWC coefficients.

Table 18.5. *Suggested Order of Runs.*

	−4	−2			−1		0			1		2			4
P	15	11	15	19	11	19	11	15	19	11	19	11	15	19	15
W	22	42	52	42	12	12	32	32	32	52	52	22	12	22	42

This ordering lacks randomization, so we randomized within the ties. We also moved one of the runs with W = 32 in a way that would not destroy the confounding, so as not to have all runs with W = 32 together, to get the ordering given in Table 18.1.

BIOGRAPHIES

Anne E. Freeny received a M.S. degree in statistics from Cornell University. She is a former member of the technical staff in the statistics and data analysis research department at Lucent Bell Laboratories in Murray Hill, NJ. Her interests include design of experiments collaborations in manufacturing and graphical methods for data presentation.

Warren Y.-C. Lai received a Ph.D. in physics from the California Institute of Technology in Pasadena, CA in 1980 and joined Lucent Bell Laboratories in Murray Hill, NJ as a member of the technical staff in VLSI research and development. His research has focused on technology integration and process development in multilevel metallization.

CHAPTER 19

USE OF EXPERIMENTAL DESIGN TO OPTIMIZE A PROCESS FOR ETCHING POLYCRYSTALLINE SILICON GATES

Fred Preuninger, Joseph Blasko, Steven Meester, and Taeho Kook

This case study describes a project to optimize a process for etching polycrystalline silicon gates in the manufacture of CMOS integrated circuits. The main objective was to reduce the variability of the electrical channel length of devices made in production lines. A new consideration is the pattern density of the circuit which affects the ultimate distribution in size of the gates. This case study illustrates the use of a robust design experiment and statistical data analysis to efficiently evaluate the capability of the process and to characterize its robustness. The range of the distribution of gate line-size, due to circuit pattern density, was reduced from 0.07 to 0.04 µm.

EXECUTIVE SUMMARY

Problem

One of the most important characteristics of integrated circuits, speed, is partly determined by the effective electrical channel length, which in turn is partly determined by the physical size and profile of the gate after etching. A batch system was being used for this poly gate etch process in production. Variability between wafers was known to be large due to wafer placement in the reactor. An alternate system that etches one wafer at a time was selected for the improved process because it would be expected to have inherently better wafer-to-wafer uniformity. A process had been developed in the new tool that met most of the requirements of the process. Upon introduction into manufacturing it was discovered that the response line-size was sensitive to the proximity of features on the wafer and to the pattern density of the circuit.

To characterize line-size variability and control in the new etcher, a fractional factorial screening experiment was performed. A region of the process parameter space was found that optimized uniformity of line-size within wafer and between wafers, and, at the same time, was insensitive to proximity of features and circuit pattern density. But results from confirmation runs at the recommended process conditions showed that there could be undesirable effects on the profile of the gate feature in some cases, namely undercut.

Known sources of variability in the size of gates patterned by etching include
- variability of the incoming material (product lot).
- variation from wafer to wafer (within a lot).
- variation from site to site on (or within) the wafer.

Two less understood sources of variability in the size of gates patterned by etching are
- the proximity of the gate features to neighboring features.
- the pattern density of the circuit.

On the other hand, the major control factors in the etch process are the pressure, power, gas flow, and gas ratio. The mean and the variability of the distribution in the size of etched features can be affected by the choices in levels for these factors.

The objective of this study was to find a set of process conditions that would
- compromise neither the cross-sectional profile of the gate features, nor any of the other previously optimized characteristics of the existing etch process.
- not increase the sensitivity of electrical channel length to different product lots, wafers, or sites within the wafer.
- reduce the variability of line-size due to proximity of features.
- reduce the sensitivity of the line-size to the pattern density of the circuits.

Solution Strategy

Although it is desirable to understand all of the sources of variability and to control them if technically possible and economically feasible, it is often more expedient to design a process that is robust against variability (see Auciello, 1989). In this case, a robust design experiment was planned to make the process robust against variability and to further understand the process latitude. A central composite design was used for the robust design experiment. Scanning electron microscopy (SEM) techniques were used to measure the line-size response and to detect undercutting in the polysilicon gate profiles.

Conclusions

A set of etch processing conditions with low power and pressure was found that minimized the effects of circuit feature proximity and circuit pattern density on line-size and profile control (undercutting). This was done by analyzing the data to find the effect of the factors on the line-sizes, and to select the important factors and their levels. Models were fit to the data using multiple regression analysis with the RS/1™ software. The functions were plotted as contours with respect to the input factors.

The gas ratio was an important factor in preventing undercutting and in controlling the target value of the gate line-size. The greatest remaining variability was due to the trade-off of either undercutting isolated lines in codes of low pattern density or having differences in line-sizes between isolated and closely spaced features in codes of high pattern density.

The results of the experiments showed that the difference between the average etch deltas for high and low densities could be reduced from 0.07 μm to less than 0.04 μm (almost half), with isolated lines being the worst case. The effect of the proximity of features could be reduced from greater than 0.03 μm to less than 0.03 μm, with high pattern density being the worst case. Assurance of the absence of undercutting could be obtained by adjusting the gas ratio and power, and the targeted size of the gates could also be adjusted with gas composition.

Process

In the manufacture of integrated circuits, structures are formed by depositing a film onto a slice or wafer of silicon and patterning it. To form metal oxide semiconductor (MOS) gates, the deposited film is polycrystalline silicon (poly). A photosensitive film (photoresist) is then deposited onto the poly and exposed to create a pattern representing the design of the circuit. This pattern is used as a protective mask during the etching of the poly. The masking layer is then removed leaving the patterned poly.

The use of plasma etching to pattern the gates in integrated circuit fabrication requires control of numerous parameters to achieve a variety of results [see (Ephrath, 1989), (Dehnad, 1989), (Van Roosmalen, 1991), (Manos, 1989), and (Voschenkov, 1992)]. The ideal requirements are illustrated in Figure 19.1a and in the scanning electron microscopy (SEM) micrograph in Figure 19.2a. An acceptable gate etch process should

- selectively remove the material of interest from the desired areas without affecting the underlying layer, silicon dioxide in this case.
- produce features with vertical cross-sectional profiles.
- produce features of the same (lateral) size as the masking layer.
- not exhibit any undercutting of the mask, as illustrated in Figures 19.1b and 19.2b, in which the size of the poly feature is smaller than that of the mask. In undercutting, both the profile and size of the feature are affected. In practice, undercutting is to be strictly avoided because of the severe effect small gates have on the function of the circuit. On the other hand, some growth in size of the gate, as illustrated in Figure 19.1c and 19.2c, can be tolerated. In practice, a target of 0.02 to 0.05 μm growth was being used to accommodate process and measurement variation.

The process must be capable of producing acceptable gate etches uniformly on features at all sites on a wafer and on all wafers in a production lot. Measurement of the size of the gates in integrated circuits is best done electrically on finished devices, and the plan for this process characterization study included confirmation of the improved process using electrically-testable device wafers. However, the details of the electrical channel length distributions are proprietary, and that part of the process improvement cycle is not discussed here. Since fabrication of complete devices is expensive and time-consuming, SEM techniques were used here to evaluate gate etching immediately following the patterning process. Cross-sectional SEMs were made of the etched profiles and measurements were made of the line-size of the features. The width of the top of the poly feature (assumed to be unchanged from the photoresist size) and the width of the bottom of the feature, were measured on photographs like those shown in Figure 19.2. As indicated in Figure 19.1d, subtracting the width at the top of the poly feature from the width at the bottom gave the delta (which is the negative of the delta usually defined in the literature). The accuracy of this technique is better than .005 μm 1 sigma, as correlated against non-destructive SEM techniques and electrical measurements.

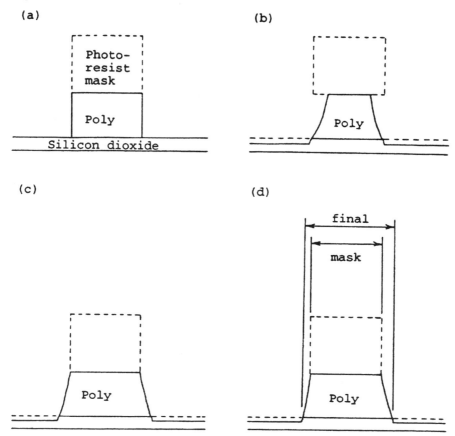

Figure 19.1. *Illustration of Etched Line-size, Profile, and Selectivity in Cross-Sectional Views of Poly Gate Structures.* (a) *shows ideal line-size control, vertical profile, and selectivity.* (b) *shows undesirable undercutting of the photoresist mask, reduction in line-size, and excessive removal of the underlying layer.* (c) *shows nonideal growth in line-size and excessive removal of the underlying layer.* (d) *shows a change in line-size (delta) from photoresist mask to final poly feature.*

The poly gate etch process was being performed in a batch system capable of processing eighteen wafers at one time. To address the issues of uniformity, a single-wafer-etcher (SWE) with a magnetically-confined plasma was chosen for the improved process. A process had been developed in the new tool that met most of the requirements. Its uniformity and selectivity for etching poly were better than in the batch system.

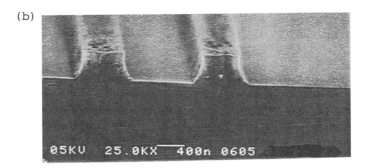
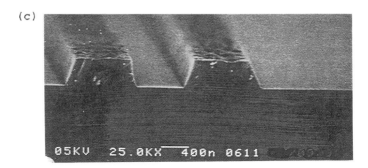

Figure 19.2. *Scanning Electron Microscopy (SEM) Photomicrograph of Typical Poly Gate Cross-Sectional Profiles.* (a) *shows an acceptable profile.* (b) *shows severe undercutting.* (c) *show excessive growth in line-size.*

However, upon introduction of the new single-wafer-etcher into manufacture, it was learned that electrical channel length distributions were somewhat degraded with respect to those of the old process. Three sources of the degradation were identified.

- The SWE had a tendency to undercut the mask.

- The SWE process gave a smaller poly gate line-size for circuit features that were close together than it did for those that were isolated. This proximity effect was about .03 to .05 μm in magnitude.
- The pattern density of the circuit considered affected the line-size, so that circuits with large areas to be etched (low pattern density of remaining poly features) gave a smaller line-size (by as much as 0.07 μm) than did circuits with high pattern density. This is the code effect.

This variability in incoming material, due to the design of the circuit itself, was not realized and not quantifiable in the earlier process development phases. Both the proximity effect and the code effect have mechanistic explanations [see (Voschenkov, 1992), (Voschenkov, 1992), and (Gottscho, 1992)] and further work is needed to relate microscopic effects on individual features to macroscopic causes and vice-versa.

The proximity effect on feature size is merely a manifestation of aspect-ratio-dependent-etching effects (see Gottscho, 1992). In this study, the nominal size of the poly features was selected as 0.6 μm. Closely spaced features were separated by about 0.9 μm. At the other extreme, a feature was considered isolated if it was more than several microns away from its neighbor.

Quantification of the code effect was more difficult and required information about the design of the integrated circuit. The pattern density was not the same in all areas of a circuit. Also, the size of the circuit affects how many circuits can be placed on a wafer, which determines the pattern density on the scale of the entire wafer. While these parameters may in fact have had more correlation to the critical responses in the poly gate etch process, the only parameter that was readily available in this study was the average pattern density of the circuit. The quantitative value of the average circuit pattern density was not used in this study, but will be designated merely as low or high.

To characterize the line-size control in terms of these new effects, a screening experiment was run using a 2^{5-1} fractional factorial design in pressure, gas composition, magnetic field, and temperature. The objectives of the experiment were to determine the important factors that affect the control of within-wafer uniformity, wafer-to-wafer line-size uniformity, and profile control, and to determine the feasibility of reducing the dependence in the line-size on the circuit density. A combination of factor settings was found that gave acceptable uniformity and control of line-size and minimized the difference in line-size for codes of extreme low and high pattern density.

The main effects and interactions of the variables indicated that low pressure of about 25 mT, low power of about 150 W, high temperature of about 40°C, and high magnetic field of about 75 Gauss gave optimal uniformity of line-size and minimal sensitivity to proximity of features and pattern density. The fifth variable, the percentage of hydrogen bromide (HBr) in the feed gas, was found to be a good independent variable to reach the target value in the line-size, and it had a strong effect on undercutting. A range of 10 to 20% HBr in chlorine at a constant total flow of 70 sccm was investigated.

The results of the screening experiment produced a process which significantly improved uniformity of line-size within and between wafers, and was robust to proximity of features and circuit pattern density. However, the new process tended to undercut. It was decided that a second experiment, the subject of this case study, should be run to carefully investigate the parameter region around the new recipe. This was done with a view to finding conditions which would achieve the line-size control afforded by the new process and eliminate the tendency to undercut. This second experiment, a robust design experiment, would also provide information on the sensitivity (or process window) in this new operating region.

DATA COLLECTION PLAN

The new robust design experiment included circuit codes that showed extremes of line-size delta and extremes of feature pattern density. A code with low pattern density and one of high pattern density were selected. Test wafers consisting of the gate oxide layer (150 Å), the poly layer (4200 Å), and the photoresist mask for the poly gate level of the appropriate code were generated. The photoresist was Hunt 238™ applied at a 1.4 µm thickness, with no post-exposure bake.

For these experiments, the single etch chamber was used. The etcher was programmed to sequentially process wafers under the various process conditions of the experimental design. An experiment of this size can easily be completed in a couple of hours, or the time it takes to normally run an entire production lot. Other studies had shown that the process is stable and repeatable within this time scale. Following etching, the wafers were processed through several steps that removed the photoresist mask. They were then measured by SEM to analyze the final line-size and profile of the poly gate.

The results of the screening experiment indicated that little would be gained by further investigation of temperature and magnetic field. A central composite experimental design (see Montgomery, 1991 for a discussion) was chosen to investigate the relationships between the responses of interest and the following reduced process parameter set:

- RF (radio frequency) power
- Pressure
- Percent Hydrogen Bromide.

The central composite design was run for both high and low pattern densities.

The nominal (center point) process was chlorine and HBr with flows at approximately 59 and 11 sccm, respectively, 15% HBr, pressure at 25 mT, and power at 150 W. The total flow was kept constant at 70 sccm, the temperature was held constant at 40°C, and the magnetic field was held constant at 75 Gauss. The central composite experimental design in the process factors and the levels of the factors are shown in Table 19.1. All of the experimental runs were made for each of the two pattern densities, and the order of the runs was random.

Two responses measured were
- line-size etch data.
- undercutting.

Samples were taken from the center, top, bottom, left, and right sides of each wafer, and line-size etch delta was measured for identical features of each type (closely spaced and isolated). Undercutting is a categorical response, either being present or absent. However, a propensity for undercutting can be inferred from negative delta values. For a process to be robust against undercutting, the delta measurements must not be close to zero. A target value of about 0.02 to 0.05 µm is acceptable (an exact value is not needed, since uniformity and control are desired, and the target value can be adjusted independently with the photoresist size).

Table 19.1. *Robust Design Experimental Settings and Measured Line-size Etch Delta. The experimental design is a central composite design in three control factors, power, pressure, and %HBr. The average delta measurements are given for two noise factors, gate spacing (closely spaced and isolated) and code pattern density (low and high).*

Row	Power	Pressure	%HBr	Low Pattern Density		High Pattern Density	
				Close-spaced	Isolated	Close-spaced	Isolated
1	180	30	20	0.00	0.01	0.03	0.06
2	180	30	10	−0.02	0.00	0.01	0.05
3	180	20	20	0.00	0.03	0.04	0.08
4	180	20	10	−0.01	0.02	0.02	0.07
5	120	30	20	−0.02	0.01	−0.03	0.01
6	120	30	10	−0.06	−0.05	−0.04	0.00
7	120	20	20	0.00	0.01	0.03	0.06
8	120	20	10	−0.02	0.01	−0.01	0.03
9	150	25	23	−0.01	0.01	0.03	0.06
10	150	25	7	−0.07	−0.03	−0.04	−0.02
11	150	33	15	−0.04	−0.04	−0.01	0.03
12	150	17	15	−0.02	−0.02	0.02	0.08
13	200	25	15	−0.03	0.02	0.04	0.06
14	100	25	15	−0.03	−0.02	0.00	0.04
15–18	150	25	15	−0.03	−0.01	0.01	0.06

DATA ANALYSIS AND INTERPRETATION OF RESULTS

Table 19.1 gives the delta measurements for each of the experimental runs for closely spaced and isolated gate features for codes with low and high pattern density. The response for within-wafer uniformity is not included, and only the averages are given for the four replicate runs (the centerpoints of runs 15 to 18).

The problem of finding an etch process which minimizes variability around a target value and which is insensitive to the proximity of features, the pattern density, and the site within a wafer falls into the robust parameter design framework of Taguchi (see Kackar, 1985 and Nair, 1992). In this framework, the noise factors are the site within a wafer, the feature type (closely spaced or isolated) and the pattern density. The control factors are the process parameters: RF power, pressure, and percent HBr. In this product array experiment, all combinations of the noise factors appear with each condition of the central composite design (the control array in the Taguchi framework) and the data could be analyzed using the inner-outer array approach of Taguchi. This analysis approach was not used in this work for three reasons:

1. Each source of variability did not have equal importance, and the investigators wanted to understand explicitly the trade-offs being made in choosing a given recipe.
2. The investigators wanted to understand how the etch deltas for the closely spaced and isolated features, and how the uniformity within a wafer, changed as functions of the process parameters and the pattern density.
3. If a process could not be found which both avoided undercutting and was sufficiently robust to pattern density, then avoiding undercut would be the overriding consideration. In this event, the different production codes could be classified as low or high pattern density, and different recipes could be used for

the two groups. This would necessitate models for the etch deltas as a function of both the process parameters and the pattern density.

The analysis performed for this experiment was to construct response surface models using RS/1™ for the following responses:
- average etch delta of closely spaced features within a wafer.
- average etch delta of isolated features within a wafer.
- the logarithm of the standard deviation of the etch delta of closely spaced features within a wafer.
- the logarithm of the standard deviation of the etch delta of isolated features within a wafer.

The explanatory variables were the process parameters and the pattern density.

To achieve robustness against pattern density, the interaction effects of pattern density with the process parameters in the response surface models are exploited to find process conditions which minimize the effect of pattern density on the etch delta.

To achieve robustness against feature spacing, the points in the process parameter region are found which minimize the difference between the average etch deltas within a wafer for the two types of features. Finding the points in the process parameter space which minimize the standard deviations within the wafer gives robustness against variability within a wafer. Furthermore, trade-offs may be quantified and weighed in the context of product requirements.

The response surface models were fit initially with the full quadratic model. Then explanatory variables that were not statistically significant predictors over the region of study were sequentially eliminated. In fact, none of the process parameters were statistically significant predictors (at the 10 percent level) for the logarithm of the standard deviation within a wafer. Thus, we took the variability within a wafer to be constant over the region investigated and did not consider this point further.

The fitted response surface models for the average etch delta within a wafer for both closely spaced and isolated features, and the tables of coefficients giving individual p-values, are given in Appendices A and B. Equation 19.1 in Appendix B is the line-size etch delta response for closely spaced features. The p-value for any predictor not appearing in a model is greater than 0.02. (This level was not used as a cutoff; it simply turned out that all predictors were either highly significant or not at all significant). Plots of the residuals did not indicate any model inadequacies. The lack of fit tests indicate some evidence (1 − p-value = 0.97) of a lack of fit for the models. However, neither variance adjustment weights nor the robust bi-square fitting algorithm in RS/1™ are helpful in this regard. The modest indication of a lack of fit for the model may be due to the fact that the variability between wafers (the pure error here) for this single-wafer-etcher is very small, and the ability of a quadratic model to approximate the response surface to the same order as for the variability between wafers is being pushed to the limit.

Understanding the sensitivity of the process responses to variations in the input parameters is made easier with graphs of the calculated line-size etch delta responses (see Figures 19.3–19.6). Since there are four variables (power, pressure, HBr concentration, and code), a two-dimensional plot can only be made by holding two variables at a chosen value and generating contour plots in the responses. Here Figures 19.3–19.6 show contours for constant values of average within-wafer etch delta as predicted by Equation 19.1 for isolated features and as predicted by Equation 19.2 for closely spaced features (see Appendix B).

In Figures 19.3–19.6, the results for the two code pattern densities are presented separately. Figures 19.3 and 19.5 are for the low pattern density, and Figures 19.4 and 19.6 are for high pattern density. The pair of figures for each pattern density shows two choices of the three remaining process parameters used as variables: one figure of the pair shows pressure and HBr concentration, while the other shows pressure and power. Within a graph, the proximity effect (difference between closely spaced and isolated features) at any combination of process conditions is apparent by simply comparing the values of the two contours. For the code effect, the value of a contour at a point in one figure must be compared to the corresponding point on the figure for the other code. Information on the sensitivity of the line-size delta response to the process conditions is obtained from the gradient or steepness with respect to a process variable.

Figure 19.3 (for the low pattern density code) and Figure 19.4 (for the high pattern density code) show contours of equal deltas predicted by equation 19.1 (see Appendix B) for isolated features, and equal deltas predicted by equation 19.2 (see Appendix B) for closely spaced features. The variables are pressure and HBr concentration; power is fixed at 100 W.

The first consideration in interpreting Figures 19.3–19.6 is to avoid regions with negative predicted values of line-size delta because they are prone to undercutting. These are in the upper left. The figures show a low pressure and high HBr concentration region exists toward the lower right-hand corner. This is where the delta values for both the closely spaced and isolated features are small and positive, as desired.

The lower right area also has minimum differences between delta contours for isolated and for closely spaced features compared to other parts of the graph. Comparing Figures 19.3 and 19.4 shows that the difference in line-size deltas for the high and low pattern density codes is also minimized in the lower right-hand corner. Since the gradient or steepness of the contours along the HBr concentration axis is greater than that along the pressure axis, the HBr concentration should be held high.

Figures 19.5 and 19.6 show the effects of pressure and power on the line-size delta. Figure 19.5 (for the low pattern density code) and Figure 19.6 (for the high pattern density code) show contours of the calculated deltas for isolated and closely spaced features predicted by Equations 19.1 and 19.2, respectively (see Appendix B), for variable pressure and power, with HBr fixed at 23%. The complicated dependence on power is evident for both codes. The curvature in the contours for the closely spaced features is due to the interaction between power and pressure. In contrast, the deltas of the isolated features simply increase with increasing power and decreasing pressure (toward the lower right corner).

Again, the first consideration in finding a region of optimal process conditions is to avoid areas with negative predicted values of delta (upper left corner).

Next, the trends in delta for pressure and power are considered. For pressure, the dependence of line-size delta is consistent with what has already been decided: delta increases at low pressure.

The situation with regard to power is more complicated. Generally, delta increases with increasing power, and regions prone to undercutting can be avoided at high power (right-hand side). High power is also favored because the delta is less sensitive to variations in power or pressure there, as indicated by the wider spacing between the contours. Therefore, at high power, line-size delta would be more robust (or less sensitive) to variations due to equipment, for example. But at high power there is more difference between closely spaced and isolated deltas than at low power, so the proximity effect is worse there.

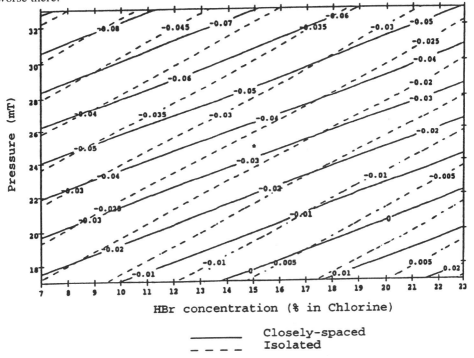

Figure 19.3. *Average Line-Size Delta Response for Low Pattern Density and* 100 W *Power using Pressure and* HBr *Concentration as Variables. The contours of equal delta shown as dashed lines are calculated from Equation* 19.1 *for the isolated features (see Appendix B). The solid lines are calculated from Equation* 19.2 *for the closely spaced features (see Appendix B). The difference between these contours at any point is the magnitude of the proximity effect.*

Thus, there is a trade-off between process robustness and avoiding undercutting at high power, and minimizing the proximity effect at low power.

The code effect is considered by comparing equivalent regions in Figure 19.5 with those in Figure 19.6, showing that the minimum difference between codes is also in the lower left corner at low pressure and low power. At higher power (at the right side) the delta for isolated features is about 0.10–0.11 µm for high pattern density, and is about 0.03–0.04 µm for low pattern density, or as much as 0.07 µm between codes.

An optimal set of process conditions is identified to be at (high) HBr concentration of about 23%, (low) pressure of about 18 mT, and (low) power of about 100 W.

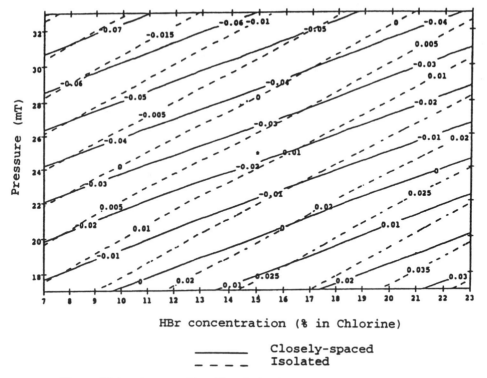

Figure 19.4. *Average Line-Size Delta Response for High Pattern Density and* 100 W *Power using Pressure and* HBr *Concentration as Variables. The contours are calculated in the same way as for Figure 19.3. The difference between these contours at any point is the magnitude of the proximity effect.*

The resulting values of the predicted line-size deltas for this process window, as read from the contour plots, are shown in Table 19.2.

Table 19.2. *Resulting Values of the Predicted Line-Size Deltas from the Contour Plots.*

	High Density	Low Density
Closely Spaced	.03 µm	.02 µm
Isolated	.04 µm	.01 µm

The magnitude of the proximity effect for each pattern density can be estimated by finding the differences between the predicted closely spaced and isolated deltas. These ideal differences are shown in Table 19.3.

The proximity effect can be minimized to less than .01 µm for both pattern densities.

The magnitude of the code effect can be estimated in the same way to be .03 µm for isolated features, and .01 µm for closely spaced features.

In practice, the process conditions may vary about their ideal values, producing some variability in the resultant delta. For total variations of about 10% in pressure and power, a process window about the ideal operating conditions can be constructed. This variability has a greater effect on the high pattern density response, as indicated by its steeper contours. The proximity effect is worse for the high pattern density codes. The worst-case point-by-point differences from the plots give an estimated proximity effect of .03 µm, and a code effect of .03 µm in this range of process conditions.

Table 19.3. *Ideal Differences Between the Predicted Closely Spaced and Isolated Deltas.*

	Proximity Effect
High Pattern Density	.01 µm
Low Pattern Density	.01 µm

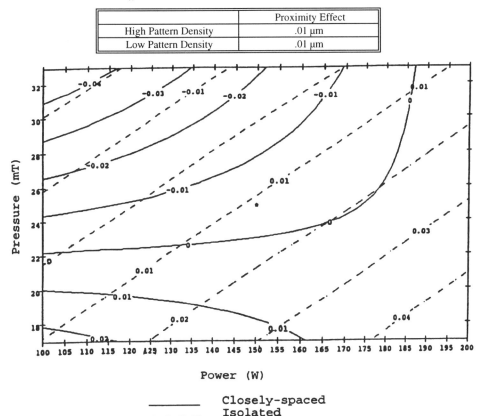

Figure 19.5. *Average Line-Size Delta Response for Low Pattern Density and 23% HBr Concentration using Pressure and Power as Variables. The contours of equal delta for the isolated features are shown as dashed lines. The solid lines are for the closely spaced features. For both types of features, the delta is largest in the lower right corner. Negative values in the upper left corner indicate possible undercutting.*

As a final practical consideration, the trade-off between high power to avoid undercutting, and low power to minimize the proximity and code effects, is important. To safely avoid undercutting, the power should not be run as low as ideally predicted, but should be chosen to give a line-size growth of at least .02 µm. Choosing the power level at about 150 W, for example, predicts the line-size deltas for both pattern densities and both feature spacings to be greater than .02 µm. But in the worst case, the delta of the isolated features of the high pattern density code increases to .08 μm, and the proximity and code effects increase to about .06 μm. In practice, however, this can be compensated for by designing the mask pattern of the isolated features to be slightly smaller than closely spaced ones.

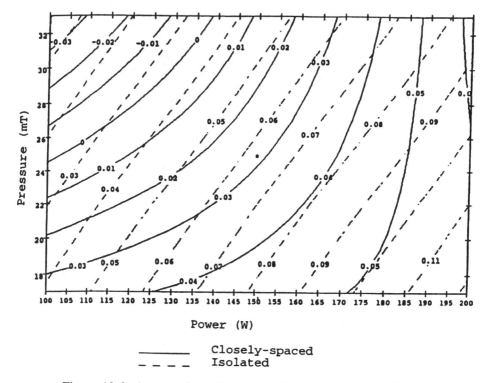

Figure 19.6. *Average Line-Size Delta Response for High Pattern Density and 23% HBr Concentration using Pressure and Power as Variables. The contours are calculated in the same way as for Figure 19.5. The difference between the contours at any point is the magnitude of the proximity effect. The difference between corresponding points on Figure 19.6 and on Figure 19.5 is the magnitude of the code effect.*

CONCLUSIONS

An optimal process, which avoids undercutting and minimizes the effects of proximity of features and the dependence on pattern density, exists at
- (high) HBr concentration of 23%.
- (low) pressure of about 18 mT.
- (low) power of about 100 W.

This choice of pressure and power is ideally predicted to have 0.01 μm difference in line-size between closely spaced and isolated features of circuits, independent of pattern density. The code effect is ideally predicted to be .01 μm at the lowest pressure and power.

When practical aspects of process control, measurement uncertainty, and manufacturing robustness are considered, a better process is chosen to be at power as high as 150 W. Here, the proximity effect is estimated to be .03 µm. With the same considerations, the code effect may be as large as 0.04 µm on isolated features.

These results are improvements over the existing process. Implementation of the new process conditions in the manufacturing line generally confirmed the improvement.

Designed experimentation has been used to find a set of optimal process conditions and to identify trade-offs for finding a practical solution. Incorporating the pattern density factor and proximity effect response explicitly into the design of a robust experiment is the first step in separating and quantifying these effects to ultimately achieve a mechanistic understanding of them. However, even in the absence of such detailed mechanistic information, robust engineering can be used to successfully minimize effects not controlled by process settings.

REFERENCES

Auciello, O. and C.L. Flamm, Eds. *Plasma Diagnostics*, Vol. 1, *Discharge Parameters and Chemistry*, New York, NY: Academic Press, 1989.

Dehnad, Khosrow. *Quality Control, Robust Design, and the Taguchi Method*, CA: Wadsworth & Brooks / Cole Advanced Books & Software, 1989.

Ephrath, L.M. "Review of Dry Etching Techniques: Recent Advances", in *Proceedings 2nd Symposium on ULSI Science and Technology*, The Electrochemical Society, ECS Proc. Vol 89-9, 394, 1989.

Gottscho, R.A., C.W. Jurgensen, and D.J. Vitkavage. "Microscopic Uniformity in Plasma Etching," *Journal of Vacuum Science and Technology B*. Vol. 10(5), pp. 2133-2147, Sept./Oct. 1992.

Kackar, R.N. "Off-Line Quality Control, Parameter Design, and the Taguchi Method (With Discussion)," *Journal of Quality Technology*, Vol. 17. pp. 176-209, 1985.

Manos, D.M. and D.L. Flamm, Eds. *Plasma Etching, an Introduction*, New York, NY: Academic Press, 1989.

Montgomery, Douglas C. *Design and Analysis of Experiments*, 4th ed. New York, NY: John Wiley and Sons, Inc., 1997.

Nair, V.J., Ed. "Taguchi's Parameter Design: A Panel Discussion," *Technometrics*, Vol. 34, pp. 127-161, 1992.

Van Roosmalen, A.J., J.A.G. Baggerman, and S.J.H. Braden. *Dry Etching for VLSI*, New York: Plenum Press, 1991.

Voschenkov, A. "Fundamentals of Plasma Etching for Silicon Technology, Part I," *International Journal of High Speed Electronics*, Vol. 1, Nos. 3,4, pp 303-345. 1992.

Voschenkov, A. "Plasma Etching Processes for Gigahertz Silicon Integrated Circuits, Part II," *International Journal of High Speed Electronics*, Vol. 2, Nos. 1,2, pp 45-88, 1992.

APPENDIX A

Model Coefficients

Table 19.4. *Least Squares Coefficients, Individual t-Statistics and Significance Levels for Etch Delta Response of Isolated Features.*

Term	Coefficients	t-value	Significance
1	−0.0370	−1.84	0.0751
Power	0.0006	6.93	0.0001
Pressure	−0.0023	14.46	0.0001
%HBr	0.0019	3.59	0.0012
Pattern Density <1 df>			0.6755
0	0.005	0.42	0.6755
1	−0.005	−0.42	0.6755
Power*Pat.Density <1 df>			0.0209
0	−0.0002	−2.44	0.0209
1	0.0002	2.44	0.0209

Table 19.5. *Least Squares Coefficients, Individual t-Statistics and Significance Levels for Etch Delta Response of Closely Spaced Features.*

Term	Coefficients.	t-value	Significance
1	0.1336	1.82	0.0790
Power	−0.0009	−1.78	0.0856
Pressure	−0.0096	−3.32	0.0024
%HBr	0.0026	5.81	0.0001
Pattern Density <1 df>			0.1418
0	0.0168	1.51	0.1418
1	−0.0168	−1.51	0.1418
Power*Pressure	0.00005	2.63	0.0136
Power*Pat.Density <1 df>			0.0051
0	−0.00022	−3.03	0.0051
1	0.00022	3.03	0.0051

APPENDIX B

Delta Calculations

The average within-wafer line-size etch delta for the isolated features was given by

$$\text{Isolated delta} = -370 + 55 \times \text{Code} + 885 \times \text{Power} - 575 \times \text{Pressure} + 285 \times \text{HBr concentration} - 315 \times \text{Power} \times \text{Code}. \quad (19.1)$$

The average within-wafer line-size etch delta for the closely spaced features is given by

$$\text{Closely spaced delta} = +1300 + 170 \times \text{Code} - 1290 \times \text{Power} - 2400 \times \text{Pressure} + 390 \times \text{HBr concentration} - 330 \times \text{Power} \times \text{Code} + 1875 \times \text{Power} \times \text{Pressure}. \quad (19.2)$$

The units for the delta responses are Angstroms ($\mu m \times 10^{-4}$). The power, pressure, and HBr concentration variables are converted from their experimental values to normalized units for direct comparison of the magnitude of their effects. For example, a power level of 150 W corresponds to 1 in the model, a pressure of 25 mT corresponds to 1 in the model, and HBr percent at 15 percent corresponds to 1 in the model. Variation about the center point of the experimental design may then be taken as percentages. The variable Code for the pattern density is taken as 1 for the low density code and −1 for the high density code. All of the coefficients of interaction terms and main effects not involved in significant interactions are significantly different from zero at the 2 % level. The individual t-statistics and significance levels are shown in Tables 19.4 and 19.5 in Appendix A.

Biographies

Fred Preuninger received a B.S. degree in Chemistry from Union College in Schenectady, New York, and M.S. and Ph.D. degrees in Chemistry from the University of Rochester, Rochester, New York. As a member of the technical staff at AT&T Lucent Technologies for sixteen years, he has worked in a variety of assignments in plasma etch process engineering and development.

Joseph P. Blasko received an A.A.S. in Electrical Engineering Technology from Pennsylvannia State University. He received a B.S. degree in Electrical Engineering from Lafayette College, Easton, Pennsylvannia, and is currently pursuing an MSEE at Lehigh University, Bethlehem, Pennsylvannia. He is currently a Member of Technical Staff at Lucent Technolgies and has worked in the plasma etch area for 13 years. Projects include 0.9, 0.5, 0.35 micron etching of poly, oxide, metal for CMOS product.

Steven Meester received the B.Sc. degree in mathematics and the M.Sc. degree in applied statistics from Simon Fraser University, and the Ph.D. degree in statistics from the University of Waterloo. Steve joined AT&T Bell Laboratories in 1991 as a Member of Technical Staff, providing statistical expertise in design of experiments and decision engineering to internal AT&T organizations. He has worked extensively in VLSI manufacturing process development and improvement, and in photonic device certification. Steve is currently a Principal Technical Staff Member at AT&T Labs where he works in credit risk management: designing systems and consulting with AT&T's business units to manage the credit risk and maximize the returns on their accounts receivable portfolios.

Taeho Kook received a B.S. degree in metallurgy from Seoul National University, Seoul, Korea, in 1974. He earned a Ph.D. degree in metallurgy and materials engineering from Lehigh University, Bethlehem, Pennsylvania, in 1985. He is currently a member of the technical staff at the VLSI laboratory of AT&T Bell Laboratories of Lucent Technologies in Orlando, Florida. His work assignments include plasma etch process development for advanced CMOS technologies.

CHAPTER 20

OPTIMIZATION OF A WAFER STEPPER ALIGNMENT SYSTEM USING ROBUST DESIGN

Brenda Cantell, José Ramírez, and William Gadson

At the request of several GCA stepper customers, a new local alignment system was recently introduced to enhance alignment of individual integrated circuits (die) within a wafer. It was also desired to determine the settings that would make the system robust (less sensitive) to process variation in the manufacturing steps prior to the alignment operation. The results of this case study showed that the new system outperforms the old MicroDFAS for metal levels. A combination of system parameters was found that makes the new system robust against process variation.

EXECUTIVE SUMMARY

Problem

The fabrication of integrated circuits on a silicon wafer requires that several layers of thin films (such as metals, silicides, and silicon dioxide) be placed on top of each other in a pre-defined order. The structure of the individual circuit is determined by patterns that are transferred from a square glass plate, referred to as the reticle mask, onto the wafer.

The functionality of the circuit depends in part on the precision with which the patterns align to levels just below it. Alignment marks are introduced on each reticle mask to help the alignment of the different thin film layers. Computer-aided alignment systems are then used to achieve the desired level of alignment precision.

A brightfield local alignment system was introduced to make alignment of individual die (local alignment) on the stepper more robust (less sensitive) to process variation in the manufacturing steps prior to the alignment operation. The brightfield system produced signals that were superior to those obtained on the darkfield (MicroDFAS) system.

However, the sensitivity of the brightfield system to variations in multi-level metal processes needed to be determined before it could replace the existing darkfield system.

Solution Strategy

As a part of the ongoing partnership activities between GCA and Digital Equipment Corporation (DEC), a joint experiment was undertaken to evaluate the new alignment system. Design of experiments is a very efficient technique for comparing the performance of different machines or systems. It can also be used to determine the settings that will make the system robust to variations introduced by factors that are difficult to control.

This study compared the alignment performance of the new alignment system with the MicroDFAS, as measured by the amount of reflected light from the alignment mark (signal strength) produced by the system when using two different mark types (segmented and bar) and two different mark polarities (clearfield and darkfield). The sensitivity of the system to the variation introduced by manufacturing steps, prior to the alignment operation, was investigated by selecting two settings, representing the typical operating ranges of a variety of GCA semiconductor manufacturing customers, from each of four different sequential process steps (oxide thickness, reflow, metal thickness, and photo resist thickness).

Since all the four mark type and polarity combinations can be placed in a wafer, only sixteen wafers were required to test all the 16 different manufacturing conditions. These sixteen wafers were then measured in the two systems to obtain the signal strength. Analysis of variance and graphical techniques were used to compare the two systems and determine the factor settings that will make the system less sensitive to variation.

Conclusions

The results of the study clearly confirmed the superiority of the brightfield alignment system over the MicroDFAS when aligning metal levels. It also provided a combination of mark type and mark polarity that is least affected by the variation in the process steps prior to the alignment operation. This information allows GCA to proceed with an alignment strategy that uses both the MicroDFAS system and the brightfield system. The new alignment system is designed to be upgradeable on current steppers, thus DEC will directly gain from upgrading the capabilities of its installed stepper base. GCA has benefitted from the use of DEC's capabilities in design of experiments to ensure the new system was adequately tested.

PROCESS DESCRIPTION

An integrated circuit is fabricated by processing one reticle pattern on top of another. Each pattern level must align in a precise manner to levels just below it. A specialized structure called an alignment mark is needed to perform this operation on very large scale integration (VLSI) circuits with many layers. The alignment marks on a wafer are aligned to a reference mark on a reticle. An example of some alignment schemes are shown in Figure 20.1(a) where the objective is to place the cross in the square. A wafer stepper is then used to expose an image of the reticle onto a position in the wafer. It systematically "steps" across the substrate until the entire wafer is exposed to the rows and columns of the integrated circuit patterns residing on the reticle (see Figure 20.1(b)).

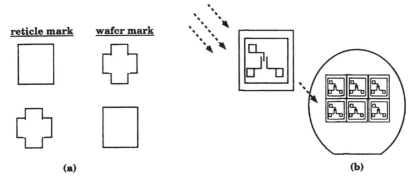

Figure 20.1. (a) *Example of Alignment Marks used to Correctly Position the Reticle Mask onto the Wafer. The objective is to place the cross in the box.* (b) *Example of a Step and Repeat Process for Exposing the Pattern on a Reticle Mask Across the Surface of the Wafer.*

Wafer steppers typically use two alignment systems, a global system that aligns the wafer to the stepper reference marks and removes wafer rotation, and a local alignment system with software algorithms that will determine the location of specific marks on a die within a wafer. This is done prior to exposing a pattern on to a wafer.

The local alignment mark consists of a series of vertical and horizontal bars (bar mark type) or a series of vertical and horizontal dots (segmented mark type) that make up lines that are either depressions (darkfield mark polarity) or bumps (clearfield mark polarity) in a wafer substrate. The stepper gets mark location data by scanning a beam of light across a mark and measuring the amount of light that is scattered. The mark can be illuminated with either darkfield illumination, which illuminates the edges of the mark only, as on the current alignment technique, or with brightfield illumination, which illuminates the entire mark, as on the new alignment technique. Each illumination system has characteristics that are best suited for a particular wafer substrate.

The quality of the alignment system is determined by the signal strength (S/N ratio, not to be confused with Taguchi's Signal-to-Noise ratios) which is the amount of light reflected from the local alignment mark. The higher the signal the easier it is to align the corresponding marks. The strength of the signal can be affected by thickness variations in the different layers of the wafer, such as oxide, metal, and photoresist.

DATA COLLECTION PLAN

The response of interest in this study is the amount of light reflected from the local alignment mark as measured by the signal strength (S/N ratio) of the alignment system. The larger the signal the better the alignment.

The explanatory variables considered in this study are divided into process factors and factors related to the alignment system. The process factors include oxide thickness, reflow, metal thickness, and photoresist thickness. These factors occur sequentially prior to alignment. The alignment system factors include alignment system, mark type and mark polarity. Although the process factors are controlled at the customer sites, variations in them can significantly affect the quality of the signal and therefore are considered to be noise[49] factors by GCA. Since GCA has control over the alignment system factors, these are called control[49] factors.

Each factor in the experiment was studied at two levels. The levels of the noise factors were chosen to be representative of the typical operating ranges for a variety of GCA semiconductor manufacturing customers, while the levels for two of the control factors (mark type and mark polarity) represent the settings most commonly used by their customers. A new alignment system, referred to as brightfield, was also compared to their existing one, "darkfield" (see Figure 20.2 and Table 20.1).

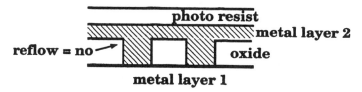

Figure 20.2. *A Simplistic View of the Multi-Level Metal Structure used in the Experiment. The four process factors are the oxide thickness, reflow (indicated by the sharpness of the oxide edges), metal layer 2 thickness, and photoresist thickness.*

Table 20.1. *Experimental Factors and High (+) and Low (-) Settings for the Experiment. The levels were selected to represent typical operating conditions in semiconductor manufacturing.*

	Factors	Levels	
		+	-
Control:	Alignment System (AS)	Brightfield	Darkfield
	Mark Type (MT)	Bar	Segmented
	Mark Polarity (MP)	Clearfield	Darkfield
Noise:	Oxide (Ox)	10,000 Å	6,000 Å
	Reflow (Ref)	Yes	No
	Metal Thickness (Met)	3 µm	1.3 µm
	Photo Resist Thickness (PR)	27,000 Å	9,000 Å

A designed experiment was carried out to obtain the information required to accomplish the objectives of the study. The 16 (2x2x2x2) different combinations for the noise factors were combined with the eight (2x2x2) different combinations of the control factors for a total of 16 x 8 = 128 test conditions. The design was generated by employing full factorials for both the control and noise variables. For the four noise factors a $2^4 = 16$ run full factorial design was used (noise array), and for the three alignment system factors a $2^3 = 8$ run full factorial design was generated (control array). The two designs were combined to give a crossed design with 16 x 8 = 128 test conditions. The design was not replicated. Table 20.2 shows the control and noise arrays.

The 16 wafers used in the experiment were processed in a split-lot manner (see Figure 20.3) according to the treatment combinations in the noise array. The 16 wafers were split

---

[49] Noise factors in a robust design experiment are those factors which are hard or impossible to control and induce unwanted variability into the process. On the other hand, control factors are those factors which can be controlled during normal production. For a more detailed description of robust design and control and noise factors, see Kackar, 1985.

into two groups to be processed through the oxidation step in two batches of eight wafers each; one batch was processed using an oxide thickness of 6,000 Å and the other using an oxide thickness of 10,000 Å. Each batch of eight wafers was then split into two groups of four wafers each, and one group of wafers was reflowed while the other group of wafers was not reflowed. Each batch of four wafers was again split into two batches to be processed using metal thickness of 9,000 Å and 27,000 Å, respectively. Finally the batches of two wafers were split one more time and given photoresist thicknesses of 1.3 µm and 3 µm, respectively.

Table 20.2. *Experimental Design Showing the 16 Combinations of the Noise Factors x Eight Combinations of the Control Factors. Each column and row combination generates an observation for a total of 128 (16 x 8) readings. The numbers refer to the order in which signal readings were obtained. Note that each wafer is measured in both alignment systems. For example, wafer 1 produced readings 1–4 and 65–68.*

			Noise Array															
		Wafer No.	1	2	3	4	5	6	7	8	9	10	11	12	13	14	15	16
		Ox	-	+	-	+	-	+	-	+	-	+	-	+	-	+	-	+
		Ref	-	-	-	-	-	-	-	-	+	+	+	+	+	+	+	+
		Met	-	-	+	+	-	-	+	+	-	-	+	+	-	-	+	+
		PR	-	-	-	-	+	+	+	+	-	-	-	-	+	+	-	+
Control Array																			
AS	MP	MT																	
-	-	-	1	5	9	13	17	21	25	29	33	37	41	45	49	53	57	61
-	-	+	2	6	10	14	18	22	26	30	34	38	42	46	50	54	58	62
-	+	-	3	7	11	15	19	23	27	31	35	39	43	47	51	55	59	63
-	+	+	4	8	12	16	20	24	28	32	36	40	44	48	52	56	60	64
+	-	-	65	69	73	77	81	85	89	93	97	101	105	109	113	117	121	125
+	-	+	66	70	74	78	82	86	90	94	98	102	106	110	114	118	122	126
+	+	-	67	71	75	79	83	87	91	95	99	103	107	111	115	119	123	127
+	+	+	68	72	76	80	84	88	92	96	100	104	108	112	116	120	124	128

All 16 wafers were first measured using the darkfield alignment system (AS = -), producing signal readings numbered 1–64 in Table 20.2, and then measured using the brightfield system, which are signal readings numbered 65–128 in Table 20.2.

There are clearly three different sizes of experimental units (see Figure 20.3) in this design. They are

1. a batch of wafers associated with the noise factors,
2. the wafer associated with the control factor alignment system,
3. the area within the wafer where the marks reside, associated with two control factors (mark type and mark polarity).

There are also restrictions in randomization: the test conditions for the noise factors were not fully randomized, and the wafers were first measured in one alignment system and then on the other. These circumstances lead to what is called a split-plot, or equivalently split-unit design.

Split-unit designs arose in agricultural experiments where plots of land (see Figure 20.3) were given, for example, a variety of fertilizers, (A1, A2, A3, A4) and then each plot was split into smaller units where, for example, different seeds (B1, B2, B3) were planted. This type of experiment has two sizes of experimental units: the main-unit (plot) to which the fertilizers are applied and a sub-unit (sub-plot) to which the seeds are applied.

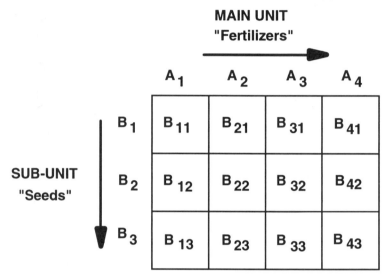

Figure 20.3. *Illustration of a Split-Unit Design Structure. The main-unit consists of the plots of land receiving the fertilizers, and the sub-unit consists of the three seed types planted in each plot of land.*

In this experiment the main unit is the batch of wafers, the split-unit is the wafer, and the split-split-unit is the area within the wafer where the marks reside. The analysis of the data should reflect the split-split-unit structure of the design and should take into account the different sizes of experimental units and restrictions in randomization. In the Appendix, we give more details about the split-split-unit structure of this experiment.

DATA ANALYSIS AND INTERPRETATION OF RESULTS

The data analysis of this experiment consisted of two steps. First, analysis of variance (ANOVA) techniques were used to identify the statistically significant control factors, the statistically significant interactions between noise and control factors, and the statistically significant interactions between control factors. Second, once these factors and interactions were identified, graphical techniques such as interaction and box plots were used to identify factor settings which were closest to the objective of the experiment (i.e., maximum signal strength and minimum sensitivity to process conditions).

A split-split-unit analysis was performed on the data. As was mentioned above, this technique was suitable for this experiment because of the way the experiment was carried out. This type of analysis requires different error terms for testing the significance of effects. The ANOVA table depicting the split-split-unit structure of the experiment is displayed in Table 20.3 in the Appendix.

One of the objectives of this study was to verify whether the brightfield alignment system performed better than the current darkfield alignment system. The ANOVA for this factor indicated that the two systems produce different signal readings. The interactions between the control and noise factors were assessed before determining which system produced a stronger signal, to insure that the best system was the one that maximized the signal and was less sensitive to process variation. The results of this analysis indicate that two interactions between the alignment systems and noise factors, AS*Ox and AS*Met*Ref, were statistically significant (see the Appendix for details).

Identifying the settings of the control factors that were least sensitive to the variations in the noise factors was easily accomplished by using a control*noise interaction plot between the noise and control factors (see Cantell and Ramírez, 1992). The interaction plot of the alignment system and oxide thickness, AS*Ox is shown in Figure 20.4. In this plot the signal strength is plotted versus the noise factor oxide thickness, and the two lines correspond to the two alignment systems. The setting for the alignment system that was least sensitive to the variation in oxide thickness was the one that produced the flattest line. This interaction is of great importance to GCA because it provides information about the performance of the alignment systems under different oxide thickness combinations; this occurs with the brightfield alignment system. Note that this setting of the alignment system also maximized the signal strength.

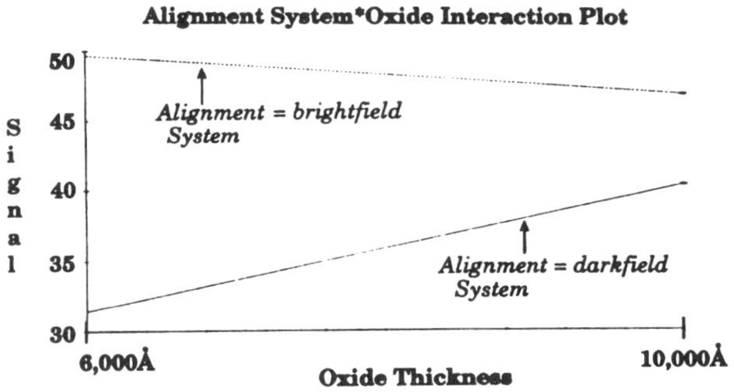

Alignment System*Oxide Interaction Plot

Figure 20.4. *Noise × Control Interaction Plot for Alignment System and Oxide Thickness. This shows that the brightfield alignment system had the least amount of variation across the levels of oxide thickness.*

The three-way noise × control interaction plot of AS*Met*Ref shown in Figure 20.5, was constructed by defining a new variable for the x-axis consisting of the four combinations of the two noise factors, metal thickness and reflow. The settings which minimize the variation in the response for the three-way interaction plot with AS*Met*Ref are not as obvious. It can be seen from this plot that the two lines are parallel except at the point (9,000 Å, yes) Met*Ref combination which suggests a mild interaction between the control factor alignment system and the noise factors. Based on Figures 20.4 and 20.5, we see that the brightfield alignment system not only produces the highest signal across the two levels of oxide and four combinations of Met*Ref, but is also least sensitive to variations in process conditions.

An ANOVA table for the remaining factors, mark polarity, mark type, their interaction, and their interactions with the alignment system and the noise factors is shown in Table 20.6 in the Appendix. The analysis indicates that the following effects are statistically significant: Mark Polarity, Mark Type, MP*MT, MP*MT*Ox, MP*AS, and MT*AS.

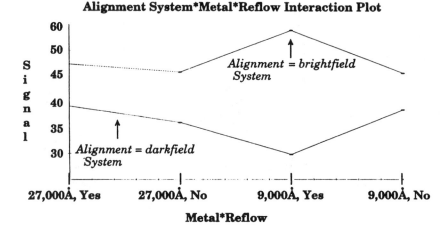

Figure 20.5. *Noise × Control Interaction Plot for AS*Met*Ref. This shows that the brightfield alignment system maximized the signal.*

The interpretation of these results begins with the three-way noise × control interaction MP*MT*Ox displayed in Figure 20.6. The preferred settings for the control factors can again be obtained by finding the flattest line. There are two lines that have the least amount of change across the noise factor oxide thickness and produce the highest signals. These lines correspond to the following settings: mark polarity = clearfield and mark type = bar, or mark polarity = darkfield and mark type = segmented.

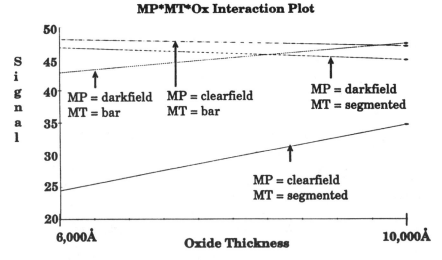

Figure 20.6. *Noise × Control Interaction Plot for MP*MT*Ox. This shows that MP=darkfield, MT=segmented or MP=clearfield, MT=bar had the least variation across the oxide thickness. They also maximized the signal.*

The remaining two-way interactions with the control factors, MP*AS and MT*AS, were interpreted by calculating the four means corresponding to each combination of these factors for each of the interactions. Since these are all control factors, we were not looking for the settings which minimized the variation in the response but those which maximized the response. For the MP*AS interaction, a setting of MP = darkfield and AS = brightfield produced the maximum average signal of 55.0625. Similarly, for the MT*AS interaction a setting of MT = bar and AS - brightfield produced the maximum average signal of 54.8125.

The results show the brightfield system to be superior to the MicroDFAS with respect to signal strength and is less sensitive to variation in the noise factors. Based on previous experimentation, GCA decided to recommend a mark polarity = darkfield and mark type = segmented. Figure 20.6 shows that this combination was less sensitive to variation in oxide thickness. This conflicts with the preferred setting for mark type from the two-way interaction MT*AS. However, a tradeoff was between increasing the signal and minimizing variation in the response.

CONCLUSIONS AND RECOMMENDATIONS

MicroDFAS, the existing local alignment system for the XLS wafer stepper aligns non-metal levels very well. MicroDFAS also aligns metal levels. However, on some multi-level metal processes, MicroDFAS was found to be not robust enough. The brightfield local alignment system was designed to provide the XLS wafer stepper with a local alignment system that is robust on wafer processes that have multiple metal levels.

This experiment tested both alignment systems for alignment performance on a simulated metal two level that was aligned back to a VIA level. The experimental design varied several process parameters (oxide thickness, reflow, metal thickness and photoresist thickness). The data clearly shows that the brightfield system is indeed better than the darkfield system for metal levels, confirming prior knowledge about the performance of the two systems.

Having an alignment system that can align metal levels with minimal variation allows GCA to use the darkfield system for non metal levels and the brightfield system for metal levels. Therefore, GCA can proceed with an alignment strategy that places both alignment systems on the stepper. This combination will make the XLS a more robust wafer stepper.

ACKNOWLEDGMENTS

The authors wish to acknowledge John Meaux of Cypress Semiconductor for providing the wafers for the experiments. They would also like to acknowledge Paul Bischoff and Norm Roberts of GCA Corporation for their assistance in gathering the data, and Charles Smith and Kevin Orvek from Digital Equipment Corporation. Ramón León, of the University of Tennessee, and the referees provided helpful suggestions that improved the readability of the final document.

REFERENCES

Box, George E. P. "Signal-to-Noise Ratios, Performance Criteria and Transformations," *Technometrics*. Vol. 30, No. 1, pp. 1-31, 1988.

Box, George E. P., and Stephen P. Jones. "Split-plot Designs for Robust Product Experimentation," *Journal of Applied Statistics*, Vol. 19, No. 1, pp. 3-26, 1992.

Cantell, Brenda S., and José G. Ramírez. "Robust Design of a Polysilicon Deposition Process Using Split-Plot Analysis," *Quality and Reliability International*, Vol. 10, pp. 123-132, 1994.

Kackar, R. N. "Off-line Quality Control, Parameter Design and the Taguchi Method," *Journal of Quality Technology*, Vol. 17, No. 4, pp. 176-209, 1985.

Hicks, C. R. *Fundamental Concepts in the Design of Experiments*, New York, NY: Holt, Reinhart, and Winston, 1973.

APPENDIX

Split-Split-Unit Design and Data Analysis

Split-Split-Unit Design

In this experiment there were three sizes of experimental units and several restrictions in randomization. The first restriction in randomization occurred during the wafer fabrication process involving the deposition and reflow of oxide, the deposition of metal, and the coating of photoresist. The experimental unit for these factors (noise array) was a batch of wafers split hierarchically into batches of wafers of smaller sizes. This type of structure is called a strip-plot or criss-cross design. These factors and their interactions are considered the main-unit of the design.

The second experimental unit, the wafer, was associated with the alignment system factor. It should be noted that the alignment system control factor resulted in another restriction in randomization; all 16 wafers were first measured using the darkfield alignment system (AS = -), test conditions 1 - 64, and then measured using the brightfield system, test conditions 65 - 128 (see Table 20.2). This factor, along with its interactions with the noise factors made up the split-unit of the design.

The third experimental unit, which receives the different combinations of mark type and mark polarity, corresponds to areas within the wafer. The four specific test locations were identified with an x-y coordinate system. The four treatment combinations were randomly assigned to the locations within the wafer. These factors and their interactions with all other factors made up the split-split-unit of the design. The entire design structure is shown in Table 20.3.

Unlike a completely randomized design, in which only one error term is used, the analysis of a split-unit design utilized more than one error term when testing for significant effects. There must be an error term corresponding to each experimental unit size. Treatments applied to a particular experimental unit should be tested using the appropriate error term. For this split-split-unit experiment, at least three error terms were needed to test the control and noise factors, along with their interactions.

Data Analysis

Typically, the data obtained from robust design experiments is analyzed using Taguchi's Signal-to-Noise ratios where the objective function would be determined by the larger-the-better scenario (see Kackar, 1985). Another approach is to analyze the eight means and standard deviations computed across all combinations of the noise array. This analysis technique provides more information than a Taguchi's Signal-to-Noise ratio analysis because factors which affect the mean, the standard deviation, or both can be identified (see Box, 1988).

The problem with these techniques is that they do not take into account the way the experiment was run (i.e., split-unit structure). Also, there is no way of estimating the interaction effects between the control and noise factors. These interactions provide useful information because they breakdown the sources of variation according to the noise factors and show how the noise factors might change the effect of the factors on the response. A better way of analyzing this type of experiment is to use an analysis that takes into account the split-unit structure of the experiment (see Box, 1992).

Table 20.3. *ANOVA Table Showing the Split-Split-Unit Structure of the Experiment. The main unit corresponds to the noise factors. The split-unit corresponds to the alignment system and its interactions with the noise factors. The split-split-unit corresponds to mark polarity, mark type and their interactions with the alignment system and the noise factors.*

Source	d.f.	
Oxide Thickness	1	**Main-unit**
Metal Thickness	1	
Photo Resist Thickness	1	
Reflow	1	
All 2-way interactions	6(6×1)	
All 3-way interactions	4(4×1)	
Ox*Met*PR*Ref	1	
Alignment System	1	**Split-unit**
AS*Ox	1	
AS*Met	1	
AS*PR	1	
AS*Ref	1	
ε_1: Error Term	5	
Mark Type	1	**Split-split-unit**
Mark Polarity	1	
MT*MP	1	
MP*Ox	1	
MP*Met	1	
MP*PR	1	
MP*Ref	1	
MT*Ox	1	
MT*Met	1	
MT*PR	1	
MT*Ref	1	
MP*MT*Ox	1	
MP*MT*Met	1	
MP*MT*PR	1	
MP*MT*Ref	1	
MT*AS	1	
MP*A5	1	
MP*MT*AS	1	
ε_2	78	

Main-Unit Analysis

The main-unit was a strip-plot design which required several replications of the experiment in order to estimate the error terms needed to test the effects of the factors. Since the experiment was not replicated, there was not enough data to estimate all of the error terms necessary to test the effects. The sums of squares (see Table 20.4) for each factor can be assessed qualitatively. Photo resist thickness appears to have the largest sums of squares among all of the noise main effects and interactions. However, as pointed out by Box (see Box, 1992), in robust type situations, we are particularly interested in the effect of the control factors and the way in which the noise factors modify the effects of the control factors. Since GCA has no control over noise factors, they do not significantly benefit from estimating the effects of noise factors such as oxide thickness.

Table 20.4. *Sums of Squares for the Effects of the Noise Factors in the Main Unit.*

Source	df	Sums of Squares
Oxide Thickness	1	279.0703
Metal Thickness	1	0.6328
Photo Resist Thickness	1	3090.9453
Reflow	1	39.3828
Ox*Met	1	3.4453
Ox*PR	1	23.6328
Ox*Ref	1	23.6328
Met*PR	1	21.9453
Met*Ref	1	51.2578
PR*Ref	1	99.7578
Ox*Met*PR	1	20.3203
Ox*Met*Ref	1	5.6953
Met*PR*Ref	1	409.6953
Ox*PR*Ref	1	10.6953
Ox*Met*PR*Ref	1	212.6953

Table 20.5. *ANOVA for Alignment System and its Interactions with the Noise Factors. Effects in bold type are significant at the 5% level.*

Source	df	SS=MS	F-Ratio	P-Value
Alignment	1	**4863.445**	**48.430**	**0.0009**
AS*Ox	1	**1110.383**	**11.057**	**0.0208**
AS*Met	1	354.445	3.529	0.1190
AS*PR	1	53.820	0.535	0.4990
AS*Ref	1	468.945	4.670	0.0831
AS*Ox*Met	1	79.695	0.794	0.4120
AS*Ox*PR	1	187.695	1.869	0.2290
AS*Ox*Ref	1	31.007	0.308	0.6016
AS*Met*PR	1	267.382	2.663	0.1638
AS*Met*Ref	1	**689.132**	**6.862**	**0.0470**
AS*PR*Ref	1	92.820	0.924	0.3815
error	5	SS = 502.10, MS = 100.4202		

Split-Unit Analysis

The second restriction in randomization was due to the way the wafers were tested on the alignment systems. The alignment system factor, along with its interactions with the noise factors made up the first sub-unit (see Table 20.5). If high order interactions are assumed to be negligible, we can pool them into an error term and perform an ANOVA for the remaining effects. The error term used to test the effects in the first subplot consists of five degrees of freedom obtained from four four-way interactions, AS*Ox*Met*PR, AS*Ox*Met*Ref, AS*Met*PR*Ref, and AS*Ox*PR*R, and the one five-way interaction,

AS*Ox*Met*PR*Ref. The results, shown in Table 20.5, indicate that Alignment system, AS*Ox, and AS*Met*Ref are significant at the 0.05 significance level.

Split-Split-Unit Analysis

The remaining factors (mark polarity, mark type, and their interactions) make up the second sub-unit (see Table 20.3). High order interactions, such as 4, 5, 6 and 7-way interactions, were assumed negligible and pooled into a 78 degree-of-freedom error term. An ANOVA table for these effects is shown in Table 20.6.

We see that the control factors (mark polarity, mark type, and their interaction) are significant. These factors also interact with the alignment system and the noise factor oxide thickness.

Table 20.6. *ANOVA for Mark Polarity, Mark Type, and Their Interactions with the Alignment System and the Noise Factors. Effects in bold type are significant at the 5% level.*

Source	df	SS=MS	F-Ratio	P-Value
Mark Polarity	1	**1519.380**	**9.28**	**0.0032**
Mark Type	1	**2389.130**	**14.59**	**0.0003**
MP*MT	1	**2784.440**	**17.00**	**0.0001**
MP*Ox	1	86.130	0.53	0.4700
MP*Met	1	1.757	0.01	0.9100
MP*PR	1	103.320	0.63	0.4297
MP*Ref	1	70.507	0.43	0.5093
MT*Ox	1	46.320	0.28	0.6016
MT*Met	1	3.445	0.02	0.8848
MT*PR	1	5.690	0.03	0.8501
MT*Ref	1	48.750	0.30	0.5896
MP*MT*Ox	1	**625.508**	**3.98**	**0.0490**
MP*MT*Met	1	150.940	0.92	0.3400
MP*MT*PR	1	46.320	0.28	0.6020
MP*MT*Ref	1	73.507	0.45	0.5000
MP*AS	1	**1505.630**	**9.15**	**0.0033**
MT*AS	1	**679.880**	**4.15**	**0.0450**
MP*MT*AS	1	625.695	3.82	0.0542
error	78	SS = 12775.859, MS = 163.793		

BIOGRAPHIES

Brenda Cantell has an M.S. in Applied Statistics, an M.S. in Industrial Engineering and a B.A. in Mathematics. Brenda is currently a Senior Statistician with Unitrode Corporation. In this position, she is responsible for providing statistical consulting in support of understanding and improving semiconductor manufacturing processes, developing and delivering statistical training, implementing more effective statistical techniques, and developing quality systems. Brenda previously worked at Digital Semiconductor as a Statistician and a Quality Engineer. Her accomplishments included deploying advanced SPC techniques to A.P.C. clean room support equipment, developing a statistical tool suite in SAS for the analysis of yield data, and providing numerous statistically related training classes. Brenda is a member of the ASA and has obtained certification in Quality Engineering from ASQC.

José G. Ramírez received a Ph.D. in statistics (1989) and an M.S. in Applied Statistics (1985) both from the University of Wisconsin-Madison, and a degree in Mathematics (1982) from Universidad Simón Bolívar in Caracas, Venezuela. He is currently a Consultant Engineer at Digital Semiconductor, where he consults and teaches statistics. Before joining Digital Semiconductor, José was at the Center for Quality and Productivity Improvement, University of Wisconsin-Madison, where he conducted research on statistical process monitoring. Prior to that he served as a statistical consultant at the Statistical Laboratory, University of Wisconsin-Madison. José is a member of the American Statistical Association, the American Society of Quality Control, and a certified Quality Engineer.

William Gadson began his work in the semiconductor industry with GCA in 1978 as a field service engineer. Since then, he has served as a wafer process engineer and as a wafer stepper applications engineer at GCA, ASM, and Nikon. He holds an A.A. in electronic engineering technology from Hutchinson Community Junior College, Hutchinson, Kansas and a B.S. in general studies from the University of Texas at Dallas. He is presently employed as the lithography supervisor for TFI, a semiconductor film and submicron lithography wafer service company.

PART 4

STATISTICAL PROCESS CONTROL

CHAPTER 21

INTRODUCTION TO STATISTICAL PROCESS CONTROL (SPC)

Veronica Czitrom

INTRODUCTION

Manufacturing processes must perform consistently over time to be capable of meeting manufacturing and design requirements. *Statistical Process Control (SPC)* is a methodology to monitor and benchmark a process to improve its variability, stability, and capability. The control chart is the main tool in this methodology. Statistical Process Control can be used to:
- monitor and reduce process variability.
- monitor and maintain the process on target.
- determine when a process needs "tweaking" (adjusting) and when it does not.
- establish process stability and detect process changes so that corrective action can be taken.
- determine the capability of a manufacturing process to make product that conforms to specifications and monitor it on-line.
- improve quality and productivity by improving the process, which reduces product inspection, scrap, and rework at the end of the line.

The case studies in this part were used to
- overcome the common practical difficulty of obtaining statistically valid control limits for control charts in the presence of natural process drift, as illustrated for a topside nitride process.
- improve manufacture of modules for surface mount technology as part of a comprehensive strategy to implement statistical process control in a printed circuit board shop.
- control semiconductor processes in wafer fabs and post-fab operations by implementing statistical process control as a standard operating procedure.

Before starting statistical process control efforts, a gauge study (Part 1) should be performed on the relevant metrology to assure reliable measurements. Initial estimates of process variability, stability, and capability for statistical process control can be obtained from a passive data collection (Part 2). A process can be actively improved using designed experiments (Part 3), and then the improved process can be monitored using statistical process control. Problems such as excessive variability that are identified using statistical process control can be addressed using designed experiments. The information obtained during statistical process control can be used to give preliminary equipment reliability estimates (Part 5).

Statistical process control concepts are used in other units. Lynch and Markle (Chapter 7) used control charts to establish process stability in a passive data collection, and Pankratz (Chapter 3) used control charts to detect changes in a gauge. Canning and Green (Chapter 26) consider process capability during a marathon to improve equipment reliability.

Control charts are introduced in the first section. The second section considers additional topics such as graphical tools, process capability, and issues regarding implementation. The third section gives a brief description of the case studies presented in the remainder of this part, and the last section gives references for further reading.

CONTROL CHARTS

Control charts are among the most important tools in Statistical Process Control. They were developed in the 1920's by Dr. Walter Shewhart, a scientist at Bell Laboratories, the research arm of American Telephone and Telegraph.

The first subsection considers basic concepts and definitions for control charts. The following subsections are on specific control charts, namely $\bar{x}$ and R charts and the statistical derivation of the $\bar{x}$ chart, $\bar{x}$ and s charts, x and MR charts, p charts, and c charts. The last subsection describes control charts in the case studies in this part.

Basic Concepts

A certain amount of variability in a process is unavoidable. When a layer of tungsten is deposited on a wafer, the thickness of the layer will vary from site to site on a wafer, and from wafer to wafer. The natural variability comes from a number of small contributors to variability such as measurement error and environmental conditions, called *common causes* or *chance causes* of variation. A process that is operating only under common causes of variation is said to be in *statistical control*. Unusual variability in the system above the natural variability or background noise is due to *assignable causes* or *special causes*. Examples of assignable causes are a candy wrapper in a chemical bath, a crack in a tube, and a broken wafer. A process that is operating in the presence of unusual events or assignable causes that increase the variability is said to be *out of (statistical) control*. If a process is out of control, corrective action should be taken to search for and reduce or eliminate assignable causes of variation.

Statistical quality control charts, Shewhart control charts, or simply *control charts*, are important tools for assessing whether a process is in control. A control chart can help separate the wheat from the chaff, the signal of an unusual event from the background noise of normal process variability. A typical control chart is illustrated in Figure 21.1. A control chart plots *values (or points)* of response data in chronological order, usually connected by straight lines. A control chart includes three horizontal parallel lines: a center line, an upper control limit above it, and a lower control limit below it. The *center line (CL)* on a control chart is a reference line about which the chart points are expected to cluster in the absence of an assignable cause. The center line is usually set at the average, the median, the mode, or the target value of the points being plotted. The *upper control limit (UCL)* line and the *lower control limit (LCL)* line define a region where most observations are expected to fall. The upper and lower *control limits* (also called *statistical control limits* or *SPC limits*) reflect the natural variability of the process, and are constructed in such a way that when the process is in control, most of the points will fall inside the control limits in a random fashion. If a point on the control chart falls above the upper control limit or below the lower control limit, the process is said to be *out of control*, and assignable causes need to be searched for and eliminated. Another indication that a process is out of control is the occurrence of a non-random pattern, such as eight consecutive points on one side of the center line. This is an example of a *run rule*.

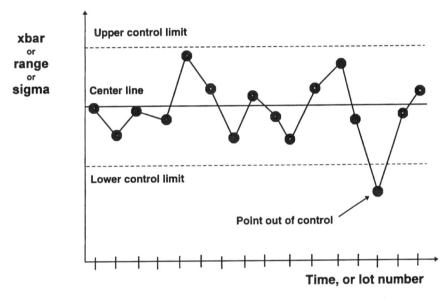

Figure 21.1. *A control chart plots values of the variable being monitored as a function of time, connected by line segments. The values being plotted (such as individual values, averages, ranges, and standard deviations) define the type of control chart. When a process is under statistical control, points will cluster around the center line (CL). A point below the lower control limit (LCL) or above the upper control limit (UCL) is out of (statistical) control.*

Many control charts have a common form. Let w be an observed process characteristic that is determined from a sample, such as the sample mean (average) $\bar{x}$. Let w have a true (population) mean μ and standard deviation σ. The center line, upper control limit, and lower control limit have the general form

$$\text{UCL} = \mu + k\sigma,$$
$$\text{CL} = \mu, \qquad (21.1)$$
$$\text{LCL} = \mu - k\sigma.$$

The upper and lower control limits are chosen to be at a distance of k standard deviations σ from the mean μ. Since μ and σ are usually not known, they are estimated from a sample. We will later give estimates of μ and σ for different types of charts. It is customary to use $k = 3$, which gives 3-*sigma control limits* that are three standard deviations away from the center line. If the distribution of w is normal, then approximately 99.73% of the values of w will fall between the 3-sigma control limits and only 0.27% of the values of w will fall outside of the 3-sigma control limits. That is, if the process is in control, a point will fall outside the control limits giving a false alarm approximately only 27 out of 10,000 times. Instead of selecting k to be 3 and finding the corresponding probability of falling between the control limits, we can select the probability and find the corresponding control limits, called *probability limits*. For example, if the probability of falling inside the control limits is chosen to be exactly 99%, then $k = 3.09$ if w is normally distributed. Lines that are two standard deviations away from the center line are called 2-*sigma warning* limits and can also be drawn on the control charts. Lines one standard deviation away from the center line are also sometimes drawn on the control chart. The 3-sigma, 2-sigma, and 1-sigma lines divide the control charts into three zones: zone A goes from the 2-sigma to the 3-sigma lines on either side of the center line; zone B goes from the 1-sigma to the 2-sigma lines on either side; and zone C goes from the 1-sigma line below the center line to the 1-sigma line above the center line.

A control chart can be constructed to plot a single observation (an "individuals" chart) or a sample statistic computed from several observations (such as the mean $\bar{x}$ that is plotted on an "$\bar{x}$ chart"). The sample statistic is computed from a *rational subgroup*, which is a sample of n observations. A single observation can be thought of as a rational subgroup of $n = 1$ observations. The rational subgroup of n (one or more) observations is taken on a regular basis, and a sample of at least 20 or 30 rational subgroups is gathered to estimate the center line and the control limits. The observations can be obtained from a passive data collection. Once the control limits and the center line have been calculated, the points corresponding to each rational subgroup can be plotted in chronological order on the control chart by operators on the shop floor.

Different criteria can be used to determine when a process is out of control. An out-of-control condition is signaled by a pattern that is very unlikely to occur if the process is in control, and is likely to occur if the process is out of control. For example, under the *Western Electric rules*, a process is considered out of control if one of the following occur:
- One point is outside the 3-sigma control limits.
- Eight consecutive points are on one side of the center line.
- Two out of three consecutive points are outside the 2-sigma warning limits on one side of the center line.
- Four out of five consecutive points are at a distance of 1-sigma or more from the center line on one side of the center line.

Control limits that are too close to each other in relation to the data give false alarms when the process is really in control, while control limits that are too far from each other in relation to the data will not give an alarm when the process really is out of control. When the control limits seem too wide or too narrow with respect to the data, it is likely that they have not been adequately determined or that the process has changed.

Control limits for *batch processes*, which are so prevalent in the semiconductor industry, need to be computed with care. There is between-batch and within-batch variability: there is variability between furnace runs as well as variability within a furnace run at different furnace locations, and there is variability between wafers and variability within a wafer at different sites on the wafer. The observations in one batch are not independent, since they share the conditions of that batch. One common mistake is to compute control limits based only on the within-batch variability in the presence of large batch-to-batch variation. This leads to too narrow control limits and a large number of false out of control points will occur. One approach to account for batch processes is to consider a sample statistic such as the batch mean $\bar{x}$ as a single "new" point in an individuals chart.

A control chart can be interpreted as a graphical representation of the test of the hypothesis that the process is in a state of statistical control. A point falling inside the control limits is equivalent to failing to reject the hypothesis of statistical control, and a point falling outside the control limits is equivalent to rejecting the hypothesis of statistical control. Two types of errors can be made: taking corrective action when it is unnecessary (concluding that the process is out of control when it is in control, or type I error), and not taking corrective action when it is necessary (concluding that the process is in control when it is not in control, or type II error). These errors can lead respectively to a needless search for an assignable cause, or to unnecessary corrective action in a manufacturing line. As the control limits are moved away from the center line, the risk of declaring the process in control when it is not in control increases, while the risk of declaring the process out of control when it is in control decreases. The risks of making the two types of errors are determined by the selection of the factor k in the control limits of equation (21.1), and by the sample sizes, mainly the number of subgroups used to calculate control limits.

There are two important types of data, continuous data and attribute data. Different control charts are used for each type of data.

Continuous data (or *variables data, measurement data,* or *interval data*), can assume any numerical value in an interval. Examples of continuous data are temperature and thickness. For continuous data, the main control chart to monitor the central tendency, or location, of the output is the $\bar{x}$, or *xbar, chart*. Control charts to monitor the variability, or dispersion, of the output are the *range (R) charts*, and the *standard deviation (s) charts*. Continuous data is usually characterized by two control charts, one for central tendency and one for variability, such as the $\bar{x}$ and R charts, or the $\bar{x}$ and s charts. The s chart is usually used when the rational subgroup sample has around $n = 10$ or more observations, and the R chart when the rational subgroup has less that 10 observations. When continuous data has only one observation in each rational subgroup, an *individuals (I, or x) chart* is used to monitor the central tendency of the process, and a *moving range (MR) chart* is used to monitor the variability of the process. Charts that monitor central tendency should not be interpreted if the associated charts for variability are out of control.

Attribute data corresponds to items that are classified as having one of several attributes, or belonging to one of several categories. Examples of attribute data for two categories are *conforming* and *nonconforming* items, or equivalently, *nondefective* and *defective* items. For example, electrical tests on chips either conform or do not conform to requirements. The *fraction nonconforming*, or *p, chart* monitors the fraction of non-conforming items, such as the fraction of nonconforming die on a wafer. *Nonconformities*, or *c (count), charts* monitor the number of nonconforming items, such as the number of particles on a wafer.

If there is an option between using continuous or attribute data to control a process, continuous data is best. It provides more information about process performance and usually leads to more efficient control procedures.

We will now introduce the following control charts: $\bar{x}$ and R charts, $\bar{x}$ and s charts, I and MR charts, p charts, and c charts. Generalizations of these control charts and other types of control charts can be found in the literature. We will give 3-sigma control limits for all the control charts. The center line and the control limits should be updated when there is a significant change in the process for reasons that are understood, such as replacement of a major piece of equipment.

Control Charts for $\bar{x}$ and R

Consider a random sample *(rational subgroup)* of n independent observations of continuous data on a process taken on a regular basis. The sample size n remains constant, and is usually around ten or smaller, and typically four or five. The $\bar{x}$, or *xbar, control chart* is used to monitor the central tendency or location of the continuous data on the process, and the *range (R) control chart* is used to monitor the variability or dispersion of the continuous data on the process. The mean $\bar{x}$ and the range R (largest minus smallest observations) are computed for the n observations in each subgroup. The average $\bar{x}$ of each subgroup is plotted in chronological order on the $\bar{x}$ control chart, and the range R of each sample is plotted in chronological order on the R chart.

Once there are some 20 or 30 subgroups for non-batch processes, the control limits for the control charts can be calculated as follows. Compute the average $\bar{\bar{x}}$ of the sample averages $\bar{x}$, and the average $\bar{R}$ of the sample ranges R. The center line and the 3-sigma control limits for the $\bar{x}$ chart are given by

$$\text{UCL} = \bar{\bar{x}} + A_2 \bar{R},$$
$$\text{CL} = \bar{\bar{x}}, \quad (21.2)$$
$$\text{LCL} = \bar{\bar{x}} - A_2 \bar{R},$$

and the center line and the 3-sigma control limits for the R chart are given by

$$\text{UCL} = D_4 \bar{R},$$
$$\text{CL} = \bar{R}, \quad (21.3)$$
$$\text{LCL} = D_3 \bar{R},$$

where A_2, D_3, and D_4 are constants that are given in tables and that depend on the number n of observations in each sample and the assumptions of normality and independence of observations (see Montgomery, 1991, Appendix VI). Around 99.73% of the points are expected to fall inside the control limits when the points come from a normal distribution and the process is under control. By the central limit theorem, this percentage is still approximately correct for the $\bar{x}$ chart for distributions that are not too far from normal. The percentage in the R chart is more sensitive to deviations from normality. Each point on the R chart gives variability within a sample (within-sample variability), and the points on the $\bar{x}$ chart indicate variability between samples (between-sample variability) and location. While the UCL and LCL of the $\bar{x}$ chart are symmetrically located around the CL at a distance of $A_2 \bar{R}$, the UCL and LCL of the R chart are asymmetrically located around the CL at different distances.

Statistical Derivation of the $\bar{x}$ Control Chart

The upper and lower control limits for the $\bar{x}$ control charts can be derived as follows. Assume that a sample of n observations is taken from a random variable X that has mean μ and standard deviation σ. The average or mean $\bar{x}$ of the observations in the sample has mean μ and standard deviation $\sigma/\sqrt{n}$, and by the central limit theorem $\bar{x}$ has an approximately normal distribution. This means that approximately $(1 - \alpha)100\%$ of the means $\bar{x}$ from samples of size n will fall between $\mu - Z_{\alpha/2}\sigma/\sqrt{n}$ and $\mu + Z_{\alpha/2}\sigma/\sqrt{n}$, where the probability that a standard normal random variable is greater than $Z_{\alpha/2}$ is $\alpha/2$. For 3-sigma limits, approximately 99.73% of the means $\bar{x}$ from samples of size n will fall between $\mu - 3\sigma/\sqrt{n}$ and $\mu + 3\sigma/\sqrt{n}$.

Assume that neither μ nor σ are known, and that we take m samples each one of size n. Denote the averages of the m samples by $\bar{x}_1, \bar{x}_2, ..., \bar{x}_m$, and the ranges of the m samples by $R_1, R_2, ..., R_m$. Let $\bar{\bar{x}}$ be the average of the m sample means, and let $\bar{R}$ be the average of the m sample ranges. Estimate μ by $\bar{\bar{x}}$. If the standard deviation s is estimated using the range $\bar{R}$, the estimate of $3\sigma/\sqrt{n}$ can be expressed as $A_2 \bar{R}$, where the constant A_2 is given in tables as a function of the sample size n. The center line and the control limits are thus given by $\bar{\bar{x}}$ and $\bar{\bar{x}} \pm A_2 \bar{R}$, respectively.

Control Charts for $\bar{x}$ and s

Consider a random sample (rational subgroup) of n independent observations of continuous data on a process taken on a regular basis. The sample size n remains constant, and is usually larger than ten. The $\bar{x}$, or *xbar*, *control chart* is used to monitor the location, and the *standard deviation (s) chart* is used to monitor the variability or dispersion, of the continuous data on the process. The control charts for $\bar{x}$ and s are derived in essentially the same way as the control charts for $\bar{x}$ and R. The mean $\bar{x}$ and the standard deviation s are calculated for each subgroup. The average $\bar{x}$ and the standard deviation s of each subgroup are plotted in chronological order on the $\bar{x}$ and s charts respectively.

Once there are some 20 or 30 samples (subgroups) for non-batch processes, the control limits for the control charts are calculated as follows. Compute the average $\bar{\bar{x}}$ of the sample averages $\bar{x}$ and the average $\bar{s}$ of the standard deviations s. The center line and the 3-sigma control limits for the $\bar{x}$ chart are given by

$$UCL = \bar{\bar{x}} + A_3 \bar{s},$$
$$CL = \bar{\bar{x}}, \quad (21.4)$$
$$LCL = \bar{\bar{x}} - A_3 \bar{s},$$

and the center line and the 3-sigma control limits for the s chart are given by

$$UCL = B_4 \bar{s},$$
$$CL = \bar{s}, \quad (21.5)$$
$$LCL = B_3 \bar{s},$$

where A_3, B_3, and B_4 are constants that are given in tables and that depend on the number of observations n in each rational subgroup and assumptions of independence and normality (see Montgomery, 1991, Appendix VI). The s chart plots within-sample variability, and the $\bar{x}$ chart tracks between-sample variability and location. As for the $\bar{x}$ and R charts, around 99.73% of the points are expected to fall inside the control limits when the observations come from a normal distribution and the process is under control. The percentage is approximately correct for the $\bar{x}$ chart even if the distribution is slightly non-normal, but the percentage for the s chart is more sensitive to deviations from normality.

Sometimes an s chart is preferable to an R chart, and sometimes an R chart is preferable to an s chart. The s chart is preferable to the R chart under two circumstances. The first is when the subgroup size n is variable. The second is when the subgroup size n is moderately large (over 10 or 12 observations) because then the standard deviation tends to be a better estimate of variability than the range. The range is calculated from only the smallest and largest observations, essentially "throws away" (ignores) the rest of the data, and is very sensitive to outliers. On the other hand, two practical reasons why the range R might be preferable to the standard deviation s are that the range is easier to interpret, and the range can be more easily calculated by hand in the absence of automation.

The control limits for the $\bar{x}$ chart computed from equations (21.4) are given in terms of the average standard deviation $\bar{s}$, and may be slightly different from the control limits for the $\bar{x}$ chart computed from equations (21.2), which are given in terms of the average range $\bar{R}$.

x and MR Control Charts

The *individuals (I), or x, control chart* measures location, and the *moving range or MR control chart* measures variability or dispersion, for continuous process data that follows a normal (or approximately normal) distribution. For a set of observations, the *moving range* corresponding to an observation is the absolute value of the difference of that observation from the previous one. The moving range cannot be evaluated for the first observation. The individual observations and the moving ranges are plotted as points on the x and MR charts respectively. To compute the control limits, let $\bar{x}$ be the average of 20 or 30 observations, and let $\overline{MR}$ be the average of their moving ranges. The center line and the 3-sigma control limits for the x chart are given by

$$UCL = \bar{x} + 2.66\,\overline{MR},$$
$$CL = \overline{MR}, \qquad (21.6)$$
$$LCL = \bar{x} - 2.66\,\overline{MR},$$

and the center line and the 3-sigma control limits for the MR chart are given by

$$UCL = 3.27\,\overline{MR},$$
$$CL = \overline{MR}, \qquad (21.7)$$
$$LCL = 0.$$

Patterns in the moving range chart should be interpreted with care, since the moving ranges are not independent (two consecutive moving ranges have one observation in common). This problem is not present in the x chart, since the individual measurements are assumed to be independent. For these reasons, some practitioners prefer not to use the moving range chart. It is possible to calculate moving ranges for more than two observations.

The widespread presence of *batch processes* makes the individuals control chart and the moving range control chart of special importance to the semiconductor industry. Sample statistics computed from each batch (such as the batch mean $\bar{x}$, the batch range R, or the batch standard deviation s) can be considered as the "individual" points of the x and MR charts. The important thing is the formula that is used for the control limit, not the name that is given to the chart. For example, the average thickness of wafers from a furnace run can be considered as the response of interest, and a control chart for individuals can be used for the furnace run average thickness.

p Control Chart

Fraction nonconforming or *p charts* are used to monitor the fraction of nonconforming or defective items produced by a process. The number p is the expected fraction of nonconforming items. It is a parameter of the binomial distribution, and it can be expressed as a percentage.

A random sample of n observations, usually around 20 or 25, is taken on a regular basis. For each random sample, the fraction $\hat{p}$ nonconforming is calculated as the number of nonconforming units in the sample divided by the number of units n in the sample. The fraction $\hat{p}$ of nonconforming values for each sample is plotted on the p control chart in chronological order.

Once there are some 20 or 30 samples, the control limits for the control charts are calculated as follows. Compute the average $\bar{p}$ of the fraction nonconforming values $\hat{p}$ from all the samples. The center line and the 3-sigma control limits for the p control chart are given by

$$\text{UCL} = \bar{p} + 3\sqrt{\bar{p}(1-\bar{p})/n},$$

$$\text{CL} = \bar{p}, \qquad (21.8)$$

$$\text{LCL} = \bar{p} - 3\sqrt{\bar{p}(1-\bar{p})/n}.$$

c Control Charts

The control chart for nonconformities, or c (count), chart is used to monitor the number of nonconformities or defects produced by a process. The expected number of nonconformities c in an inspection unit is the parameter of a Poisson distribution. The nonconformities in an inspection unit are assumed to be independent. Inspection units are examined on a regular basis, and the number of nonconformities is plotted in chronological order on the control chart. Let $\bar{c}$ be the average number of nonconformities in a group of 20 or 30 inspection units. Since the variance of a Poisson distribution is the same as its mean, the center line and the 3-sigma control limits for the c chart are given by

$$\text{UCL} = \bar{c} + 3\sqrt{\bar{c}},$$

$$\text{CL} = \bar{c}, \qquad (21.9)$$

$$\text{LCL} = \bar{c} - 3\sqrt{\bar{c}},$$

The c charts would seem ideally suited for particle counts in the semiconductor industry. Unfortunately, they are not. The basic problem is that the occurrence of individual particles are not independent events: particles tend to come in clumps. The lack of independence means that particle counts do not follow a Poisson distribution, and the c charts work only moderately well. Particle counts are not too well approximated by the normal distribution either (the distribution of particles is often asymmetrical, with a high number of low particle counts and occasional high particle counts), and even transformations like taking the logarithm to try to normalize the data don't always work well.

Control Charts in the Case Studies

The case studies illustrate the use of different types of control charts. Marr used an $\bar{x}$ control chart on a nitride process. Watson studied contributors to module thickness variability in surface mount technology, using $\bar{x}$ and R charts for contributors with several observations per sample, and I and MR charts for contributors with individual observations per sample. Joshi and Sprague used $\bar{x}$, s, and individuals charts for critical dimension measurements in a lithographic process.

Marr (Chapter 22) initially used control limits transferred from the development organization. A passive data collection from 25 lots was planned to find statistically valid control limits. A clear drift was uncovered in the data, and identification of the assignable cause led to the change of a mass flow controller. Marr gives a methodology to remove process drift, and illustrates how it was used on the silicon nitride film thickness data to set valid control limits without further delay. This is important since if the drift is not removed from the data, the estimate of variability is too large. This leads to control limits that are too wide, and a process that appears to be in control when in reality it is not.

Watson (Chapter 23) calculated $\bar{x}$ and R charts for the heat up rate of lamination presses by measuring temperature at five openings on a press. He initially used the temperature readings at the five openings in calculating $\bar{R}$. Since this estimate of $\bar{R}$ incorporated variability at each opening as well as variability between openings, the value of $\bar{R}$ was quite large, which resulted in $\bar{x}$ and R charts with overly wide (with respect to the data) control limits. To correct the problem, temperature was later measured at only one opening.

Joshi and Sprague describe how, for an $\bar{x}$ chart, using standard formulas for the control limits (and not taking into account the fact that the data came from a batch process) led to specification limits that were too narrow, and to too many points out of control. Joshi and Sprague present two methods of computing control limits for batch processes: through the use of an "effective sample size," and by using variance components. They illustrate the use of individuals control charts to plot the individual values of $\bar{x}$ on one chart, and the individual values of s on another chart.

ADDITIONAL TOPICS IN SPC

The first three subsections will describe additional graphical tools, process capability, and implementation issues. Other statistical techniques that aid in statistical process control include analysis of variance to estimate and reduce components of variability (see Part 2 on passive data collection), and design of experiments to actively improve the process (see Part 3). The last subsection describes additional topics in the case studies.

Additional Graphical Tools

In addition to control charts, a comprehensive statistical process control program will use graphical tools such as:
- *histograms*, to see the shape, location or central tendency, and spread or variability, of the data.
- *capability graphs*, to see the shape, location, and spread of the data, and to relate them to process specifications.
- *pareto charts*, to identify the most important or frequently occurring assignable causes for points out of control.
- *cause and effect or fishbone diagrams*, to systematically identify the causes for points out of control, often through brainstorming.
- *scatter diagram*, to identify potential relationships between two variables.

Additional useful graphical tools include box plots, time series plots, multiple variable plots, and flow charts. Graphs are described in the glossary at the end of the book.

Process Capability

The *specification limits* of a process variable reflect the region where "good" product is manufactured, and are set by engineers, management, product designers, and customers. The observations of the process variable should be between the *lower specification limit*

(LSL) and the *upper specification limit (USL)*. A specification interval goes from the lower specification limit to the upper specification limit.

Specification limits and control limits are not the same. Specification limits reflect the region where good product is produced and may be set by customers, while control limits reflect the natural variability of the process and are calculated using process data. Specification limits should be wider than control limits. Specification limits are given for individual observations, while control limits are often given for summary statistics computed from the individual observations, such as the mean $\bar{x}$. Specification limits can be added to individuals control charts, but should not be added to $\bar{x}$ control charts because they can be misleading. The spread of individual observations is always larger than the spread of sample averages, so while the averages may be inside the specification limits, individual observations could be far outside of specification limits.

The *capability* of a process to meet specifications can be assessed (see Part 2) once the process is in statistical control. A graphical assessment of process capability is provided by a *capability graph*, which is a histogram of the individual observations of the process variable to which the lower and upper specification limits are added. If the observations in the capability graph have reduced variability compared to the distance between the specifications, and are centered around the target value, the process is capable. The *capability indices* C_p and C_{pk} are numerical measures of process capability, but only C_{pk} considers whether the process is centered around the target.

The *capability index* C_p is a measure of process capability that compares the natural process variability to the process specifications. C_p is estimated by

$$C_p = \frac{\text{USL} - \text{LSL}}{6s}.$$

Here s is the standard deviation from a sample of values of the process variable. If the process variable is normally distributed, 99.7% of the observations are expected to fall within six standard deviations (6s, or "six sigma") of the mean. If C_p equals one, the observations are just inside specifications, provided they are centered around the target value. The higher the value of C_p, the more capable the process.

The capability index C_{pk} is a measure of process capability that compares the natural process variability to the process specifications and also takes into account whether the process is centered around the target value. C_{pk} is estimated by

$$C_{pk} = \min\left[\frac{\text{USL} - \bar{x}}{3s}, \frac{\bar{x} - \text{LSL}}{3s}\right],$$

where $\bar{x}$ is the average and s is the standard deviation from a sample of values of the process variable, and *min* indicates the smallest of the two quantities in parenthesis. The larger the value of C_{pk}, the more capable the process. A value of 1.3 or 1.5 is often considered satisfactory for observations that come from a normal distribution. C_p measures how capable the process *could be* (if it were centered around the target), while C_{pk} measures how capable the process is. Since the estimate of C_{pk} is highly variable, a confidence interval for C_{pk} can be given instead of the value of C_{pk}, or C_{pk} can be computed from one hundred or more observations for greater precision. If specification limits have not been defined for a process variable (which is often the case for non-critical parameters) engineering limits may be used instead.

Process capability at the end of the manufacturing process is often measured by the number of *parts per million* (ppm) out of specification. The number of parts per million out of specification can be estimated by checking all the parts produced in a given period of time and finding the number that is out of specification, by selecting a sample of all the parts produced and finding the fraction that is out of specification, or by finding the probability that a part is out of specification for a given probability distribution for the observations.

When the process capability is inadequate, it may be possible to improve it by reducing the natural process variability using designed experiments (Part 3). However, there is a minimum variability inherent in the performance of a given process on a given piece of equipment below which the variability cannot be reduced. In this case it may be possible to change the specification limits to a more appropriate value, or it may be necessary to change the equipment or the process itself to improve the capability. Changing the specification limits should only be changed through consultation with the customer and *not* just to make the C_{pk} look better.

A word of caution. A high value of C_{pk} does not automatically mean the process has high capability. The value of C_{pk} can be strongly affected by the choice of sampling plan. If observations are only taken at the center of the wafer, the variability may appear to be low and C_{pk} may appear to be high for a process that is not capable due to high variability across the wafer.

Implementing SPC

We have considered a number of statistical tools that are used for statistical process control. Actual implementation of SPC or other statistical techniques requires more than statistical tools. Understanding human behavior is essential for the successful implementation of a new SPC program. Some practical issues to be aware of when implementing statistical process control are:

- *Obtain buy-in.* Support from management and cooperation from the operators and engineers doing the work is vital for success. Human behavior issues are critical and can make or break an SPC implementation program.
- *Educate personnel.* Operators, engineers, and management need to be aware of the benefits of using SPC and of the tools used.
- *Empower operators.* The operator or manufacturing engineer who is closest to the process should be trained and given responsibility for collecting data, maintaining control charts, and interpreting results. Engineers or managers can be consulted in more difficult situations.
- *Create cross-functional SPC teams.* Operators doing good work on a few carefully selected issues which are given high visibility will generate enthusiasm. Nothing succeeds like success.

- *Simplify implementation.* Ideally, use computer software that is easy to use, has visual displays, is on-line to be able to take quick action, and is conveniently close to the work station.
- *Establish a corrective action system.* There should be clear documented guidelines so that operators, engineers, and managers know what action to take when the process is out of control. Such a corrective action system should be included in any comprehensive statistical process control program. Corrective actions should be recorded. A general rule followed by many companies is that no control chart should be implemented unless there is an accompanying out of control action plan.
- *Begin with critical upstream processes.* To avoid using control charts as wall paper that no one looks at, select processes that are important to control. Select upstream processes, since SPC will have greater impact on quality if the root causes at the beginning of a manufacturing process are addressed than if the finished or almost finished product is inspected.
- *Update processes.* Change the processes and the responses that are monitored based on current needs.
- *Define metrics, set goals, monitor progress.*
- *Assign duties to personnel.*
- *Check gauges.* Determine the performance of the measurement equipment by performing gauge studies (Part 1).
- *Make plans for data collection.*

Additional Topics in the Case Studies

Watson (Chapter 23) used four graphs (two control charts, a histogram, and a time series) for each one of six contributors to module thickness variability. He used MR and I charts for individual observations, and R and $\bar{x}$ charts for samples with more than one observation. The charts indicate the 3-sigma, 2-sigma, and 1-sigma control limits. The histograms are for the individual observations, and the addition of specification limits permits visualization of process capability. The time series plot gives the individual observations in chronological order, and includes the specification limits. The plot of the data on the time series plot is identical to the plot of the data on the corresponding I charts, but is different from the $\bar{x}$ charts for samples with more than one observation.

Watson found that poor caliper (gauge) resolution caused some measurements of multilayer thickness range to appear to be zero. In addition, it appeared that multilayer thickness specifications may have been incorrect, since the multilayer process has a high C_{pk}, yet contributes most of the variability to the module thickness process that has a low C_{pk}.

Joshi and Sprague (Chapter 24) describe how before statistical process control was implemented, the natural variability of the process was not recognized, and process "tweaking" was used to try to get the product to meet specifications. Initial implementation of control charts not taking into account the batch nature of many semiconductor processes led to a large number of points outside control limits (and many false alarms), which created resistance to their use. Different strategies were used to reduce the initial resistance, and eventually to implement statistical process control throughout the fab. On the technical side, an "effective sample size" was used to take into account the batch nature of some processes, and recalculation of the control limits reduced the number of points outside control limits. On the organizational side, management took an active role to support implementation of SPC and other continuous improvement methods in the fab. An SPC group was formed to teach engineers SPC methods, assist in implementing control charts at key steps, form groups in the fab to foster use of SPC, and help automate the calculation and use of control charts.

Introduction to Case Studies

This section gives a brief overview of the case studies presented in the remainder of this part of the book.

Chapter 22. Removing Drift Effects When Calculating Control Limits. *Ray L. Marr*. When the mean of a process drifts over time, control charts tend to have overly wide control limits, making the process appear to be in control when in reality it is not. This case study gives a procedure to remove the effect of such a drift, in order to establish statistically valid process control limits. The procedure is illustrated for a topside nitride process where a silicon nitride film is deposited on silicon wafers by plasma-enhanced chemical vapor deposition. The root-half-mean-square-successive-difference method gives appropriate control limits in the presence of small drifts, distributional outliers, and shifts in the data by removing their effect from the estimate of the standard deviation, which reduces it and narrows the control limits. The method was used in an automated system that was instrumental in reducing scrap by 80% within two and a half months.

Chapter 23. Implementation of a Statistical Process Control Capability Strategy in the Manufacture of Raw Printed Circuit Boards for Surface Mount Technology. *Ricky M. Watson*. This case study describes the use of statistical process control to analyze and reduce module thickness variability in a surface mount process in a printed circuit board shop. The variability reduction was part of a strategy to define current manufacturing capabilities and to enhance future manufacturing capabilities by ensuring that all key processes are monitored using statistical process control. The need for the strategy arose when Texas Instruments started to implement the Motorola Six Sigma philosophy, and design groups requested information relating production capabilities to proposed designs. It became necessary to expand the previous statistical process control effort on process variables to include product variables. Large components of variability in the module thickness were identified, and corrective action was taken.

Chapter 24. Obtaining and Using Statistical Process Control Limits in the Semiconductor Industry. *Madhukar Joshi* and *Kimberley Sprague*. This case study considers key techniques that led to the successful implementation of statistical process control in fabs at DEC. Key elements in the implementation of SPC were active support from management, and the formation of groups to help transfer the methodology, implement it in key process steps, and automate its use. A method and supporting formulae to obtain control limits for control charts for batch processes were developed. An "effective sample size" technique was derived in response to a large percentage of points falling outside of the calculated control limits on the control charts for batch processes. In processes for which within-batch variability is less than between batch variability, the technique is equivalent to using individuals charts for the within-batch averages. The methodology is illustrated with data on critical dimensions in a photolithographic process.

SUGGESTED READING

All SEMATECH statistical publications that are available to the general public can be accessed from the following Internet URL: http://www.sematech.org/

Books that contain introductory chapters on statistical process control are:

    Ryan, Thomas P. *Statistical Methods for Quality Improvement*, John Wiley and Sons, Inc., 1989.

    John, Peter W.M. *Statistical Methods in Engineering and Quality Assurance*. New York: John Wiley and Sons, Inc., 1990.

A book on statistical process control that includes a description of control charts for batch processes is:

    Leitnaker, Mary G., Sanders, Richard D., and Hild, Cheryl. *The Power of Statistical Thinking— Improving Industrial Processes*. Reading, MA: Addison–Wesley Publishing Co., Inc., 1996.

Four books on statistical process control are:

    Montgomery, Douglas C. *Statistical Quality Control*, second edition, New York: John Wiley and Sons, Inc., 1991.

    Wetherill, G. Barrie, and Brown, Don W. *Statistical Process Control, Theory and Practice*, London: Chapman and Hall, 1991.

    Wadsworth, Harrison M. Jr., Stephens, Kenneth S., and Godfrey, A. Blanton. *Modern Methods for Quality Control and Improvement*, New York: John Wiley and Sons, Inc. 1986.

    Western Electric. *Statistical Quality Control Handbook*, Indianapolis: Western Electric Co., Inc, 1956.

CHAPTER 22

REMOVING DRIFT EFFECTS WHEN CALCULATING CONTROL LIMITS

Ray L. Marr

This case study illustrates a common practical difficulty in setting control limits from drifting data and presents a remedy. It further discusses experience with a procedure that uses this method and interfaces with the existing SPC charting system for comprehensive analysis of control charts in AMD Fab XIV. Such methods may obviate a strict process capability study by giving sound control limits despite small drifts, distributional outliers, and shifts (deliberate or otherwise) in the data. In one module, its use was instrumental in reducing scrap by 80% within two and a half months.

EXECUTIVE SUMMARY

Problem

When a process mean drifts, the usual (Root-Mean-Square-Error, RMSE) method of estimating the standard deviation is inappropriate, and the customary 3-sigma control limits based on centerline $\pm 3\hat{\sigma}$ will give overly wide control limits. But all is not lost. There is a way to correctly estimate the standard deviation for the mean while virtually subtracting out the effect of a drift. As will be seen below, this method can often produce reliable results in the absence of a completely passive process capability study.

Solution Strategy

In the case of a "new" topside nitride process, we were operating with the control limits of the transferring organization until sufficient data were gathered to calculate our own statistical control limits. It was desired to establish statistically valid control limits on the process, but a clear drift was present (see Figure 22.1). Changing a mass flow controller corrected the drift problem, but it was undesirable to wait another 25 to 30 runs before setting statistically valid control limits. So the Root-Half-Mean-Square-Successive-

Difference (RHMSSD) method (described in the Statistical Basis for the Common and Proposed Methods section) was used to get statistically calculated control limits right away.

Conclusions

When preceded by a robust screening for distributional outliers and when the drift between successive sampling events is small, the RHMSSD is an effective way to establish control limits uninflated by cumulative drift over a long baseline.

PROCESS

A silicon nitride film opaque to ultra-violet (UV) light is deposited on silicon wafers by plasma-enhanced chemical vapor deposition (PECVD) at the wafer surface when silane gas (SiH_4) reacts with dry ammonia (NH_3) in a plasma under suitable conditions of temperature, RF power, and pressure. The resulting film, though opaque to UV light, is transmissive within the visible range; hence a refractive index can be used as a sensitive measure of film stoichiometry.

DATA COLLECTION PLAN

To monitor the PECVD process for each lot, the refractive index (among other things) was measured and recorded for each lot. Refractive index was measured on a prism coupler with a 3-mm spot size at wafer center for one wafer; an average over so large an area (a circle of 3-mm diameter) was believed reasonably normal in distribution and very much constant within the lot. I had recommended the engineer collect 25 data points for calculation of reliable control limits, but indicated that we could get reasonable, though less reliable, control limits with fewer points.

The engineer set provisional control limits at the control limits used by the organization that developed the original process and began passive collection of data from 25 lots, it being decided to display the control charts only on violation of the inherited control limits. The data presented in Table 22.1 have all been linearly coded to map the inherited control limits into the values 0 and 100. It should be understood that the actual differences in the refractive index are magnified greatly in this case-study, but even very small changes in refractive index are indicative of assignable process changes that ought to be taken seriously.

STATISTICAL BASIS FOR THE COMMON AND PROPOSED METHODS

Suppose y_t represents an independent normal SPC plot point for each t (t = 1, 2,..., n), collected while the process is subject to a slight drift over time. Hence the true mean (μ_t) may vary slowly with time (t), though the variance (σ^2) about the drifting mean is constant. If we used the usual Root-Mean-Square-Error (RMSE) method to estimate the standard deviation of σ for the control charts,

$$\text{RMSE} = \sqrt{\frac{\sum_{t=1}^{n}(y_t - \bar{y})^2}{n-1}} = \sqrt{\frac{n\sum_{t=1}^{n}y_t^2 - \left[\sum_{t=1}^{n}y_t\right]^2}{n(n-1)}} = \sqrt{\frac{\sum_{t=1}^{n}y_t^2 - n\bar{y}^2}{n-1}}.$$

where $\bar{y}$ is the mean of the y's, we would have an inappropriate estimate of σ because it is highly inflated in the presence of drift. To estimate the lot-to-lot standard deviation while subtracting out the effect of the drift, we calculate the Root-Half-Mean-Square-Successive-Difference (RHMSSD) estimate of the true standard deviation σ:

$$\text{RHMSSD} = \sqrt{\frac{\sum_{t=2}^{n}(y_t - y_{t-1})^2}{2(n-1)}}.$$

The expected value of the Half-Mean-Square-Successive-Difference (HMSSD) is given by

$$\text{E(HMSSD)} = E\left\{\frac{\sum_{t=2}^{n}(y_t - y_{t-1})^2}{2(n-1)}\right\} = \sigma^2 + \frac{\sum_{t=2}^{n}(\mu_t - \mu_{t-1})^2}{2(n-1)}. \quad (22.1)$$

When the drift between consecutive sampling events is fairly small (compared with the standard deviation σ), the second term is negligible, and the expected value of HMSSD is approximately σ^2. So we see that the Half-Mean-Square-Successive-Difference (HMSSD) is a good estimator of the true underlying variance, even when the process is subject to a slight drift, and its square root (the RHMSSD) is a good estimator of the underlying standard deviation in the presence of a slight drift. No particular assumption about the functional form of the drift need be assumed to use this method, in contrast to a regression method (which might also be reasonable, though less robust).

On the other hand, the usual Mean-Square-Error (MSE) estimate of σ^2 would have expected value

$$\text{E(MSE)} = E\left\{\frac{\sum_{t=1}^{n}(y_t - \bar{y})^2}{n-1}\right\} = \sigma^2 + \frac{\sum_{t=1}^{n}(\mu_t - \bar{\mu})^2}{n-1}, \quad (22.2)$$

which can differ from σ^2 by a very large amount indeed over n runs if the cumulative range of the drift is large.

The Appendix includes additional comments on the Half-Mean-Square-Successive-Difference (HMSSD) control limits, including serially correlated random errors.

ANALYSIS AND INTERPRETATION OF RESULTS

With the nitride process, we had data from before and after the installation of the new mass flow controller at observation #30. The coded data are presented numerically in Table

22.1 and graphically in Figure 22.1. Successive differences of observed means are reported in the column labeled "Diff." and are just the change of the current observation from the previous observation.

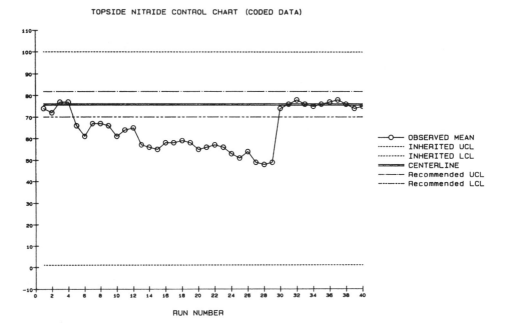

Figure 22.1. *After successfully transferring the nitride process to Fab XIV (note runs #1–4), the engineer installed the inherited control limits as well, with intent to passively gather 25 additional observations for a capability study before establishing new control limits. Aghast at the drift evidenced at observation #29, the engineer took immediate action. He replaced the mass flow controller and the (coded) refractive index returned again to its former level. Perhaps the inherited control limits had been calculated earlier by the usual Root-Mean-Square-Error method oblivious to evidence of a bad mass flow controller. Unfortunately, ineffective control limits are often self-perpetuating. Permitting the process to meander through the whole range of those control limits before troubleshooting the process virtually guarantees a large Root-Mean-Square-Error on recalculation.*

After taking successive differences, the outlying nature of the difference of 25 in run 30 is apparent (see Table 22.1), and it should be omitted from estimation of the standard deviation since changing the mass flow controller was expected to change the mean. The difference in run 5, where the mass flow controller began its sojourn south, is also an outlier because it exceeds the 3σ-limits on the successive differences. The 3 σ -limits on the differences are $3\sqrt{2} \approx 4.24$ times the RHMSSD. The centerline may be set at the process target value, or alternatively, at the mean of observations 30 through 40 made after the drift was corrected, or even at the mean of observations 1 through 4 and 30 through 40, if we trust our eyes. It is frequently useful in guaranteeing long-term stability to establish a target as the centerline. Since no target value had yet been established as appropriate, the centerline had to be determined from the mean of points not influenced by the drift. Four different values of the control limits are now calculated.

Table 22.1. *Mean Refractive Index over a 3-mm Spot and Differences Between Consecutive Values. Actual measurements have been linearly coded.*

Run No.	Observed Mean	Diff.	Run No.	Observed Mean	Diff.	Run No.	Observed Mean	Diff.	Run No.	Observed Mean	Diff.
1	74	...	11	64	3	21	56	1	31	76	2
2	72	−2	12	65	1	22	57	1	32	78	2
3	77	5	13	57	−8	23	56	−1	33	76	−2
4	77	0	14	56	−1	24	53	−3	34	75	−1
5	66	−11	15	55	−1	25	51	−2	35	76	1
6	61	−5	16	58	3	26	54	3	36	77	1
7	67	6	17	58	0	27	49	−5	37	78	1
8	67	0	18	59	1	28	48	−1	38	76	−2
9	66	−1	19	58	−1	29	49	1	39	74	−2
10	61	−5	20	55	−3	30	74	25*	40	75	1

Control Limits Based on RMSE

Mean = 64.5, RMSE = 9.87 ⇒ Control Limits = 64.5 ± 29.6 (very misleading!).

Control Limits Based on RHMSSD

Mean for runs 30 to 40 = 75.9.

RHMSSD = $\sqrt{\frac{1035}{2 \times 39}}$ = 3.64 ⇒ Control Limits = 75.9 ± 10.9 (less misleading).

RHMSSD(deleting diff. run 30) = $\sqrt{\frac{410}{2 \times 38}}$ = 2.32 ⇒ Control Limits = 75.9 ± 7.0 (good).

RHMSSD (delete diff. runs 30, 5) = $\sqrt{\frac{289}{2 \times 39}}$ = 1.98 ⇒ Control Limits = 75.9 ± 5.9 (better).

Note that the denominator for the RHMSSD method is always twice the number of squared successive differences summed in the numerator. Note further that an outlier in an observed value should be deleted before differencing, as otherwise it would typically give rise to two outliers in the successive differences that should be deleted. On the other hand, a large shift (caused by a change of setting or radical maintenance) would necessitate deleting only one successive difference; screening for such shift outliers is essential when data are not flagged in the database following radical maintenance or changes in equipment set-up. Finally note that the centerline is estimated from the datapoints that manifest stability, or it is set at the target (or a historically justified value). Thus different numbers of terms might properly be summed to estimate the mean and standard deviation for the control limits.

Potential Automation Procedure for RHMSSD Control Limits

The following procedure might be used to automate the calculation of these control limits:
1. Delete all evidently misrecorded observations; form differences of successive remaining observations, but delete those differences that straddle a process shift known from a priori engineering considerations, such as radical maintenance or adjustment.
2. Average the remaining squared successive differences, divide by 2, and take the square root to get the re-calculated RHMSSD.
3. Delete all successive differences outside $\pm 3\sqrt{2} \times$ RHMSSD.
4. Re-calculate RHMSSD as in step 2.
5. Set control limits at Target or Mean $\pm 3 \times$ RHMSSD, where the Mean may be either a historically derived (and still valid) mean or the mean of points that manifest stability.

Step 1 permits removal of data entry errors and of distortion arising from pooling process runs with very different means; the differencing palliates the effect of small drifts; and step 3 removes the effect of large shifts of the mean.

I should emphasize: the "screening" is not used to remove points from the SPC charts, but is only used to better determine the maximum reasonable deviation from the centerline that might be encountered when the process is really operating in control and at the centerline.

CONCLUSIONS AND RECOMMENDATIONS

This case study illustrates a method of establishing reasonable control limits in the presence of drift. Although it can screen for severe outliers, it is recommended that a good, robust screening replace the non-robust screening illustrated above. Since this work four years ago, the author has refined this method and packaged it with certain others to create an engineering tool for process control analysis with English-language diagnostics, resulting in software that interfaces well with AMD's on-line data management and SPC system. The results have been impressive: tight, realistic, statistically appropriate control limits have replaced subjectively or incorrectly set action limits. In addition, since large shifts are screened, the recommended control limits are generally valid in the absence of a strict capability study, provided the active changes occur in only a small number of datapoints (say, less than 25%). This is a real necessity in the diffusion area of a manufacturing fab, where passive data collection in a capability study might be a costly enterprise. In one module, it was instrumental in reducing scrap by 80% within two and a half months, with no backsliding since then.

Acknowledgments

Successive difference methods date from ballistics testing in the late 19th century, where they were used to provide effective estimates of dispersion in the presence of drift caused by heat, wind, and barrel wear. Von Neumann and Hart in the early forties really developed and popularized the methods (see Read, 1988 or Von Neumann et al., 1941). These methods were all published before I was born; the purpose of this paper is to introduce them into quarters of an industry where they are needed but not known. My only claim to originality is in assessing the needs of one fab and knitting together a coherent procedure for analysis of process control charts that realistically handles at least 95% of the control charts seen in this fab, where realism dictates a reasonable amount of ruggedness against the common violations of the ideal assumptions of independent observations, coming from a single normal population with fixed mean and variance.

Moore (1955) discusses the impact of nonnormality on the relative efficiency of the HMSSD as compared with the MSE for estimating σ^2. For large samples, the HMSSD estimator in effect discards a fraction σ^4/μ_4 of the information compared with the usual MSE estimator, where μ_4 is the fourth moment about the mean.

Thanks are due to Bob Anderson and to the many other fine engineers of AMD Fab XIV who have shared their problems and their practical experience with me in the course of statistical consultations.

References

Kotz, S. and N.L. Johnson. "Three Sigma (3σ) Rule," *Encyclopedia of Statistical Sciences*, 9:236, New York, NY: John Wiley and Sons, Inc., 1988.

Moore, P.G. "The Properties of the Mean Square Successive Difference in Samples from Various Populations." *Journal of the American Statistical Association*, 50, 434-456, 1955.

Read, C.B. "Successive Differences," *Encyclopedia of Statistical Sciences*, 9, 65-68, S. Kotz and N.L. Johnson, Eds. New York, NY: John Wiley and Sons, Inc., 1988 (This excellent reference includes a useful bibliography. But note that the author's $d_1 = d\sqrt{\pi}/2$, that "n-4" should read "n-r" in equation (6) and that $\sqrt{n}\,(z - \sigma)/\sigma$ is asymptomatically $N(0, (2\pi/3) + \sqrt{3} - 3)$ as $n \to \infty$.

Von Neumann, J., R.H. Kent, H.R. Bellinson, and B.I. Hart. "The Mean Square Successive Difference," *Annals of Mathematical Statistics*, 12, 153-162, 1941.

Appendix

How does the RHMSSD method (Centerline ± 3 × RHMSSD after deleting outlying differences) of establishing control limits compare with a modified method[50] (Centerline ±3.145 × median $\{|y_t - y_{t-1}|\}$)? Since both methods in effect use absolute values of successive differences (sometimes called "moving ranges"), both methods would be suitable to establish control limits in the presence of a slight drift. The modified method is easier to apply (no need to delete outliers and recalculate!) and more resistant to shift outliers, but at

---

[50] This is a modified Clifford's method.

the cost of some efficiency.[51] There are 39 moving ranges: the 19th smallest is 1 and the 20th smallest is 2, so the median of the ranges is 2. The actual control limits using the modified method are Centerline ± 6.30, which is quite reasonable.

My concern with the modified method where moving ranges are only available to one significant digit (most of the values being 0, 1, or 2) is how dramatically the control limits would change (from Centerline ± 6.290 to Centerline ± 3.145) if only one moving range of value 2 were changed to a value 1. This might be considered more a shortcoming of excessive round-off than of the median moving range, but let it serve as a warning that virtually all methods based on order statistics (like the median) or rank statistics are sensitive to over-zealous round-off.

In summary, the RHMSSD with the indicated screening (or robust pre-screening at, say, $\text{median}_i X_i \pm 4.5 \,\text{median}_j |X_j - \text{median}_k (X_k)|$) should be somewhat more efficient than a median moving range, though less convenient; median moving ranges are more susceptible to problems of excessive round-off, but either method can establish workable control limits for process data contaminated by drift. A mean moving range could also be used at the price of a little efficiency (60% rather than 67% if the observed data have no outliers or departures from normality).

On closer examination, the successive differences before the mass flow controller's replacement are seen to have larger magnitudes than those afterward. Thus the limits should be re-calculated after another 15 to 20 observations. But meanwhile far tighter control is maintained over the process with the RHMSSD method than would be the case with the original control limits transferred to our fab or the control limits that might be incorrectly calculated from the usual RMSE calculation.

Some non-practitioners may balk at the several outlier screens on the grounds that the resulting control limits may be somewhat less than 3-sigma because of the resultant underestimation of the standard deviation. The answer to this objection is twofold: (1) use of "exact" 3-sigma control limits is traditional rather than sacrosanct; in a production environment, the *right* frequency of spurious signals to accept will depend on the risks, but typically should be less than 5% (the Central Limit Theorem, Chebyshev Inequality, the Camp-Meidell Inequalities, or the Three-Sigma Theorems of Vysochanskii and Petunin (see, e.g., Kotz and Johnson, 1988) will usually ensure this for many common non-normal distributions) ; (2) a series of screenings, one for distributional outliers and one for shifts, each unlikely to filter out data under the ideal distribution, would ordinarily bias the standard deviation by a negligible amount in the case that these data were actually part of the population of interest, unavoidable, tolerable, and with stable frequency over time; but in the more common case that the data were glitches of an inconsistent, erratic "non"-population, the screenings prevent dramatic overestimation of control limits.

There *are* situations where the RHMSSD method is inappropriate and misleading:

---

[51] The efficiency of an estimator is the fraction of the information, or equivalent fraction of observations, effectively thrown away when the population really is ideal, by estimating a parameter using an unnecessarily rugged procedure rather than the procedure optimal for the ideal distribution.

1. When there is a small number of plot points and drift effects are not expected, the loss in effectively discarding 33% of the information is not offset by any advantage.
2. When the data have strong serial correlations, as for similar data collected at many successive instants within a processing step, say using a SECS II interface. Negatively autocorrelated data tend to yield overly wide control limits; positively autocorrelated data, the reverse.

In the presence of stationary autocorrelated *random* errors, the σ^2 term of Equation (22.1) would have to be multiplied by $(1 - \rho)$, where ρ is the true correlation between consecutive errors. But modeling a drift by means of a (slowly) changing non-random mean is probably more realistic than modeling it with autocorrelated random errors unless the true consecutive drifts $\mu_t - \mu_{t-1}$ are large in magnitude relative to σ (say, $|\mu_t - \mu_{t-1}| > \sigma$ for a typical t), or the drift is unavoidable and allows no compensation, or the sampling interval is such that nearly the whole population is sampled. Indeed, the first condition represents a greater excursion than what is usually considered a "slow" drift (we would probably want to sample more frequently in such a case and then the RHMSSD method might again become viable); the case when the process is in a constant state of flux and virtually every unit is measured, begs for the autocorrelated random errors model. Frankly, under such conditions one wonders whether a workable process has really been found and whether more effort should not be put into ascertaining the causes of the process variability than into optimal calculation of control limits.

A common case for which the time-order of the data is questionable is illustrated by wafer electrical test data. There are so many process steps affecting the results, each with its own time ordering; a drift in the data plotted in order of time of electrical testing is more likely a result of drift in tester than in product performance. In such cases successive-difference methods still provide valid estimates of the true standard deviation, but in the absence of tester drift, apparent product drift effects are either "random" or difficult to find the cause of, and estimates of standard error (the standard deviation of a summary statistic) may suffer from the inefficiency inherent in successive difference methods. Robust methods should be preferred if there is danger of distributional outliers.

BIOGRAPHY

Ray L. Marr received a Ph.D. in statistics in 1988 from North Carolina State University after many years of graduate work in mathematics and mathematical statistics at Indiana University (masters degrees in 1969 and 1980, Ph.D. candidacy in 1975). These years cover extensive work in applying statistics to the manufacturing process in the pharmaceutical industry in New York and North Carolina, as well as applying it to electronic equipment manufacturing in Florida. As Fab XIV statistician from 1988 to 1994, Ray supported AMD's largest volume fab in experimental design and analysis, statistical process control, statistical process troubleshooting, and the statistical equipment control project. In 1995, he was recalled from a SEMATECH assignment to derive a calculus for computing mathematically rigorous error bounds for transcendental functions in various AMD floating point units. Current research interests include correlation, graphical methods of simultaneous inference, and development of practical statistical methods to improve semiconductor manufacturing.

CHAPTER 23

IMPLEMENTATION OF A STATISTICAL PROCESS CONTROL CAPABILITY STRATEGY IN THE MANUFACTURE OF RAW PRINTED CIRCUIT BOARDS FOR SURFACE MOUNT TECHNOLOGY

Ricky M. Watson

A strategy was developed at a Texas Instruments' printed circuit board shop to define present manufacturing capabilities, enhance future manufacturing capabilities, ensure that all key processes are monitored by the use of statistical process control (SPC), and provide a focus for the six sigma teams in place. Detailed data analysis of the product quality variable "surface mount technology (SMT) module thickness" is discussed in relation to the capability strategy.

EXECUTIVE SUMMARY

Problem

The Texas Instruments circuit board shop in Austin, Texas faced a problem when company design groups asked for capability data relating production capabilities to proposed designs. The Defense Systems and Electronics Group at Texas Instruments had adopted the Motorola Six Sigma philosophy in 1991, which requires a close relationship between design and production. The goal was to attain a six sigma level of quality by the year 1995 (six sigma, as defined by Motorola, is equivalent to a Cpk of 1.5, or 3.4 parts per million).

Solution Strategy

Most of the previous SPC effort in the board shop was focused on process variables. Product variables needed to be determined and SPC implemented for those variables. Quality function deployment (QFD) was the tool chosen to determine the product variables. QFD was selected because of its excellent ability to determine all levels of customer care-abouts (those qualities most important to the customer, or user of the circuit boards). A group of "product specialists" was selected for their broad knowledge of the production process. The product specialists constructed the QFD model. The initial list of customer care-abouts was at a level higher than usually considered in the production shop, including items such as high input output, impedance, solder joint life, etc. The next level included items to be measured and analyzed with SPC at the product level, such as board thickness, hole location, circuit line width, etc. The final level was the process variables that directly supported and affected the product variables, such as temperatures, pressures, and chemical concentrations, etc. This study chronicles SPC implementation of the surface mount technology (SMT) module thickness product variable identified from the QFD model. The module thickness is the total thickness of two multilayer circuit boards bonded to a substrate to make a module. The module is a package requiring less space than two separate multilayer boards connected with a connector.

Data was collected and SPC was used to monitor and improve the module thickness. Process variables that affect the module thickness (multilayer bonding thickness, incoming substrate thickness, lamination press temperatures, lamination core thickness, and lamination adhesive thickness) were also monitored using SPC.

Conclusions

The strategy to improve manufacturing was successfully implemented. For the module thickness process, data collected over a six month period indicated that the process was incapable. Efforts to understand the sources of variability led to an important discovery. Multiple part numbers were processed, and each one corresponded to a unique design. The design engineers were given the list of part numbers that corresponded to large deviations from the target thicknesses, so that they could incorporate that information into future designs. Work with the supplier to reduce variation in the substrate was initiated.

PROCESS

Layers of different circuit patterns, with individual sheets of adhesive material, are placed in a stack. The stack is positioned with external pins to ensure good layer-to-layer registration. The stack is placed in a lamination press, which uses high temperature and pressure to "melt" the adhesive and laminate the layers of circuits together to form a multilayer board. Some process variables that are crucial to attaining the correct board thickness are

- thickness of incoming layer material
- number of adhesive sheets
- constant temperature

- constant pressure
- length of time that the stack is under high pressure and temperature.

Once laminated, a multilayer board is completed by the following process steps:

1. Drill: Holes are drilled through the board to act as a pathway for electricity to pass from one side of the board to the other.
2. Hole plate: Copper is electroplated into the holes to act as the electrical conductor in the holes.
3. Image: The external image is placed onto the outside of the board.
4. Circuit plate: Tin/lead solder is electroplated onto the surface of the board and in the holes to act as the surface for component soldering and as a protective covering for the copper conductor underneath.
5. Reflow: Tin/lead solder is heated to its melting point as a quality check for the solder quality.
6. Profile rout: The board is routed (mechanically removed) from its surrounding material framework.

Once these processes are completed, two multilayer boards are pinned together with the correct number of sheets of adhesive and a substrate to make one module. The multilayer boards are laminated in a similar fashion as that of the circuits that form the multilayer boards. The substrate acts as a heat sink, drawing heat from the boards, and as a connective device, eliminating large external connectors. This completed module is measured and compared to the customer's specification.

DATA COLLECTION PLAN

Three passive data collections were performed to study the three thicknesses of interest: multilayer, module, and substrate thickness. The data that is presented is coded for proprietary reasons. Data for the multilayer thickness was collected by measuring two boards from each lot processed at the profile rout machine. The boards were selected at random from each lot, and the sample was taken at a random corner. The measurement device used was a digital caliper with 0.67 resolution, well within the required amount of resolution (2 minimum). $\overline{X}$ and R charts were constructed. Subgroups of size two were selected from each lot to be able to calculate within-lot variability as well as between-lot variability.

The second thickness variable, module thickness, was handled differently. Based on the results of the multilayer thickness data, it was felt that the within-lot variation would be very low. One sample board was taken from each lot at a corner selected at random. The sample was taken at the profile rout operation. X (individuals) and moving range charts were utilized because there was no logical way to group the data.

The last thickness variable was the substrate thickness. This was the thickness of the substrate as it arrived from the supplier. One substrate from each lot was sampled using a micrometer similar to the one used to measure the multilayer and module thicknesses. X and moving range charts were again used since the data points were unrelated.

Three lamination variables were measured: the heat up rate for the lamination presses, lamination core thickness, and lamination adhesive thickness. Five openings are located in each press. The heat up rate for each opening was sampled. $\overline{X}$ and R charts were used with subgroups of size five from the five readings taken at the same time.

Data Analysis and Interpretation of Results

All graphs were constructed using deviations from the nominal value, rather than the actual values. This is a common practice in our board shop because it is much easier to understand how something differs from its target than it is to understand the actual value. For example, it is more useful to describe how far from the target an arrow hit, than it would be to state the actual location of the arrow on an X-Y co-ordinate system.

General Notes Regarding the Charts Used in this Case Study

All figures will be analyzed in the following order. All range and moving range charts are in the top left and will be analyzed first. All X and $\overline{X}$ charts are in the top right. All histograms are in the bottom left. They show the specifications and will be analyzed for spread, location, and shape. Also, all histograms contain ten cells across the scale, a limit of our software. The bottom right chart is a chart of individual points plotted against the specification. All charts have the oldest data on the left and the latest on the right. All control limits on control charts are calculated based only on the data shown. The SPC software we use automatically plots three zones on either side of the center line of the $\overline{X}$ and X charts. Each zone is one third of the distance from the control limits to the center line. All zones on the R charts are to be disregarded, and all zones on the X and $\overline{X}$ charts indicate the zones "A" (outside), "B" (middle), and "C" (closest to center line). Western Electric Run Rules are applied to all X and $\overline{X}$ charts. The values of C_{pk}, the standard deviation, and the mean are given in the middle of each figure. When a process is out of control, the value of C_{pk} does not have much meaning.

Module Thickness

The first variable measured was the module thickness product variable (see Figure 23.1). The moving range chart shows that there is one data point above the upper control limit, and there are several data points that are very close to zero. This indicates that the process is at times able to attain a low range. Our conclusion was that the difference from lot-to-lot (point to point) could be much lower and stable than was exhibited. The X chart exhibited one point above the upper control limit, and one area in which the second to tenth points from the right are all in zone C. This violates a run rule in which an out of control situation occurs if eight points in a row are in one of the zones closest to the center line. This indicates that the process has changed, because the probability of eight points in a row in the C zones (closest to the center line) is very unlikely. This chart indicates that the module thickness was out of control. The histogram is centered, but it goes beyond the specification limits. Finally, the C_{pk} is 1.03 or 4.6 sigma (1.03 C_{pk} × standard deviations), taking into account the 1.5 sigma shift used by Motorola. Since our goal was at least a 1.5 value of C_{pk}, the module thickness process was incapable. Overall, the module thickness measurement is not good. These results made it necessary to examine the causes of module thickness variability.

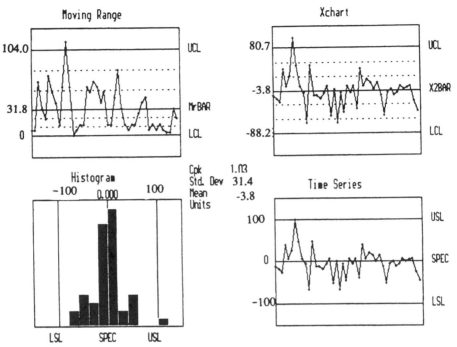

Figure 23.1. *Results of the Final Module Measurements. The data was collected over a six month period, and indicates an incapable process. The C_{pk} value of 1.03 (4.6 sigma) is lower than the desired minimum value of 1.5.*

Multilayer Thickness

The next variable analyzed was the multilayer thickness, a component of module thickness (see Figure 23.2). The range chart shows four points above the upper control limit and several points equal to zero. Notice that many points are in steps at the same height. This is due to poor resolution of the caliper used to take measurements. The $\overline{X}$ chart is frequently out of control. The majority of the variation for multilayer thickness appears to be lot-to-lot variation since the average range ($\overline{R}$) is only 6.7, indicating that the average difference between boards in the same lot is only 6.7. However, the $\overline{X}$ between-lots control limits are wider than 18. This becomes more obvious when the time series chart is examined. The observations come in pairs corresponding to the two boards in one lot, which indicates that the within-lot variability is smaller than the lot-to-lot variability. The histogram is centered on the specification. Its spread is not excessive when compared to the upper and lower specification limits, and the distribution is bell-shaped. The C_{pk} is very good at 2.22.

Since the module thickness is unacceptable, we must search for the biggest contributor to its variability. The percentage of the final module variability that is due to the multilayer thickness variability is calculated as follows. The standard deviation of the multilayer thickness (13.9) is squared to give the variance (standard deviations cannot be combined) then doubled (two multilayers are combined to make one module) and divided by the square of the module standard deviation (31.4) and multiplied by 100), the resulting percentage is 39%:

$$(13.9)^2/(31.4)^2 \times 100 \times 2 = 39\%.$$

This says that one of the greatest sources of variation in the final module (39%) is multilayer thickness, and we learned earlier that the greatest source of variation in multilayer thickness is lot-to-lot variation. Therefore, if multilayer lot-to-lot variation is decreased, the module thickness should become more capable (higher C_{pk}).

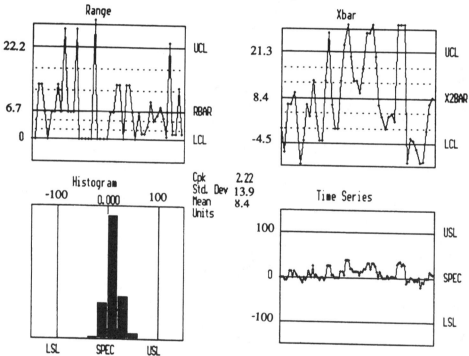

Figure 23.2. *Multilayer Thickness, the First Component of Module Thickness. The data was collected over a six month period. The multilayer C_{pk} is 2.22, even though the range and $\overline{X}$ charts are out of control.*

Incoming Inspection Substrate Thickness

Another component of module thickness is the incoming inspection substrate thickness as it arrives from the supplier (see Figure 23.3). Again, the poor resolution of the measurement device, the caliper, is seen in the observations with zero range. The graph does not look unusual. It is in control. The X chart shows two areas at either end that appear to be staying on one side of the center line, indicating a lower degree of variation and a change in centering. The histogram shows very poor results: the location is not clear, the shape is not normal, and the spread is too wide. The C_{pk} is too low at 0.97. This variable is clearly in need of improvement. Its contribution to the total variation is

$$(7.6)^2/(31.4)^2 \times 100 = 6\%.$$

The substrate thickness is clearly a minor contributor to module thickness variability. The total contribution to module thickness variability from multilayer and substrate variability is 45% (39% + 6%).

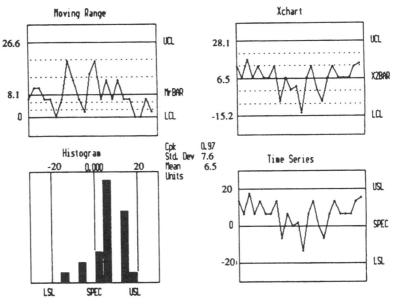

Figure 23.3. *The Second Component of Module Thickness, Incoming Inspection Substrate Thickness. After five months of data, a capability problem is indicated by the low C_{pk} of 0.97 (4.5 sigma).*

The remaining contributors to module thickness variability are in the lamination process. The process engineering members of the team indicated that there are three main contributors to lamination process variability: the heat-up rate of the lamination presses, the core thickness, and the adhesive thickness. These will now be considered.

Lamination Temperature

The heat up rate of the lamination presses was suspected to be the primary variable to monitor in the lamination process. Temperature data was collected following a fixed length of time after every press closure and pressure was applied. The temperatures at all five openings on the press were recorded and charted. Figure 23.4 shows the results.

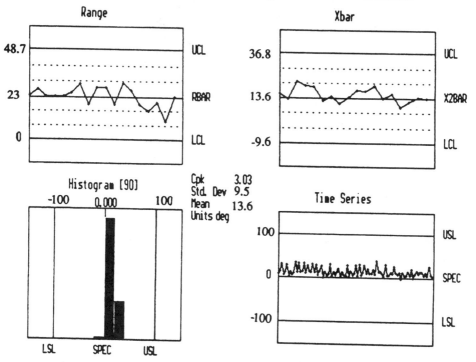

Figure 23.4. *Lamination Temperature. Data was collected following a fixed amount of time after every press closure and pressure was applied. All five openings on the press were recorded and charted. Notice the unusual range chart.*

The range chart looks very unusual. All points but one are hugging the center line. This indicates a high but tightly grouped set of ranges. The $\overline{X}$ chart looks similar. The problem is that the average range $\overline{R}$ is too large because it was calculated using the data from the five openings. For example, for the $\overline{X}$ chart,

$$\overline{\overline{X}} \pm A_2 \overline{R} = \text{control limits.}$$

When $\overline{R}$ is large, it drives the control limits for the $\overline{X}$ chart out. This problem was resolved by using data from only one opening in later charts for lamination temperature. The histogram shows that the observations are located close to specifications, with a tight spread, and the shape of the distribution is unclear. Our philosophy regarding the shape of the distribution in the histogram is that if the value of C_{pk} is high (for example, 3.03 in this case), then it doesn't matter if the shape of the distribution is not bell-shaped (normal).

Core Thickness and Adhesive Thickness

The last two components of module thickness which are part of the lamination process are the core thickness (see Figure 23.5) and the adhesive thickness (see Figure 23.6). The core is the part of the board that has the layer images. The adhesive is the material that binds to the core, allowing the board to be laminated.

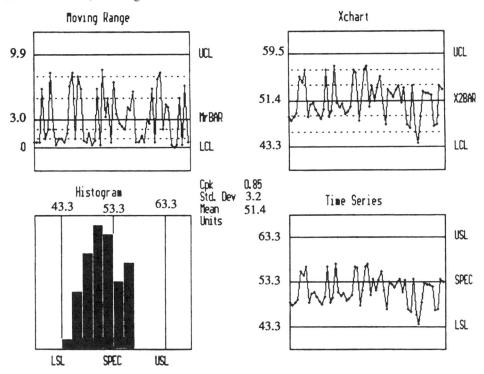

Figure 23.5. *Lamination Core Thickness.*

A typical multilayer board may contain five cores. The core graphs are X and moving range charts. $\overline{X}$ and range charts were not used because the data could not be grouped together logically. The charts in Figure 23.5 indicate that the process is not doing well. The moving range chart shows data that is close to zero and data that is close to the upper control limit, indicating instability. The X chart shows a similar pattern. The histogram is not very bell-shaped, slightly off-center, and the high spread reaches the lower spec limit. The C_{pk} is only 0.85.

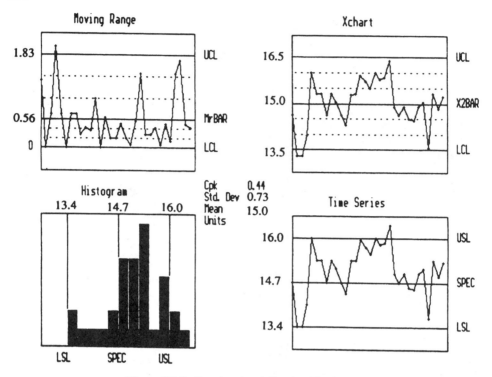

Figure 23.6. *Lamination Adhesive Thickness.*

The last variable is the adhesive thickness which also uses X and moving range charts. The moving range chart looks in control, except for the fourth point from the left. The X chart does not look as promising. Two points are outside the control limits. The chart has several places where run rules are violated (eight points in a row on one side of the mean). This chart is out of control. The histogram is far from bell-shaped, slightly off-center, and spread from one specification to the other. The C_{pk} is abysmal at 0.44.

CONCLUSION

The team concluded that the module thickness had to be improved. A C_{pk} of only 1.03 translates to thousands of defects per million. The fact that virtually zero defects were seen prior to shipment was attributed to inadequate sampling methods, and to the fact that the production volume was low.

The greatest source of variation in the module thickness is the multilayer process. The multilayer variable showed many control problems in spite of its high C_{pk}. The multilayer thickness may have inappropriate specifications since it has a high C_{pk}, yet contributes most (39%) of the variation to the module process that has a low C_{pk}.

Efforts to understand the source of the multilayer variation resulted in a very important discovery. The wild swings on the $\overline{X}$ and range charts were due to the processing of multiple part numbers. Each part number was unique in design compared to all other part numbers, and therefore each part number was laminated at a different average. The shop generally made little contribution to the design of a board. We felt that this information would provide useful feedback to the design group concerning the manufacturability of their designs. We immediately began giving the design engineers a list of part numbers that appeared to deviate more than usual from the targeted thicknesses. We continue to provide this information today.

The substrate thickness presented a problem in the sense that it was controlled by the supplier, not by our shop. Efforts were initiated to work with the supplier to decrease the variation of the substrate.

The only work scheduled for the temperature rate of the lamination press heat up is to graph data from one opening only on a moving range and X chart, eliminating the other four opening data points. The reason for this change is that poor calibration techniques led to very large ranges when the five openings were considered, which resulted in exceedingly wide control limits for the $\overline{X}$ charts. In spite of the out-of-control situations, the C_{pk} is sufficiently high to allow us to focus our efforts on variables that have worse capability. The team did not feel that the slight variation in opening temperature was a large cause of thickness variation.

Efforts have been initiated to work with the suppliers of the adhesive and core to find ways to reduce variation. Our future success for multilayer and module thickness rests with the improvement of the supplier's adhesive and core.

BIOGRAPHY

Ricky M. Watson joined Tracor, Inc. in Austin, TX in 1994 and is currently the SPC/Team Facilitator for all of Tracor Aerospace, where he is responsible for all SPC and team training, producibility for all programs and projects, and SPC and team implementation. He was previously with Texas Instruments, Inc. for 18 years as a manufacturing manager and Facilitator. He received his bachelor's degree in biology from the University of Texas at Austin in 1985. Rick has completed one textbook, published numerous articles on SPC, and conducted many speeches, including at the World Quality Congress in Helsinki, Finland in 1993. He currently sits on the Board of Examiners for the Austin Quality Award and Texas Quality Award.

CHAPTER 24

OBTAINING AND USING STATISTICAL PROCESS CONTROL LIMITS IN THE SEMICONDUCTOR INDUSTRY

Madhukar Joshi and Kimberley Sprague

There are two distinct but related objectives to this case study:
1. illustrate the method and formulae first proposed in 1986 for use in Digital Equipment Corporation's wafer fabs to obtain workable statistical process control (SPC) limits;
2. discuss key SPC implementation techniques that led to successful control of the semiconductor processes in the wafer fabs.

The first of these two objectives focuses on describing the method used for finding workable control chart limits. This method was required because when the text book approach using Shewhart constants (d_2, c_4, etc.) was used to compute SPC limits, a large percentage of the plotted points fell outside the calculated control limits. Such inappropriate limits resulted from not accounting for the fact that many wafer processing steps are batch processes.

The second objective of this case study is to discuss the resistance to initiating SPC charts in the wafer fabs that was present at the start of this project, and the methods used to overcome that resistance. The concept of "effective sample size" (N_{eff}) allowed process engineers to implement SPC charts with workable limits on virtually any semiconductor process - both within and beyond the wafer fabs. It was a key ingredient in overcoming the resistance to SPC. Management's active support of SPC was the other key ingredient. When these two aspects were applied simultaneously, they led to successful implementation of SPC, just-in-time manufacturing (JIT), design of experiments (DOE) and related continuous improvement methods across all groups within Digital's semiconductor facility in Hudson, MA.

EXECUTIVE SUMMARY

Problem

In 1986, all wafer fabs, as well as post-fab operations (probe, back prep, wafer saw, packaging, burn-in, and electrical test before and after burn-in, etc.), were run on the premise that any output within specification limits was perfect and all else was worthless. Each individual wafer lot at each fab operation, for instance, was judged strictly on the basis of whether or not it met the "tech file," Digital's specification document provided by the chip designers, for that operation. There was no concept of tracking the history of any operation and using it to judge whether the process of making similar wafer lots was stable and predictable. This mode of operation, called "spec mentality" within Digital, led to numerous tweaks to a process with little regard to the natural variability of that process. This tended to further increase process variability and led to numerous fire drills. Managers who had heard of Dr. Deming's SPC message asked process engineers to bring SPC to semiconductor manufacturing. When engineers tried using the text book approach, the results were disastrous: too many charted points outside of the apparent SPC limits— contrary to what they had heard from SPC leaders such as Dr. Deming and Dr. Juran. Consequently, engineers wanted nothing more to do with what they thought was SPC.

Solution Strategy

An SPC focus group, formed expressly to understand and resolve the root causes of the above systemic problem, realized that there were technical as well as managerial causes for the difficulties encountered. Specifically, unlike one-widget-at-a-time production, semiconductor manufacturing has numerous batch processes. Consequently, statistical methods listed in the vast majority of readily available SPC publications in 1986 did not apply directly.[52] The SPC focus group developed the concept of effective sample size, N_{eff}, so that the rest of the text book SPC formulae could be used by the process engineers. Management then initiated the statistical process, analysis, control and enhancement (S.P.A.C.E.) group to help implement SPC using this new concept. The S.P.A.C.E. group educated wafer process engineers, assisted in implementing SPC charts at key process steps so that the concept of effective sample size could be pilot tested, formed SPC user groups to foster the use of SPC in the wafer fabs, and helped automate the process of obtaining SPC limits and checking additional applicable Western Electric rules.

Conclusions

Presently, SPC is standard operating procedure within Digital's wafer fabs. As SPC became a proven technology within fabs, management of post-fab operations (wafer probe, electrical test, back prep, package assembly, final assembly, etc.) implemented SPC in their operations as well. Finally, these successes were followed by other continuous improvement initiatives such as just-in-time (JIT) manufacturing for all semiconductor operations and institution of designed experimentation (DOE). Key ingredients of this success were:

    1. listening to clients to understand their problems in their own language

---

[52] Very few of the publications that were in the public domain in 1986, for instance, those referenced by Woodall and Thomas (1993), discussed the batch problem in SPC computations. Engineers, therefore, tended to equate SPC methods with those used for monitoring one-widget-at-a-time manufacturing. By 1992, a number of good references were available. See, for instance, Hahn and Cockrum (1987), Montgomery (1991), Neuhardt (1987), Pence (1991), Relyea (1990), and Yang and Hancock (1990).

2. the use of batch-to-batch control limits that were not based on within-batch variability
3. insistence that no continuous improvement technique be used blindly
4. management's philosophy that continuous improvement experts must be integral members of manufacturing groups
5. management's active support (through persuasion rather than dictation) of continuous improvement explorations.

Structure of the Case Study

This case study is divided into four major sections. The Technical Problems section presents the technical problem and identifies its root cause. The second section, Effective Sample Size (N_{eff}), discusses solutions to that technical problem. The third section, Implementation of SPC, describes some of the methods used to implement SPC. The Conclusions section lists some reasons why we think our methods were successful.

THE TECHNICAL PROBLEM

Question: When is the information from three data points equal to the information from only one data point?

Answer: When the data corresponds to a batch process where each batch contains material that is relatively homogeneous compared to the variability between batches.

In this section, we will provide the details associated with the above dialogue. The technical discussion will be deliberately limited to considering a situation with only two sources of variation: within batch and between batches.

When reading this section, it is important to remember that this case study started in 1986. At Digital's semiconductor facility in Hudson, MA., circa 1986, fab processes were "controlled" by comparing process outputs to specifications generated by the design group. Management felt that there had to be a better way to run the manufacturing areas than to tweak a machine if an out-of-spec condition occurred. Efforts to use the SPC techniques were widely publicized in 1986, but unfortunately they failed miserably. This led many fab process engineers to state that "SPC cannot be applied in wafer fabs. Unlike one-widget-at-a-time manufacturing, wafer manufacturing is a high tech operation." Some engineers who made honest efforts to set up what they thought were SPC charts felt frustrated by the excessive number of out-of-control flags associated with their charts. Intuitively they felt that something was wrong; how could they have so many special-cause flags?

What was wrong? Integrated circuits are made from many die produced on a wafer (a "batch" of integrated circuits is manufactured on each wafer), wafers are batch-processed in diffusion tubes (a batch of 200 wafers is simultaneously processed in a furnace), etc., and the data is not independent. One individual die on a wafer is more likely to be defective if its neighbors are defective than if they are not; one individual wafer in a chemical vapor deposition furnace is more likely to be too thick if other wafers in that furnace run are too thick than if they are not. This led to incorrectly tight limits when unsuspecting users applied the standard one-widget-at-a-time SPC chart formulas to the "obvious" subgroup size (one batch), especially for cases where many measurements were recorded.

Figure 24.1 illustrates this problem for the $\overline{X}$ chart of a photolithographic process. Appendix A lists the data used in Figure 24.1. Three critical dimensions (CDS) were measured at different sites (top, center, bottom) on a product wafer. When the standard $\overline{\overline{x}} \pm A_3 \overline{s}$ formulas were used (see Appendix B), incorrect[53] SPC limits were found for a subgroup size of three, 11 of 74 data points ($\approx 15\%$) on the mean chart were found to be out of control beyond the control limits. This was only one of many such charts in the Digital fabs and by no means the one with the largest percentage of points outside of the calculated SPC limits! Since so many points were outside control limits when applying only one rule, no one dared suggest using the full set of Western Electric rules.

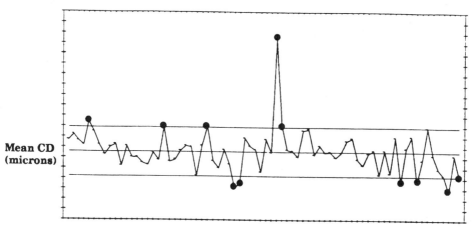

Figure 24.1. $\overline{X}$ *Chart for a Photolithography Process. Three critical dimension readings were measured on each wafer. SPC limits were calculated using a subgroup size of 3. For the mean chart, 15% (11 of 74) data points failed the "one point beyond the control limits" rule.*

Aside from the typical reluctance to implement a new tool due to a lack of experience using it and the need for a change in philosophy (e.g., stopping a process based on SPC results instead of on specification limits), there was certainly little incentive for process engineers to use an approach indicating that the process needed to be shut down every seven runs! Given the variety of fab processes and the problems engineers were having implementing SPC charts, it was essential to develop a method that was straightforward. In the next section, we will describe the concept of "effective sample size" that was devised to remove the technical problems associated with using the one-widget-at-a-time SPC methods in the semiconductor industry. In the last section, we will describe the implementation approach that successfully moved us from resisting SPC to making it standard operating procedure.

---

[53] "Incorrect" based on hindsight. At the time, the cause of the problem was not understood.

Effective Sample Size (N_{EFF})

The photolithography example in Figure 24.1 illustrates how SPC limits computed without taking into account the batch structure of the data led to a large number of points (15%) falling outside the apparent 3-sigma limits. This seems to contradict the theory which suggests that if the process is "in control", i.e., operating with only common causes of variation, then only 0.27% of the charted points will fall outside the 3-sigma limits. To resolve this apparent inconsistency and to recommend an appropriate SPC method, the management of Digital's semiconductor operations formed an SPC focus group. This group was comprised of a statistician, a quality representative, and process engineers from a number of key areas of the wafer fabs (photo, diffusion, etch, etc.). The group spent three months studying the issues and testing solutions.

A Cause That Was Not

The technical cause of the excessive number of $\overline{X}$ values outside the apparent 3-sigma limits turned out to be the inappropriate application of SPC theory to semiconductor manufacturing.

Most of the literature on SPC theory assumes that products are manufactured one unit at a time, and that each sample of N units produced at a given time can be considered *random* and *independent* production units. Under these conditions, the Central Limit Theorem applies to the average ($\overline{X}$) of individual readings, and $\overline{X}$ follows the normal distribution fairly well. Based on this assumption, the standard deviation $\sigma_{\overline{x}}$ of the sample means is given by[54]

$$\sigma_{\overline{x}} = \sigma_X \div \sqrt{N} \qquad (24.1)$$

in terms of the standard deviation σ_X of the individual readings. There are many formal statements of this theorem, such as the statement on page 244 in William Feller's classic text (see Feller, 1970) which is based on a paper by J.W. Lindberg (see Lindberg, 1922).

Traditionally, one would plot the $\overline{X}$ values for a specified time window (e.g., K = 30 consecutive $\overline{X}$ values) on an $\overline{X}$ control chart. The upper and lower control limits (UCL and LCL) for the $\overline{X}$ control chart would be based on the estimates of standard deviation within each sample. For instance, one could estimate σ_X as the square root of the average of the within-sample variances, or one could compute $\overline{s}$ as the average of the within-sample standard deviations and then estimate $\sigma_X = \overline{s}/c_4$ (c_4 is a function of N – sample size and has been well documented. See, for instance, ASQC, 1983). Equation (24.1) would then be used to estimate $\sigma_{\overline{x}}$. The $\overline{X}$ control limits would be $3\sigma_{\overline{x}}$ units away from the process mean. Consequently, in a stable and normally distributed process, only a very small fraction (0.27) of the $\overline{X}$ values would fall outside these control limits.

---

[54] This formula holds for any distribution, normal or not. The 99.73% probability associated with 3-sigma limits is a result of normality.

As already illustrated in Figure 24.1, the fraction outside the control limits was much larger than 0.27 in the early attempts to construct SPC charts. Some people conjectured that this huge disparity might be a consequence of non-normality. However, Wheeler and Chambers (see Wheeler, 1992) have studied the problem of dealing with non-normal data and have shown that even if the original data were non-normal and even if $\overline{X}$ values were computed from small samples (e.g., N = 4), at least 99% of such values would fall within properly computed 3-sigma limits. Even if we plot individual readings from non-normal data, Wheeler and Chambers (see Wheeler 1992) have documented that at least 98.2% of the charted points would lie inside the properly computed 3-sigma limits. Thus, lack of normality was ruled out as a possible cause of over 15% of $\overline{X}$ values falling outside the computed 3-sigma limits for the photolithographic example.[55]

Another proposed reason for the large number of points outside the control limits was that the process was truly out-of-control. Engineers familiar with the process reviewed the data and noted that one point did look like an outlier but that, for the rest of the data points, there were as many points just within the control limits as fell just beyond them. This left only one alternative for the cause of the problem: the 3-sigma limits had not been computed properly! We will take up this issue in the rest of this chapter.

Batch Processes

Semiconductor operations, especially wafer fabrication areas (fabs), include numerous processes. For example,
1. chips (the true "production units") are generally not built one at a time; instead, there are many chips on each wafer (batch of chips).
2. a number of steps in the wafer production, such as diffusion, process a "boat" or a "run" of wafers at once— introducing further "batchness."

How do batches affect SPC limits?[56] If widgets were produced independently, one-at-a-time, then an $\overline{X}$ based on N observations would have $\sigma_{\overline{x}} = \sigma_X \div \sqrt{N}$. For a batch process, however, the observations used to compute $\overline{X}$ are not independent; in fact, they are correlated with each other.[57] Consequently, $\sigma_{\overline{x}}$ is larger than $\sigma_X \div \sqrt{N}$. Not surprisingly, the LCL and UCL computed under the assumption of independent individual readings is too tight for the actual $\overline{X}$ values and led to an excessive number of charted points falling outside the apparent 3-sigma limits.

Computing Effective Sample Sizes

The SPC focus group devised the method described in this section of using an effective sample size to address the batch structure of the data and the consequent correlation

---

[55] Dr. Wheeler provided many of his results to the SPC focus group before their official publication.
[56] See Appendix C for the 1992 statistical "answer" to this question.
[57] The problem discussed here cannot be resolved by choosing the sites at random, either. As documented in a separate paper by Joshi and Fang (see Joshi, 1987), a number of electrical parameters on 99 wafers, each with 49 sites, were analyzed to identify site clusters that were uncorrelated with each other so as to obtain a statistically better selection of uncorrelated sites. For each electrical parameter, the clusters did form clearly explainable patterns but there were a maximum of two such clusters for any one parameter.

between observations, thereby obtaining more reasonable control limits. This method was used for all batch process steps that had the potential for correlated data. It drastically reduced the number of points on the control charts that fell outside the computed control limits, and was a key to the successful implementation of SPC in Digital's wafer fabs.

Instead of assuming, for example, that the site level readings on a wafer form a set of N independent random readings on each wafer, we posed the question in reverse, namely:

> If the data were in fact randomly and independently selected from a process with mean μ and standard deviation σ_X, how many independent random readings would be in that sample?

We called this computed number the *effective sample size* (N_{eff}).

Consider the process data represented in Table 24.1, with K wafers (subgroups) and N readings taken on each wafer. The mean and sample standard deviation for each wafer are shown in columns headed "$\overline{X}$" and "Standard Deviation," respectively. Each reading has two suffixes: the first suffix identifies the wafer (subgroup) number (1 to K) and the second suffix identifies the observation number (1 to N) in that subgroup. For instance, $x_{3,2}$ is the second reading on the third wafer. The subgroup means and sample standard deviations are shown with a subscript identifying each subgroup. Finally, the average values for the $\overline{X}$ and the standard deviation columns are shown at the bottom of the table.

Table 24.1. *A Table for Computing N_{eff} for a batch process.*

Subgroup Number	Data Values				$\overline{X}$	Standard Deviation
1	$X_{1,1}$	$X_{1,2}$	---	$X_{1,N}$	$\overline{X}_1$	s_1
2	$X_{2,1}$	$X_{2,2}$	---	$X_{2,N}$	$\overline{X}_2$	s_2
3	$x_{3,1}$	$X_{3,2}$	---	$X_{3,N}$	$\overline{X}_3$	s_3
---	---	---	---	---	---	---
---	---	---	---	---	---	---
K – 1	$X_{K-1,1}$	$X_{K-1,2}$	---	$X_{K-1,N}$	$\overline{X}_{K-1}$	s_{K-1}
K	$X_{K,1}$	$X_{K,2}$	---	$X_{K,N}$	$\overline{X}_K$	s_K
					$\overline{\overline{X}}$	$\overline{s}$
	$s^2 = \Sigma\, s_k^2 \div K$ estimates σ_x^2. Alternatively, σ_x can be estimated as $\overline{s} \div c_4$.					

If each $\overline{X}$ were the mean of a random independent sample of size N_{eff} from a population with mean μ and standard deviation of σ_x, then the standard deviation of $\overline{X}$ would be $\sigma_{\overline{x}} = \sigma_x \div \sqrt{N_{eff}}$. Hence, we decided to compute N_{eff} as the ratio

$$N_{eff} = \sigma_x^2 / \sigma_{\overline{x}}^2 \tag{24.2}$$

The unbiased estimate of $\sigma_{\overline{x}}^2$ is $s_{\overline{x}}^2$, where $s_{\overline{x}}$ is the sample standard deviation of the $\overline{X}$ values in the dataset. Symbolically:

$$\hat{\sigma}^2_{\overline{x}} = (s_{\overline{x}})^2. \tag{24.3}$$

The unbiased estimate of σ_x^2 is the average of variances of individual readings in each wafer. Symbolically:

$$\hat{\sigma}_x^2 = \sum s_k^2 \div K. \qquad (24.4)$$

Equivalently, σ_x of Equation 24.4 can be estimated by the ratio:

$$\hat{\sigma}_x = \bar{s} \div c_4, \qquad (24.4a)$$

where one uses the c_4 value corresponding to sample size N. Combining Equations 24.2, 24.3 and 24.4, we get:

$$N_{eff} = \frac{\sum s_k^2 \div K}{(s_{\bar{x}})^2}. \qquad (24.5)$$

Using Equation 24.4a, Equation 24.5 becomes,

$$N_{eff} = \frac{(\bar{s} \div c_4)^2}{(s_{\bar{x}})^2}. \qquad (24.5a)$$

In either case, the computed value is always *rounded upwards*. For instance, if N_{eff} of Equation 24.5 is 0.648, then we would use $N_{eff} = 1$. The value for N_{eff} determines which one-widget-at-a-time Shewhart constants to use when computing LCL and UCL.[58]

N_{eff} can be thought of as the quotient of within-wafer variability (See Equation 24.2) to between-wafer variability. In many semiconductor processes, within-batch variability is smaller than between-batch variability, so that the quotient is less than 1, and N_{eff} is rounded up to 1. When N_{eff} is 1, each $\overline{X}$ is treated as a single observation in an individuals control charts.

This approach had several benefits.

1. It widened the control limits to more realistic values and thus reduced the number of points out-of-control on the control charts. Process engineers were more willing to respond to the associated special-cause flags.
2. It allowed the use of one-widget-at-a-time standard tables (see ASQC, 1983) and procedures provided N_{eff} was used in place of the actual sample size. For instance, we could continue to invoke the Western Electric special-cause rules wherever applicable. The ability to invoke the ASQC tables provided a "comfort zone" to management as well as to process engineers.
3. The calculations were easy to do and the theory was understandable. As processes improved over time, the die-to-die (within-wafer) variation became smaller than the wafer-to-wafer variation, so N_{eff} became 1 and the SPC limits for all of the $\overline{X}$ charts were based on the moving range of those $\overline{X}$ values. This led to even further standardization of local SPC software.

---

[58] Process engineers continued taking all N readings in each wafer for engineering reasons, such as "to study wafer uniformity," the engineering equivalent of studying within-wafer variability.

Application to a Single Wafer Process

The use of effective sample sizes successfully resolves the problem of obtaining workable SPC limits for any process step that handles one wafer at a time. Photolithography exposure and dry etch are steps of this type. An example is shown in Figure 24.2, where critical dimension measurements of die on a wafer are taken at fixed sites on a wafer.

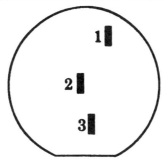

Figure 24.2. *Data Sites on a Wafer. Actual number of sites and their locations are Digital confidential information.*

An example of the instructions for determining N_{eff} is shown next for the data in Appendix A. We have data on $N = 3$ sites per wafer for each of $K = 74$ lots. To compute Neff, begin by obtaining $\overline{X}$, the wafer mean for the product wafer measured in every lot, as shown in the next-to-last column in Appendix A. Next, compute the standard deviation $s_{\overline{x}}$ of these wafer means, given by 0.0981. The square of this number is the denominator of Equation 24.5. To obtain the numerator, compute the standard deviation s_x for each wafer, as shown in the last column of the table in Appendix A. Square s_x for each wafer, sum all of the s_x^2 values ($\Sigma s_x^2 = 0.4616$) and divide that total by the number of lots ($K = 74$). This gives the numerator of Equation 24.5. The effective sample size is given by:

$$N_{eff} = (0.4616 \div 74) \div (0.0981)^2 = 0.6482,$$

which gives $N_{eff} = 1$ by rounding up to the nearest integer. Also, $\overline{\overline{X}} = 3.1276$.

Figure 24.3 shows the associated $\overline{X}$ chart for data of Appendix A based on $N_{eff} = 1$. The control limits for the $\overline{X}$ chart, where each average $\overline{X}$ is treated as one point on an individuals control chart, are given in Appendix B. Now, only 1 (of 74) data points falls outside of the control limits.

When the effective sample size was first implemented, N_{eff} was either 1 or 2—much smaller than the actual number of sites measured on a wafer. As SPC theory and actions by process owners helped improve single-wafer processes, N_{eff} converged to 1.

Figure 24.4 gives the s chart for the data given in Appendix A. The control limits for the s chart, where each standard deviation s is treated as one point on an individuals control chart, are given in Appendix B.

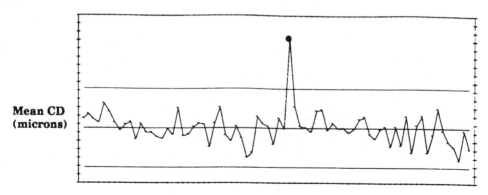

Figure 24.3. *The photolithography process example using the data of Appendix A—SPC limits are now calculated using the effective sample size, $N_{eff} = 1$. Only 1 data point out of 74 (1.4%) fails the "point(s) beyond the control limits" rule.*

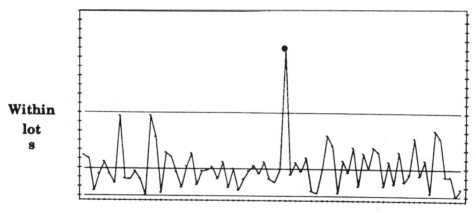

Figure 24.4. *S Control Chart for Within-Lot Variability for the Photolithography Process. The chart treats each standard deviation s as a point on an individuals chart. This is based on a moving range of s values with one-time iteration to eliminate points beyond initial control limits before recalculating workable limits. Note: all data points are shown in this figure although one (Lot 40) was removed in the first iteration.*

Site-to-Site Correlations

We have often mentioned that the observations at different sites on a wafer correlated. Table 24.2 gives the correlations between the three wafer sites for the data of Appendix A, where three sites are measured on a product wafer for each of 74 consecutive lots. If the site readings had been independent, the off-diagonal correlations would have all been zero.[59] In this case, the correlations are 0.668, 0.548, and 0.425.

Table 24.2. *Sample Correlation Between Readings.*

	Site 1	Site 2	Site 3
Site 1	1.000	0.668	0.578
Site 2	0.668	1.000	0.425
Site 3	0.578	0.425	1.000

For most parameters, Neff in 1986 was 1; occasionally it was 2. In the latter case, the sites formed two clusters that had clear engineering interpretations, e.g., center versus edge effects. As process control began to work, Neff became 1 for every parameter. At that stage, we stopped computing Neff formally and began using a "nested" model described in Appendix C where the control limits for the sample means at the highest level of nesting are always computed as if each sample mean were an individual reading.

IMPLEMENTATION OF SPC

Having an appropriate implementation system for SPC is as important as having workable SPC limits. This section of the case study describes the progression of Digital's semiconductor manufacturing area from a fab reacting only to specification limits in 1986 to one in 1992 where SPC charts are used in all facets of semiconductor manufacturing. In order to give some perspective on how long it took to evolve to this point, time frames are also mentioned.

In order for SPC to be successful, the implementation effort had to be supported by production and engineering management. *This is the key*: Without engineering management support, the engineers would have continued to react to out-of-spec situations; without production management cooperation, the SPC charts would have been merely wallpaper.

At Digital (circa 1986), SPC implementation was initially targeted for one area (photolithography) as a pilot to show the tool's use to other areas as well as to work out any implementation "bugs" before fab-wide implementation was attempted. We began implementing SPC by spending the first several months of the program training the engineering and production people who would be maintaining and reacting to the SPC charts. Topics covered in various formal SPC classes included: a brief history of SPC, computing means and ranges, plotting points on a paper chart, computing SPC limits from the data using the effective sample size, and SPC setup and special cause rules.

The courses included workshop exercises so that everyone could apply the material learned to actual case studies. Any misunderstandings in the method and its application were addressed in class.

---

[59] If we assume that the site-to-site correlations were the same for any pair of sites, then using the Neuhardt formula (see Neuhardt, 1987), the point estimate of that common correlation would have been

$$\rho = \sigma^2_{lots} \div (\sigma^2_{lots} + \sigma^2_{sites}) = .007548 \div (.007548 + .006237) = 0.548,$$

which is similar to the correlation between Sites 1 and 3.

Armed with this information, engineers initiated data collection activities. This allowed them to not only have relevant data but also signaled the beginning of a new continuous improvement effort. In a sense, the data collection effort helped change the mindset from "we react to out-of-spec conditions" to "we stop the process based on out-of-control occurrences." Finally, the engineers put manual paper charts on line. While this approach is not feasible for paperless manufacturing areas, we found that the people who have to actually calculate means and ranges, plot them, and then apply the SPC special-cause rules have a much better understanding of how SPC works.

An important aspect of the implementation was good communication among functional groups (process engineering, manufacturing and equipment engineering) and among each of three shifts. SPC user groups were formed and weekly meetings were held to discuss SPC problems and findings. This was to support the formal training with real-time solutions to specific implementation issues. The meetings usually lasted 45 to 60 minutes with 15 to 20 people in attendance. In the beginning, procedural issues, such as the correct way to calculate means and ranges or how to fill out the log sheets and plot points, were discussed. As those logistical items were resolved, user group focus shifted to the methods for addressing out-of-control points. The pilot program was expanded to other areas and people from these areas joined the user meetings; the initial users became more familiar with SPC and they no longer needed to attend meetings. As people became comfortable with the theory and practice of SPC, the user group meetings evolved into one-on-one consulting with the S.P.A.C.E. group, a small group of continuous improvement methods experts who reported to the semiconductor manufacturing engineering manager. At that stage, roughly six months after the user group meetings had begun, formal group meetings were discontinued.

In 1987, the next step of the implementation was to move from manual to automated data plotting. "Pop-up" control charts allowed manufacturing technicians to enter data as they normally would and then the data collection system would plot data points on a control chart that was displayed on a terminal for the technician. Rule violations would flag operators. However, for the initial automated SPC charts, the charts and limits had to be set up by the computer system managers. In 1991, a more flexible, automated SPC setup system was implemented so that engineers could directly set up SPC charts by specifying the SPC limits and special-cause rules that they wanted used on a specific process parameter. We are now working to move most of the setup and response duties from the process engineering groups to the on-line production people.

In 1989, after two to three years of experience in obtaining SPC limits, identifying applicable Western Electric rules that went with each chart, responding to the associated SPC flags, and using the menu-driven system, both the engineers and mangers had enough faith in the usefulness of SPC in the wafer fabs to ask for a "management metric" of process instability. The instability index (INS) was the agreed-upon solution. For each SPC chart, INS was computed as the percentage of charted points that violated the applicable Western Electric rules. Charts with high INS were subject to review for identification of special or common causes, verification of those causes through special studies (designed experiments of the appropriate variety), identification and verification of corrective actions for each cause and implementation of those actions.

As Digital's semiconductor site began to develop a reputation for the success of its SPC efforts, several engineers from other Digital sites visited to experience our system firsthand.

In 1990, one such visitor documented how he saw SPC implementation in action in his Digital-internal trip report (see Miller, 1990). Some key points of his report are noted below.

1. The SPC banner has been picked up by the local "Yield Enhancement Group" who conduct weekly SPC chart review meetings.[60] These meetings are attended by engineers from process, yield, probe, equipment support, the defect reduction team, and quality. The meetings are usually chaired by an engineering manager.
    a. Each process engineer presents control charts for his/her critical parameters. Any out-of-control points are discussed. All process critical parameters are charted, but only those requiring attention are presented in the meeting.
    b. The yield and probe engineers present control charts of yields and electrical parameters. They have a method of re-sorting their data by the time that a particular process step took place. This allows them to tie their test measurements back to the process steps that control them.
    c. The quality engineer presents control charts of parameters related to product quality and reliability issues.
    d. Throughout the meeting the engineering manager asked questions, assigned action items, and requested reports on previous action items.
2. Most of the engineers in this meeting are of senior level and are responsible for a small group of people. These people are treated with respect and confidence that empowers them. This fosters a healthy work environment where people take ownership and the job gets done.
3. Most of the emphasis is put on the process being in statistical control and not on C_{pk}. They use a statistic called INS to measure and compare the relative instability of their process parameters. C_{pk} is mentioned, but it is statistical control that is critical.
4. There is little room for emotions when meetings are conducted in this manner. Process stability is of central concern. Issues are surfaced and resolved with the support of the proper data. Emotions don't play a role in the decision making process.
5. Each piece of process equipment has a control chart that is updated from qualification runs each day or each shift. These charts help to improve the communication between process engineers and equipment support engineers.
6. The site management and the engineering management at Hudson are committed to a quality program that includes total quality control (TQC), JIT, and SPC. This commitment has resulted in a group within engineering for consulting and training on SPC and DOE. This group is called the statistical process analysis control and enhancement group (or S.P.A.C.E. group). Its members consist of statisticians and senior level engineers with process backgrounds.

---

[60] Note: These meetings are review meetings. All out-of-control flags are responded to in real time.

7. Much of the success of Hudson's SPC program can be traced back to the efforts of the S.P.A.C.E. group. They have done the fundamental research and ground work that made it possible to successfully apply SPC to semiconductor processes. They have developed techniques for conducting process capability studies that can be carried out by the typical process engineer. They have been instrumental in specifying requirements for software used in generating pop-up control charts on-line.
8. The impact of designed experiments is ever apparent in Hudson. DOE goes hand-in-hand with SPC. As a problem is identified with a control chart, an experiment is run to determine how to correct the problem. The control chart is then used to monitor the effect of any process changes. The S.P.A.C.E. group plays an active role in training and consulting in experimental design. They offer a DOE class to their engineers that is of the highest caliber.

During 1991 to 1992 our improvements have focused on software enhancements so that standard methods of calculating SPC limits and managing SPC charts would be available to all of the process engineers. As confidence in the use of SPC has increased, local software experts have developed:
1. software to help analyze the data for unimodality, symmetry, etc., prior to "releasing" any SPC chart to production.
2. standardized control-charting software that enables the engineers to automatically check the Western Electric rules that the software in Step 1 above suggested be applied to a particular SPC chart.
3. menu-driven SPC data updating and flag-raising software.

As of September 1992, SPC was a standard operating procedure within Digital's wafer fabs. Over 400 charts exist in each wafer fab. About 125 of these were judged critical enough to be followed religiously. Production groups have developed decision trees and action plans for each critical chart. Other charts are studied primarily by engineering.

Conclusions

SPC is now standard operating procedure within Digital's wafer fabs. This was achieved by supplying the managerial support and technical expertise that allowed and encouraged process engineers to implement SPC on their processes. As SPC became a proven technology within wafer fabs, management of post-fab operations (wafer probe, electrical test, back prep, package assembly, final assembly, etc.) began implementing SPC in their operations as well. The acceptance of SPC as a useful continuous improvement technique was rooted in the experiences we had in the wafer fabrication areas.

The SPC success was followed by other continuous improvement initiatives such as JIT for all semiconductor operations and institution of designed experimentation. Showing that SPC could be customized for the semiconductor environment made it easier to convince management, production, and engineering that other continuous improvement tools can be modified to work with batch processing outputs. Indeed, JIT and DOE programs have been implemented more quickly because of the SPC history of continuous improvements, and these programs have resulted in more than a 50% process cycle time reduction (from unprocessed wafers to finally assembled packages).

Key ingredients of this overall success were:
1. listening to clients to understand their problems in their own language
2. insisting that none of these techniques be used blindly
3. management's philosophy that continuous improvement experts must be integral members of manufacturing groups
4. management's active support (through persuasion rather than dictation) of continuous improvement explorations.

REFERENCES

American Society for Quality Control (ASQC), *Glossary and Tables for Statistical Quality Control*, Milwaukee, WI: ASQC, 1983.

Feller, W. *An Introduction to Probability Theory and Its Applications*, New York, NY: John Wiley and Sons, Inc., 1970.

Hahn, G.J. and M. B. Cockrum. "Adapting control charts to meet practical needs: A chemical processing application," *Journal of Applied Statistics*, vol. 14, pp. 33-50, 1987.

Joshi, M. and P. Fang. "Statistical mapping of silicon wafers," Digital Equipment Corporation internal publication, May, 1987.

Lindberg, J.W. "Eine neue herleitung des exponentialgesetzes in der wahrscheinlichkeitsrechnung," *Mathematische Zeitschrift*, vol. 15, pp. 211-225, 1922.

Miller, R. A. Digital Equipment Corporation internal memorandum, April, 1990.

Montgomery, D.C. *An Introduction to Quality Control*, 2nd Ed. New York, NY: John Wiley and Sons, Inc., 1991.

Neuhardt, J.B. "Effects of correlated sub-samples in statistical process control," *IIE Transactions*, pp. 208-214, 1987.

Pence, H. "Standardized variance components control charts and control of processing for multiple products," *5th Annual Symposium on Statistics and Design of Experiments in Automated Manufacturing*, Arizona State University, AZ, 1991.

Relyea, D.B. *Practical Application of SPC in the Wire & Cable Industry*, White Plains, NY: Quality Resources, 1990.

Wetherill, G.B. and B. W. Brown. *Statistical Process Control: Theory and Practice*, London: Chapman and Hall, 1991.

Wheeler, D.J. and D. S. Chambers. *Understanding Statistical Process Control*, Knoxville, TN: SPC Press, Inc., 1992.

Yang, K. and W. M. Hancock, "Statistical quality control for correlated samples," *Industrial Journal of Production Research*, vol. 28, pp. 595-608, 1990.

APPENDIX A

Data for Photolithography Process Example

Critical dimension measurements at three sites per wafer for one wafer in each one of 74 lots. The last two columns give the mean and the standard deviation of the three sites on one wafer, respectively

LOT #	Site 1	Site 2	Site 3	Xbar	Stdev
LOT 1	3.26	3.09	3.20	3.183	0.0862
LOT 2	3.27	3.24	3.12	3.210	0.0794
LOT 3	3.19	3.17	3.18	3.180	0.0100
LOT 4	3.21	3.15	3.12	3.160	0.0458
LOT 5	3.32	3.19	3.31	3.273	0.0723
LOT 6	3.25	3.25	3.17	3.223	0.0462
LOT 7	3.18	3.17	3.13	3.160	0.0265
LOT 8	3.25	2.92	3.17	3.113	0.1721
LOT 9	3.19	3.14	3.12	3.150	0.0361
LOT 10	3.20	3.16	3.13	3.163	0.0351
LOT 11	3.00	3.09	3.09	3.060	0.0520
LOT 12	3.19	3.12	3.15	3.153	0.0351
LOT 13	3.10	3.10	3.10	3.100	0.0000
LOT 14	3.00	3.30	3.00	3.100	0.1732
LOT 15	3.22	3.00	3.00	3.073	0.1270
LOT 16	3.07	3.06	3.06	3.063	0.0058
LOT 17	3.15	3.02	3.20	3.123	0.0929
LOT 18	3.14	3.13	2.99	3.087	0.0839
LOT 19	3.24	3.30	3.20	3.247	0.0503
LOT 20	3.06	3.09	3.09	3.080	0.0173
LOT 21	3.14	3.02	3.11	3.090	0.0624
LOT 22	3.17	3.03	3.20	3.133	0.0907
LOT 23	3.17	3.13	3.17	3.157	0.0231
LOT 24	3.19	3.09	3.17	3.150	0.0529
LOT 25	2.95	3.05	3.04	3.013	0.0551
LOT 26	3.17	3.21	3.09	3.157	0.0611
LOT 27	3.28	3.21	3.26	3.250	0.0361
LOT 28	3.00	3.12	3.13	3.083	0.0723
LOT 29	3.06	3.06	3.03	3.050	0.0173
LOT 30	3.17	3.17	3.07	3.137	0.0577
LOT 31	3.06	3.08	3.06	3.067	0.0115
LOT 32	2.91	2.97	2.97	2.950	0.0346
LOT 33	3.00	3.02	2.92	2.980	0.0529
LOT 34	3.24	3.12	3.22	3.193	0.0643
LOT 35	3.17	3.10	3.19	3.153	0.0473

Appendix A (continued)

LOT #	Site 1	Site 2	Site 3	Xbar	Stdev
LOT 36	3.10	3.09	3.22	3.137	0.0723
LOT 37	2.99	3.04	3.06	3.030	0.0361
LOT 38	3.17	3.17	3.22	3.187	0.0289
LOT 39	3.06	3.14	3.17	3.123	0.0569
LOT 40	3.81	3.89	3.30	3.667	0.3201
LOT 41	3.21	3.25	3.30	3.253	0.0451
LOT 42	3.15	3.20	3.06	3.137	0.0709
LOT 43	3.10	3.19	3.10	3.130	0.0520
LOT 44	3.14	3.01	3.16	3.103	0.0814
LOT 45	3.22	3.23	3.24	3.230	0.0100
LOT 46	3.24	3.24	3.23	3.237	0.0058
LOT 47	3.16	3.05	3.13	3.113	0.0569
LOT 48	3.01	3.26	3.20	3.157	0.1305
LOT 49	3.20	3.00	3.17	3.123	0.1079
LOT 50	3.12	3.13	3.13	3.127	0.0058
LOT 51	3.11	3.02	3.17	3.100	0.0755
LOT 52	3.07	3.13	3.17	3.123	0.0503
LOT 53	3.30	3.11	3.13	3.180	0.1044
LOT 54	3.17	3.20	3.21	3.193	0.0208
LOT 55	3.01	3.19	3.07	3.090	0.0917
LOT 56	3.13	3.03	3.03	3.063	0.0577
LOT 57	3.17	3.00	3.19	3.120	0.1044
LOT 58	3.10	3.06	3.24	3.133	0.0945
LOT 59	3.04	3.00	3.01	3.017	0.0208
LOT 60	3.16	3.05	3.19	3.133	0.0737
LOT 61	3.00	3.02	3.05	3.023	0.0252
LOT 62	3.23	3.09	3.27	3.197	0.0945
LOT 63	3.01	2.98	2.95	2.980	0.0300
LOT 64	3.17	3.17	3.09	3.143	0.0462
LOT 65	3.30	3.24	3.06	3.200	0.1249
LOT 66	3.03	2.96	2.95	2.980	0.0436
LOT 67	3.07	3.17	3.02	3.087	0.0764
LOT 68	3.25	3.24	3.24	3.243	0.0058
LOT 69	3.27	3.06	3.00	3.110	0.1418
LOT 70	3.10	3.13	2.90	3.043	0.1250
LOT 71	3.00	2.98	3.06	3.013	0.0416
LOT 72	2.97	2.89	2.95	2.937	0.0416
LOT 73	3.11	3.11	3.11	3.110	0.0000
LOT 74	3.02	3.00	2.99	3.003	0.0153

Appendix B

Computations Used in 1986

The control limits for three control charts, two $\overline{X}$ charts and one s chart, are computed for the photolithographic data given in Appendix A.

Some summary statistics that are used in the formulas are:

$$\overline{\overline{X}}\, \overline{sx} = 3.128 \qquad s_{\overline{x}} = 0.0981$$

$$\overline{MR} = 0.0884 \qquad \overline{s} = 0.0618$$

The control limits for the $\overline{X}$ chart, without taking into account the fact that the data comes from a batch process and using a subgroup size of N = 3, are:

$$\text{UCL} = \overline{\overline{X}} + A_3 s = 3.128 + 1.954 \times 0.0618 = 3.245$$

$$\text{LCL} = 3.128 - 1.954 \times 0.0618 = 3.010$$

These (inappropriate) limits for the X chart are plotted in Figure 24.1. The limits are too narrow, so there are too many points out of control.

The control limits for the $\overline{X}$ chart, using the effective sample size that acknowledges the fact that the data comes from a batch process, will now be computed. The effective sample size is:

$$N_{\text{eff}} = \left(\sum s_k^2 / k\right) / (s_{\overline{x}})^2 = (0.4616/74)/(0.0981)^2 = 0.6477.$$

Rounding up, $N_{\text{eff}} = 1$, the control limits for the $\overline{X}$ chart, where each average $\overline{X}$ is treated as one point on an individuals chart are[61]:

$$UCL_{\overline{X}} = \overline{\overline{X}} + 2.66\,\overline{MR} = 3.128 + 2.66 \times 0.0884 = 3.363,$$

$$LCL_{\overline{X}} = 3.128 - 2.66 \times 0.0884 = 2.893.$$

These limits for the $\overline{X}$ chart are shown in Figure 24.3.

The control limits for an s chart, where each standard deviation s is treated as one point on an individuals control chart, are:

$$UCL_s = \overline{s} + 2.66\,\overline{MR}_s = 0.0618 + 2.66 \times 0.0521 = 0.200$$

$$\text{LCL}_s = 0.0618 - 2.66 \times 0.0521 = -0.094 \text{ which is set to 0.}$$

We allowed "deleting points outside control limits" once before setting workable limits. Since s for the 40th lot was 0.3201, limits were recomputed ignoring that value of s. Thus, workable s chart limits were

---

[61] This multiplier, 2.66, is a two-digit approximation of 3, d_2 (= 3 ÷ 1.128 = 2.6596), which is applied to $\overline{MR}$ to find the 3-sigma limits for an Individuals chart.

$$UCL_s = 0.0583 + 2.66 \times 0.0459 = 0.180.$$

$$LCL_s = 0.0583 - 2.66 \times 0.0459 = -0.064 \text{ which is set to } 0.$$

These limits are shown in the s chart of Figure 24.4.

APPENDIX C

Computations Used in 1992 Using Variance Components

Using Proc GLM of SAS® with sites nested within lots provides the following estimates of the variance components for the photolithographic data in Appendix A:

$$\hat{\sigma}^2_{sites} = MS_{sites} = 0.006237$$
$$(= s_x^2 \text{ from 1986})$$

and

$$MS_{lots} = 0.02888 = 3\hat{\sigma}^2_{lots} + \hat{\sigma}^2_{sites}$$

so that

$$\hat{\sigma}^2_{lots} = (0.0288 - 0.006237) \div 3$$
$$= 0.007548$$

and

$$s_{\bar{x}}^2 = 0.0288 / 3 = 0.009627,$$

which is the same as $s_{\bar{x}} = 0.0981$ (from 1986 computations).

We can plot $\overline{X}$ of each lot as points on an individuals chart (this is equivalent to $N_{eff} = 1$) and then compute the control limits. Using the Wheeler and Chambers approach (see Wheeler, 1992), we get

$$UCL_{\bar{x}} = 3.128 + 2.66 \times 0.0884 = 3.363$$

and

$$LCL_{\bar{x}} = 3.128 - 2.66 \times 0.0884 = 2.893.$$

These limits are shown in Figure 24.3. Using the Wetherill and Brown approach (see Wetherill, 1991), and using $3 \times s_{\bar{x}}$ as the "margin of error" would have led us to

$$UCL_{\bar{x}} = 3.128 + 3 \times 0.0981 = 3.425$$

and

$$LCL_{\bar{x}} = 3.128 - 3 \times 0.0981 = 2.831.$$

These limits are close to those obtained using $N_{eff} = 1$. However, since $s_{\bar{x}}$ is affected by trends in the data, the limits based on $s_{\bar{x}}$ tend to be wider than those based on using moving ranges of $\overline{X}$ values. Therefore, we chose to use the Wheeler and Chambers moving range method for computing $\overline{X}$ limits.

Ordinary s chart limits based on the one-widget-at-a-time constants (see ASQC, 1983) can be used to monitor site-to-site variability within each wafer. For $N = 3$, these computations are

$$UCL_s = B_4 \bar{s} = 2.568 \times 0.0618 = 0.159$$

and

$$LCL_s = B_3 \bar{s} = 0.$$

These limits are shown in Figure 24.5 and are comparable to those shown in Figure 24.4 where we eliminated the "above the initial UCL" s value from the computations. We chose to use the moving range approach for both the $\overline{X}$ and s charts to provide one set of instructions for all types of charts.

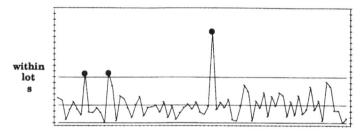

Figure 24.5. *The Control Chart for Within-Lot Variability for the Data of Appendix A. This is based on ASQC one-widget-at-a-time constants (see ASQC, 1983) for s chart limits.*

BIOGRAPHIES

Madhukar Joshi is the S.P.A.C.E. group manager in Digital's semiconductor facility in Hudson, MA. This group is responsible for designing, customizing, and implementing continuous improvement techniques for semiconductor operations. These techniques include (but are not limited to): JIT, SPC, DOE, as well as reliability and risk analysis. The S.P.A.C.E. group develops and implements associated educational programs. Madhukar has developed a workshop-based DOE text and is compiling a book of DOE case studies in the semiconductor industry.

Kimberley Sprague is a principal engineer in the statistical process, analysis, control, and enhancement group at Digital Equipment Corporation's semiconductor manufacturing facility in Hudson, MA. There she works with individuals and cross-functional teams in the wafer fab and probe, assembly, and test areas to design and implement continuous improvement initiatives. Ms. Sprague graduated from MIT in 1984 with a B.S. degree in chemical engineering and an M.S. degree in chemical engineering practice.

PART 5

EQUIPMENT RELIABILITY

CHAPTER 25

INTRODUCTION TO EQUIPMENT RELIABILITY

Veronica Czitrom

INTRODUCTION

Equipment reliability has direct impact on productivity and manufacturing costs. If the equipment is not available for manufacturing, material cannot be processed and expensive equipment is underutilized. Costs are particularly high in the semiconductor industry where one piece of equipment can cost millions of dollars, and the costs continue to escalate.

One of the main thrusts of SEMATECH has been to organize equipment reliability improvement efforts throughout the U.S. semiconductor manufacturing equipment supplier community. It has set up numerous Equipment Improvement Programs (EIPs) and Joint Development Programs (JDPs), each of which might involve personnel from SEMATECH, a semiconductor equipment manufacturing company, and a semiconductor manufacturing company.

The case studies in this unit are used to
- improve the reliability of a wafer stepper using a marathon in which the tool was exercised almost continuously for three weeks, saving an estimated $7 million in development costs and one year in development time.
- improve the reliability of an automated memory repair system using a statistically designed experiment to make the wafer handling sub-system insensitive to customer usage.
- maximize equipment availability by performing preventive maintenance on manufacturing equipment and replacing or restoring components subject to failure before they fail.

A study of equipment reliability should be preceded by gauge studies (Part 1) if process measurements are taken. Relevant information from a passive data collection (Part 2), designed experiment (Part 3), statistical process control (Part 4), or any other source may prove useful.

The next section of this introduction to reliability presents some of the definitions of basic terms that are given in the SEMI E10 standards. The following section gives an overview of the case studies in the remainder of this part, and the last section gives suggestions for further reading.

SEMI E10 Standards Definitions

The SEMI E10 Standards provide guidelines for measuring the performance of semiconductor equipment in a manufacturing environment. Some definitions from E10-96 are:

- *Assist*— an unplanned interruption that occurs during an equipment cycle where all three of the following conditions apply:
  * The interrupted equipment cycle is resumed through external intervention (e.g., by an operator or user, either human or host computer).
  * There is no replacement of a part, other than specified consumables.
  * There is no further variation from specifications of equipment operation..
- *Failure (equipment failure)*—any unplanned interruption or variance from the specifications of equipment operation other than assists.
- *Interrupt* - any assist or failure.
- *Maintenance* - the act of sustaining equipment in a condition to perform its intended function.
- *Cycle (equipment cycle)*—one complete operational sequence (including product load and unload) of processing, manufacturing, or testing steps for an equipment system or subsystem.
- *Total time*—all time (at the rate of 24 hours/day, 7 days/week) during the period being measured.
- *Non-scheduled time*—a period when the equipment is not scheduled to be utilized in production.
- *Operations time*—total time minus non-scheduled time.
- *Uptime (equipment uptime)*—the hours when the equipment is in a condition to perform its intended function.
- *Downtime (equipment downtime)*—the time when the equipment is not in a condition, or is not available, to perform its intended function.

Reliability, availability, and maintainability (RAM) are measures of equipment performance which have been used widely in industry for decades. Equipment performance (RAM) should be tracked with regard to time and/or equipment cycles. Equipment reliability concentrates on the relationship of equipment failures to equipment usage (productive time and equipment cycles). Some of the frequently used RAM statistics defined by the E10-96 standards are:

EQUIPMENT RELIABILITY—the probability that the equipment will perform its intended function, within stated conditions, for a specified period of time. The time when the equipment is performing its intended function is the only appropriate time to consider in equipment reliability calculations.

- *MTBF = Mean (Productive) Time Between Failures*—the average time the equipment performed its intended function between failures, given by productive time divided by the number of failures during that time.
- *MTBA = Mean (Productive) Time Between Assists*—the average time the equipment performed its intended function between assists, given by productive time divided by the number of assists during that time.

- *MTBI = Mean (Productive) Time Between Interrupts*—the average time the equipment performed its intended function between interrupts, given by productive time by the number of interrupts during that time.

EQUIPMENT AVAILABILITY—the probability that the equipment will be in a condition to perform its intended function when required.
- *Equipment Dependent Uptime*—the percent of time the equipment is in condition to perform its intended function during the period of operations time minus the sum of all maintenance delay, out-of-spec input downtime, and facilities-related downtime.
- *Supplier Dependent Uptime*—the percent of time the equipment is in a condition to perform its intended function during the period of operation time minus the sum of user maintenance delay, out-of-spec input downtime, and facilities-related downtime.
- *Operational Uptime*—the percent of time the equipment is in a condition to perform its intended function during the period of operations time.

EQUIPMENT MAINTAINABILITY—the probability that the equipment will be retained in, or restored to, a condition where it can perform its intended function, within a specified period of time.
- *MTTR = Mean Time To Repair*—the average time to correct a failure and return the equipment to a condition where it can perform its intended function.

EQUIPMENT UTILIZATION—the percent of time the equipment is performing its intended function during a specified time period.
- *Operational Utilization*—the percent of productive time during operations time.
- *Total Utilization*—the percentage of productive time during total time.

Assumptions that underlie use of an MTBF metric are that a failure is fixed when it occurs, that the tool generates randomly distributed failures, and that when a failure occurs neither the failure event nor the act of fixing it affect other components and subsystems. In real systems these assumptions aren't strictly true, and serious violation complicates the analysis.

As improvements are made to the system, increases in reliability may be modeled. For example, a *growth model* in MTBF can be used to describe the increase more precisely, and to estimate MTBF with confidence in periods of active improvement. E10-96 describes how to use the well known AMSAA model (or the Power Law Non-Homogeneous Poisson Process Model) for analyzing reliability growth data.

The *marathon run*, or *manufacturability demonstration*, is a simulation of a manufacturing environment to test the operation of manufacturing equipment and estimate its reliability by using it continuously. A marathon run is a structured procedure for data collection and analysis from which reliability, availability, and maintainability (RAM) statistics can be estimated. The results of a marathon can be used to prioritize use of resources to improve equipment reliability. A marathon run is the reliability test most frequently used at SEMATECH. Canning and Green (Chapter 26) provide an example of a marathon on a wafer stepper.

The *Cost Of Ownership (COO) model* employs utilization statistics to estimate the cost of processing one wafer on a given equipment system. The COO model allows a better understanding of manufacturing costs, and it can be used as a decision tool for purchasing equipment for manufacturing. The SEMI Standard E35 describes the COO model.

INTRODUCTION TO CASE STUDIES

This section gives a brief overview of the case studies related to equipment reliability that are presented in the remainder of this part.

Chapter 26. Marathon Report for a Photolithography Exposure Tool. *John T. Canning and Kent G. Green.* This case study describes a marathon run on a deep ultraviolet (DUV) wafer stepper during a SEMATECH Joint Development Program (JDP). It is estimated that the disciplined methodology of the JDP and the marathon saved over one year in the development cycle and $7 million in development costs. Once the stepper was installed and an acceptance test was performed, a five-day passive data collection was performed to test the new tool's stability, and a three-week marathon test in a manufacturing environment was performed to estimate production performance. The SEMATECH marathon test plan was designed to consider a process-related response (test stepper overlay on three product level film stacks), and equipment-related responses (reliability and productivity). Analysis of the data indicated the need to reduce software failures, improve throughput, and reduce downtime. The results gave a clear understanding of the required technical and operational improvements, and the supplier implemented a corrective action plan to address them.

The marathon data on wafer registration overlay was collected as follows. The stepper was used continuously to align and expose wafers with three different film stacks 24 hours a day, five days a week, for three weeks. Wafer registration was measured at 130 sites per wafer. Reliability information was collected in accordance with Semi E10-90 standards.

Analysis of wafer overlay for each one of the three film types (as in a passive data collection, Part 2), included:
- *capability analysis* using c_{pk} values and capability plots, where histograms of all the data collected during the marathon were compared to the specification limits. The observations were expected to represent typical data distributions encountered in a manufacturing environment.
- *analysis of variance (ANOVA)* to quantify the sources of variability, in this case run-to-run, wafer-to-wafer, field-to-field, and site-to-site. The major contributors were targeted for reduction.
- *box plots* to help visualize the components of variance.

Analysis of productivity and reliability used the three film types simultaneously and included:
- *estimation* of productivity (tool throughput, operational utilization, mean number of wafers per day, utilization capability, productivity capability), and reliability (hardware mean time between failures, software mean time between failures, mean time to repair, mean time between assists, mean wafers between interrupts, supplier dependent uptime, equipment dependent uptime).
- *pareto graphs* to identify major reasons for assist failures, given an unexpectedly high assist rate.

Chapter 27. Experimentation for Equipment Reliability Improvement. *Donald K. Lewis, Craig Hutchens, and Joseph M. Smith.* Semiconductor equipment must perform reliably in different customer environments. This case study illustrates the use of designed experiments to improve reliability in the robotic wafer transfer and alignment sub-system of an automated memory repair system at a customer site. It resulted in a reduction of the average number of wafer handling assists per 1000 system hours from seven to zero. The equipment presented a serious reliability problem in the operating environment of a particular customer. The problem was traced to assists that were attributed to the manner in which the customer operated the equipment. The urgent need to solve the problem made it evident that it would be necessary to use a structured improvement methodology that would provide valid results. A robust design experiment was conducted to improve the operation of the wafer handling sub-system in order to improve its performance and avoid the effects of environmental factors that reflected usage of the system at the customer site.

The case study describes a fundamentally different assessment of Mean Time Between Failures (MTBF) by an equipment supplier and a Japanese customer. After looking at the customer log book, the equipment supplier traced the difference to the fact that the customer's definition of failure included assists. The robust design experiment was planned to reduce the number of assists.

Chapter 28. How to Determine Component-Based Preventive Maintenance Plans. *Stephen V. Crowder.* Preventive maintenance (PM) can often increase equipment availability by replacing or restoring components subject to failure before they fail, instead of by fixing them after they break as is often done in American industry. This case study provides guidance in establishing preventive maintenance plans for manufacturing equipment to maximize equipment availability. For preventive maintenance to be beneficial, unexpected or emergency maintenance of a component must be more expensive than the scheduled preventive maintenance. This will often be the case in the semiconductor industry, since unexpected equipment failure will result in costly process downtime for repairs, while scheduled PM's may result in less process downtime. Preventive maintenance policies that maximize equipment availability can be established using knowledge of past equipment performance and relative times required for scheduled and unscheduled maintenance actions. Two preventive maintenance policies are considered: age replacement policies in which a component is replaced at some fixed age if it has not failed, and condition-based policies in which a degradation variable is monitored and a component is replaced when this variable crosses a specified threshold. The condition-based policy attempts to provide preventive maintenance only when necessary, but requires more equipment or component knowledge than the age replacement policy. These policies are illustrated with two examples, one on mass flow controller failure data, and the other on ion implanter source filament degradation. Both models are based on simple, but useful, statistical models.

SUGGESTED READING

All SEMATECH statistical publications that are available to the general public can be accessed at the following Internet URL: http://www.sematech.org/

Some books on reliability are:

O'Connor, Patrick D.T. *Practical Reliability Engineering*, third edition, New York: John Wiley and Sons, Inc., 1991.

RADC Reliability Engineers Toolkit: An Application Oriented Guide for the Practicing Reliability Engineer, Griffiss Air Force Base, New York: Systems Reliability and Engineering Division, Rome Air Development Center, Air Force Systems Command, July 1988.

Kececioglu, Dimitri. *Reliability Engineering Handbook*, Vols. 1 and 2. New Jersey: Prentice Hall, 1991.

A Reliability Guide to Failure Reporting, Analysis, and Corrective Action Systems, Milwaukee, WS: American Society for Quality Control, 1977.

Tobias, Paul A. and David C. Trindade. *Applied Reliability, Second Edition,* New York: Van Nostrand Reinhold, 1995.

Klinger, David, Yoshinao Nakada, and Maria A. Menendez. *AT&T Reliability Manual*, New York: Van Nostrand Reinhold, 1990.

Asher, Harry and H. Feingold. *Repairable Systems Reliability Modeling, Inference, Misconceptions and Their Causes*, New York: Marcel Dekker, Inc., 1984.

Ireson, W. Grant and Clyde F. Coombs, Jr. editors. *Handbook of Reliability Engineering and Management*, New York: McGraw-Hill, Inc., 1988.

Semiconductor Equipment and Materials International (SEMI) standards may be ordered from the SEMI Standards Web page accessed from the URL http://www.semi.org/. The standards referred to in this chapter are:

SEMI E10-96, Standard for Definition and Measurement of Equipment Reliability, Availability, and Maintainability (RAM), SEMI 1996.

SEMI E35, Cost of Ownership for Semiconductor Manufacturing Equipment Metrics, SEMI 1995.

Government military standards for reliability can be obtained from the Department of Defense, Washington, D.C. Some government standards are:

MIL-HDBK-217E. *Reliability Prediction of Electronic Equipment*, Griffiss Air Force Base, New York: Rome Air Development Center, October 1986.

MIL-HDBK-189, *Reliability Growth Management*.

MIL-STD-756B, *Reliability Modeling and Prediction*.

MIL-STD-1629A, *Procedures for Performing a Failure Mode, Effects, and Criticality Analysis*.

MIL-STD-1635(EC), *Reliability Growth Testing*.

CHAPTER 26

MARATHON REPORT FOR A PHOTOLITHOGRAPHY EXPOSURE TOOL

John T. Canning and Kent G. Green

This chapter presents the purpose, procedure and results of a marathon test performed on a wafer stepper developed as a part of the SEMATECH joint development program (JDP) with GCA, a unit of General Signal Corporation. The marathon was designed to test three key aspects of the exposure tool: overlay, productivity, and reliability. The data from the marathon was analyzed to identify possible technical and operational improvements. It is estimated that the focus of the JDP and SEMATECH Qualification Plan (Qual Plan) marathon process saved over one year in the development cycle and $7 million in development costs.

EXECUTIVE SUMMARY

Problem

The purpose of the joint development program (JDP) with GCA, a unit of General Signal, was to accelerate the development of domestic world-class steppers at both I-line and deep ultraviolet (DUV) wavelengths for SEMATECH Phase 2 and 3 applications (0.6-0.35 µm geometries). The goals of the JDP required that the DUV stepper meet Phase 3 overlay requirements on specific product level film stacks, and that the stepper exceed the productivity and reliability requirements. The initial challenge was to design a stepper that could meet these aggressive requirements. The next challenge was to devise a means to collect and analyze appropriate data to verify that the stepper met the JDP requirements and to identify improvements. The final challenge was to incorporate these improvements in the stepper for commercialization.

Solution Strategy

GCA met the design challenge by developing the XLS, a new-generation stepper featuring wide field lenses at both I-line and DUV wavelengths, a high performance XYZ stage with a six-axis interferometer system and a new system architecture with improved internal metrology (see Figure 26.1). The first prototype XLS DUV stepper (system #2104) incorporating these design features was delivered to SEMATECH in March 1991.[62]

SEMATECH performed an acceptance test, a passive data collection, and a marathon on the prototype stepper to identify and address any technical or operational shortfalls. GCA used the data from these tests to incorporate improvements in the commercial XLS stepper, thus addressing the final challenge.

The commercial XLS DUV Stepper (system #2105) was delivered to SEMATECH in January, 1992. This stepper went through the following qualification plan.

1. a three-day acceptance test procedure (ATP) that covered major tool specifications
2. a five day passive data collection (PDC) test in which no changes were made to the tool during the data collection period in order to test the stability of the tool
3. a three week marathon test in a manufacturing environment to determine production performance.

This case study considers the results of the three-week marathon on the commercial stepper. The marathon test addressed three major areas of tool production performance in a manufacturing environment:

1. overlay on product level film stacks
2. productivity capability
3. tool reliability.

To meet project timelines, the marathon test was run for three weeks. This was ample time to provide sufficient data for overlay and productivity. To achieve higher confidence in the reliability data, a six week marathon would have been preferred. The key results are shown in Table 26.1.

Conclusions

The conclusions drawn from the results of the three-week marathon on the commercial XLS DUV system #2105 are as follows:

Overlay
- The SEMATECH Phase 3 overlay requirements were met on all levels except Metal 2/Via. GCA developed an alternative brightfield alignment system that resolved this problem. It also improved the throughput on this level.

Productivity
- The tool throughput criteria was achieved, and the average productivity (wafers per day) was limited by the operational methodology.

---

[62] See SEMATECH Statement of Work #89060049C-MEM.

Table 26.1. *GCA XLS DUV Stepper Key Marathon Results. The SEMATECH Phase 3 overlay requirements were met on all levels except Metal 2/Via. GCA developed a brightfield alignment system to correct this problem. Tool throughput met requirements, and the average productivity was limited by the 70% utilization. The software reliability was well below goals, and GCA corrected the error sources with a software revision. Supplier dependent uptime and mean time to repair exceeded the goals due to fast recovery from the software failures.*

Parameter	Goals	Results
Overlay (Mean+3σ)		
Contact/Poly	120 nm	110 nm
Via/Metal 1	150 nm	133 nm
Metal 2	150 nm	178 nm
Productivity		
Tool Throughput (200 mm)	35 wafers per hour	38 wafers per hour
Operational Utilization	NA	70%
Wafer Per Day (Average)	NA	472
Reliability		
Hardware Mean Time Between Fail	329 hours	329 hours
Software Mean Time Between Fail	329 hours	82 hours
Mean Time to Repair	5	2
Supplier Dependent Uptime	90%	95%

Reliability
- The failure rate did not meet the specification due to software. GCA understood these problems, and resolved them with Version 1.2 software.
- Supplier dependent uptime and mean time to repair exceeded the JDP specification due to fast recovery from the software failures.

In summary, the SEMATECH qualification plan and statistical analysis of data are excellent methodologies for verifying that semiconductor equipment meet manufacturing requirements. In this case study, a clear understanding of required improvements was demonstrated, and a supplier corrective action plan was implemented.[63]

PROCESS DESCRIPTION

Figure 26.1 illustrates the DUV stepper system configuration that was used for the JDP.

The photolithography process is critical to the fabrication of a working integrated circuit. In this process, a photoactive chemical (resist) is spin-coated onto a silicon wafer. The wafer is then aligned and exposed with a light source in the exposure tool. Finally, the wafer is developed, leaving the wafer surface exposed in some areas and covered with resist in others (see Figure 26.2). The purpose of this process is to allow the implant or deposition of dopants into designated areas of the wafer surface or to define metal interconnect and isolation regions by etching. To save money in the marathon process, these downstream processes were eliminated, and the wafers were reused by stripping the exposed resist and recoating for subsequent align/expose cycles.

---

[63] GCA went out of business on May 2, 1993 due to insufficient demand for their products.

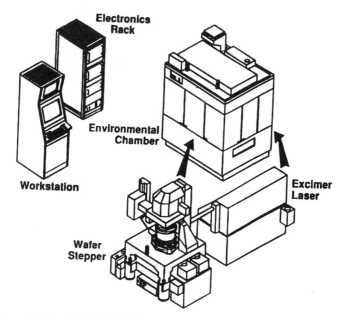

Figure 26.1. *GCA XLS DUV Stepper System Configuration. The main system unit is the wafer stepper which contains the optical system (the 4X reduction lens and the illuminator) for transferring the reticle pattern onto the wafer and the wafer stage which moves and aligns the wafer for each exposure. The environmental chamber controls the system temperature and directs air flow through the stepper. The Excimer laser provides the DUV light source for exposure. The electronics rack houses all the control electronics, and the workstation contains the system computer and the operator interface.*

The photolithography process is repeated several times in the building of an integrated circuit, and it is critical that each level properly overlay to the previous level. It is also imperative that the geometries be within predetermined critical dimension (CD) specifications. Both of these aspects of the photolithography process are determined primarily by the capability of the exposure tool.

One purpose of this marathon was to determine the overlay (alignment) capability of the XLS system on production film stacks. Three different film stacks were used: Contact/Poly, Via/Metal 1 and Metal 2/Via (see Figure 26.3). The film stacks were produced by short loop processing to the required film topology and reflectivities of full flow wafers, i.e., the Metal 2/Via stack is produced by putting a blanket Metal 1 film on the silicon wafer, patterning and etching the Metal 1 followed by a blanket dielectric layer which is patterned and etched to form the vias and then the Metal 2 film is deposited to complete the film stack.

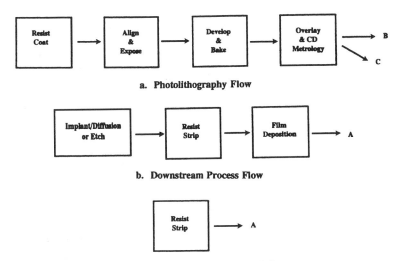

a. Photolithography Flow

b. Downstream Process Flow

c. Rework Cycle for Marathon Wafers

Figure 26.2. *Generic Photolithography Process Flow. In a, the wafers are resist coated, aligned/exposed and developed/baked. Then a sample is measured for overlay and CD control. In b, after the photo process, the wafers go to implant/diffusion or etch to define specific structures. Next, the resist is stripped and another film is deposited in preparing the wafers for another loop through the photoprocess. Part c depicts the rework process used in the marathon to save time and money by recycling wafers for each level.*

The XLS system aligns wafers in the following way:
1. The reticle management system (RMS) selects the chosen reticle from the RMS library and loads, aligns, and secures the reticle on the platen chuck of the lens with a vacuum for use during wafer exposure.
2. The automatic wafer handler (AWH) carries the wafer from the send carrier to the pre-align chuck, where the major flat (or notch) of the wafer is pre-aligned. After pre-alignment, the AWH moves the wafer from the pre-align station to the wafer stage where the wafer is loaded and secured to the wafer chuck with the vacuum.
3. The wafer stage moves to position the wafer under the reduction lens, and the entire wafer is leveled and brought into focus.
4. A wafer exposed at a previous level contains two types of alignment marks, global and local, which are used to align the wafer to the XLS system. Once the wafer has been leveled, the automatic wafer aligner (AWA) performs a two point global alignment which aligns the wafer in preparation for local alignment.

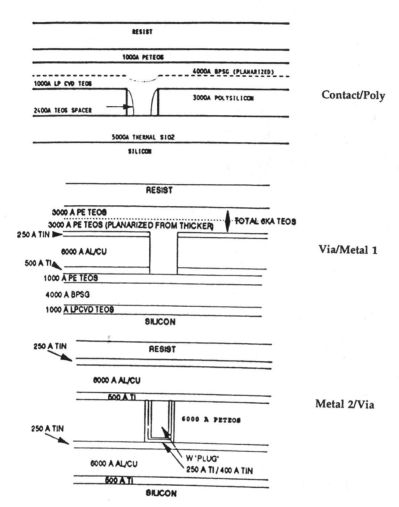

Figure 26.3. *Product Level Film Stacks. The film stacks are produced by short loop processing that produces film topologies and reflectivities equivalent to full flow wafers. As an example, the Metal 2/Via stack is produced by putting a blanket Metal 1 film on the silicon wafer. Patterning and etching of the Metal 1 is followed by a blanket dielectric layer, which is patterned and etched to form the vias. Then the Metal 2 film is deposited to complete the film stack.*

5. After global alignment takes place, user specified alignment sites (typically ten exposure fields randomly selected across the wafer) are measured to determine precise alignment data using the micro dark field alignment system (DFAS). The alignment software uses micro DFAS as a measuring device to calibrate grid errors (translation, scale, orthogonality and rotation errors) for the wafer being measured and adjusts the XLS system calibration to accommodate the wafer. After the measurement corrections have been applied, the wafer is exposed.
6. Once the wafer is exposed, the stage moves to the load position and the AWH removes the exposed wafer from the wafer chuck, loads a new wafer onto the wafer chuck, and then places the exposed wafer in the receive cassette of the AWH. This process is repeated for every wafer in the send cassette of the AWH.

DATA COLLECTION PLAN

For the three-week marathon test, a 24-hour per day, five-day per week schedule was followed. During the available tool time, the tool was continually used by manufacturing to align and expose the three production film stacks sequentially in lots of 25 wafers each (i.e. 25 wafers of stack A, 25 of stack B, 25 of stack C, 25 of stack A, 25 of stack B, etc.). Full-flow production lots were also exposed when available. The sampling plan called for the first and 25th wafer from the first lot of each of the three production film stacks to be measured for overlay on each shift. Each wafer was measured at thirteen fields radially selected around the center of the wafer. Each field was measured at the center and four corners in both the x and y-axis for a total of 130 registration measurements per wafer. In practice, about 90% of the wafers were actually measured. The aggregate of these measurements was then used to determine the overlay capability of the tool.

Productivity data was collected from all the wafers run during each shift: short flow production film stack wafers, full-flow production wafers, and SPC test wafers. The throughput data was collected in the stepper's activity log. It is dependent on the number of fields exposed and the exposure time per field for each product level. The operational utilization is based on data from the steppers activity log, the manufacturing system log (Comets) and an independent manual log maintained by the operators during the marathon. Comets and the manual logs accounted for job set-up time, engineering use, and other non-productive activities. The reliability data was also collected in accordance with Semi E10-90 standards by all three logs to minimize disagreements between manufacturing, engineering and the supplier.

DATA ANALYSIS AND INTERPRETATION OF RESULTS

Typically, photolithography exposure tool overlay performance is characterized and evaluated by several methods of data analysis. The first method is the capability analysis which is an overall bottom line evaluation of all the data from the marathon test. Figures 26.4– 26.6 are capability plots generated for each product level. They are the distribution of all data collected in the marathon and represent the drift and variation expected in a long term manufacturing environment. Of prime interest is the ability of a tool and process to consistently meet the product specifications. The measure of this is called the process capability index or C_{pk} and is the ratio of the process specification divided by the process variation. Values greater than 1.0 are acceptable, and 1.33 is a common target.

In Figure 26.4, the Contact/Poly C_{pk} is 1.35 in the X axis and 1.10 in the Y axis, meeting the requirement that C_{pk} be greater than one. Likewise, in Figure 26.5 for the Via/Metal 1 level, the C_{pk} values are 1.48 and 1.13. The Metal 2/Via performance in Figure 26.6 did not meet the requirement with C_{pk} values of 0.84 and 0.86. The causes of this poor performance are the inherent difficulty of the alignment system to detect alignment marks on metal films and the unusually poor quality metal films used in this marathon (It is analyzed further in subsequent sections).

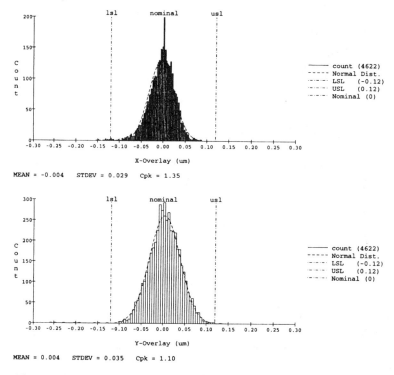

Figure 26.4. *Capability Plots of Contact/Poly Overlay. The Cpk is 1.35 in X and 1.10 in Y, meeting manufacturing requirements for SEMATECH phase 3 product. They represent the variation expected in a manufacturing environment.*

The next method of analysis is to perform an analysis of variance (ANOVA) on the data for each product level. This analysis determines the variance for the major contributors of variability to the stepper and the tool/process performance. In this method, the variance is determined for site-to-site, field-to-field, wafer-to-wafer and run-to-run sources of variability. Site-to-site is the variation across one exposure field of the stepper (also called intrafield error) and is caused primarily by lens magnification errors, reticle rotation, reticle manufacturing errors, and wafer alignment. Wafer alignment includes alignment mark detection, stage scaling, and orthogonality and translation errors. Field-to-field is variation across one wafer and is caused by wafer-stage repeatability and wafer-process induced distortions. Wafer-to-wafer variance is a measure of the short term stability of the stepper (normally one eight-hour shift). In the case of the XLS, an internal metrology system calibrates and resets the system "baseline" (lens magnification, reticle rotation, stage scaling, and orthogonality) at the beginning of each shift. Run-to-run is a measure of the long term stability of the tool and process. In this case, it includes the shift-to-shift and the day-to-day drift and variation.

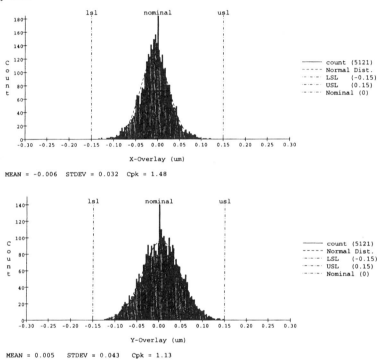

Figure 26.5. *Capability Plots of Via/Metal Overlay. The Cpk is 1.48 in X and 1.13 in Y, meeting manufacturing requirements for SEMATECH phase 3 product. They represent the variation expected in a manufacturing environment.*

Table 26.2 shows the result of the ANOVA for this marathon. For the Contact/Poly and Via/Metal 1 levels where the Cpk met manufacturing requirements, the predominant sources of variation are site-to-site (~48%) and run-to-run (~28%). The site-to-site variations for these two levels are shown in Figures 26.7 and 26.8. Site C is the center of the field and TL, TR, BL, and BR are the top left, top right, bottom left and bottom right corners respectively. Examining the mean values (see Figure 26.9) shows some reticle rotation on the Contact/Poly level but not on the other two levels. This data indicates no systematic errors in reticle rotation due to the stepper or in lens magnification. Normal reticle manufacturing errors are in the range of 20 - 30 nm and are systematic, i.e. they directly add to these mean values. In the Contact/Poly level, the rotation could be caused by misplaced alignment marks on the reticle. In the other levels, the mean errors could be attributed to random reticle errors. The interquartile ranges (IQR) for sites TL, TR, BL and BR are not significantly different from site C, also indicating no large random lens magnification or reticle rotation errors. These IQRs (except the Y-axis Via/Metal 1) are consistent with good wafer alignment performance.

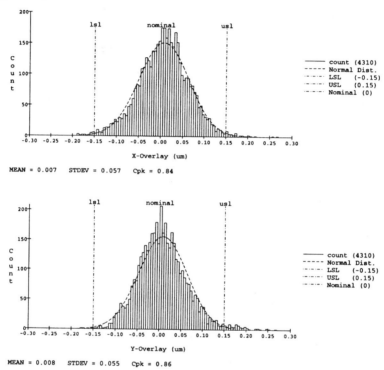

Figure 26.6. *Capability Plots of Metal 2/Via Overlay. The Cpk is 0.84 in X and 0.86 in Y, <u>not</u> meeting manufacturing requirements for SEMATECH phase 3 product. The causes of this poor performance are the inherent difficulty of the alignment system to detect alignment marks on metal films and the unusually poor quality metal films used in this marathon.*

The run-to-run variations for Contact/Poly and Via/Metal 1 levels are shown in Figures 26.10 and 26.11. The mean ± 3 sigma variation for the first and 25th wafer in each lot is plotted for each shift. The means are fairly well distributed around zero indicating good calibration of the tool and normal tool/process drift due to slight changes in temperature and pressure. The variation appears to increase in the last week of the test. These increases may have been due to the effectiveness of the resist strip process used in recycling the wafers. Subsequent experience in running daily SPC wafers demonstrated non-uniform build-up of resist at the alignment sites due to incomplete resist strip. It manifested itself as a degradation of overlay due to incorrect measurement of the alignment mark.

Table 26.2. *XLS Stepper Overlay Analysis of Variance. For the Contact/Poly and Via/Metal 1 levels, the predominant sources of variation are site-to-site (~48% due to reticle error) and wafer alignment and run-to-run (~28% due to tool/process drift over time and incorrect wafer alignment from a resist build-up on the recycled wafers). The low values for field-to-field and run-to-run indicate good stage repeatability and short term stepper stability. For the Metal 2/Via level, the major source of variance is field-to-field (~39% due to poor compensation for wafer process scaling errors when the alignment system selected sites in close proximity to each other to handle the > 30% film reflectivity on these wafers). The next major source is site-to-site (~31% due to the inherent difficulty of the alignment system to align to metal films compounded by the poor quality fields). GCA developed a brightfield alignment system to correct this problem.*

Source of Variance	Percent of Variance					
	Contact/Poly		Via/Metal		Metal 2/Via	
	X	Y	X	Y	X	Y
Run to Run	31.5	30.9	18.1	29.6	0.0	23.3
Wafer to Wafer	12.0	16.7	9.2	11.8	20.7	16.9
Field to Field	8.5	14.6	11.5	11.8	52.9	24.5
Site to Site	48.0	37.8	61.2	46.7	26.5	35.3

Also, there is a very large excursion at the date 0902-B1 which was attributed to an evacuation in the fab during that shift due to a chemical spill. Air flow to the bay was increased to clear chemical fumes, and it is unknown if there were temperature fluctuations during this time that may have caused some drift on the tool. In addition, only 33 points (versus 65) were read by the metrology tool (KLA 5015), suggesting that it was having problems reading the wafers on that day.

The low values for field-to-field variance on Contact/Poly and Via/Metal 1 are indicative of the very good short term stage repeatability ($X = 24$ nm and $Y = 29$ nm, 3 sigma) and low process induced distortion. The low values for wafer-to-wafer also indicate very good short term stepper stability.

The ANOVA (See Table 26.2) on the Metal 2/Via level showed a large field-to-field variation (~39%, average X, Y) along with a ~31% site-to-site variation. This large field-to-field is attributed to the poor quality of the Metal 2 films used in this marathon. Normally, the local alignment sites are randomly selected across the entire wafer. In this case, the metal film reflectivity and grain size variation (> 30% across the wafer) caused the alignment sites to be located close together in the bottom half of the wafer. The result was poor compensation for wafer-process-induced scaling errors manifesting itself in large field-to-field variation.

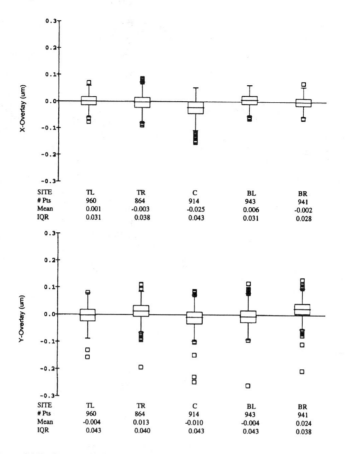

Figure 26.7. *Contact/Poly Overlay Site-to-Site Variation. Exam-ination of the mean values (plotted in Figure 26.9) shows some reticle rotation attributed to misplaced alignment marks on the reticle. The Interquartile ranges (IQR) for sites TL, TR, BL, and BR are not significantly different from that for site C, indicating no large random lens magnification or reticle rotation errors.*

The large IQR (see Figure 26.12) on Metal 2/Via site-to-site variation was attributed to the inherent difficulty of the alignment system to align to metal levels. In the darkfield alignment system used on the XLS, the grain boundaries of a metal film create noise in the alignment detection system which makes it difficult for the stepper to determine the true location of the alignment mark. The > 30% film variations compounded the problem and led to incorrect alignment adjustments and poor overlay. Based on this data, GCA developed a brightfield alignment system which was less sensitive to noise on metal levels. Beta site testing at other customer sites confirmed the improved performance equivalent to dielectric levels.

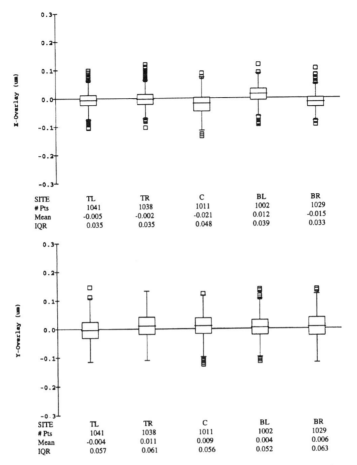

Figure 26.8. *Via/Metal Overlay Site-to-Site Variation. The mean values as plotted in Figure 26.9 are random and are attributed to random reticle errors. Within each axis, the IQRs are relatively equivalent, indicating no large random lens magnification or reticle rotation errors. The Y IQR is significantly larger than the X IQR, suggesting a Y-axis alignment issue.*

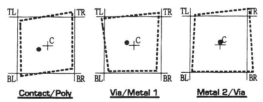

Figure 26.9. *Mean Value Plots for Site-to-Site Variation. The mean values for contact/poly show some reticle rotation attributed to misplaced alignment marks on the reticle. The mean values for via/metal 1 and metal 2/via are random and attributed to random reticle errors at these sites.*

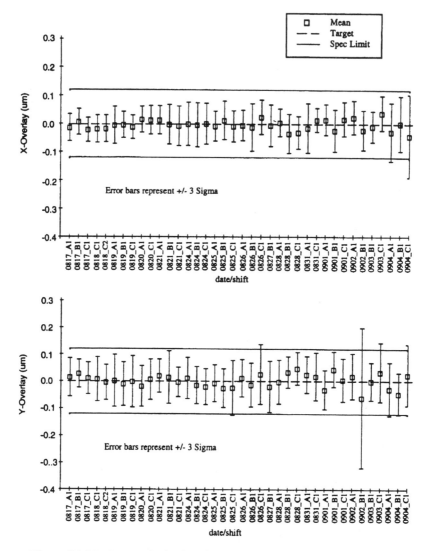

Figure 26.10. *Contact/Poly Overlay Run-to-Run Variation. The mean values are fairly well distributed around zero, indicating good tool calibration and normal tool/process drift due to slight changes in temperature and pressure. The variation increases in the last week of the test and is attributed to incorrect measurement of the alignment mark due to resist build-up from incomplete resist strip in the wafer recycling process. The large excursion at date 0902_B1 was attributed to increased air flow at the tool due to a fab evacuation.*

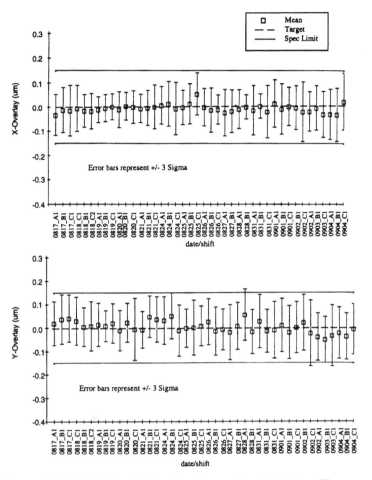

Figure 26.11. *Via/Metal 1 Overlay Run-to-Run Variation. The mean values are fairly well distributed around zero, indicating good tool calibration and normal tool/process drift due to slight changes in temperature and pressure. The variation increases in the last week of the test and is attributed to incorrect measurement of the alignment mark due to resist build-up from incomplete resist strip in the wafer recycling process. The individual run variation is greater in the Y-axis as seen in Figure 26.8, suggesting a Y-axis alignment issue.*

Figure 26.13 also clearly demonstrates the issue with metal level alignment. The mean shift appears to increase in the last week of the test suggesting a problem with resist stripping on this level also.

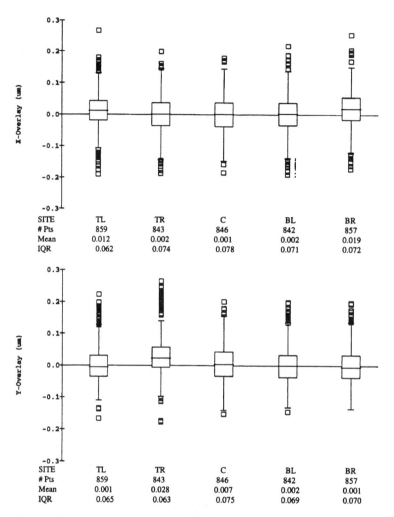

Figure 26.12. *Metal 2/Via Overlay Site-to-Site Variation. As plotted in Figure 26.9, the mean values are random and attributed to random reticle errors. The IQRs are relatively equivalent, indicating no large random lens magnification or reticle rotation errors. The large IQRs are attributed to the inherent difficulty of the alignment system to align to metal levels and the poor quality (>30% reflectivity variation) of the metal films.*

Table 26.3 shows the marathon productivity results. The tool throughput goals were based on the JDP specification of 88 exposure fields and 0.2 sec/exposure (80 laser pulses at 400 Hz). It also included a step and settle time per exposure of 0.5 sec., an alignment/load overhead of 30 seconds per wafer and a 10% contingency. The production throughput goals were based on 93 exposure fields and an actual exposure time of 0.47 seconds. The actual results were comparable to the goals (except for Metal 2/Via), which indicates that the throughput parameters were well understood and predictable. The Metal 2/Via was under target due to the excessive time needed to align to the poor quality metal film.

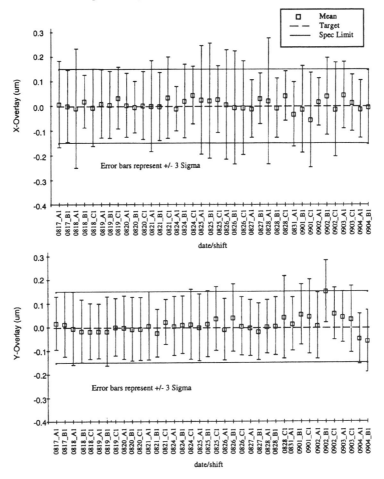

Figure 26.13. *Metal 2/Via Overlay Run-to-Run Variation. The mean values are fairly well distributed around zero, indicating good tool calibration and normal tool/process drift due to slight changes in temperature and pressure. The variation increases in the last week of the test and is attributed to incorrect measurement of the alignment mark due to resist buildup from incomplete resist strip in the wafer recycling process. The large excursion at date 0902_B1 was attributed to increased air flow at the tool due to a fab evacuation. The random run-to-run variation is significantly greater in the X-axis, suggesting an unique issue with X-axis alignment.*

Table 26.3. *XLS Stepper Marathon Productivity Results. Throughput results were comparable to the goals (except for Metal 2/Via), indicating that throughput parameters are well understood. The Metal 2/Via required excessive time to align to the poor quality metal film. A total of 6,839 wafers were run averaging 456 wafers per day. The operational utilization was about 70% due mainly to job set-up and engineering use. Productivity capability is 620 wafers per day.*

Parameter	Goals	Results
Throughput		
Tool Throughput	35 wafers per hour	38 wafers per hour
Product Throughput		
Contact/Poly	30 wafers per hour	31 wafers per hour
Via/Metal 1	30 wafers per hour	31 wafers per hour
Metal 2/Via	30 wafers per hour	26 wafers per hour
Productivity		
Total Wafers Run	NA	6,839
Wafers Per Day (Average)	NA	456
Operational Utilization	NA	69.8%
Utilization Capability	≥ 90%	94.9%
Productivity Capability	NA	620 wafers per day

Table 26.4. *XLS Stepper Marathon Reliability Results. The software reliability was well below the goals, and the error sources were corrected in the Version 1.2 software. The assist rate was unexpectedly high due mainly to Z-stage, AWH, and X-stage intermittent hardware problems. They were corrected during and after the marathon. The resultant mean-wafers-between-interrupt was quite low due to the high assist rate. The mean-time-to-repair and the supplier-dependent uptime exceeded the goals due to fast recovery from the software failures.*

Parameter	Goal	Results
Hardware Failures	--	1
Software Failures	--	4
Hardware Mean Time Between Fail	329 hours	329 hours
Software Mean Time Between Fail	329 hours	82 hours
Mean Time to Repair	5 hours	2 hours
Assists	--	62
Mean Time Between Assists	--	5.3 hours
Mean Wafers Between Interrupt	--	102 wafers
Supplier Dependent Uptime	90%	94.9%
Equipment Dependent Uptime	95%	97.2%

A total of 6,839 wafers were run during the 15 day marathon, averaging 456 wafers per day. The operational utilization was 69.8%, meaning that the tool was actually stepping wafers about 70% of the total operational time (24 hours x 15 days). The utilization capability was 94.9% (i.e., the percentage of time the tool could have been stepping wafers if product was run 100% of the time that the tool was available. Productivity capability was 620 wafers per day (i.e., the number of wafers per day that could have been produced at full utilization capability). The unproductive time was attributed to job set-up time, engineering use, unplanned maintenance, and waiting test results.

Table 26.4 summarizes the marathon reliability results. There was one hardware failure. Z-stage cables which worked intermittently (contributing to 22 assists) failed on the last day of the test. Four software failures each required a reboot of the control system. The low mean-time-to-repair of two hours was the result of fast recovery from the software failures.

The assist rate was unexpectedly high and a pareto chart is shown in Figure 26.14. The 22 TZF assists were attributed to faulty Z-stage piezo cables and limited Z travel range. They were corrected at the end of the test by replacing the piezo cables and adjusting the Z-motors to provide a full range of travel. The 20 AWH assists were largely resolved after the marathon with a hardware upgrade and a complete preventive maintenance (PM). A parallel beta site test of the AWH hardware upgrade demonstrated a 6X improvement in assist rate. The 13 XY-stage assists were caused by a faulty X-stage motor and were eliminated when the motor was replaced during a PM. The 6 RMS assists were caused by aligner timeouts and were to have been addressed with new RMS firmware in 1Q93. The single µDFAS assist had no assignable cause.

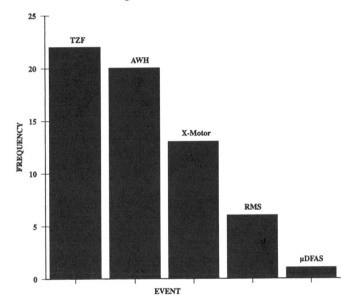

Figure 26.14. *XLS Marathon Assist Pareto Chart from Reliability Results. The assist rate was unexpectedly high. The three major contributors were 22 TZF errors due to faulty Z-stage piezo cables and limited Z travel, 20 AWH errors due to hardware and set-up errors, and 13 XY-stage errors due to a faulty X-stage motor.*

The resultant mean wafer between interrupt (MWBI) of 102 wafers was quite low due to the high assist rate. Overall supplier dependent uptime of 94.9% was quite high due to fast recovery from the failures and good maintenance procedures.

GCA used the results from this marathon and other customer tests to prioritize their engineering resources on product overlay, software reliability, AWH and RMS upgrades, and focus control improvements. The marathon test procedure provided a very disciplined and critical measure of the stepper performance and readiness for production. The first XLS marathon was completed in October 1991 on the prototype i-line system and identified overlay performance (150-350 nm) and reliability (MWBI = 15) as the key issues. In less than 10 months, the marathon was completed on the commercial system which incorporated significant hardware and software improvements. It is estimated that the focus of the JDP and Qual Plan marathon process saved over one year in the development cycle and $7 million in development costs.

Conclusions and Recommendations

The photolithography exposure tool program presented the challenges of design, evaluation and improvement. The SEMATECH Qualification Plan, with corresponding statistical data analysis, is an excellent methodology for evaluating exposure tool performance and identifying required improvements. In this case study, a clear understanding of required improvements was demonstrated and a supplier corrective action plan was implemented. This process is recommended for all equipment programs.

Specifically, the 3-week marathon demonstrated that the overlay capability of the XLS stepper was sufficient for Phase 3 dielectric films and inadequate for metal films. The process capability index (C_{pk}) was 1.1 for contact and via levels and 0.8 for Metal 2. The analysis of variance (ANOVA) showed that the predominant sources of variation were site-to-site (~48%) and run-to-run (~28%) for dielectric levels. It showed the significance of reticle errors on site-to-site variance and the need for stable, well characterized marathon processes. This latter point was demonstrated by the effects of resist strip on alignment of recycled wafers in the run-to-run variance and the effect of poor quality metal films on the field-to-field variance. The large with-in site variations (IQR) on the Metal 2/Via clearly demonstrated the weakness of the dark field alignment system on metal films. The ANOVA also demonstrated the relatively good short term stability and stage repeatability of the stepper. These results led GCA to develop an alternate bright field alignment system, evaluate the impact of reticle errors on alignment off-sets, and develop software for local alignment using two points per field. These improvements were successfully demonstrated in beta site testing at other customer sites.

The productivity statistics reinforced the need to improve throughput and minimize downtime. As part of the overlay improvement, GCA optimized the alignment marks to reduce job setup time. The reliability data confirmed the need to drive software failures to zero, to improve stage cabling, and to implement the AWH and RMS improvement. GCA had projects in all these areas and was on a path to implement them when the company went out of business.

In summary, the SEMATECH marathon test plan provided a very disciplined and critical measure of stepper performance and readiness for production. A clear understanding of required improvements was demonstrated, and a supplier corrective action plan was implemented. The focus of the JDP and Qual Plan process is estimated to have saved over one year in the development cycle and $7 million in development costs. This process is recommended for all equipment programs.

REFERENCES

SEMATECH. *SEMATECH Qualification Plan Overview*, 91050538B.GEN, Austin, TX: SEMATECH, 1993.

SEMATECH, *GCA XLS Stepper Joint Development Contract Statement of Work*, 89060049C, MEM, Austin, TX: SEMATECH.

SEMI, E 10-92. *Guidelines for Definition and Measurement of Equipment Reliability, Availability, and Maintainability (RAM)*.

BIOGRAPHIES

John T. Canning is manager of advanced lithography at SEMATECH. Prior to this assignment, he managed the GCA stepper joint development program and the lithography start-up team as an IBM assignee to SEMATECH. John's previous experience at IBM includes strategic planning, mask equipment development, and semi-conductor packaging. He has Master of Engineering and Bachelor of Engineering degrees in mechanical engineering from Stevens Institute of Technology.

Kent G. Green received a Bachelor of Science degree in chemical engineering from Kansas State University in 1987. He started working for Texas Instruments (TI) the same year as a photolithography engineer. Kent joined the SEMATECH photolithography department as a TI assignee in August, 1991. His work experience has included g-line and deep ultraviolet (DUV) photolithography processing. His statistical background includes undergraduate and graduate level courses in statistics as well as several work sponsored courses on statistics and the design of experiments.

CHAPTER 27

EXPERIMENTATION FOR EQUIPMENT RELIABILITY IMPROVEMENT

Donald K. Lewis, Craig Hutchens, and Joseph M. Smith

To be fit for use by a customer, semiconductor process and test equipment must perform reliably under real-world sources of variation in wafer parameters, setup modalities, and environmental conditions. Using design of experiments methodology, an experiment was conducted on the robotic wafer transfer and alignment sub-system of an ESI Model 9000 automated memory repair system. The experiment identified design factors related to a serious reliability problem in the operating environment of a particular customer. The results demonstrated how the wafer handling sub-system could be improved to avoid the effects of environmental factors representing customer usage of the system. This case study describes the experimental design, data analysis, and implementation of the results which achieved a reduction from an average of seven wafer handling assists per 1000 system hours prior to the experiment to zero occurrences after implementation of a design change.

EXECUTIVE SUMMARY

Problem

In December 1990 a key Electro-Scientific Industries (ESI) customer in Japan informed ESI of their dissatisfaction with the mean-time-between-failure (MTBF) performance of their Model 9000 systems. In the customer's estimation, ESI's competitor provided superior reliability performance; therefore, improvement in MTBF was required if ESI was to remain a vendor of choice for automated memory repair equipment. ESI was given six months to respond with a solution. The customer's MTBF claim surprised ESI personnel, as ESI data indicated the system to be more reliable than the competition's. However, ESI had failed to recognize that their assessment of MTBF differed fundamentally from the way their Japanese customer measured MTBF. The customer considered a failure to be any event that caused the system to suspend processing and that could not be corrected by a system operator, who (for this customer) had little technical skill or experience with the equipment. Scrutiny of system problem logs at this account confirmed the superior hardware reliability of ESI's installed systems, using ESI's standard definition of MTBF, but also indicated that the criterion of mean-wafers-between-assists (MWBA) of concern to the customer needed to be addressed.

The issue was clear: using the customer's definition of MTBF, which included assists, ESI had to characterize the performance of the wafer handling sub-system to explain the cause of assists, as well as identify system design variables which would de-sensitize the wafer handling sub-system to variation in this customer's environment. Due to the customer's expectation that demonstrable improvement had to be achieved without delay, it became evident that a structured improvement methodology was called for if valid results were to be provided in the prescribed time frame.

Solution Strategy

A team, comprised of ESI design engineers skilled in the operation of the equipment and a consulting statistician, was assembled to address the problem. After gaining a common understanding of the nature of the problem, the team decided that assists were more likely a result of the manner in which the customer operated the equipment than functionally oriented. Various theories of causation related to customer use were presented and debated but all were related to variables that were realistically outside of the control of ESI; that is, the customer could not be expected to operate the equipment in the manner suggested by some of the hypothesized solutions. Further discussions made it clear that design changes to the wafer handling system to compensate for the manner in which the customer used the equipment would have to be investigated.

A robust design experiment was designed and conducted. Its purpose was to determine how eight factors relating to the wafer handling sub-system design affected the performance of the pattern recognition system. Specifically, the experiment looked at how this system performed under conditions that varied according to three uncontrollable environmental factors. The experiment utilized a two-level fractional factorial design of 64 combinations of the eight design and three environmental factors. The data was analyzed to determine which of the eight design factors caused variations in the effects that the three environmental factors had on system performance. Essentially, the experiment sought to identify interactions between design and environmental factors. The purpose of this was to exploit them to attenuate (and remove) the effect of the uncontrollable environmental factors.

Obtaining the resources necessary to conduct the experiment was a challenge encountered at the time of the project. During that period, attention was focused so intently on finding a solution in the field (Japan) that no one was immediately available to run the experiment at ESI's headquarters in Portland, Oregon. In addition, the project was slowed by the lack of a Model 9000 system available solely for conduct of the experiment.

Conclusions

The experiment demonstrated that definite deterioration in vision system reliability occurred due to variations in the environment in which the Model 9000 system was being operated, particularly, the intensity of the ambient light. However, the effects of ambient light variation (which casts a shadow on the wafer that makes pattern recognition difficult) are avoided by starting the pattern recognition search in a region of the wafer away from the shadow. Implementation of this design change solved the assist problem to the complete satisfaction of the customer.

PROCESS

As digital technology becomes more powerful and less expensive, the demand for memory chips continues to grow. Electro-Scientific Industries (ESI) is the leading producer of completely automated memory repair systems for improving production yields for such chips. ESI's Model 9000 uses a diode-pumped laser with programmable spot size to process an increasingly diverse mix of memory devices. Memory repair by laser link cutting is a micro-machining process used in production to correct faulty integrated circuits. Because high density devices are very susceptible to yield loss due to failed circuit elements, redundant elements are being increasingly designed into these circuits. The laser disconnects failed circuit elements by cutting appropriate link structures. Additional link cuts substitute on-board spares for faulty elements. The spot size, or diameter, of the laser beam doing the cutting is between 2.5 and 13 microns, equal to 1/100 of the width of a human hair. In the proper configuration, the system can make up to 1000 laser link cuts per second.

The system used to perform the link cutting task is a highly sophisticated combination of electronic, mechanical, optical, and laser technology. Off-line memory testers send the repair information through a network so that the corrections are made while the memory chips are still in wafer form prior to packaging. The wafers are carried in cassettes which are loaded and unloaded by the system's on-board robot. With the aid of a machine vision sub-system, the robot also does the primary alignment of the wafer to the X/Y stage of the repair system.

Each wafer must be pre-aligned on the wafer handler before it is moved to the system chuck for processing. The first step in this alignment is to perform a scene angle alignment. This alignment determines a global orientation of the wafer to allow the system to search for either corner target along the primary flat. Enough positional information is provided to place the wafer on the chuck for processing. These alignments require that the visual image be learned, or trained, by the optical recognition system. Since the camera is fixed, the corner target placement must be in the field of view of the camera. And, within the field of view, a window is defined over which the video image is learned. A difficult alignment problem with this sub-system led to the experimentation described in this case study.

Data Collection Plan

The objective of the experiment was to determine whether changes in the design of the optical recognition system would improve the quality of the flat find (FF). The FF is measured, on a scale of zero to 1000, as the degree of correlation between the image (provided by the pattern recognition system) of a corner of the primary flat and the learned (expected) image. A maximum value of FF is desirable. While other qualities of the optical "read" were observed in the experiment, FF was the critical response and will be the subject of this case study.

Figure 27.1 gives a graphical representation of the process. The following eight factors related to the design of the optical recognition system were investigated in the experiment. (Their names, along with their settings in the experiment, are described in Appendix A.)

Design and Environmental Factors Depicted in Figure 27.1

A. train box size: size of the window over which the video image is learned
B. corner orientation: two corners of the major wafer flat
C. binary threshold: parameter for the video sub-system which determines at which signal the threshold pixels are designated as black or white
D. illumination level: intensity of illumination
E. illumination angle: angle of light striking the wafer
F. illumination uniformity: uniformity of illumination
G. teach scene angle: orientation of the scene angle of the video image done before finding the corner target
H. train condition: set up of pre-align targets.

Brief descriptions of each design factor and the factor settings for the experiment are given in Appendix A. Three factors representing the customer's environmental usage of the system were varied in the experiment along with the design factors.

J. ambient light: intensity of the ambient light external to the system
K. wafer position: initial wafer displacement
L. wafer angle: initial wafer orientation angle before the wafer handling robot moves the wafer into the system for pre-alignment.

By including these factors appropriately in the experiment design, the experimenters were able to determine if changes in the environment cause FF to vary, and if so, whether any of the design factors could attenuate these effects.

The eight design factors were varied at each of two levels in a fractional factorial design in sixteen treatment combinations. The matrix of combinations is listed in Appendix B. This design is a member of the family of Resolution IV fractional factorial designs, which allow for the independent estimation of the main effects of the eight factors and groups of confounded two-factor interactions.[64] At each of the treatment combinations of design

---

[64] The confounding pattern follows the defining relations presented in Diamond (1989), p. 126, for a sixteen treatment combination, eight factor fractional factorial. They are: I = ABDE, I = ABCF, I = BCDG, I = ACDH. All main effects are therefore confounded with three-factor and higher-order interactions, which are assumed to be negligible in this experiment. This confounding scheme is not unique; alternative schemes for constructing a

factors, a matrix of environmental factors was run in another fractional factorial of four treatment combinations (see Appendix C). This arrangement follows closely the inner array/outer array methodology of Taguchi (1986). By arranging the design in this fashion, clear estimation of each of the design x environment interactions is possible.[65] The resulting experimental design, known as a split-plot design, utilized 16 x 4 = 64 combinations of eleven design and environmental factors.[66]

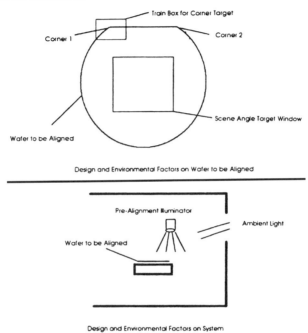

Figure 27.1. *Diagram of Design and Environmental Factors.*

At each of the 64 combinations, two wafers were cycled through the system five times each, and FF was determined. Relative to the variation in FF between treatment combinations, both wafer-to-wafer and within-wafer variation were generally small; therefore, the ten replications were averaged into one value of FF.[67] The data appear in Appendix D.

---

design with equivalent properties can be found in other sources, such as Box, Hunter, & Hunter (1978). The availability of a computer program, based upon the Diamond (1989) text, which designs and analyzes two-level factorial experiments, dictated the choice of this particular confounding scheme.

[65] By combining defining relations it can be demonstrated that all two-factor design x environment interactions are confounded with five-factor and higher-order interactions involving the remaining factors. These interactions can safely be ignored.

[66] The design has a split-plot structure because treatment combinations of the eight design factors ("whole plots") were varied from setup to setup of the equipment while the four treatment combinations of environmental factors were varied within setups ("split plots"). For a discussion of split-plot arrangements in two-level fractional factorial designs, see Box and Jones (1992) and Daniel (1976).

[67] Eight of the sixty-four treatment combinations did suffer from extremely large between-wafer and/or within-wafer variation. In six of these cases, the mean FF was near or below the borderline reading of 900. Not

Data Analysis and Interpretation of Results

Data Analysis

Two analytical approaches can be followed for assessing the effects of the eight design and three environmental factors in the experimental data. First, a regression model (which would include all main effect and two-factor interaction terms specified above) can be fitted to the 64 data (taking into account the split-plot structure of the design). An alternative approach is to further reduce the data by summarizing (calculating averages and standard deviations) the four values of FF at each of the sixteen design factor treatment combinations. While the former analysis could be preferred because the effects of the eleven factors are more completely described, the latter analysis is simpler and, in the end, was the basis for the decisions that were eventually made. Note that by following the latter approach, the standard deviation (the second to last column of Appendix D) is a measure of the overall effects of the environmental factors at each of the sixteen design conditions, and the mean is a measure of the overall quality of the flat find across the environmental conditions.

To determine the main effects of the design factors on the two responses, Mean and Log Standard Deviation of FF, the differences between the average responses at the lo (minus sign) and hi (plus sign) levels in each of the eight columns of the matrix in Appendix B were calculated. The main effects represent the change in the average response from the lo to the hi level of a factor. In addition, the seven remaining interaction effects provided by the design (each effect representing four confounded two-factor interactions) were analyzed in the same fashion. These results are provided in Appendix E.

Appendices F and G present plots of the main effects for mean FF and Ln(S), respectively.

In order to identify which of the fifteen effects were statistically significant (beyond random variation), the effects were plotted on normal probability paper (see Figures 27.2 and 27.3). The plotted points which fall off the line formed by the remainder of the points are the likely active (significant) effects. For a description of this graphical technique, see Box, Hunter, & Hunter (1978).

Interpretation

Analysis of Mean FF

The two design factors which dominate the mean FF are B and D. In fact, there is an apparent interaction[68] between them, which is plotted in Figure 27.4. The interaction suggests that FF deteriorates at only one combination of B (+: "find corner 2 of the wafer first") and D(-: "set illumination a lower than normal"). As long as this design combination is avoided, mean FF will be fairly high, regardless of the environmental factor conditions.

---

surprisingly, the pattern recognition system lacks repeatability in correlating the learned and observed image when the average reading is poor.

[68] In reality the BD interaction is confounded in the design with the AE, CG, and FH interactions. However, the interpretation of the BD interaction is substantively much more straightforward than the others. Moreover, the main effects of factors A, C, E, G, and H were negligible; therefore, the existence of these interactions is less likely.

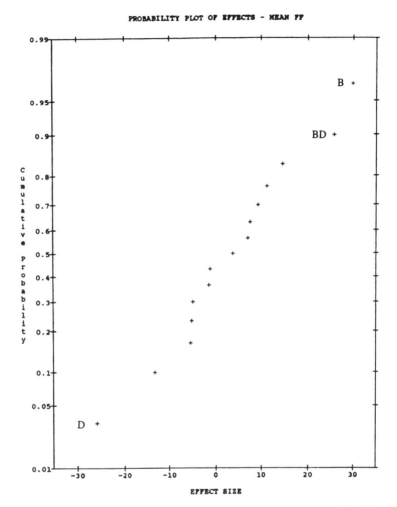

Figure 27.2. *Normal Probability Plot of Mean FF Effects. The three plotted points at the extremes, representing main effects B and D and interaction effect BD, appear to fall off the line created by the remaining thirteen points, which suggests that these three effects are significantly different from the remainder.*

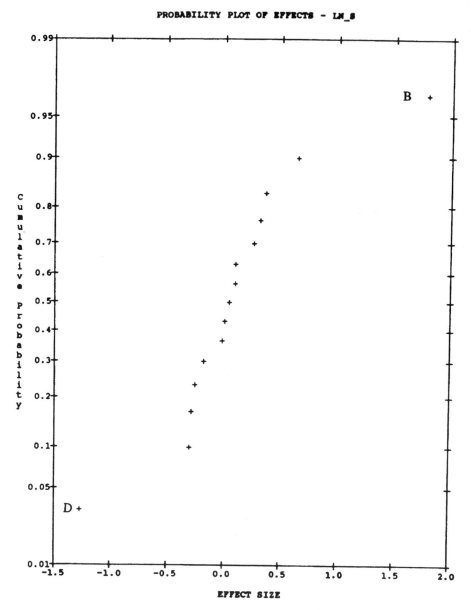

Figure 27.3. *Normal Probability Plot of Ln(S) Effects. The two plotted points at the extremes, representing the B and D main effects, appear to fall off the line created by the remaining fourteen points, which suggests that these two effects are significantly different from the others.*

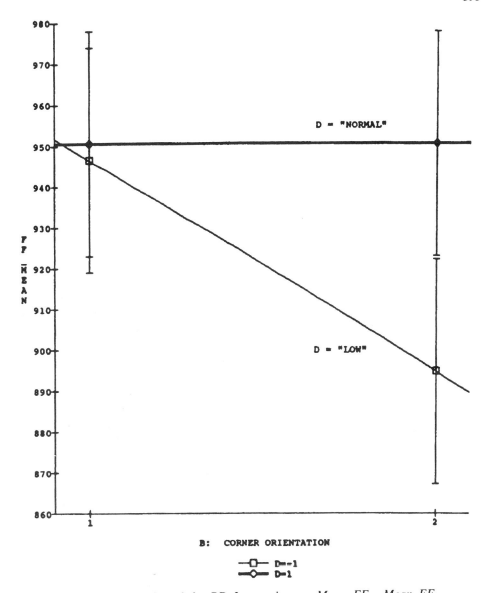

Figure 27.4. *Plot of the BD Interaction on Mean FF. Mean FF varies with illumination level but only if the target is corner 2. Similarly, mean FF varies with the choice of target corner but only if illumination level is lower than normal. By choosing corner 1 as the target, the effect of illumination is eliminated, and the mean FF is maximized.*

Analysis of FF Standard Deviation

From the main effects plots in Appendix G, the normal probability plot in Figure 27.3, and the significance of the effects in Appendix E, the two design factors that dominate the logarithm of the standard deviation (and hence the standard deviation) are B and D. Since the standard deviation (and its logarithm) measure the variation in FF across the four conditions of the three environmental factors, this means that FF is sensitive to the customer's env'ronment, and that the design of the optical recognition system can be set to attenuate the effects of the environment.

At the low setting of B ("find corner 1 first") and high setting of D ("set illumination at normal setting"), the effects of the environmental factors are reduced (low values of the standard deviation of FF in the table in Appendix D.) Upon closer inspection, the environmental factors are dominated by factor J: the four largest standard deviations in Appendix D correspond to values of FF that are much lower when J (ambient light) is high (-) than when it is low (+). In other words, the optical recognition system has the greatest difficulty recognizing the pattern when the ambient light is high, the interior illumination level is low, and the search begins at the far corner (three factor interaction).

In hindsight, the explanation of this phenomenon is straightforward. The arm of the system robot for the interior casts a shadow on the far corner. This shadow on the far target is strongest when the interior illumination is low and the ambient light is high, making it difficult for the optical recognition system to recognize the pattern. This shadowing was not easily detected by visual inspection; however, it did have a measurable effect on the performance of the optical recognition system.

This understanding of the problem proved to be the key to alignment reliability improvement. The problem was addressed by a modification of the alignment software to search for the target only in regions away from the shoulder of the robot with the robot arm extended. The modification incurred a potentially modest increase in wafer handling time to allow for the search of all four quadrants of this target area.

CONCLUSIONS

The primary focus of the project was to increase the MWBA for the ESI Model 9000. The experiment tested the system under varying environmental, target wafer, and system setup conditions. The wafer alignment process used a robot to position the wafer under the vision system for capturing either one of two targets, typically located on each edge of the major wafer flat. Using the two targets required only two wafer placements to ensure one of the targets was in the capture range of the vision system. It appeared that one of the targets was unreliable but the cause was not obvious prior to the experiment. As a result of the experiment, this step in the wafer alignment process was found to be sensitive to variation in the illumination and the choice of alignment target. By modifying the alignment software to only search for the target in the areas away from the shadow cast by the robot, a key part of a comprehensive package of improvements for increasing MTBF was achieved. The goal was to increase the MTBF of 68 hours to a MTBF of 240 hours by mid-1991. After all improvements had been installed, the measured MTBF for systems at the customer site had surpassed 540 hours.

The primary benefit of the project to the customer was the reduction in processing cost per wafer output. A secondary benefit was improved customer confidence in the ESI product (as well as problem-solving expertise), which is beyond measure. The customer also saved time that had otherwise been spent investigating or re-optimizing the setup parameters. Of significant benefit to ESI was a reduction in field service and engineering requests. Previously, after each failure, the problem was temporarily solved by optimizing setup criteria at the customer site. However, reliability would then deteriorate as the external conditions changed, prompting a call for field service or engineering support once again. Finally, the approach determined by this case study efficiently determined the sensitivity of the process in an experiment conducted in the factory. By understanding the nature of the problem, a potentially expensive redesign of the illumination and/or vision sub-systems was avoided.

In the design of systems which may be subject to the effects of environmental conditions, design of experiments is a vital approach for identifying which inputs have impact on system performance. In particular, this approach is much less resource-intensive than alternative methods of investigation. It can be performed in-house, under controlled conditions, utilizing a small set of data, as compared to testing in a production environment at multiple customer sites, where the boundaries of expected operating and environmental conditions may not be experienced.

ACKNOWLEDGMENTS

The authors thank Rodger Dwight of ESI and the reviewers for helpful suggestions in the writing of this case study.

REFERENCES

Box, G.E.P., S. Hunter, and W. Hunter, *Statistics for Experimenters*, New York, NY: John Wiley and Sons, Inc., 1978.

Box, G.E.P. and Stephen Jones, "Split-Plot Designs for Robust Product Experimentation," *Journal of Applied Statistics*, Vol. 19, No. 1, pp. 3-26. 1992.

Daniel, C. *Applications of Statistics to Industrial Experimentation*, New York, NY: John Wiley and Sons, Inc., 1976.

Diamond, W. *Practical Experiment Designs for Engineers and Scientists*, 2nd edition, New York, NY: Von Nostrand Reinhold, 1989

Taguchi, G. *Introduction to Quality Engineering*, White Plains, NY: UNIPUB, 1986.

APPENDIX A

Names, Levels, and Short Description of Experiment Factors

Design Factors

Factor Name	Levels Lo (-)	Hi (+)	Description
A - Train Box Size	Small	Large	Size of the trained window for corner target
B - Corner Orientation	Corner 1	Corner 2	Corner 1 is the near target / Corner 2 is the far target
C - Binary Threshold	40%	60%	Binary threshold for pixel analysis
D - Illumination Level	Low	Normal	Illumination setting
E - Illumination Angle	Current	New	Illumination angle
F - Illumination Uniformity	Non-unif.	Uniform	Uniformity of the light striking the target
G - Teach Scene Angle	Small/non-centered	Large/centered	Method for learning the Scene Angle target
H - Train Condition	Off-center	Normal	Set-up of pre-align targets

Environmental Factors

Factor Name	Levels Lo (-)	Hi (+)	Description
J - Ambient Light	High	Low/off	Level of ambient light
K - Wafer Position	Back	Normal	Initial position of wafer in cassette
L - Wafer Angle	+/-20 deg.	0 deg.	Angle of wafer flat to cassette top

APPENDIX B

Matrix of Design Factors

16 Treatment Combinations; The first four columns describe a full factorial in the factors A, B, C, and D. The remaining four columns are generated from the first four by utilizing the defining relations: I = ABDE, I = ABCF, I = BCDG, and I = ACDH.

T.C.	A	B	C	D	E	F	G	H
1	-	-	-	-	-	-	-	-
2	+	-	-	-	+	+	-	+
3	-	+	-	-	+	+	+	-
4	+	+	-	-	-	-	+	+
5	-	-	+	-	-	+	+	+
6	+	-	+	-	+	-	+	-
7	-	+	+	-	+	-	-	+
8	+	+	+	-	-	+	-	-
9	-	-	-	+	+	-	+	+
10	+	-	-	+	-	+	+	-
11	-	+	-	+	-	+	-	+
12	+	+	-	+	+	-	-	-
13	-	-	+	+	+	+	-	-
14	+	-	+	+	-	-	-	+
15	-	+	+	+	-	-	+	-
16	+	+	+	+	+	+	+	+

Appendix C

Matrix of Environmental Factors.

Experimental Design in 3 Environmental Factors in 4 Treatment Combinations. The first two columns describe a full factorial in the factors J and K. The third column is generated by utilizing the defining relation: I = JKL.

T.C.	J	K	L
1	-	-	+
2	+	-	-
3	-	+	-
4	+	+	+

Appendix D

Experiment Design

Experimental Design in 8 Design Factors and 3 Environmental Factors in 64 Treatment Combinations, with the Values of FF. The last three columns summarize the values of FF across the four combinations of the environmental factors.

	DESIGN FACTORS							ENVIRONMENTAL FACTORS						
								J: - K: - L: +	J: + K: - L: -	J: - K: + L: -	J: + K: + L: +			
T.C.	A	B	C	D	E	F	G	H					Mean	Std. Dev.	Ln(s)
1	-	-	-	-	-	-	-	-	916	934	912	922	921	9.6	2.26
2	+	-	-	-	+	+	-	+	967	964	959	972	966	5.4	1.69
3	-	+	-	-	+	+	+	-	906	934	880	926	911	24.1	3.18
4	+	+	-	-	-	-	+	+	856	967	822	959	901	73.0	4.29
5	-	-	+	-	-	+	+	+	939	952	926	943	940	10.8	2.38
6	+	-	+	-	+	-	+	-	957	954	964	961	959	4.4	1.48
7	-	+	+	-	+	-	-	+	879	941	847	935	900	45.3	3.81
8	+	+	+	-	-	+	-	-	738	973	776	972	865	125.4	4.83
9	-	-	-	+	+	-	+	+	944	948	948	948	947	2.0	0.69
10	+	-	-	+	-	+	+	-	956	963	957	958	958	3.1	1.13
11	-	+	-	+	-	+	-	+	957	954	964	961	959	4.4	1.48
12	+	+	-	+	+	-	-	-	941	972	953	969	959	14.5	2.67
13	-	-	+	+	+	+	-	-	932	939	935	940	937	3.7	1.31
14	+	-	+	+	-	-	-	+	961	960	960	956	959	2.2	0.79
15	-	+	+	+	-	-	+	-	925	951	934	957	942	14.8	2.69
16	+	+	+	+	+	+	+	+	972	971	929	974	961	21.7	3.08

APPENDIX E

Factor Effects

Effects of the Factors on Flat Find Mean and Ln(S). The effects which are labeled with an asterisk are considered to be statistically significant. The remaining effects do not appear to differ significantly from zero (See Figures 27.2 and 27.3).

Effect		Effect Size	
		Mean	Ln(s)
Average		935.3	2.42
Main			
	A - Train Box Size	11.4	0.16
	B - Corner Orientation	-26.1*	1.90*
	C - Binary Threshold	-4.9	0.26
	D - Illumination Level	29.9*	-1.15*
	E - Illumination Angle	14.4	-0.35
	F - Illumination Unif.	-1.4	0.16
	G - Teach Scene Angle	9.1	-0.10
	H - Train Condition	7.6	-0.06

Effect		Effect Size	
		Mean	Ln(s)
Interactions			
	AB/DE/CF/GH	12.9	-0.55
	AC/BF/DH/EG	-5.1	-0.16
	AD/BE/CH/FG	6.6	-0.01
	AE/BD/CG/FH	26.1*	-0.18
	AF/BC/DG/EH	-5.6	0.21
	AG/BH/CE/DF	-1.6	0.10
	AH/BG/CD/EF	3.9	-0.01

APPENDIX F

Main Effects of Mean FF

Plots of Main Effects and 95% Confidence Intervals about the Means - Mean FF. The larger the magnitude of the effect, the greater the slope of the line from the value zero. A positive slope indicates an increasing effect from lo to hi on Mean FF. A negative slope indicates a decreasing effect on Mean FF.

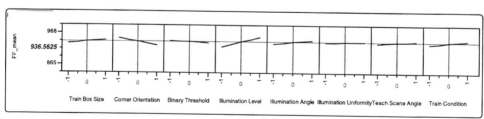

APPENDIX G

Main Effects of Ln(s)

Plots of Main Effects and 95% Confidence Intervals about the Means - Ln(s). The larger the magnitude of the effect, the greater the slope of the line from the value zero. A positive slope indicates an increasing effect from lo to hi on Ln(s). A negative slope indicates a decreasing effect on Ln(s).

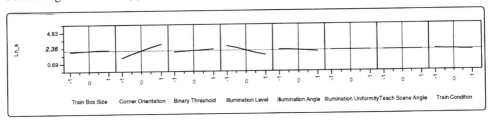

BIOGRAPHIES

Donald K. Lewis, Ph.D., is Principal of Lewis & Lewis, Portland, Oregon, a consulting firm specializing in the dissemination and implementation of modern strategies and methods of continuous improvement. He also serves as an Adjunct Assistant Professor at the Oregon Graduate Institute of Science & Technology. Don received his B.A. in mathematics from Claremont McKenna College and Ph.D. in biostatistics from the University of North Carolina at Chapel Hill.

Craig Hutchens is business manager for ESI's Link Processing Systems Business Unit. Craig has been with ESI since 1981 and has been involved in laser-based memory repair since its early beginnings at AT&T Bell Laboratories. He previously held various management and senior technical positions in engineering and customer support at ESI. During the time of this experiment, Craig was the project engineer for the Model 9000.

Joseph M. Smith received a B.S. degree in electrical engineering from Texas A&M University in 1983. From 1984 to 1991, he worked as a systems applications engineer for semiconductor products at ESI. Since 1991 he has been a member of the software engineering group at ESI, where his efforts have been involved in software design, test, and quality assurance.

CHAPTER 28

HOW TO DETERMINE COMPONENT-BASED PREVENTIVE MAINTENANCE PLANS

Stephen V. Crowder

This case study provides guidance on establishing preventive maintenance (PM) plans for manufacturing equipment. From the knowledge of past equipment performance and relative times required for scheduled and unscheduled maintenance actions, PM policies that maximize equipment availability can be determined. This case study addresses age replacement policies (in which a component is replaced at some fixed age if it has not failed) and condition-based policies (in which a degradation variable is monitored and a component is replaced when this variable crosses a specified threshold). Emphasis is placed on assumptions associated with different PM strategies and on what data are necessary to implement those strategies. All the techniques discussed are illustrated with examples. This case study should prove useful for equipment engineers and others who are faced with the problem of scheduling PM. References listed at the end of this case study provide information on the basics covered here, as well as information on more complicated techniques for scheduling PM.

EXECUTIVE SUMMARY

Problem

Manufacturing equipment reliability depends greatly upon the quality built in during the design and production stages. Equipment reliability, especially in a manufacturing environment, also depends on the quality of equipment operation and maintenance. A simple argument for preventive maintenance (PM) is that equipment availability may often be increased by replacing or restoring components subject to failure before they fail. The strategy of only fixing it when it breaks will often lead to a very low equipment availability which could be increased through the scheduling of PMs. On the other hand, too frequent PMs can also result in low availability. For any component which ages or wears out over time and for which the time required to repair or replace an in-service failure results in more downtime than a scheduled replacement, the availability can be increased through PM. This case study deals with the problem of determining how to appropriately schedule PMs to maximize equipment availability.

Solution Strategy

This case study deals with PM in the sense of replacement or renewal of components subject to failure. For PM to be beneficial, there must be some advantage to avoiding an unexpected failure in service. The scheduled PM of a component must be less costly than an unexpected or emergency maintenance (EM). This will often be the case in industries such as semiconductor manufacturing, as unexpected equipment failure will result in costly process downtime for repairs, while scheduled PMs may result in less process downtime. The first major section of the case study addresses the age replacement policy, where a component is replaced or restored to as good as new condition based solely on its age, or amount of time in use. Both model-based and empirically-based approaches to maximizing an availability criterion are outlined. An example using mass flow controller failure data illustrates both approaches. The second major section covers the case in which more is known about the component than simply its age. A measure of wear, or degradation, of the component is available, and the resulting PM is based on tracking this degradation level to determine when a component should be replaced. The approach discussed in this case is based on empirical observation of how the component degrades as a function of use. An example involving the degradation of an ion implanter source filament illustrates this PM policy (referred to as a condition-based policy).

Conclusions

The traditional approach to PM in American industry has been to remove a component from service when it fails, or simply fix it when it breaks. An approach which would increase equipment availability is to remove the component from service at predetermined intervals. This is the so-called age replacement policy. A technique which offers even more potential for increased efficiency over either of these approaches is the condition-based policy. Rather than replacing only at failure or at predetermined intervals, this policy attempts to provide PM only when necessary. This approach requires more knowledge about how the component or equipment works, what state variables need to be monitored, and having the appropriate diagnostics built into the equipment to allow easy monitoring of the key variables. Each of the strategies for scheduling PM discussed in the case study is based on simple, but useful, statistical models. The key assumptions associated with each model are emphasized, as are the data requirements and estimation techniques necessary for implementing the models and plans. The material presented should provide a starting point for semiconductor manufacturing engineers wishing to establish PM plans for their equipment or other such systems. The references listed provide information on more complicated models and techniques for scheduling PM.

PREVENTIVE MAINTENANCE

Introduction

Preventive maintenance (PM) has been defined (see Gertsbakh, 1977) as "the total of all service functions aimed at maintaining and improving reliability" of equipment and includes such things as "the replacement and renewals of elements, inspections, testing, and checking of working parts during their operation." This case study deals with PM in the sense of replacement or renewal of components subject to failure.

For PM to be beneficial, there must be some advantage to avoiding an unexpected failure in service. The scheduled PM of a component or system must be less costly than an unexpected or emergency maintenance (EM).

Designing PM plans for equipment requires knowledge of that equipment's likely time to failure, which can be statistically modeled if appropriate data are available. This case study presents PM strategies based on simple, but useful, statistical models, with emphasis on the key assumptions associated with each model, as well as data requirements and estimation techniques necessary to implement the strategies. To establish a PM schedule, one of two criteria is typically considered:
1. expected maintenance cost per unit of operating time
2. equipment availability, defined as the percent of time the equipment is available for use when needed.

This case study will consider the problem of scheduling PM to maximize equipment availability. The approach which seeks to minimize cost is entirely analogous and is not covered. The availability ratio for the PM models discussed is defined by

$$A = \text{Expected Uptime} / (\text{Expected Uptime} + \text{Expected Downtime}) \quad (28.1)$$

for a cycle, where a cycle is defined as the time between successive maintenances (preventive or emergency). A cycle thus includes operating time plus downtime for either PM or EM. Expected values are used to define the availability ratio because the cycle length will depend on the random component time to failure. Throughout the case study, optimum PM schedules will be determined by maximizing the above availability ratio.

Overview

In the remainder of the case study, the following are covered:
- the age replacement policy for components, in which a component is replaced after a fixed amount of operating time;
- the condition-based policy, in which PM is based on some measure of degradation or wear;
- additional comments and concluding remarks on PM;
- brief discussion of the reference list;
- appendices with selected technical details.

The discussion deals primarily with a few specific useful models that can be applied to the problem of scheduling PM. The material presented should provide a starting point for manufacturing engineers wishing to establish PM plans for their equipment or other such systems. The references listed provide information on more complicated models and techniques for scheduling PM.

AGE REPLACEMENT POLICY

Statement of Age Replacement Policy

The age replacement policy is a PM policy for components, such as equipment parts, that may fail in operation if not replaced in time. The objective is to determine a replacement time which is neither so short as to cause an excessive number of replacements, nor so long as to unnecessarily risk failure. The policy is simply stated:
- Perform PM (replacement) after t_0 hours of operation without failure.
- If the component fails before t_0 operating hours have elapsed, perform EM at the time of failure.

The PM plan is thus determined by the choice of the constant t_0. Figure 28.1 illustrates the application of this policy.

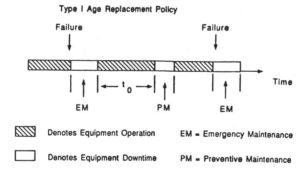

Figure 28.1. *Perform PM after t_0 hours of operation without failure, or if the component fails before t_0 operating hours have elapsed, perform EM at the time of failure. The PM plan is thus determined by the choice of the constant t_0.*

Key Assumptions Associated with Age Replacement Policy

Several important assumptions need to be met before the age replacement policy will be appropriate. They are
- Failed components are replaced by components which are as good as new, or are restored to as good as new condition.
- The component times to failure are independent and are from the same statistical distribution.
- The time to perform an EM is greater than the time to perform a PM.
- Component failure results in system (equipment) failure, i.e., loss of availability.

Availability Criterion for Age Replacement Policy

The availability ratio for the policy is, expressed as a function of t_0:

$$A(t_0) = \frac{\int_0^{t_0}(1-F(x))dx}{\int_0^{t_0}(1-F(x))dx + T_1 F(t_0) + T_2(1-F(t_0))}, \tag{28.2}$$

where

- T_1 = Expected time to perform an EM (in this context, replacement of a component at failure);
- T_2 = Expected time to perform a PM ($T_2 < T_1$);
- F = Cumulative distribution function (cdf) of the time-to-failure data.

Expression (28.2) is just the expected uptime during a cycle divided by the total expected cycle length, as defined earlier. This result is derived in Appendix A. Maximization of (28.2) will yield the optimal t_0 for the age replacement policy. Presumably, values for T_1 and T_2 can be determined from historical maintenance records. Then if the cdf in (28.2) is known, finding t_0 is simply a maximization problem, requiring a numerical integration and optimization routine. In addition to finding the maximum of the function $A(t_0)$, it is strongly recommended that $A(t_0)$ be plotted as a function of t_0. Plotting the entire curve provides more information than simply computing the maximum. Examining the entire curve will demonstrate how sensitive the availability is to the choice of t_0. It should also be pointed out that both the availability and the value of t_0 maximizing $A(t_0)$ will be sensitive to the constants T_1 and T_2. These points will be illustrated in the Example using Mass Flow Controller Failure Data section.

Data Requirements and Estimation Techniques for Age Replacement Policy

In practice, to establish an age replacement schedule, times of operation to failure (or censoring times) and times required for EM and PM are needed. The times of operation to failure may not be routinely tracked by plant computer systems, so special data collection or a separate life test of components may be necessary. For the age replacement policy, define $X_i = i^{th}$ time to failure (or censoring time) for the component of interest.

As mentioned above, it is assumed that the X_i's are independent and have the same statistical distribution with cdf F, say.

Typically, the function F will not be known and must be estimated from the data. To do so, assume a distributional form (i.e., Weibull) for F and estimate its parameters, or estimate F using distribution-free techniques. Techniques for life data analysis (when a distributional form or model can be assumed) are outlined by Lawless (1982) and Nelson (1982, 1983). Often the technique of maximum likelihood estimation is used with a distributional form to estimate F. Nelson (1983) has compiled a how to booklet on the topic of analyzing reliability data, which provides guidelines for model-based estimation of F.

The distribution-free technique estimates F by the empirical cdf

$$\hat{F}_{emp}(t) = [1/n] \times [\text{number of } X_i\text{'s} \leq t],$$

which is the sample fraction of the n observations which are less than or equal to t. This estimate is used because it is a consistent estimator of F, and it leads to a simple distribution-free estimate of the availability ratio. If a distributional form for F is known or can be identified, model-based estimation is preferable. The distribution-free approach has advantages in terms of simplicity of estimation and the search for a maximum, and will be just as good with large sample sizes.

Once an estimate of F is obtained and values for T_1 and T_2 are determined, the ratio (28.2) can be maximized. The steps necessary to determine an age replacement PM plan for components are thus

1. Collect times to failure (and censoring times) for the component of interest under normal operating conditions. These data may be available from plant maintenance or equipment development records. If they are not, an experimental life test may be necessary. Lawless (1982) discusses life test plans, and MIL-STD-781C discusses and tabulates widely used procedures.

2. Determine the unknowns in the availability ratio (28.2). These include estimates of T_1, T_2, and the cdf, F. Estimates of T_1 and T_2, the times for an EM, and scheduled maintenance, respectively, can be obtained from historical maintenance records. The cdf F can be estimated using either a model-based or empirical approach. Nelson (1983) demonstrates how to analyze this type of data and how to estimate F using both graphical and numerical techniques. This includes discussion of various computer programs available to do the analyses.

3. Substitute the estimates of T_1, T_2, and F from Step 2 into the expression for $A(t_0)$, to yield an estimated availability ratio, $\hat{A}(t_0)$. If a model-based estimate of F is used, $\hat{A}(t_0)$ will probably have to be evaluated numerically. If the empirical cdf is used, the expression for (t_0) can be evaluated easily using a calculator.

4. Evaluate and plot $\hat{A}(t_0)$ as a function of t_0, for $t_0 > 0$. These evaluations and the plot will indicate what value of t_0 maximizes $\hat{A}(t_0)$, and how sensitive the availability is to the choice of t_0.

5. Having determined the value of t_0, say $\hat{t}_0$, that maximizes $\hat{A}(t_0)$, the resulting PM plan is
   - Perform PM (replacement or restoration) after $\hat{t}_0$ hours of operation without failure, or
   - If the component fails before $\hat{t}_0$ hours have elapsed, perform EM at the time of failure.

These steps are illustrated in the example below.

It should be noted that $\hat{A}(t_0)$ and $\hat{t}_0$ are estimates, subject to sampling error. That is, other samples of times to failure would lead to different estimates of the cdf F, different estimates of $A(t_0)$, and hence different $\hat{t}_0$ values. Calculations of approximate confidence limits for the parameters of the model-based estimate of F are too complicated to give here, but they appear in Nelson (1982). Given confidence intervals for these parameters, the user could choose different values for each parameter within its own confidence interval, and substitute these values into the expression for A(t₀). Then the values of t₀ maximizing these resulting expressions give an idea of how much variability is associated with the estimate $\hat{t}_0$. The point to remember is that $\hat{t}_0$ should be viewed as an estimate of the value which maximizes A(t₀) rather than as a deterministic value.

Example using Mass Flow Controller Failure Data

The techniques for establishing an age replacement policy is here illustrated using simulated mass flow controller failure data (see Table 28.1). Table 28.1 presents fifty times to failure, ordered from smallest to greatest. Fifty data points will typically be enough to obtain good estimates of the parameters of a distribution.

Table 28.1. *Mass Flow Controller Time-to-Failure Data. It is ordered from least to greatest. These data are used in establishing a condition-based replacement policy.*

i	X_i(Hrs)	i	X_i(Hrs)
1	7.54	26	82.8
2	12.0	27	82.9
3	12.9	28	83.0
4	14.9	29	86.9
5	19.4	30	92.4
6	22.4	31	94.5
7	36.9	32	95.1
8	40.4	33	96.4
9	41.4	34	97.8
10	49.2	35	106
11	49.7	36	107
12	50.1	37	108
13	51.0	38	112
14	58.9	39	116
15	59.4	40	117
16	59.9	41	118
17	69.1	42	123
18	70.0	43	137
19	70.5	44	148
20	73.7	45	153
21	77.8	46	163
22	79.9	47	165
23	80.2	48	180
24	80.4	49	198
25	81.3	50	240

Suppose that based on these lifetime data, we wish to establish a PM plan for the replacement of a mass flow controller. A reasonable model to consider for this type of data is the Weibull model. Fitting a Weibull model, which has cdf

$$F(t) = 1 - \exp\{-(t/\alpha)^\beta\},$$

to these data, we obtain maximum likelihood estimates of $\hat{\alpha} = 98.1$ and $\hat{\beta} = 1.82$. The result that $\hat{\beta} > 1$ suggests an increasing hazard function (See Lawless, 1982), making this component a candidate for PM. Roughly speaking, an increasing hazard function means the component ages with use, or has an increasing probability of failure as it ages. Assume, for sake of argument, that $T_1 = 10$ hours and $T_2 = 1$ hours, so that $T_1 / T_2 = 10$. That is, an in-use mass flow controller failure causes the loss of 10 hours of availability of the system for use, while a scheduled replacement causes the loss of only 1 hours of availability. Then the optimum age replacement policy is determined by t_0 maximizing

$$\hat{A}(t_0) = \frac{\int_0^{t_0} \exp\{-\left(\frac{x}{98.1}\right)^{1.82}\} dx}{\int_0^{t_0} \exp\{-\left(\frac{x}{98.1}\right)^{1.82}\} dx + 10(1 - \exp\{-\left(\frac{t_0}{98.1}\right)^{1.82}\}) + \exp\{-\left(\frac{t_0}{98.1}\right)^{1.82}\}} \quad (28.3)$$

where (28.3) is obtained from (28.2) by substituting in the values for T_1, T_2, and the estimate of F. Figure 28.2 shows the estimated availability curve, $\hat{A}(t_0)$, for this example, for t_0 ranging from 5 to 300 hours. Note that a unique maximum occurs at $\hat{t}_0 = 35$, determined by numerical maximization, and that the estimated maximum attainable availability is $\hat{A}(35) = 0.94$, or 94% system availability with respect to this component.

The optimum PM policy, based on these data, is thus to replace the component after 35 hours of operation without a failure. If a failure occurs before 35 hours of operation, the component should be replaced and PM rescheduled. Notice from Figure 28.2 the effect of too frequent PMs. If PMs were scheduled, say, for every 5 hours of operation, the resulting availability would be only 83%. Too infrequent PMs, on the other hand, would also decrease the availability, since the curve decreases for $t_0 > 35$ hours. With no PM ($t_0 = \infty$), the availability decreases to 0.90. Thus the curve indicates not only the maximum possible availability, but also the sensitivity with respect to the choice of t_0. To show the importance of the constants T_1 and T_2, assume now that $T_1 = 2$ hours and $T_2 = 1$ hour, so that $T_1 / T_2 = 2$. In this case, (28.2) is maximized for $\hat{t}_0 = 125$, and $\hat{A}(125) = 0.98$, or 98% availability. So reducing T_1 / T_2 to 2 has a dramatic impact on the optimal time between scheduled PMs, and reducing T_1 increases the availability, as would be expected.

In case we have no knowledge of the family of distributions to which F belongs, the empirical, or distribution-free estimators can be used in (28.2). If the empirical cdf, $\hat{F}_{emp}(t)$, is used in (28.2), the numerator becomes (see Ingram and Scheaffer, 1976)

$$\int_0^{t_0} (1 - \hat{F}_{emp}(x)) dx = \frac{1}{n} \sum_{i=1}^{n} \min(X_i, t_0),$$

and the availability ratio, for a given t_0, is thus estimated by

$$\hat{A}(t_0) = \frac{\frac{1}{n}\sum_{i=1}^{n}\min(X_i,t_0)}{\frac{1}{n}\sum_{i=1}^{n}\min(X_i,t_0) + T_1 \hat{F}_{emp}(t_0) + T_2\left[1 - \hat{F}_{emp}(t_0)\right]}$$

(28.4)

For a fixed t_0, and a given data set (the X_i's), this quantity is easily computed using a calculator. It is also the case (see Ingram and Scheaffer, 1976) that equation (28.4) is necessarily maximized for t_0 equal to one of the X_i's, so the search for the maximum is greatly simplified.

Returning to the mass flow controller example, we will illustrate maximization of (28.4), again with $T_1 = 10$ hours and $T_2 = 1$ hour. This approach requires evaluation of $\hat{A}(t_0)$ (28.4) at each of the ordered data points, ranging from $X_1 = 7.54$ to $X_{50} = 240$. Substituting $t_0 = X_1, X_2, \ldots, X_{50}$ (the ordered X_i's) into (28.4) leads to $\bar{A}(X_1) = 0.86$, $\bar{A}(X_2) = 0.90$, ... , $\bar{A}(X_{50}) = 0.90$. Figure 28.3 shows the empirical availability plot for this example. The maximum value occurs at $\hat{t}_0 = X\ 10 = 49.2$, with an estimated availability of 0.94. The optimum PM policy, using these data with the empirical approach, is thus to replace the component after 49 hours of operation without a failure.

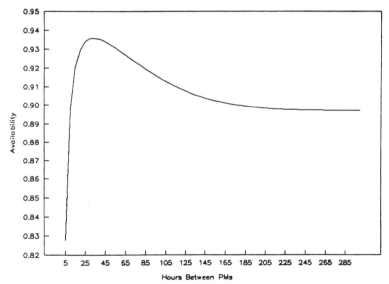

Figure 28.2. *Availability Curve for the Age Replacement Policy for Mass Flow Controllers Assuming a Weibull Distribution. The curve suggests an optimal policy with t_0 approximately equal to 35, with an availability of 94%.*

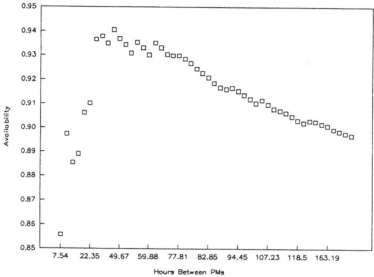

Figure 28.3. *Availability Plot for the Age Replacement Policy for Mass Flow Controllers with No Distributional Assumption. The curve suggests an optimal policy with t_0 approximately equal to 49, with an availability of 94%.*

In this particular case, the two approaches gave roughly the same answer. Although the model-based estimate was $\hat{t}_0 = 35$ hours and the empirical-based estimate was $\hat{t}_0 = 49$ hours, the estimated availability is essentially the same for any choice of t_0 in this interval. It should be noted, however, that for small samples (less than size n = 25), the model-based and distribution-free estimates can be quite different. With larger samples, the estimates should be comparable if the chosen model adequately represents the data.

In summary, the age replacement policy may be useful in situations where the component ages or wears out over time and the time required to repair or replace an in-service failed component results in more downtime than a scheduled replacement. If the form of the cdf F can be identified, the model-based approach should be used to establish the PM schedule. The distribution-free estimator of F can be used when the form of F is not known, but caution should be exercised with small sample sizes. The distribution-free approach does have the advantage in terms of simplicity and computations required to estimate the best t_0. More theoretical details of this particular PM plan are given by Barlow and Hunter (1960), Barlow and Proschan (1996), and Gertsbakh (1977), while application oriented papers include Barlow (1978), Bergman (1977), and Glasser (1969).

Condition-Based Policy

Statement of Condition-Based Replacement Policy

A second type of PM policy for components is called the condition-based policy. For the age replacement policy in the previous section, we assumed our only knowledge of the component was whether it was operating or not. The time to failure distribution depended only on the age of the component. In some situations, more knowledge regarding the state of the component may be available. For example, in a mechanical system, some measure of fatigue or wear in a component may be available. In an electronic system, the state of the system may be reflected by measuring a component output which changes due to some gradual process of deterioration or wear. The condition-based policy uses this additional information to establish the PM schedule. This policy is also sometimes referred to as a control limit replacement policy. Rather than deciding the schedule for PMs based solely on the age of the component, the PM is scheduled based on the condition, or state of the component. An appropriately chosen state variable is monitored to determine when a PM is necessary. A simple example is what type of strategy to use for the replacement of an automobile tire. An age replacement policy would require replacement of the tire after a pre-determined number of miles of use, while a condition-based strategy might be to monitor the amount of tread left on the tire and replace the tire when the amount of tread reaches some minimum. Use of this additional information about the component can increase its availability.

The condition-based policy is
- perform a PM (replacement) whenever the monitored state variable reaches a critical limit, or
- if a failure occurs before the state variable reaches the critical limit, perform EM at the time of failure.

The PM plan is thus determined by identifying an appropriate state variable to monitor and choosing a critical limit for that variable. Figure 28.4 illustrates the application of this policy.

Key Assumptions Associated with the Condition-Based Policy

The assumptions associated with the Age Replacement Policy section also apply to the condition-based policy. Additional assumptions associated with use of a state variable are
- The state variable is monotonic (i.e., accumulated wear).
- The probability of failure in the short time interval $(t, t+\Delta t)$ depends only on the value of the state variable at time t. Physically, this means that only the actual level of the variable being monitored influences the failure probability in a future short time period.

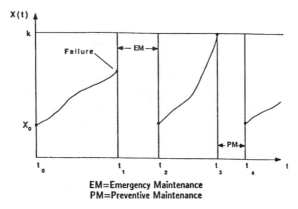

EM=Emergency Maintenance
PM=Preventive Maintenance

Figure 28.4. *Scheme of PM using State Variable X(t). Perform a PM whenever the monitored state variable reaches a critical limit k, or if a failure occurs before the state variable reaches the critical limit, perform EM at the time of failure. The PM plan is thus determined by identifying an appropriate state variable to monitor and by choosing a critical limit for that variable.*

- The probability of failure in the time interval $(t,t+\Delta t)$ is an increasing function of the monotonic state variable. That is, for a monotonically increasing state variable, increases in the state variable increase the probability of failure in the interval $(t,t+\Delta t)$. And for a monotonically decreasing state variable, decreases in the state variable also lead to an increase in the probability of failure in the interval $(t,t+\Delta t)$.

Availability Criterion for Conditioned-Based Policy

The availability ratio for the condition-based policy is, expressed as a function of the control limit k,

$$A(k) = \frac{E\{\min(L_i, T_i(k))\}}{E\{\min(L_i, T_i(k))\} + T_1 \Pr(L_i \leq T_i(k)) + T_2(1 - \Pr(L_i \leq T_i(k)))} \quad (28.5)$$

where

L_i = Time to failure for i_{th} component,

$T_i(k)$ = Time at which the state variable for the i_{th} component reaches the limit k,

T_1 = Expected time to perform an EM (in this context replacement of a component at failure),

T_2 = Expected time to perform a PM ($T_2 < T_1$).

Expression (28.5) is, as before, the expected uptime during a cycle divided by the total expected cycle length. This result is derived in Appendix B. Maximization of (28.5) will yield the optimal k for the condition-based replacement policy. To find the maximum, it is again recommended that A(k) be evaluated and plotted as a function of k. As with the age replacement policy, the value of k maximizing A(k) will be sensitive to the constants T_1 and T_2.

The availability ratio (28.5) does not lead directly to finding an optimal limit k because of the difficulty in specifying a model for the probability $Pr(L_i\ T_i(k))$. However, a simple empirical approach can be used to estimate the optimal k. Derivation of the empirical estimate assumes very little about the form of the probability model, making only the assumptions listed above in the Key Assumptions Associated with Age Replacement Policy section.

Data Requirements and Estimation Techniques for Condition-Based Policy

In practice, to establish a condition-based replacement policy, the data required are sample traces of the state variable, denoted by $X_i(t)$, from the time the component is placed in service until the time of failure. Figure 28.5 gives a hypothetical example of a trace of a state variable.

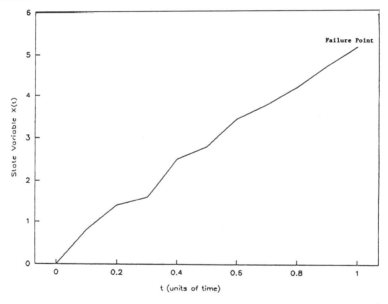

Figure 28.5. *Example of the Trace of a State Variable. It is monitored from the time the component is placed in service until the time of failure. Failure occurs after one unit of time.*

Suppose Figure 28.5 represents the first of a sample of n traces. Then $L_1 = 1.0$ (Time to failure), and if we choose, for example, k = 3, then $T_1(3) = 0.53$, the time at which the state variable $X_1(t)$ reached 3. Also needed are the times required for EM (T_1) and PM (T_2). If traces of state variables are not routinely collected by plant computer systems, special data collection or a separate experimental life test of components may be necessary to identify an appropriate state variable.

Given these data, the numerator in (28.5) can be estimated empirically by

$$\hat{E}(k) = \frac{1}{n}\sum_{i=1}^{n} \min(L_i, T_i(k)).$$

$Pr(L_i \leq T_i(k))$ can be estimated empirically by

$$P(k) = 1/n \text{ [number of cycles for which } L_i \leq T_i(k)].$$

$P(k)$ is just the sample fraction of times that the component fails before the state variable reaches the limit k. Note that each sample trace contributes one data point to the estimates $\hat{E}(k)$ and $\hat{P}(k)$. That is, each sample trace contains a single time to failure or time to reach the limit k. The availability ratio (28.5) can then be estimated by

$$\hat{A}(k) = \frac{\hat{E}(k)}{\hat{E}(k) + T_1\hat{P}(k) + T_2(1-\hat{P}(k))} \tag{28.6}$$

Once values for T_1 and T_2 are determined, the ratio (28.6) can be maximized, by evaluating and plotting $\hat{A}(k)$ as a function of k. Because of the form of $\hat{A}(k)$, its maximum within the range of the data must occur at a value of k for which $k = X_i(L_i)$, for some i. Thus the search for the optimal limit using this empirical estimator is straightforward.

The steps necessary to determine a condition-based PM plan for components are
1. Collect sample traces of the state variable from the time the component is placed in service until the time of failure. If such data are not readily available, an experimental life test may be necessary to identify an appropriate state variable and obtain component times to failure.
2. Determine the unknown time constants T_1 and T_2 in the availability ratio (28.5). Estimates of T_1 and T_2 can be obtained from historical maintenance records.
3. Using the empirical estimates $\hat{E}(k)$ and $\hat{P}(k)$, evaluate and plot $\hat{A}(k)$ as a function of k. These evaluations and the plot will indicate what value of k maximizes $\hat{A}(k)$, and how sensitive the availability is to the choice of k.
4. Having determined the value of k, say $\hat{k}$, that maximizes $\hat{A}(k)$, the resulting condition-based PM plan is
   - Perform PM (replacement or restoration) when the state variable first crosses the threshold value $\hat{k}$
   - If the component fails before the state variable crosses the threshold, perform EM at the time of failure.

These steps are illustrated in the example below.

Example Using Ion Source Failure Data

The techniques for establishing a condition-based replacement policy are illustrated using ion source failure data. An important component of an ion implanter in semiconductor manufacturing is the ionization chamber. Inside the chamber, electrons are created from a hot filament source, which is a wire of tungsten or tantalum that is heated by an electric current (see Figure 28.6).

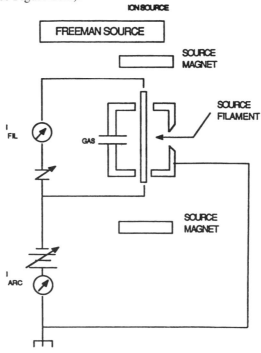

Figure 28.6. *Schematic Diagram of an Ion Source Chamber. This is an important component of an ion implanter used in semiconductor manufacturing. Inside the chamber, electrons are created from a hot filament source, a wire of tungsten or tantalum heated by an electric current. During the ionization process, the source filament is subject to deterioration and eventual breakage, causing downtime for replacement of the filament.*

At a certain temperature, electrons boil off the filament and ionize any source gas molecules they collide with. During the ionization process, the source filament is subject to deterioration and eventual breakage, which causes downtime for replacement of the filament. This is a frequent source of downtime for the ion implanter. An age replacement policy would use only time-to-failure data for the filaments, model the lifetime distribution, and optimize the PM schedule using expression (28.2). In practice, however, the current through the filament is closely monitored. As the filament wears, with a fixed voltage, the current through the filament steadily decreases until breakage occurs. Thus, a possible state variable to use in setting up a PM plan for source filament replacement would be the variable current through the filament as a function of operating time. A sample trace in this example is a plot of the decreasing current from the time of replacement with a new filament to the time of breakage. Figure 28.7 illustrates six hypothetical traces used in this example. In practice, more traces (say 20 - 25) should be used to establish a condition-based policy. Table 28.2 lists the times to failure and the values of the state variable at failure for each trace.

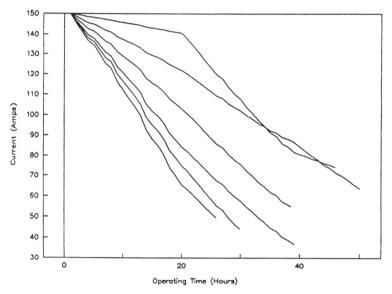

Figure 28.7. *Sample Traces of the Variable Current. This is used to monitor the state of the ion source filament. The current is monitored from the time the new source filament is placed in service until the time it breaks.*

Table 28.2. *Ion Source Failure Data. It includes time to failure and amps at failure. These data are used in establishing a condition-based replacement policy.*

i	L_i(Hours)	$X_i(L_i)$(Amps at failure)
1	26	50
2	30	45
3	39	37
4	38	50
5	50	64
6	46	75

Expression (28.6), the estimated availability ratio expressed as a function of k, is used to optimize the condition-based PM schedule. It is assumed that $T_1 = 4$ hours is required for emergency replacement, while $T_2 = 3$ hours is required for a scheduled replacement. As mentioned earlier, within the range of data, only the $X_i(L_i)$ values need to be considered as possible choices for k. First, the computation of $\bar{A}(k)$ for $k = X_1(L_1) = 50$ will be illustrated, using expression (28.6). In the numerator of (28.6),

$$\hat{E}(50) = \frac{1}{6}\sum_{i=1}^{6}\min(L_i, T_i(50))$$

$= 1/6 \ [\min(L_1, T_1(50)) + \min(L_2, T_2(50)) + \min(L_3, T_3(50)) + \min(L_4, T_4(50))$
$\quad + \min(L_5, T_5(50)) + \min(L_6, T_6(50))]$

$= 1/6 \ [\min(26,26) + \min(30,28) + \min(39,34) + \min(38, > 38) + \min(50, >50)$
$\quad + \min(46, > 46)]$

$= 1/6 \ [26 + 28 + 34 + 38 + 50 + 46]$

$= 37.$

In the derivation of $\hat{E}(k)$, times for the state variable to reach the control limit k may be unknown (censored) if failure occurs before the control limit is reached. In this example, the time required for the fourth observation of the state variable to reach 50 amps ($T_4(50)$) is not known exactly. It is only known that it is greater than 38 hours. So, $\min(L_4, T_4(50))$ is represented by $\min(38, > 38) = 38$ hours. In the denominator of (28.6),

$\hat{P}(50) = 1/6$ [number of traces for which $L_i \leq T_i(50)$]

$= 1/6$ [number of traces for which failure has occurred by 50 Amps]

$= 1/6 \ [4]$

$= 0.667.$

Thus $\hat{A}(50)$ is estimated by

$$\hat{A}(50) = 37/(37 + 4(.667) + 3(.333)) = 0.9098$$

Continuing in this fashion for each value of $k = X_i(L_i)$,

$\hat{A}(37) = 0.9052, \qquad \hat{A}(56) = 0.9118,$
$\hat{A}(45) = 0.9073, \qquad \hat{A}(64) = 0.9100,$
$\hat{A}(50) = 0.9098, \qquad \hat{A}(75) = 0.9059.$

Figure 28.8 shows the empirical availability curve, for values of current from 35 amps to 75 amps. In general, more data points will help to clearly define an availability curve.

For this small data set, the maximum availability occurs at $\hat{k} = 56$ amps, with an estimated availability of 91.18%. So the best condition-based policy, with empirical estimation of A(k), is to replace the source filament when the current decreases to 56 amps, or at failure, whichever occurs first.

By comparison, the age replacement policy would not use the sample traces in Figure 28.7, but only the times to failure listed in Table 28.2. Using the availability ratio (28.4), the optimum age replacement policy has $\hat{t}_0 = X_3 = 38$ hours with $\hat{A}(38) = 0.9083$, or 90.83% availability. With no preventive maintenance, which in this case corresponds to k = 0, the estimated availability is

$$\hat{A}(0) = \overline{L} / (\overline{L} + 4) = 38.17 \ 38.17 + 4 = 0.9051,$$

where $\bar{L}$ is the sample average time to failure. Note that all the availability ratios in this example are close to one because the replacement times are small relative to the expected operating time until failure. If the state variable is properly chosen, the condition-based policy will result in greater availability than the age replacement policy, which will be an improvement over no PM at all.

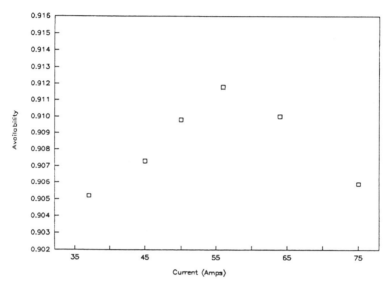

Figure 28.8. *Availability Plot for the Condition-Based Replacement Policy for Ion Source Filaments with no Distributional Assumptions. The plot suggests an optimal policy with a control limit of k approximately equal to 55 amps, with an availability of 91.2%.*

In summary, the condition-based policy is useful in situations where more is known about a component than just its operating age. The motivation for the condition-based policy is that replacements will be made only as needed, rather than according to a fixed schedule. The key assumption is that a state variable can be used to predict an upcoming failure. Indeed, the main difficulty in applying this type of PM policy may be in identifying an appropriate state variable. The effectiveness of the condition-based policy will depend on how closely the current value of the state variable is related to the probability of component failure. If a state variable chosen for this type of PM policy has nothing to do with the probability of failure, any resulting condition-based policy could be improved by simply using the time-to-failure data and constructing an age replacement policy using the techniques found in the Age Replacement Policy section.

In the development stages of manufacturing equipment, a thorough equipment characterization should include a study of causes of failure and identification of those state variables closely related to component failure. A correctly chosen state variable will allow determination of a condition-based PM policy which will increase the equipment availability over that of a simple age replacement policy. Values of appropriate state variables should routinely be monitored, with automated data collection of traces so that an optimal PM schedule can easily be determined, or an existing PM schedule can easily be reevaluated. It is also important to consider, in the development stages, the ease of monitoring the key state variables. Bergman (1978), Hertzbach (1967), and Gertsbakh (1977) provide theoretical details associated with this approach. Bergman (1977) and Geurts (1983) present applications of the condition-based policy, although Geurts' approach is different from the one presented here.

SUMMARY AND CONCLUSIONS

Further Comments on PM Plans

The traditional approach to PM in American industry has been to remove a component or system from service when it fails or to simply fix it when it breaks. One commonly used approach which will increase equipment availability is to remove the component from service at predetermined intervals. This is the so-called age replacement policy. A technique which offers even more potential for increased efficiency over either of these approaches is the condition-based policy. Rather than replacing only at failure or at predetermined intervals, this policy attempts to provide PM only when necessary. This approach requires more knowledge about how the component or equipment works and what state variables need to be monitored. It also requires having the appropriate diagnostics built into the equipment to allow easy monitoring of the key variables. If an appropriate state variable can be identified, the condition-based policy will result in greater availability than the age replacement policy, which is an improvement over no PM at all.

A key problem in establishing efficient PM schedules is collecting the appropriate data for estimating model parameters. Plant data collection systems, either automatic or manual, typically do not keep track of operating times between failures, which must be obtained to estimate time-to-failure distributions. Data systems should be designed to routinely collect such data and to establish new PM plans or to evaluate existing plans. Also, the expected times to complete EMs and PMs need to be well estimated because the optimal PM schedules for the models reviewed here are sensitive to these times.

Comments on the Reference List

An extensive literature exists on the subject of PM, primarily in the operations research area, with relatively little appearing in the traditional statistical quality control literature. Comprehensive literature reviews are given by McCall (1965), Pierskalla and Voelker (1976), Sherif and Smith (1981), and Valdez-Flores and Feldman (1989). These references provide a starting point for information on more complicated models and techniques which are beyond the scope of this case study. Excellent texts on PM include Barlow and Proschan (1996) and Gertsbakh (1977), although the treatment of the subject is somewhat mathematical. Useful texts describing how to estimate parameters of time-to-failure distributions include Lawless (1982) and Nelson (1982, 1983) for i.i.d. failure data. Lau (1983) provides a bibliography of other books on the subject of reliability and maintainability.

REFERENCES

Barlow, R. "Analysis of Retrospective Failure Data Using Computer Graphics," *Proceedings 1978 Annual Reliability and Maintainability Symposium*, pp. 113-116, 1978.

Barlow, R. and L. Hunter,. "Optimum Preventive Maintenance Policies," *Operations Research*, Vol. 8, pp. 90-100.

Barlow, R. and F. Proschan, *Mathematical Theory of Reliability,* Philadelphia, PA: Society for Industrial and Applied Mathematics, 1996.

Bergman, B. "Some Graphical Methods for Maintenance Planning," *Proceedings 1977 Annual Reliability and Maintainability Symposium*, pp. 467-471, 1977.

Bergman, B. "Optimal Replacement Under a General Failure Model," *Advances in Applied Probability*, V. 10, pp. 431-451, 1978.

Gertsbakh, I. *Models of Preventive Maintenance*, New York, NY: North-Holland, 1977.

Geurts, J. "Optimal Age Replacement Versus Condition-based Replacement: Some Theoretical and Practical Considerations," *Journal of Quality Technology*, Vol. 15, No. 4, pp. 171-179, 1983.

Glasser, G. "Planned Replacement: Some Theory and its Application," *Journal of Quality Technology*, Vol. 1, No. 2, pp. 110-119, 1969.

Hertzbach, I. "Preventive Maintenance Using Prognosis Parameter," *Engineering Cybernetics*, No. 1, pp. 54-63, 1967.

Ingram, C. and R. Scheaffer,. "On Consistent Estimation of Age Replacement Intervals," *Technometrics*, Vol. 18, No. 2, pp. 213-219, 1976.

Lau, H. (1983), "Reliability, Maintainability and Cost-Effectiveness: A Bibliographical Note," *Microelectronics Reliability*, Vol. 23, No. 1, pp. 21-40, 1983.

Lawless, J. *Statistical Models and Methods for Lifetime Data*, New York, NY: John Wiley and Sons, Inc., 1982.

McCall, J. "Maintenance Policies for Stochastically Failing Equipment: A Survey," *Management Science*, Vol. 11, No. 5, pp. 493-524, 1965.

Nelson, W. *Applied Life Data Analysis*, New York, NY: John Wiley and Sons, Inc. 1982.

Nelson, W. "How to Analyze Reliability Data," *ASQC Basic References in Quality Control*, Volume 6, 1983.

Pierskalla, W. and J. Voelker, "A Survey of Maintenance Models: The Control and Surveillance of Deteriorating Systems," *Naval Research Logistics Quarterly*, Vol. 23, pp. 353-388, 1976.

Sherif, Y. and M. Smith, "Optimal Maintenance Models for Systems Subject to Failure: A Review," *Naval Research Logistics Quarterly*, V. 28, pp. 47-74. 1981.

Valdez-Flores, C. and R.A. Feldman, "Survey of Preventive Maintenance Models for Stochastically Deteriorating Single-Unit Systems," *Naval Research Logistics*, Vol. 36, pp. 419-446, 1989.

MIL-STD-781, *Reliability Testing for Engineering Development, Qualification, and Production*.

APPENDIX A

Availability Ratio For Age Replacement Policy

For the age replacement policy, the availability ratio is given by

$$A(t_0) = \frac{E\{\min(x_i, t_0)\}}{E\{\min(x_i, t_0)\} + T_1 F(t_0) + T_2\{1 - F(t_0)\}}. \qquad (28.7)$$

Defining a cycle to be the random time between successive component replacements (including either an emergency or PM), then (28.7) can be interpreted as the ratio

Expected Uptime / (Expected Uptime + Expected Downtime)

for a cycle. The numerator in (28.7) is simply the expected operating time during the ith cycle. It is equal to X_i (the ith time to failure) if failure occurs before t_o. It is equal to t_o otherwise. In the denominator of (28.7), the expected downtime is equal to T_1 if failure occurs before the component reaches the age t_o, which happens with probability $F(t_o)$. It is equal to T_2 if the component reaches the age to without failure, which happens with probability $(1-F(t_o))$. It can be shown that

$$E\{\min(x_i, t_0)\} = \int_0^{t_0}(1-F(x))dx$$

so that $A(t_o)$ can be expressed as

$$A(t_0) = \frac{\int_0^{t_0}(1-F(x))dx}{\int_0^{t_0}(1-F(x))dx + T_1 F(t_0) + T_2(1-F(t_0))}.$$

APPENDIX B

Availability Ratio For Condition-Based Policy

The long-run availability ratio for the condition-based policy is given by:

$$A(k) = \frac{E\{\min(L_i, T_i(k))\}}{E\{\min(L_i, T_i(k))\} + T_1 \Pr(L_i \leq T_i(k)) + T_2(1 - \Pr(L_i \leq T_i(k)))} \quad (28.8)$$

The ratio in (28.8) can be interpreted as the ratio

Expected Uptime / (Expected Uptime + Expected Downtime)

for a cycle. The numerator in (28.8) is the expected value of the operating time during a random cycle. It is equal to L_i if failure occurs before the state variable reaches the limit k, and it is equal to $T_i(k)$ if the variable reaches k before the component fails. In the denominator, the expected downtime is equal to T_1 if failure occurs before the state variable reaches k, which happens with probability $\Pr(L_i T_i(k))$. It is equal to T_2 if the state variable reaches k before failure occurs, which happens with probability equal to $1- \Pr(L_i T_i(k))$.

BIOGRAPHY

Stephen V. Crowder is a senior member of technical staff in the statistics and human factors department at Sandia National Laboratories in Albuquerque, New Mexico. He received a Ph.D. in statistics from Iowa State University in 1986 and worked as a statistician in industry before joining Sandia in 1989. His current responsibilities include consulting in statistical process control, experimental design, and data analysis with engineers and scientists at Sandia. Current research interests include preventive maintenance using degradation measures and statistical process control for low volume manufacturing.

PART 6

COMPREHENSIVE CASE STUDY

CHAPTER 29

INTRODUCTION TO COMPREHENSIVE CASE STUDY

Veronica Czitrom

This case studies book has used the SEMATECH Qualification Plan as a framework for implementing statistical methods for process and equipment improvement. Each part has focused on a particular statistical method that supports a critical element of that Qualification Plan. The book would not be complete without a comprehensive case study that "puts it all together."

The Qualification Plan for an Equipment Improvement Project (EIP) at SEMATECH might consist of gauge studies to establish the capability of the measurement tools, a passive data collection to establish the process and equipment baseline, designed experiments to characterize and optimize the process and establish process capability, and a marathon run to improve equipment reliability. In a comprehensive case study, Reece and Shenasa (Chapter 30) describe the application of these four parts of the Qualification Plan during a furnace EIP.

Reece and Shenasa's case study describes the redesign of a new type of vertical furnace used to deposit thin films on silicon wafers. The process they considered is chemical vapor deposition (CVD) of silicon nitride. The project considered equipment reliability as well as process improvement. A brief description of some of the highlights of their application of the SEMATECH Qualification Plan follows.

(1) A *gauge study* was used to establish the capability of the gauge to measure the thickness of a thin silicon nitride film. The gauge was found to be capable of performing the measurements, since the measurement error was very small compared to the 1100 Ångstrom thickness that was measured.

(2) A *passive data collection* was used to identify major sources of variability in the silicon nitride deposition process and to establish process stability. The sampling plan consisted of 26 furnace runs with one wafer at the top of the furnace and another at the bottom of the furnace. *Control charts* indicated that the process was stable, and an initial estimate of process capability gave a C_{pk} value of 1.9. A later furnace run with eleven wafers equally spaced along the furnace indicated that wafer thickness along the furnace followed a sinusoidal curve. This meant that the original sampling plan was inappropriate, since it did not account for variability along the furnace in each run, and that the initial estimate of process capability was unduly optimistic.

(3) A screening *designed experiment* was used to identify the major factors and interactions affecting the deposition of silicon nitride. The experiment was a fractional factorial design of resolution IV in seven factors in sixteen runs with five center points. It was not possible to perform some of the experimental points due to pump capacity limitations, so five additional runs were selected using the D-optimal criterion. Four responses were considered: film thickness, and three measures of film thickness variability (total variability, within-wafer variability, and wafer-to-wafer variability), from which capability potential was also modeled. Data analysis included modeling the responses using regression, and examining graphs such as residual plots, interaction graphs, and contour plots. The center points were used to detect the presence of curvature in several responses, although the factors that caused it could not be identified without further experimentation.

(4) *Confirmation runs* and a second *passive data collection* were performed on a new furnace with improved pump capacity.

(5) A *marathon* was used to measure the utilization capability of the furnace under conditions approximating a manufacturing environment. The furnace was exercised 24 hours a day, seven days a week, for over five weeks. Equipment dependent uptime was found to be 91%.

(6) A second *designed experiment* was performed to establish a process window and process sensitivity to changes in process inputs. A central composite design in a subset of five of the seven factors was used, which allowed fitting a quadratic model (with all two-factor interactions and quadratic terms) that would identify the factors causing curvature in the response. The experiment verified the robustness of the operating conditions.

CHAPTER 30

CHARACTERIZATION OF A VERTICAL FURNACE CHEMICAL VAPOR DEPOSITION (CVD) SILICON NITRIDE PROCESS

Jack E. Reece and Mohsen Shenasa

Members of the SEMATECH consortium work with suppliers of processing equipment for the semiconductor industry to help those suppliers regain world leadership in the manufacture of new tools suitable for advanced semiconductor processing. SEMATECH requires aggressive use of applied statistics methods in the execution of the projects it undertakes to assure the member company community that the information reported is complete and reliable. This case study describes the integrated application of those statistical techniques using the SEMATECH Qualification Plan throughout a study to characterize a new type of vertical furnace used for the deposition of thin films on silicon wafers. The process illustrated uses chemical vapor deposition (CVD) techniques to deposit approximately 1100 Å of silicon nitride (Si_3N_4) on batches of 125 150-mm wafers.

EXECUTIVE SUMMARY

Problem

SEMATECH Equipment Improvement Program (EIP) investigations have a variety of objectives:
   1. characterizing the process capability of a specific process tool, including identifying and minimizing its sources of variability and establishing boundaries for its process window;
   2. establishing estimates of the cost of ownership of a processing tool;
   3. estimating the reliability of the tool in a manufacturing environment.

The investigation of the chemical vapor deposition (CVD) silicon nitride (Si_3N_4) process described was part of a cooperative effort between SEMATECH, the tool supplier, and National Semiconductor Corporation. The project involved redesign of the process tool and considered reliability as well as process issues. This case study considers only the process characterization portion of the project.

Solution Strategy

The investigation followed the SEMATECH Qualification Plan. This plan assures that investigators collect the proper information using sound statistical techniques and subject processing tools to the same levels of testing and scrutiny, and that all studies supported by SEMATECH report the same types of information.

This case study illustrates measurement capability analyses, passive data collection (PDC) studies, experiments (DOE) to quantify the contribution of major process inputs, and experiments to characterize fully the process window of the tool. It also describes the errors than can occur due to improper sampling plans (PDC) and due to invalid assumptions made in defining variable settings in an experiment (DOE).

Conclusions

The total measurement error associated with the tool used for this project (thickness measurements only) was 0.67 Å (one standard deviation). In addition, the measurement tool showed no appreciable instability nor sensitivity to different operators, etc. Since the films being deposited were approximately 1100 Å thick, this level of measurement error was of no concern.

An initial passive data study (26 furnace runs under constant conditions, two wafers per run) suggested that the existing process had a Cpk of around 1.9. Variation in thickness within the monitor wafers accounted for about two thirds of the process variability. However, later investigations using 11 wafers per run showed that the sampling plan used initially (a wafer at each end of the furnace) did not detect sinusoidal variation of film thickness from one end of the furnace to the other. Therefore, the original estimate of Cpk was extremely misleading and overestimated process capability by a factor of two or three.

A screening experiment involving seven process variables helped identify a process space with less total variability. This experiment contained combinations of total gas flow and system pressure which the conductance of the system could not support. Augmenting the original design with additional experimental runs that comprehended the constraints in the process tool removed the confounding present in the broken original matrix. Interpretation of the results from this experiment with additional information provided by the tool supplier identified potential operating conditions. A second passive data study following the screening experiment (30 runs, six wafers per run to capture within run variability more accurately) produced a Cpk of around 1.8. Full characterization of this window using response surface methodology and quadratic models identified still further possible improvements in the process which led to a Cpk greater than 2.0.

The complete project provided the tool supplier with valuable information regarding the design and performance of its tool. The process characterization demonstrated that this furnace design matched or exceeded the performance of any comparable tool available in world markets.

INTRODUCTION

This paper illustrates the sequential application of elements of the SEMATECH Qualification Plan in the characterization of a chemical vapor deposition (CVD) process. It includes
1. an initial capability analysis of the film thickness measurement tool,
2. a passive data study intended to identify major sources of variation within the process and to demonstrate process stability,
3. an initial screening experiment to identify and quantify major effects and interactions among process inputs,
4. a final response surface experiment to identify a process window and to quantify the sensitivity of the process outputs to minor changes in process inputs.

A marathon was also performed, but it is only mentioned in this case study.

ESTABLISHING MEASUREMENT TOOL CAPABILITY

The measurement tool studies estimated the short term measurement error (repeatability error) and the longer term error involving several days and different operators (total measurement or reproducibility error). The investigators made no effort to isolate the reproducibility error since the numbers were small enough that they had no effect on interpreting the process. The analysis techniques used represent approximations of more rigorous analysis of variance methods commonly reported by other investigators and agree very closely with the more rigorous methods, particularly with balanced data structures (equal numbers of observations in all cells). In the absence of a suitable traceable standard, the investigators relied on representative artifact wafers; therefore, no estimate of measurement bias was possible.

Repeatability Error

To estimate repeatability, the engineers measured the same site 30 times on each of three wafers without removing the wafer from the measurement device between measurements. The three wafers chosen spanned the range of expected variation of wafer thicknesses for the process being investigated. Figure 30.1 is a box and whisker plot illustrating the variation found for each set of measurements. Obviously the repeatability error is relatively independent of nominal film thickness over this range of values (the interquartile ranges (IQR) of the wafers are quite similar).

Table 30.1. *Standard Deviation and P/T by Wafer, Measurement Repeatability Study.*

Nominal Thickness	StDev Thickness	P/T Repeat
777	0.13257	0.51
1024	0.111675	0.33
1178	0.224505	0.57

Table 30.1 illustrates the standard deviations found for the repeated measurements on each wafer and the P/T ratio associated with each.

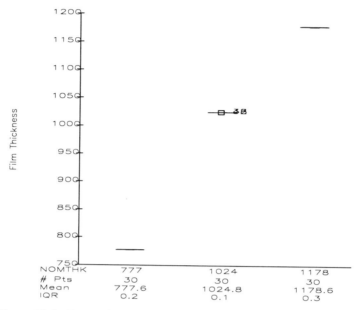

Figure 30.1. *Box and Whisker Plot Illustrating Repeatability Error as a Function of Nominal Wafer Thickness.*

P/T$_{rpt}$ is the ratio of 6 times the repeatability measurement error divided by the tolerance of the process with the result multiplied by 100:

$$\text{P/T}_{rpt} = \left(\frac{6 * \hat{\sigma}_{rpt}}{\text{USL} - \text{LSL}} \right) * 100 .$$

Since the process specification offered a process window of ±10% around the target value, the denominator in each calculation in Table 30.1 is 20% of the observed nominal thickness. Having P/T$_{rpt}$ less than 5.0 is highly desirable in a measurement tool as it represents the inherent error in the measurement tool relatively independent of external influences.

Total Measurement Error

Table 30.2 summarizes the calculations for total measurement tool error. To estimate this error, three operators measured five sites on the same wafer twice a day for 10 days. Pooling the variances found at each wafer site provides a useful approximation of the total measurement error. This approximation blends contributions from operators, days, time of day, and any possible interactions and assumes that all variables in the study are crossed. In addition, it treats wafer site as a fixed effect associated with the wafer and not the measurement tool. All other variables are random effects. The P/T_{tot} ratio calculation is analogous to that just illustrated: the numerator is six times the estimated total measurement error, and the denominator is 220 Å since the target value of the film thickness is 1100 Å±10%. The acceptibility criterion for P/T_{tot} is $P/T_{tot} \leq 30$.

Table 30.2 *Estimation of Total Measurement Error and Associated P/T Ratios.*

Wafer Site	Measurement Count	Measurement Variance
1	60	0.183898
2	60	0.181921
3	60	0.488983
4	60	0.251695
5	60	0.864124
Pooled Variance		0.394124
Measurement Error $\hat{\sigma}_{measerror}$		0.627793
P/T (Tot Meas Error)		1.712163

Conclusions from the Measurement Capability Study

Both measurement studies and their associated P/T ratios demonstrated that the measurement device was suitable for its intended purpose and that further analyses need not include an adjustment for measurement error. Had the P/T ratio for total measurement error exceeded 30%, then the investigators would have applied a correction factor based on $\hat{\sigma}_{measerror}$ to any variation of measurements seen at the lowest level of hierarchy—within wafer.

PRELIMINARY PASSIVE DATA STUDIES

At the start of this project, the sampling plan recommended by the supplier placed one wafer at the top of the furnace load, one at the bottom, and one in the middle. A few initial trials suggested that the film thickness on the middle wafer was not practically different from the thicknesses observed at either end. Therefore, the initial passive data study included only the top and the bottom wafers. Five measurements were made on each wafer.

Figure 30.2 illustrates the control chart produced from this preliminary study involving 26 furnace runs under constant conditions. The chart includes the mean thickness and the standard deviation from each run. Trend analysis of the individual chart using the Western Electric Trend Rules suggests that Run 14 was out of statistical control. The signal given (Rule 3) indicates that four of the last five points were above one sigma. SEMATECH does not regard this as a serious problem, choosing instead to react only to points above or below three sigma or to situations in which eight consecutive points are on one side of the center line. Therefore, particularly since no pattern continued in the individuals chart beyond Run 14, the investigators considered the process stable and under statistical process control at this point. The standard deviation chart uses an estimated subgroup size three (two wafers per run, five measurements per wafer) based on Satterthwaite's approximation[70] of degrees of freedom for nested variances (see Satterthwaite, 1946). Trend analysis of this chart indicates no runs or other out-of-control features.

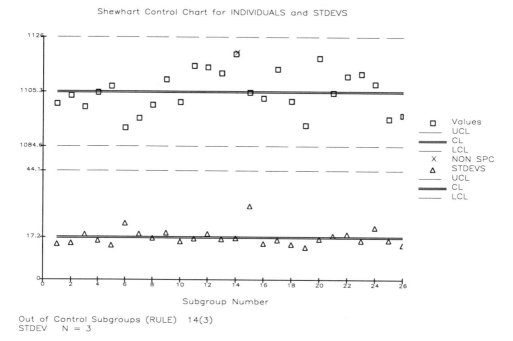

Figure 30.2. *Individual Chart for Run Means and Run Standard Deviation from an Initial 26-Run Passive Data Study Combined.*

Variance component analysis of the data from the initial passive data study (assuming nested random variables) produced Table 30.3. The table suggests that the majority of the variability in the data is due to within-wafer-contributions. Figure 30.3 is a bar graph further illustrating the relative magnitudes of the standard deviations and their sources.

Table 30.3. *Variance Components Analysis of Initial Passive Data Study. By convention, a negative estimate of variance has been set to 0.*

Variance Source	Standard Deviation	Variance Component	% of Total Variance
RUN	0	-16.916	0
WFtoWF	10.326	106.622	30.244
WITHIN WAF	15.686	245.918	69.756
TOTAL	18.776	352.541	100

---

[70] The approximation implemented here was supported by a custom program running under RS/1 software.

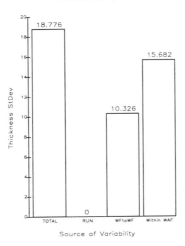

Figure 30.3. *Relative Contributions of Sources of Variation in the Initial Passive Data Study Expressed as Standard Deviations.*

The negative variance component due to run to run variation indicates it was very small compared to the others. The method of estimation often finds negative answers for such components; convention sets them to 0.

Figure 30.4 is a capability graph of the data gathered during the passive data study. It indicates that the process is very near target with mean = 1105.3 and standard deviation = 18.3. This number differs slightly from that computed using variance components. The reason they are different is that they are estimated using different formulas.

While the results of this study were certainly encouraging, given the relatively high capability statistics found, subsequent analysis with a revised sampling plan showed that they were extremely misleading and that the original sampling plan had missed a significant source of variation within the process. As the project team was discussing the design of experiments to begin characterizing this process (see the next section), the author (JR) insisted that each treatment combination (run) should include enough monitor wafers to give a good representation of the variation of film thickness from one end of the furnace to the other. The compromise plan placed monitor wafers in every 15th slot, starting with slot 5 (usable wafer slot numbers range from 5 to 155 from bottom to top of the furnace). Therefore each run contained 11 monitor wafers with dummy wafers placed in the empty slots. Four to five dummy wafers preceded and followed the first and last monitor wafers.

Before starting the screening experiment, the team ran one additional batch using the new sampling plan and the same operating conditions as the initial passive data study. Figure 30.5 is a scatter plot illustrating the new results. The symbols on this graph represent the mean thickness found at each slot; the vertical lines are locations of monitor wafers in the original sampling plan recommended by the supplier.

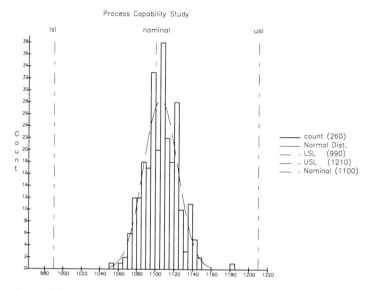

Figure 30.4. *Capability graph from initial passive data collection study. Process Mean = 1105.3 Å; Standard Deviation = 18.3 Å. Capability potential (Cp) = 2.0; Cpk = 1.9.*

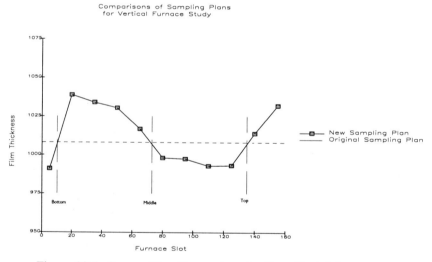

Figure 30.5. *Scatter Plot Comparing the True Variability within a Furnace Run with the Original Sampling Plan in the Initial Passive Data Collection Study.*

Figure 30.5 supports the team's original decision to use only two monitors of the three originally suggested by the tool supplier. However, it also clearly indicates how an incorrect sampling plan can produce misleading information. Clearly the between wafer (or within run) variation contributes significantly to process variability, but it escaped detection in the original plan.

Conclusions Supported by the Initial Passive Data Study

The team elected to move to the next phase of the investigation -- to understand and to quantify contributions of major process variables using a screening design for two reasons:
1. The initial passive data study indicated a stable process under statistical process control.
2. Time and cost considerations made repeating the entire PDC study with more wafers per run unattractive, particularly since similar studies following the experimental design phase would characterize the actual sources of variation in the process.

SCREENING EXPERIMENTS

While the initial passive data study produced misleading information regarding within run variability, it did demonstrate that the current process was stable and suitable for experimentation to reveal the important variables available to modify the process.

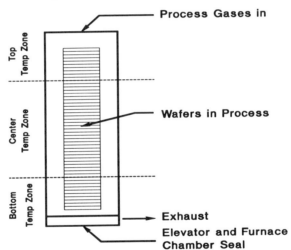

Figure 30.6. *Schematic of Vertical Furnace Reactor.*

Figure 30.6 is a highly schematic representation of the furnace used in this investigation. The furnace provides slots for up to 160 wafers in a quartz boat supported on an elevator. The elevator raises the wafers from atmospheric conditions into the furnace chamber then seals the furnace. System controls maintain the desired pressure at a constant flow within the furnace using a closed loop approach. This method introduces nitrogen gas just upstream of the vacuum pump running at constant speed. Adjusting the amount of nitrogen introduced controls the pressure in the system, once the reactive gases enter the furnace.

A mixture of reactive gases enters the top of the furnace and travels to the bottom through three temperature-controlled zones. Because the top entry is the only source of reactants, the process depletes the gases as they pass through the heated reaction chamber. To counter the effects of this depletion on the deposition rate of the film from top to bottom in the furnace, a typical recipe maintains a higher temperature at the bottom of the reactor than at the top. The rate of silicon nitride deposition is higher at higher temperatures.

The separation between wafers within the furnace chamber (boat pitch) can also influence the process. Larger pitch values (in 0.001 inch) may allow better diffusion of the process gases between adjacent wafers. However, increasing this spacing means that each batch contains fewer wafers and throughput suffers. Table 30.4 lists the factors used in this study and their settings. These factors will now be described.

Table 30.4 *Factors in the First Experiment and Their Settings.*

Factor	Abbrev	Units	Settings
Top Offset	TPOFF	Deg C.	-30 to -10
Center Temp	CTMP	Deg C.	750 to 810
Bottom Offset	BTOFF	Deg C.	10 to 30
Total Flow	TFLOW	SCCM	200 to 400
Ammonia/Dichloro-silane Ratio	ASR		2 to 10
Closed Loop Pressure	CLP	MTorr	150 to 350
Boat Pitch	BP		0.14, 0.2, 0.28

The metrics describing the gas composition flowing through the chamber (Total Flow and Ammonia/ Dichlorosilane Ratio factors) reflect a concern for the chemistry of the reaction. Alternatively, one could use the raw flows of the gases ammonia and dichlorosilane as factors. The author (JR) has found that describing the variables as indicated in Table 30.4 usually produces more meaningful regression models, particularly when the gases react with each other. The alternative factor metrics

$$TFLOW = (Flow\ NH_3) + (Flow\ SiCl_2H_2),$$
$$ASR = (Flow\ NH_3) / (Flow\ SiCl_2H_2)$$

can be solved for the two gas flows, namely

$$Flow\ NH_3 = (ASR)*(Flow\ SiCl_2H_2),$$
$$Flow\ SiCl_2H_2 = (TFLOW) / (1 + ASR).$$

In a similar fashion one could establish temperature ranges of interest for the top, center, and bottom zones of the furnace. Treating the center temperature zone as the controlling zone and slaving the other two zones to it using relative offset temperature settings (as in Table 30.4) assures that the gas entry zone will always remain cooler than the center and that the exit zone will always remain warmer than the center.

The range of boat pitch settings reflects whether or not the operators placed a wafer in every slot or in every other slot. Because of this a center point setting for this boat is not available. Fortunately, another boat with a 0.020 inch pitch was available, so the investigation used that pitch for a center point in the design matrix.

Responses

Table 30.5 lists the responses described in this example. Each treatment combination (run) in the experiment contained 11 wafers on which the operators measured 9 sites. Film thickness is the average of those 99 measurements; thickness standard deviation is the standard deviation of those measurements and is a blend of two sources of variation—that contributed by wafer-to-wafer variation within each batch and that contributed by within or across wafer variation on each wafer in each batch. This discussion illustrates the results of estimating these sources of variation using Method 1 below because it involves simpler calculations.

Table 30.5. *Responses for the Initial Screening Experiment.*

Response Name	Abbrev
Film Thickness	THK
Thickness Standard Deviation	THKSD
Wafer-to-Wafer Variance	WTWVAR
Within-Wafer Variance	WIWVAR

Method 1 overestimates the contribution of wafer-to-wafer variance somewhat, so it is a relatively conservative approach. Method 2 also overestimates this contribution, but not as much as Method 1.

Method 1

To estimate the variability contributed by variations in film thickness across each wafer in each run (within-wafer variability), compute the variance of the measurements of the nine sites on each wafer in a particular run, then pool them to provide within-wafer variance (WIWVAR) for each run:

$$\sigma^2_{wiwvar} = \frac{df_{waf1}*\sigma^2_{waf1} + df_{waf2}*\sigma^2_{waf2} + \ldots + df_{waf11}*\sigma^2_{waf11}}{df_{waf1} + df_{waf2} + \ldots + df_{waf11}}.$$

In the equation above, *df* is the degrees of freedom associated with each variance estimate; in this case it is one less that the number of measurements across a wafer.

To estimate the variability contributed by variations in the average film thicknesses on the 11 wafers in each run (wafer-to-wafer variability), compute the average thickness on each wafer and calculate the variance of those averages; this value is the wafer-to-wafer variance (WTWVAR) for each run.

Method 2

An inherent assumption in this calculation is that the sites within wafers and wafer positions within runs are random nested effects. This assumption is not literally true since each run had wafers in the same positions and since operators measured the same sites on each of those wafers. Technically these are fixed effects, and one can argue that they are not nested. However, the purpose of the investigation is to make the film thickness produced by the process robust to the location of a wafer within the furnace and to the location of a measurement point within a particular wafer. Therefore, one may argue that the wafers chosen and the points on those wafers are representative of the entire population of wafers and sites available. Therefore, the values obtained are not true variance components as a classical statistician might compute; rather they are effective estimates of the sources of dispersion active within a particular treatment combination. Calculations using this approach require special software routines capable of estimating expected mean squares based on a model definition. See Appendix A for comments regarding software. Figure 30.7 is a graphical illustration of the nested structure.

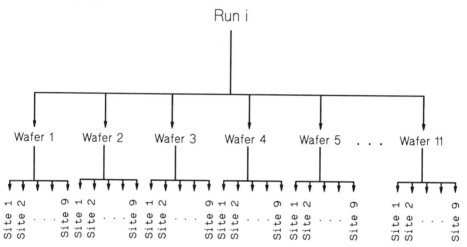

Figure 30.7. *Nested Structure Assumed for* 11 *Wafers Within a Furnace Run and Nine Measurements per Wafer.*

Another approach to modeling the performance of the process is to use the estimated capability potential (ECP). Clearly one of the abuses of capability potential calculations is their premature application to any process. However, a computation of an estimated capability can provide models and predictions regarding factor settings which might maximize that measurement.

Each treatment combination in this investigation included 99 measurements of thickness. Typical specification limits for processes of this type require maintaining the film thickness within ±10% of some target value. If the mean thickness obtained in each run is the target, then a calculation of estimated capability potential becomes

$$Cp_{est} = \frac{0.2 * THK}{6 * \sigma_{thksd}}.$$

Technically, this response blends two other responses: film thickness and thickness standard deviation. As such, it is a derivative of another statistic called coefficient of variation, which is the ratio of the standard deviation to the mean (a commonly used name for coefficient of variation is uniformity). Many statisticians, including the author (JR) warn against such a combination as it can lead to meaningless models or to models extremely difficult to comprehend.

They will argue instead that one should model each response separately, then combine the two models later into a formula that computes the desired metric. Obviously creating and exploiting such a formula requires special software utilities not available in all packages (see Appendix A). This paper uses this approach in reporting estimated capability potential.

Design Matrix

The investigation used a resolution IV fractional factorial design which required 16 treatment combinations and included five center point replicates. In this design matrix, third order terms are confounded with factor main-effects and at least one two-factor interaction (second order term) is confounded with another two-factor interaction. Table 30.6 summarizes the randomized worksheet used.

Table 30.6. *Randomized Worksheet Based on a Resolution* IV *Design for Seven Factors in* 16 *Runs with 5 Center Point Replicates. In this table, the abbreviation TC stands for treatment combination.*

TC	TPOFF	CTMP	BTOFF	TFLOW	ASR	CLP	BP
1	-20	780	20	300	6	250	0.20
*2	-30	810	10	400	10	150	0.28
3	-30	750	10	400	2	350	0.28
4	-20	780	20	300	6	250	0.20
*5	-10	750	30	400	2	150	0.28
6	-30	750	30	200	10	350	0.28
7	-10	810	30	400	10	350	0.28
8	-10	750	30	200	2	350	0.14
9	-10	750	10	200	10	150	0.28
10	-30	750	10	200	2	150	0.14
11	-20	780	20	300	6	250	0.20
12	-10	750	10	400	10	350	0.14
13	-10	810	10	200	2	350	0.28
14	-10	810	30	200	10	150	0.14
15	-20	780	20	300	6	250	0.20
16	-30	810	30	200	2	150	0.28
*17	-30	750	30	400	10	150	0.14
*18	-10	810	10	400	2	150	0.14
19	-20	780	20	300	6	250	0.20
20	-30	810	30	400	2	350	0.14
21	-30	810	10	200	10	350	0.14

Initial Results: Repairing the Design

Prior to setting up the design matrix, the investigators checked the pumping capacity (conductance) of the furnace system. Unfortunately, those pumping experiments used only nitrogen gas, rather than the actual mixture of ammonia and dichlorosilane. This produced an overestimation of its conductance under operating conditions. The treatment combinations in Table 30.6 marked with * were not achievable; the system could not provide 150 MTorr pressure when the total flow of gases was 400 SCCM. The engineer originally leading this investigation did not attempt to find another low pressure that was useful under these conditions; rather, he immediately substituted the high setting of pressure in each case (350 MTorr). This produced considerable collinearity between pressure and total flow and introduced some uncertainty into the analysis of the data.

Figure 30.8a (shaded portion) indicates the approximate region of inaccessible combinations of total flow and pressure. To correct this problem, the author (JR) augmented this design matrix with additional treatment combinations provided by a D-optimal algorithm. The symbols in Figure 30.8a represent a new grid of candidate treatment combinations created for these two variables to allow the design generation algorithm to closely approach the inaccessible region in selecting new treatment combinations. The following equation was used to exclude inaccessible treatment combinations from consideration using RS/1 software:

$$CLP < 200 \text{ and } TFLOW > 275 \text{ or } BP <> 0.14.$$

A better way to define an exclusion condition requires graphing settings of one input variable against settings of the other close to the boundary of the inaccessible region and estimating an equation for a line connecting those points. Figure 30.8b illustrates such a graph. An alternative equation for the exclusion becomes:

$$TFLOW < (2.6*CLP - 120) \text{ or } BP <> 0.14.$$

The original modification added six runs to the experiment, but time constraints allowed execution of only the first five. The new runs excluded those combinations of pressure and total flow that the furnace could not produce and used only the 0.14 boat pitch. Preliminary analysis of the data from the broken experiment indicated that the 0.14 pitch was the most desirable setting for this factor. Table 30.7 illustrates the final worksheet, including the five new runs and the results.

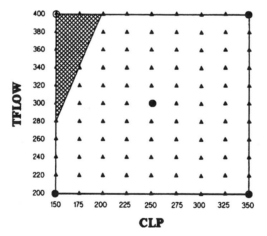

Figure 30.8a. *Estimated Inaccessible Region (Shaded Portion) of Total Flow and Pressure Settings.*

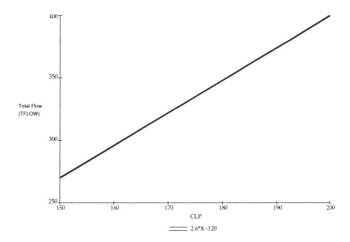

Figure 30.8b. *Plot of Total Flow versus Closed Loop Pressure Near the Boundary Conditions.*

Table 30.7. *Final Randomized Worksheet for Initial Screening Experiment with Results.*

TC	TPOFF	CTMP	BTOFF	TFLOW	ASR	CLP	BP
1	-20	780	20	300	6	250	0.20
2	-30	810	10	200	10	150	0.28
3	-30	750	10	400	2	350	0.28
4	-20	780	20	300	6	250	0.20
5	-10	750	30	200	2	150	0.28
6	-30	750	30	200	10	350	0.28
7	-10	810	30	400	10	350	0.28
8	-10	750	30	200	2	350	0.14
9	-10	750	10	200	10	150	0.28
10	-30	750	10	200	2	150	0.14
11	-20	780	20	300	6	250	0.20
12	-10	750	10	400	10	350	0.14
13	-10	810	10	200	2	350	0.28
14	-10	810	30	200	10	150	0.14
15	-20	780	20	300	6	250	0.20
16	-30	810	30	200	2	150	0.28
17	-30	750	30	200	10	150	0.14
18	-10	810	10	200	2	150	0.14
19	-20	780	20	300	6	250	0.20
20	-30	810	30	400	2	350	0.14
21	-30	810	10	200	10	350	0.14
22	-30	810	30	400	10	200	0.14
23	-30	750	30	200	2	350	0.14
24	-10	810	30	200	2	350	0.14
25	-10	810	10	200	10	350	0.14
26	-10	750	10	400	2	350	0.14

Table 30.7 (continued).

TC			MNTHK	THKSD		WTWVAR	WIWVAR
1			1143	119		15215	284
2			1039	55		2952	321
3			1153	218		51459	216
4			1124	120		15887	357
5			745	122		15997	62
6			689	197		42014	150
7			2104	93		7622	1729
8			842	127		17452	248
9			524	32		1050	30
10			567	71		5502	70
11			1128	118		15465	239
12			674	20		331	102
13			2357	187		36834	1292
14			806	101		10915	260
15			1136	110		12881	316
16			1755	264		61873	14173
17			393	96		9950	30
18			1314	184		36220	573
19			1169	116		14453	283
20			2406	165		25406	4203
21			886	174		32304	1062
22			1497	101		8907	2331
23			815	155		25860	251
24			2032	238		56180	5801
25			959	286		89556	1074
26			1028	51		2363	433

Analysis of the Screening Experiment

Regression Coefficients from Least Squares Regression

Table 30.8 summarizes the least squares regression coefficients found for each response listed earlier in Tables 30.5 and 30.7. The ~ before each term indicates the factor settings were scaled -1 to +1 for the regression analysis.

Table 30.9 provides additional information regarding the terms in Table 30.8.

As originally constructed, the design assured that interaction terms did not alias main effects (resolution IV). The original design model contained only main effects and the constant term (represented by 1 in Table 30.8). The original design matrix confounded interaction terms with other interaction terms, and the augmented design still contains those confoundings, but at a lesser level. The interactions included in Table 30.8 represent good engineering guesses about which interactions in particular alias chains might be active. Inclusion of these interactions does introduce some risk since alternative interpretations of the data using other interactions are certainly possible. However, if one concentrates primarily on the main effects, then clear signals emerge that can lead to further experiments to complete the characterization.

Table 30.8. *Summary of Regression Coefficients and Statistics for Responses in Tables 30.5 and 30.7.*

Term	FILM THICK	THKSD	WIWVAR	WTWVAR
1	1163.1	98.92	343.2	9629
~TPOFF	33.1	-24.76	-81.2	-6279.9
~CTMP	417.3	31.64	417.4	8103.4
~BTOFF	50.5	19.95	95.9	5270
~TFLOW	255.7	-30.16		-7634.6
~ASR	-254	-35.76	-147.7	-9946.1
~CLP	53.6	29.74	156.1	7285.5
~BP	149.3	10.57	28.3	7982.3
~TPOFF*CTMP		22.92		
~TPOFF*BTOFF				-5391.4
~CTMP*BTOFF			102.7	
~CTMP*ASR	-46.2			
~CTMP*BP		-25.49		-7427.6
~BTOFF*CLP		-19.64		-5044.8
~BTOFF*BP			70.2	
~CLP*BP			-75.9	
~TFLOW*CLP	-98.9			
~TFLOW*BP				10803.7
TRANSFORM*	LOG	LOG	LOG	LOG
R-SQUARE*	0.99	0.87	0.94	0.91
REL PRESS*	0.97	0.57	0.78	0.59
LOF SIG.*	0.007	0.0002	0.01	0.0007
COND NUM.*	3.7	2.1	1.5	2.7

*See Table 30.9 for definitions of these terms.

Residual Diagnostics

The residual distributions produced by each regression model generally satisfied the requirement that they should be normally and independently distributed with mean 0 and constant variance s^2. Several of the models did exhibit one or more outliers, but no assignable cause existed to eliminate any of them. Figures 30.9a and 30.9b, respectively, illustrate the normal probability plot and case order graph from the model for thickness standard deviation. The approximately linear plot of the residuals in Figure 30.9a suggests they have a normal distribution. Figure 30.9b shows that the residuals meander around 0 with no obvious pattern, therefore, they exhibit no obvious time dependence. Residuals from other responses produced similar graphs.

Figures 30.10a and 30.10b illustrate other residual plots. Figure 30.10a is a graph of the residuals from the model for wafer-to-wafer variance versus the settings of the center temperature. If the model is adequate, then one expects a random scatter about 0 in this graph. The line in this graph is a smoothing function which suggests that the residuals are generally higher at the center values than at the extremes. Figure 30.10b is an adjusted y graph. The dark line in this figure represents the expected behavior of wafer-to-wafer variance as a function of center temperature using the model found by regression analysis.

If the model is adequate, then all the points plotted along this line should scatter randomly about it. The dotted line is a smoothing function which traces the model residuals added to the adjusted y plot. Both Figure 30.10a and 30.10b suggest that describing the behavior of wafer to wafer variance requires quadratic terms in the model. This is further indication of lack of fit. The design matrix used can detect curvature since it contains center points, but it lacks the degrees of freedom to model curvature with respect to any predictor. In other words, it is not possible to tell which of the predictors (factors) is creating the curvature. Similar graphs result for each predictor.

Table 30.9 *Definition of Terms in Table* 30.8.

Transform	The software used for this analysis recommended transforming each response to its natural logarithm to improve the fit of the model. Another software utility converted the transformed coefficients back to an approximation of their natural metric to make interpretation easier. Therefore, even though the least squares regression used the natural logarithm of each response, the coefficients appear as untransformed values.
R-Square	R-Square is a measure of how much of the variability in the response is explained by the regression model.
Relative PRESS	Relative PRESS is a measure of the ability of the regression model to predict observations not used to create it. Close agreement between R-Square and Relative PRESS indicates a very useful model. Discrepancies between the two values indicate that the model needs additional terms to predict one or more cases.
Lack of FIT (LOF) Significant	LOF Significance is the p-value associated with the LOF mean sum of squares in an analysis of variance table. Very low values, such as those in Table 30.8, indicate that the regression models may lack important terms to describe fully the variability in the data. Typically a model requiring higher-ordered terms such as quadratic effects will have LOF. Alternatively, the estimate of the pure error sum of squares could be artificially low.
Condition Number	A perfectly-balanced orthogonal design matrix has condition number = 1.0. The further the condition number is from 1, the less balanced or orthogonal the matrix is. Originally this investigation used a classical fractional factorial design for 7 factors in 16 treatment combinations (with center points) which had condition number » 1. The small deviation from 1.0 was due to the use of a center point not exactly at the center of the settings for boat pitch. Repairing the design with the D-optimal algorithm as described in the Initial Results: Repairing the Design Section introduced additional treatment combinations which produced an unbalanced design matrix. The higher the condition number, the more collinearity exists between supposedly independent factor settings. Condition numbers less than 5 are highly desirable and do not materially affect the interpretation of the results.

Locating an Operating Region

The software used allows optimizations of the regression models found according to specifications set by the user. For example, the user can maximize a response which holds others within a specified range while using only selected factor settings or ranges. Table 30.10 summarizes a few of the optimization attempts made using a capability potential estimated by using formulas based on the model for thickness and combinations of models for process dispersion. The formulas used were

$$\text{ECP}: \frac{0.2*\text{THK}}{6*\text{THKSD}},$$

$$\text{ECPVAR:} \frac{0.2*\text{THK}}{6*\sqrt{(\text{WTWVAR}+\text{WIWVAR})}},$$

where THK represents the regression model for film thickness, THKSD the model for the total standard deviation of a run, WTWVAR the model for wafer to wafer variance, and WIWVAR the model for within wafer variance.

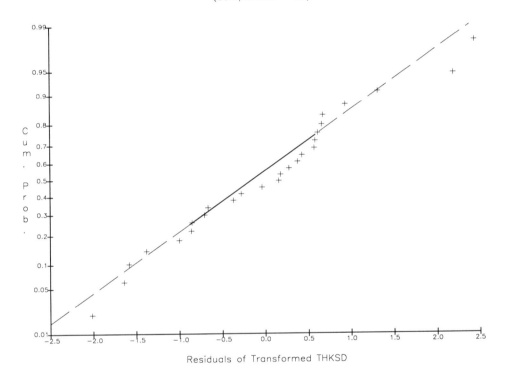

Figure 30.9a. *Normal probability plot of residuals from regression model for thickness standard deviation.*

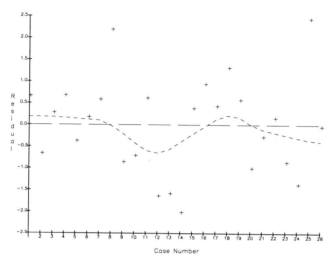

Figure 30.9b *Case order plot of residuals from regression model for thickness standard deviation.*

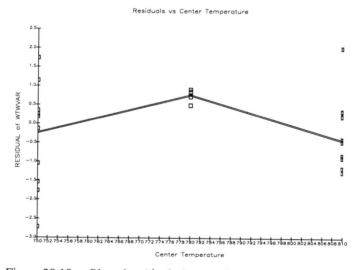

Figure 30.10a. *Plot of residuals from wafer to wafer variance versus settings of center temperature.*

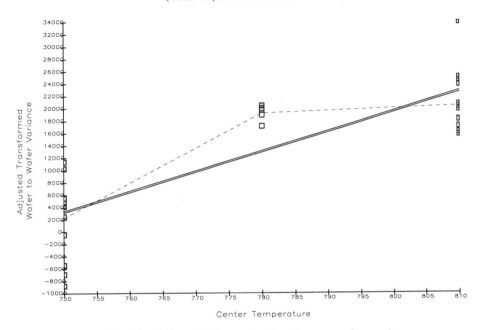

Figure 30.10b. *Adjusted Y graph of wafer to wafer variance vs. settings of center temperature.*

Note that all of the models generally converge to approximately the same region of the process space: high total flow, low closed loop pressure, high ratio of ammonia to dichlorosilane, boat pitch = 0.14, and minimum temperature offsets with respect to the center temperature. Unfortunately, this study demonstrated that high flow with low pressure was not achievable. Furthermore, running the process with a large excess of ammonia was also undesirable, as doing so places considerable stress on the vacuum pumps and on the exhaust scrubbers (to remove the excess ammonia). Because the models all contained considerable lack of fit, the investigators used these results as indications of trends, rather than final solutions.

Table 30.10. *Summary of Optimizations. Estimated capability potential is used from formulas based on models for film thickness and thickness standard deviation (ECP), and on film thickness and variance estimates (ECPVAR).*

Factor, Response or Formula	Range	Initial Setting	MAXIMIZE ECP	MAXIMIZE ECPVAR
Factors				
TPOFF	-30 to -10	-20	-10.014	-10
CTMP	750 to 810	780	750.04	761.6
BTOFF	10 to 30	20	10	17.332
TFLOW	200 to 400	300	400	400
ASR	2 to 10	6	9.9909	9.9997
CL_LP_PRS	150 to 350	250	150.03	156.06
BTPTCH	0.14,0.2,0.28		0.14	0.14
Responses				
MNTHK			741.28	873.1
THKSD			12.378	20.77
WIWVAR			21.832	31.252
WTWVAR			119.96	228.38
Formulas				
ECP			1.9963	1.4012
ECPVAR			2.0751	1.8062

Visualizing the Operating Region

Since all the estimates of process capability converged at the conditions indicated above, that convergence fixes the components of the major interactions shown in Table 30.8, except for those between the settings of the top offset and bottom offset temperatures. Figure 30.11 illustrates this interaction at three levels for each factor for the response WTWVAR; the curvature in the graph is due to the logarithmic transform used for the response. The settings for the other factors in the model approximate those found in the optimizations.

Figure 30.11 indicates that the level of WTWVAR is relatively independent of the top offset value when the bottom offset is 10. The rate of change of WTWVAR with change in top offset increases as the bottom offset increases. Therefore, the best results should occur with minimum values in both offsets, and the process should be somewhat robust to minor variations in top offset conditions. This graph also suggests that even smaller offsets might produce further improvement in the process variability.

Figures 30.12 and 30.13 are contour graphs illustrating the dependence of film thickness and the estimated capability potential on gas ratio, center temperature, and total flow, given settings for the other four factors. In both figures, the best capability potential occurs at relatively high gas ratios. In addition, a total flow at 400 SCCM (see Figure 30.13) is more desirable than 300 SCCM (see Figure 30.12). However, reasonable process uniformity is possible at lower gas ratios (3:1, for example), provided some method exists to allow higher flows at low pressure.

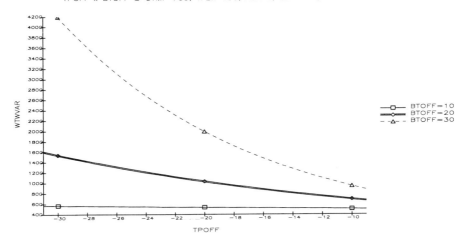

Figure 30.11. *Interaction Graph for TPOFF x BTOFF and Response WTWVAR. Curvature of the lines is due to the log transform used to model WTWVAR.*

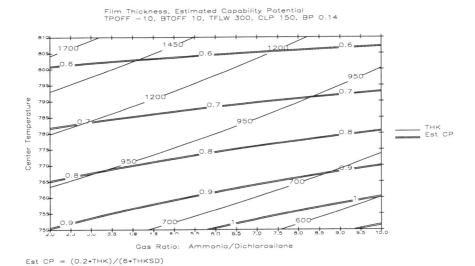

Est CP = (0.2*THK)/(6*THKSD)

Figure 30.12. *Contour Graph Showing Dependence of Estimated Process Capability Potential as a Function of Center Temperature and Gas Ratio at Fixed Settings for Other Five Factors. This graph illustrates a total flow of 300 SCCM.*

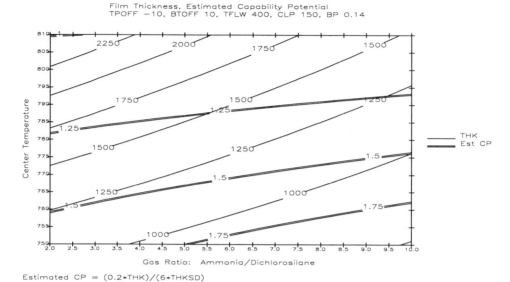

Figure 30.13. *Contour Graph Showing Dependence of Estimated Process Capability Potential as a Function of Center Temperature and Gas Ratio at Fixed Settings for Other Five Factors. This graph illustrates a total flow of 400 SCCM.*

Conclusions and Lessons Learned from the Screening Experiments

Certain combinations of gas flows and pressures called for in the design matrix proved unattainable in the actual process, despite the fact that engineers had tested the system with inert gas before starting the work. Using a gas mixture with different flow properties than those of the actual process gases produced misleading information. Augmenting a broken design with new treatment combinations using a computer-generated design algorithm (D-optimal in this case) allowed complete analysis of the results.

The significant lack of fit obtained for each model suggested that the behavior of the responses as functions of the input variables using simple linear models with interactions was inadequate. Graphs of residuals versus factor settings and adjusted y graphs suggested considerable curvature was probably present. This indicates that the process is not fully characterized at this point and that defining the relationships between inputs and outputs more precisely requires additional experiments.

CONFIRMATORY TRIALS AND SECOND PASSIVE DATA COLLECTION

The investigation described through the Screening Experiments Section of this case study occurred at a single factory site and used a particular version of the furnace tool. At the end of this phase of the study, the project moved to another site and began using a new version of the furnace which included improvements identified in the initial studies. Particularly, the new furnace had improved conductance and allowed lower pressures at higher gas flows than the previous model. Runs to confirm the results of the screening experiments were run in the new process tool.

A complicating factor that influenced the confirmatory runs was an initial faulty setup of a pressure sensor on the new furnace. As a result of this setup, the original confirmatory runs did not support the conclusions drawn from the designed experiment because the pressure in the furnace was actually twice as high as intended. Identifying and correcting this fault required considerable time and, to keep the study on schedule, shortened the time available for further experiments. As a result, the investigators decided to modify an existing recipe furnished by the supplier of the tool following the findings of experimental design. Table 30.11 illustrates the final adopted recipe, based on graphical data analysis, engineering judgment of where to set critical parameters, and additional supplier inputs.

Table 30.11. *Final Recipe Parameters for Post Experiment Verification and Marathon Runs.*

Factor	Value
Center Temperature:	782
Top Offset:	-17
Bottom Offset:	11
Closed Loop Pressure:	150
Gas Ratio:	3
Boat Pitch:	0.14
Total Gas Flow:	254

Of the conditions listed in Table 30.11, the only one not anticipated by the results of the designed experiment is the total gas flow. Given the lack of fit detected in the models for all the responses (possibly due to curvilinear relationships), this is actually excellent agreement. Or the modifications made to the new furnace improved its flow characteristics somewhat such that lower total flows would produce good results.

Figure 30.14 is a capability graph derived from 31 confirmatory runs using the recipe in Table 30.11. The sampling plan for this passive data study used six wafer slots chosen to capture as much of the within-tube variability as possible. This was based on studying the box and whisker plots from runs made using the same 11-wafer sampling plan used for the designed experiment. In conducting the designed experiment, the engineers placed wafers in slots 5, 20, 35, 50, 65, 80, 95, 110, 125, 140, and 155. The confirmatory trials used slots 10, 25, 55, 72, 119 and 134. Figure 30.14 indicates that the process performed extremely well during these confirmatory trials, producing a Cpk of 1.8. The reduced sampling plan used in these trials may have underestimated the total variability in the system, however. Control charts for these confirmatory runs (not included in this paper) also showed a process in a state of statistical process control.

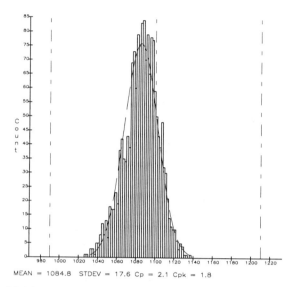

Figure 30.14. *Capability graph of* 31 *Confirmatory Runs (Six Wafers per Run, Nine Measurement Sites per Wafer). This follows the designed experiment and uses the recipe in Table* 30.11.

EXTENDED FURNACE TRIALS (MARATHON)

The Introduction Section stated that a major objective of equipment studies conducted in cooperation with SEMATECH is to measure the utilization capability of a processing tool under conditions approximating a manufacturing environment. Following the confirmatory trials, the project team exercised the furnace 24 hours a day, 7 days a week and logged over 200 production-like runs during over 900 hours of operation. During this period they demonstrated an equipment dependent uptime of 91%. To conserve monitor wafers they reduced the sampling plan further to include only slots 10, 72, and 134. Choosing these slots captured most of the variability seen in the confirmatory trials completed earlier. With this sampling plan, Cpk remained approximately 1.8, once about 15 runs had been removed for assignable causes due to equipment or operator errors.

CHARACTERIZATION OF THE PROCESS WINDOW

In the normal execution of the SEMATECH Qualification Plan, full characterization and refinement of the process window occur before one starts the manufacturing simulation (marathon). However, in this case the desire to generate as much useful data as possible from this furnace for estimation of utilization capability statistics delayed this step until near the end of the project. Analysis of the screening design suggested that curvature was present in one or more responses. The original two-level screening design with center points could detect that curvature, but could not model it. Therefore, a more complex experiment involving a quadratic model was necessary.

Factors

Table 30.12. *Factors Used in the Sensitivity Experiment and Their Settings.*

Factor	Abbrev	Units	Settings
Top Offset	TPOFF	Deg C.	-10 to -20
Center Temp	CTMP	Deg C.	765 to 795
Bottom Offset	BTOFF	Deg C.	10 to 20
Total Flow	TFLOW	SCCM	200 to 400
Closed Loop Pressure	CLP	MTorr	150 to 300

The confirmatory trials following the initial screening experiment and the marathon study demonstrated that the 0.14 boat pitch and a 3:1 ratio of ammonia to dichlorosilane gave completely acceptable results. The gas ratio is also the same as that commonly reported by other engineers using this type of process. Therefore, the next experiment involved only five factors, and the engineers adjusted their ranges to bracket the conditions used during the final stages of the project in order to determine process sensitivity to minor changes or variations in these factors. Table 30.12 summarizes the settings used for this experiment.

Responses

Table 30.13. *Responses (Outputs) for the Second Experiment.*

Response Name	Abbrev
Film Thickness	THK
Thickness Variance	THKVAR
Wafer-to-Wafer Variance Component	WTWVC
Within Wafer Variance Component	WIWVC

The same responses apply to this experiment as applied to the first described in the Responses Section of the Screening Experiments. However, the analysis concentrated on only two forms of variation -- the variance of the 99 thickness measurements collected per run and the approximate variance components based on the assumed random nested model illustrated earlier under Method 2 in the Responses Section of the Screening Experiments. Table 30.13 summarizes the responses and their abbreviations.

Design Matrix

The original screening experiment (see the Screening Experiments Section) suggested that a number of two-factor interactions were active in this process. In addition, the regression models produced had considerable lack of fit, further suggesting that one or more quadratic effects could be active as well. To accommodate these possibilities, the project team selected a Box-Wilson Central Composite Design for the second experiment. Designs of this type provide five levels of each input variable and allow estimation of all two-factor interactions as well as quadratic terms.

Normally in constructing such a design matrix, one creates at least a resolution V factorial or fractional factorial matrix and augments it with star points located outside the factorial space. The loci of the star points and factorial points approximately describe a sphere (or hyper-sphere, depending on the number of factors). In this case, the ranges selected for the factor settings were the limits of those settings that the team wished to explore, so creating new points beyond them was not desirable. In the randomized worksheet illustrated in Table 30.14, the limits of the factor settings have become the star points surrounding computed fractional factorial factor settings such that the design matrix maintains the spherical symmetry. The factor settings used bracketed the conditions currently in use on the furnace, but did not formally contain them. Treatment combinations 33 and 34 represent added runs under process nominal conditions.

Table 30.14 *Factor Settings for Second Experiment. Treatment combinations 33 and 34 are added runs representing process nominal and were not part of the design matrix.*

TC	CTMP	TPOFF	BTOFF	CLP	TFLOW
1	780	-15	15	225	300
2	780	-15	15	225	400
3	765	-15	15	225	300
4	772.5	-12.5	12.5	187.5	250
5	772.5	-12.5	12.5	262.5	350
6	772.5	-12.5	17.5	262.5	250
7	772.5	-12.5	17.5	187.5	350
8	780	-15	15	225	300
9	772.5	-17.5	12.5	187.5	350
10	772.5	-17.5	12.5	262.5	250
11	772.5	-17.5	17.5	187.5	250
12	772.5	-17.5	17.5	262.5	350
13	780	-10	15	225	300
14	780	-15	10	225	300
15	780	-15	15	300	300
16	780	-15	15	150	300
17	780	-15	15	225	300
18	780	-15	15	225	200
19	780	-15	20	225	300
20	780	-20	15	225	300
21	780	-15	15	225	300
22	787.5	-12.5	12.5	187.5	350
23	787.5	-12.5	12.5	262.5	250
24	787.5	-12.5	17.5	262.5	350
25	787.5	-12.5	17.5	187.5	250
26	787.5	-17.5	12.5	262.5	350
27	787.5	-17.5	12.5	187.5	250
28	787.5	-17.5	17.5	262.5	250
29	787.5	-17.5	17.5	187.5	350
30	780	-15	15	225	300
31	795	-15	15	225	300
32	780	-15	15	225	300
33	782	-17	11	150	248
34	782	-17	11	150	229

Table 30.15. *Results from the Second Experiment.*

TC	THK	WTWVC	WIWVC	THKVAR
1	1322	1691	279	1831
2	1496	2244	508	2569
3	1070	2683	136	2601
4	1040	883	97.2	908
5	1363	1455	315	1652
6	1179	3263	240	3237
7	1184	3684	195	3578
8	1330	1685	294	1841
9	1345	1472	269	1621
10	1154	1253	177	1318
11	1043	2374	104	2284
12	1358	7380	326	7104
13	1365	8295	303	7853
14	1328	1248	275	1421
15	1494	2050	546	2429
16	1137	1130	147	1184
17	1339	1599	327	1795
18	1096	1063	143	1119
19	1345	5330	415	5310
20	1312	2955	285	2999
21	1342	1636	350	1853
22	1493	5181	1238	5996
23	1439	10363	444	9961
24	1698	6112	716	6330
25	1278	5740	457	5729
26	1657	496	583	1039
27	1251	851	286	1068
28	1433	1200	438	1540
29	1476	2243	440	2500
30	1340	1160	378	1443
31	1633	7509	1450	8346
32	1338	1320	351	1564
33	1047	320	111	404
34	1046	303	105	383

Analysis, Interpretation of Results, and Conclusions

Analysis

Stepwise regression analysis of each response (using backwards stepwise methods) produced the lists of regression coefficients summarized in Table 30.16. The ~ symbol before a term indicates conversion of the factor settings to an orthogonal scale before the regression analysis. Where a response required a transformation for best fit, matched scaling has been applied to that response. This technique allows the benefits of the response transformation, but the reported coefficients are approximations of their natural *untransformed* values.

Table 30.16. *Summary of Regression Coefficients and Regression Statistics for the Second Experiment.*

Term	THK	THKVAR	THKVAR (ROBUST)	WTWVC	WTWVC (ROBUST)	WIWVC
1	1339.0	1644.73	1666.95	1553.29	1467.5	322.04
~CTMP	264.94	772.99	949.49	484.3	720.6	331.6
~TPOFF	6.67	1167.91	980.18	1259.4	995.51	23.66
~BTOFF	-2.08	1185.62	1332.32	1300.53	1441.46	29.59
~PRS	160.62	503.49	650.52	375.22	510.29	191.79
~TF	214.33	603.31	407.41	448.18	167.73	228.94
~CTMP**2		1655.39	1718.64	1572.84	1676.95	-74.96
~CTMP*TPOFF	57.38	3210.39	2652.15	3464.04	2719.36	92.33
~CTMP*BTOFF	48.02	-1505.13	-993.24	-1243.27		
~CTMP*PRS	74.58					-274.00
~CTMP*TF		-898.72	-1452.58	-1002.42	-1753.01	-252.72
~TPOFF**2		1710.12	1791.75	1706.29	1853.97	
~TPOFF*BTOFF		-889.54	-1358.27	-1112.08	-1624.72	145.06
~TPOFF*PRS	40.75					
~TPOFF*TF	-41.25	-758.32				
~BTOFF**2		589.7	719.97		780.46	
~BTOFF*PRS	33.75					
~BTOFF*TF	-53.76			845.52		-142.87
~PRS**2						-111.47
~PRS*TF		-967.67	-1353.95	-1225.52	-1516.18	-209.86
~TF**2	-31.89					-117.44
Transform	NONE	LOG	LOG	LOG	LOG	INV
Reg Type	LS	LS	BISQUARE	LS	BISQUARE	LS
R-Square	0.990	0.961	0.958	0.934	0.924	0.985
Rel PRESS	0.955	0.799	N.A.	0.756	N.A.	0.945
LOF Signif.	0.006	0.043	0.262	0.058	0.251	0.337

Figures 30.15a and 30.15b, respectively, show the normal probability plots of the residuals for the least squares fit of thickness variance and for the robust bisquare fit. The initial least squares fit of the model gave each observation equal weight. Figure 30.15a shows that several points near the extreme left of the figure may be outliers. The curved line on Figure 30.15b is the weighting curve associated with each residual. Projecting a vertical line from a residual to this curve followed by projecting the intersection of that line horizontally to the right hand axis indicates the weight assigned to a particular observation. In Figure 30.15b, the point at the far left of the graph receives very little weight. This is equivalent to removing the observation from the data without physically doing so.

Residual plots for the least squares model for WTWVC and for the robust bisquare model for WTWVC had similar appearances. The outliers present in the least squares fits for both responses are the probable explanation for the lack of fit detected in these responses.

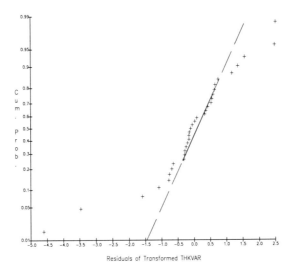

Figure 30.15a. *Normal Probability Plot of Residuals from Least Squares Fit of THKVAR.*

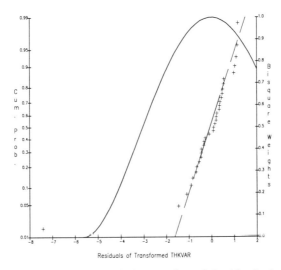

Figure 30.15b. *Normal Probability Plot of Residuals from Robust Bisquare Fit of THKVAR.*

Figures 30.16a and 30.16b are graphs of the residuals versus the settings of center temperature for the wafer to wafer variance component model and the adjusted y graph for that response as a function of center temperature, respectively. These graphs are analogous to Figures 30.10a and 30.10b produced during the residual analyses of the screening experiment. Notice in Figure 30.16a how the residuals vary randomly around 0. The single point near the bottom of that graph is the outlier given very little weight (refer to Figure 30.15b) by the robust regression technique. Similarly the solid line in Figure 30.16b illustrates the expected behavior of the response with respect to center temperature, adjusted for all other predictors. Notice how all the points cluster closely about this curved line. The single exception is the outlier down-weighted by the robust regression technique. The smoothing curve on this graph is almost obscured by the actual model line.

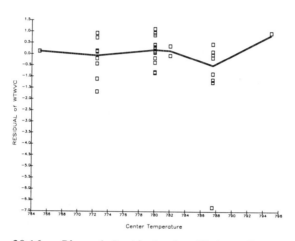

Figure 30.16a. *Plot of Residuals for Wafer-to-Wafer Variance Component versus Settings of Center Temperature.*

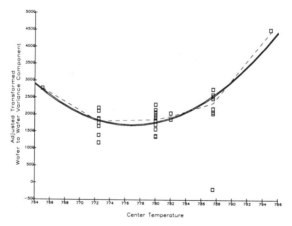

Figure 30.16b. *Adjusted y Graph of Wafer-to-Wafer Variance Component Model.*

Interpretation of Results

Table 30.16 clearly shows extremely large influences on both THKVAR and WTWVC from the interaction CTMP*TPOFF and from the quadratic terms $CTMP^2$ and $TPOFF^2$. The relatively large R-Square values for these models and the behavior of the residuals for each of them indicate a reasonable degree of reliability for making predictions regarding process performance.

However, the model for film thickness in Table 30.16 still has significant lack of fit despite excellent R-square and Relative PRESS statistics. No outliers that could account for this lack of fit were present in this model as they were in the others. To understand how this lack of fit signal could result, let us consider the origin of this statistic.

Analysis of variance separates the variability of data into its several sources:

Total (Corr) Sum of Squares	This is an expression of the total variation of a group of observations about their grand mean: $$TOTss = \sum_{i=1}^{n}(y_i - \bar{y})^2$$
Regression Sum of Squares	This is an expression of the amount of variation in a group of observations explained by the model: $$REGss = \sum_{i=1}^{n}(\hat{y}_i - \bar{y})^2$$
Residual Sum of Squares	This is an expression of the amount of variation in a group of observations not explained by the model: $$TOTss - REGss = RESss = \sum_{i=1}^{n}(y_i - \hat{y}_i)^2$$ The residual sum of squares is due to one or both of two causes: the model lacks a required term or there is random error.

If, as in this case, an experiment contains replicated trials, then analysis of variance can separate the residual sum of squares into its components: pure error and lack of fit. Statistical tests then determine whether or not the lack of fit mean square (LOFms = Lack of Fit Sum of Squares /degrees of freedom) is significantly different from the error variance. If it is, then significant lack of fit exists, and the model may be incorrect.

The statistical test involves forming the ratio of the LOFms to the error variance and comparing that ratio to standard tables. A significant test results when the observed ratio (Fisher's F-ratio) is unusually large, given the degrees of freedom associated with the numerator and the denominator. A large ratio can result because the model lacks an important term (larger numerator) or because the estimate of error is artificially small (small denominator).

In this example, the project team elected to run all the replicated trials in sequence, rather than in random order as indicated in the worksheet. Doing so minimized setup errors associated with readjusting furnace temperature and flows from run to run and probably minimized the variation between runs. This is probably the source of the observed lack of fit, rather than assuming some important term is missing from the model.

Conclusions

Figure 30.17 is an interaction graph displaying the relationships among three levels each of center temperature and top offset on the thickness variance. The curvature in the lines is due both to the transform used in the regression model and to the contributions of the quadratic terms. Note that in the lower range for center temperature that a top offset of -15 produces superior results, while in the higher range for center temperature a top offset of -20 produces better results.

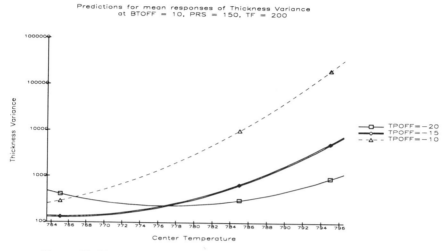

Figure 30.17. *Interaction Graph Illustrating the Influence of Center Temperature and Top Offset on the Response THKVAR. The vertical axis of this graph is logarithmic. The curvature in the lines is due both to the transform applied to the response and to the contributions of quadratic terms for both center temperature and top offset.*

Table 30.17 illustrates the results of some optimizations of the response estimated process capability potential (Cp) as well as optimizations of computed Cp values using models for THKVAR and combinations of models for WTWVAR and WIWVAR similar to those illustrated earlier in the Screening Experiments Section. In addition, the investigators wished to find conditions that produced acceptable variability with a deposition rate near 20 Å/min. The formulas computed based on the regression models are as follows:

ECP1 (Based on thickness variance)	$\dfrac{0.2 * \text{THK}}{6 * \sqrt{\text{THKVAR}}}$
ECP2 (Based on variance components)	$\dfrac{0.2 * \text{THK}}{6 * \sqrt{\text{WTWVC} + \text{WIWVC}}}$
Deposition Rate	$\dfrac{\text{THK}}{50}$

In Table 30.17, TRUE value for the EXCLUSION formula indicates that the optimum found was outside the actual space investigated in the design (it is an extrapolation). Values of 18 for the deposition rate in the last two columns reflect a constraint requiring a value between 18 and 22 Å/minute.

Table 30.17. *Summary of Optimizations for Capability Potential Estimates. ECP1 is the estimate based on the model for film thickness and film thickness variance; ECP2 is the estimate based on the model for film thickness and the models for WTWVC and WIWVC. A TRUE value for EXCLUSION means that the projected optimum value is an extrapolation outside the actual region covered by the experimental design. The values in the last two columns reflect a constraint on the formula for deposition rate to 18 to 22 Å/minute.*

Factor, Response, or Formula	Range	Initial Setting	MAXIMIZE ECP1 (Thk Var)	MAXIMIZE ECP2 (Var Comp)	MAXIMIZE ECP1 (Thk Var) (Constrain Dep Rate)	MAXIMIZE ECP2 (Var Comp) (Constrain Dep Rate)
Factors						
CTMP	765 to 795	780	791.13	795	774.37	778.29
TPOFF	-20 to -10	-15	-20.00	-19.96	-17.30	-17.57
BTOFF	10 to 20	15	10	10.81	10	10
PRESS	150 to 300	225	299.99	299.98	150.45	150
TFLOW	200 to 400	300	399.98	400	209.24	200.66
Responses						
WTWVC			128.61	117.89	169.12	155.94
WIWVC			286.01	211.64	63.59	70.55
THK			1871.7	1933.5	900.	900.
THKVAR			533.87	606.33	184.9	197.5
Formulas						
EXCLUSION			TRUE	TRUE	TRUE	TRUE
DEPRATE			37.43	38.67	18	18
ECP1			2.70	2.62	2.21	2.13
ECP2			3.06	3.55	1.97	1.99

In the absence of any constraint on desired deposition rate, the optimization of either model predicts superb process capability. But that capability comes at the price of relatively large gas flow rates and relatively high variability within the film thickness. The high gas flows would tax the pumping systems and could lead to premature failures due to the volume of gas pumped. Determining whether or not that capability prediction would remain true at much lower deposition times would require additional experiments in that region. Arguably the higher deposition rates predicted could enhance throughput of the system. But the total system cycle time was about six hours and included loading and unloading the wafers, bringing the wafers to process temperature and down again, for example. The actual deposition time was only 50 minutes in that six-hour period. Therefore, cutting that time in half would offer little increase in throughput. Notice that the estimation of process capability from the models for the variance components produces slightly more conservative estimates.

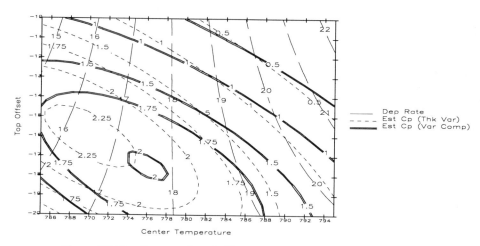

Figure 30.18. *Contour Graph Showing the Relationship of Estimates of Process Capability with Center Temperature and Top Offset at Constant Settings of the Other Variables: BTOFF 10; CLP 150; TOTFLOW 200. The process is robust to minor changes in center temperature and top offset.*

Figures 30.18, 30.19, and 30.20 are contour graphs showing the variation in predictions of three estimates of process capability and deposition rate as functions of center temperature and top offset with bottom offset, pressure, and total flow fixed. Among the figures, total flow varies from 200 to 250 SCCM. The graphs show a process relatively robust to minor changes in center temperature and top offset. In addition, they suggest that reducing the total flow from the process nominal setting of 254 to about 200 SCCM could produce an improvement in process capability (less variation between wafers and within each wafer) with a modest loss of deposition rate.

This relationship of process capability to flow rate contradicts the results from the screening experiment. However, this contradiction is due to the fact that the more complicated model supported by this design identified relatively strong interactions and quadratic terms involving total flow that the lower resolution screening experiment did not detect. Notice also that the model for capability based on thickness variation predicts a slightly different set of optimum conditions compared to the same model based on variance components. Determining which model is predicting performance the most precisely will require additional experiments in each predicted region and comparison of the results.

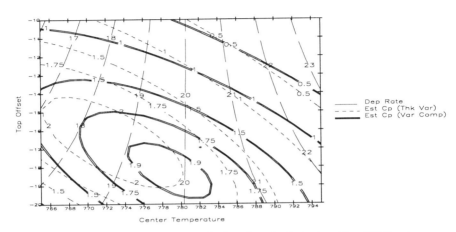

Figure 30.19. *Contour Graph Showing the Relationship of Estimates of Process Capability with Center Temperature and Top Offset at Constant Settings of the Other Variables: BTOFF 10; CLP 150; TOTFLOW 225. The process is robust to minor changes in center temperature and top offset.*

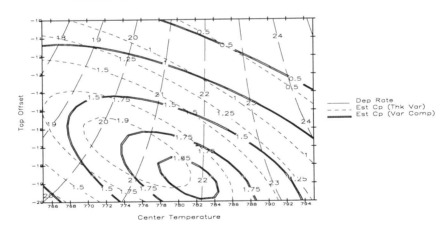

Figure 30.20. *Contour Graph Showing the Relationship of Estimates of Process Capability with Center Temperature and Top Offset at Constant Settings of the Other Variables: BTOFF 0; CLP 150; TOTFLOW 250. The process is robust to minor changes in center temperature and top offset.*

Summary

This chapter illustrates the major steps in the use of statistical methods to characterize and to understand the process capability of a semiconductor manufacturing tool—a furnace used for the low pressure chemical vapor deposition (LPCVD) of silicon nitride (Si_3N_4).

Characterization of measurement tool capability was an important first step to assure that measured responses were not misleading due to the variability of a measurement tool or to its drift.

Establishing the baseline capability of the processing tool assured that it was stable enough so that its inherent variation would not confuse experimental results. A key element in this portion of the study was the use of a sampling plan adequate to capture the representative variability within the process. In this study, the initial sampling plan failed to detect serious within-run variability, but it did provide adequate indication that the process was inherently stable.

Methods of statistical experimental design are relatively robust to operational errors as occurred here. When the errors do occur, mechanisms exist to augment a damaged design matrix with additional treatment combinations to reduce any collinearity among otherwise orthogonal terms induced by the mistake.

Overinterpretation of limited screening experiments can produce misleading results, particularly when regression models have significant lack of fit perhaps due to one or more undetected two-factor interactions or active higher order terms. In this case, the resolution IV screening experiment identified large main effects for both pressure and total flow that suggested one should operate this process at low pressure and high total flow to minimize variability. Not estimated in this experiment was the interaction between these two factors. The investigators chose an alternative interaction somewhat confounded with the pressure x total flow interaction in the initial interpretation. A later sensitivity (response surface) experiment correctly identified the main effects, interactions, and quadratic effects active in this process.

Modeling and interpreting the variability in a batch process requires careful consideration. Statistics such as capability potential are attractive options since they have received such wide use in manufacturing industries. With a relatively large number of observations in each run, estimating Cp for each run from its mean and standard deviation was one option. However, because this statistic is actually a combination of those two responses, directions given by the interpretation of models based on it may be unreliable. The approach demonstrated modeled dispersion and location separately, then interpreted a combination of them.

Deciding which statistic or group of statistics to use in modeling variability was still another issue. This case study illustrates the use of the run standard deviation, the run variance, and estimates of variance from two sources (within run and within wafer) derived in two ways. Method 1 computed the pooled variance across the wafers in a run and the variance of the wafer means in a run. Method 2 assumed that locations on a wafer and locations of a wafer within the furnace were nested random variables and then computed estimates of variance components. All of these methods provide an adequate estimate of the variation in a process; the last two separate the total variability into components to allow studying them separately.

Finally, the response surface experiment verified the robustness of the process nominal settings used in the marathon tests and indicated that minor improvements in process capability might occur if one shifted the operating conditions to slightly lower temperatures and lower total flow. Lowering both inputs does reduce the deposition rate about 10%, however.

ACKNOWLEDGMENTS

This project involved more than 20 engineers, technicians, and support personnel drawn from SEMATECH and National Semiconductor Corporation, as well as the suppliers of the processing tool and the gases used in the process. Specifically, the authors wish to acknowledge Richard Wilkes and William O'Toole (SEMATECH assignees who served as project managers for SEMATECH on this activity), Mansour Moinpour of National Semiconductor Corporation (who acted as chief process engineer, and Richard Lynch (a SEMATECH assignee who preceded Jack Reece as statistical consultant for the project).

REFERENCES

Satterthwaite, F. E., "An Approximate Distribution of Estimates of Variance Components," *Biometrics Bulletin*, Vol 2, 1946.

APPENDIX A

The major software tool used during this investigation was the RS/1 family[71] of statistical software provided by Domain Solutions Corporation, Cambridge, MA. Included were the parent software RS/1, the advanced data analysis advisor RS/Explore, and the experimental design module RS/Discover.

The multiple regression utility with RS/Explore provides convenient routines for creating formulas based on response models and for optimizing and visualizing these response models or formulas.

Variance component analyses initially used the *proc nested* routine within SAS[72] provided by the SAS Institute, Cary, NC. Analyses in the Characterization of the Process Window Section used a SEMATECH custom program running under RS/1 which assumes a balanced and nested random effects model and uses the method of moments to compute Type I Sums of Squares.

BIOGRAPHIES

Jack E. Reece, Ph.D., retired as a Fellow of the SEMATECH statistical methods group in June, 1996. In his capacity as a statistician at SEMATECH he generated training

---

[71] RS/1, RS/Explore, and RS/Discover are trademarks of Domain Solutions Corporation, 150 Cambridge Park Drive, Cambridge, MA 02140.
[72] SAS is a trademark of the SAS Institute, Cary, NC.

materials to help engineers understand statistical methods, particularly design of experiments, and provided consulting and training in the use of applied statistical methods across all semiconductor fab operations. Dr. Reece has more than 25 years' experience as a process engineer/statistician in industries such as petrochemicals, photographic products, coating technology, and semiconductors.

Currently Dr. Reece is a private consultant based in Lake George, CO. He may be reached on the Internet at jreece@pcisys.net.

Mohsen Shenasa is a staff process engineer in the process/device development group of National Semiconductor Corp. His responsibilities include process characterization of nitride, poly and gate oxide for 0.5 µm technologies. Mohsen received a Ph.D. degree (1988) in physics from University of Wisconsin. He has been a contributing author to several technical papers.

APPENDIX: INTRODUCTION TO INTEGRATED CIRCUIT MANUFACTURING

Ronald J. Schutz

Integrated circuits, or *chips*, are the heart of the electronics industry. Integrated circuits are used in computers, telephones, televisions, watches, cars, and airplanes. The electronics industry creates more manufacturing jobs in the United States (2.6 million) than the steel, automobile, and aerospace industries combined.

The manufacture of integrated circuits has recently experienced an explosive growth cycle. To understand some basic principles of integrated circuit manufacturing, it is first important to realize that virtually all integrated circuits are made from *silicon*. In fact, this whole technology depends on the electrical properties of silicon, the main constituent of common sand.

Silicon is a *semiconductor*. As the name infers, semiconductors do not conduct electricity well like copper and aluminum, nor do they totally halt its flow like rubber and glass. Instead, silicon's electrical conductivity can vary over a broad range. Most importantly, its conductivity continuously increases as small amounts of impurities, or *dopants*, are added to its structure. Some dopants, such as phosphorous and arsenic, add "conducting" electrons to the silicon structure. The resultant material is said to be *n-type* silicon, because negatively charged electrons travel through the material. On the other hand, dopants such as boron remove electrons, leaving a *hole* for every boron-type dopant atom. This material is called *p-type* silicon because the flow of positively charged holes also results in electrical conductivity.

A piece of "lightly" doped silicon is in a very tenuous state of conductivity. When such material exists between two contact points, it can be switched between a conducting state and a non-conducting state by a third contact placed between the two. This switching, on and off, occurs only if p-type and n-type regions are properly positioned between the two contacts (*source* and *drain*) and the correct signal is applied to the third intermediate contact (the *gate*). A silicon device that is switched on and off by an applied signal voltage is called a *transistor*.

The invention of the transistor at Bell Laboratories in 1947 was an important technological milestone. Perhaps an equally important milestone was the invention of the *integrated circuit*. This device is made up of many transistors created on the flat, polished surface of a single piece or *chip* of silicon. Not only does the process of fabricating integrated circuits include steps to form these transistors, it also contains steps to form thin wires used to connect them together into a complete, self-contained circuit. Before integrated circuits, such circuits were made from many individual transistors and other circuit elements wired together on a large printed circuit board. Figure A.1 shows a cutaway drawing of a transistor in an integrated circuit, and the basic steps to create it.

470

1. The transistor's insulating layer of silicon dioxide (blue) is grown in a hot furnace on a positively doped silicon substrate on a wafer.

2. A layer of photoresist (red), sensitive only to ultraviolet light, is deposited on the wafer.

3. Light shining through a mask of chromium (black)-patterned glass chemically changes the exposed photoresist.

4. A chemical solvent washes away the exposed photoresist, leaving a patterned layer of photoresist above the silicon dioxide.

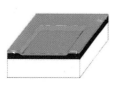

7. A layer of polysilicon (green) is deposited on the wafer. It will form the transistor gate for transmitting electrical current.

8. A new layer of photoresist is deposited on the polysilicon layer.

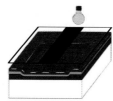

9. Light shining through a new patterned mask chemically changes the photoresist that is exposed to the light.

10. A chemical solvent removes the exposed photoresist, leaving a patterned layer of photoresist above the polysilicon.

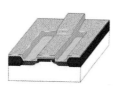

14. A new layer of silicon dioxide is deposited. It will insulate the transistor structure, except for metal contacts that will be added.

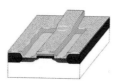

15. A layer of photoresist is deposited on the wafer, in preparation for metal contacts to the source, drain, and gate of the transistor.

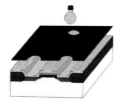

16. Light shining through a patterned mask chemically alters most of the photoresist, except three small areas for metal contacts.

17. A chemical solvent removes the exposed photoresist, opening access to three small areas of silicon dioxide.

20. Aluminum (blue-green) is deposited over the surface and inside the three shafts. It will provide electrical connections.

21. A fourth layer of photoresist is deposited on the wafer.

22. Light shining through the fourth patterned mask chemically changes the exposed photoresist.

23. A chemical solvent removes the exposed photoresist, creating a photoresist pattern above the aluminum.

471

Figure A.1. *The creation of a transistor in an integrated circuit. The figure illustrates the basic steps for making a transistor.*

5. A layer of silicon dioxide is etched away by a gas plasma, leaving only a thin insulating layer.

6. A chemical solvent washes away the remaining photoresist, leaving an uneven silicon dioxide surface.

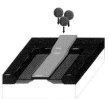

11. Polysilicon and a thin layer of silicon dioxide are removed by etching, exposing the underlying silicon substrate (white).

12. A chemical solvent washes away the remaining photoresist, leaving the polysilicon transistor gate structure (green).

13. Phosphorous atoms are implanted in the positively doped silicon substrate, forming the source and drain as negatively charged wells (orange).

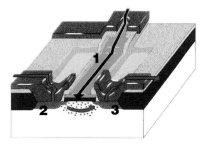

18. Dry etch removes the exposed silicon dioxide, opening shafts to the negatively doped substrate and the polysilicon.

19. A chemical solvent removes the photoresist. The negatively doped source and drain (orange) and the polysilicon gate (green) are opened to metal contacts.

The finished transistor operates as a switch. If no electrical charge is applied to the polysilicon gate (**1**), then no current flows from the negatively charged source (**2**) to the negatively charged drain (**3**). If a positive electrical charge is applied to the gate, it acts through the thin insulating layer of silicon dioxide (blue), creating a channel of negative charge. This channel allows current to flow from source to drain, and on through the aluminum connectors to other parts of the integrated circuit.

24. The exposed aluminum is etched away, leaving aluminum "wires" that will carry current to and from the transistor.

25. The photoresist is removed with a chemical solvent, completing the transistor. Many transistors are created simultaneously.

The integrated circuit fabrication process builds all the transistors on a chip simultaneously. Therefore, the number of transistors in the circuit is limited only by the size of the chip and how small the transistors can be made. Today, it is not uncommon for integrated circuits approximately one square centimeter to contain millions of transistors. Very large scale integration (VLSI) refers to putting over a hundred thousand components on an integrated circuit. In such circuits, the size of a single transistor element can be as small as 0.35 microns (micrometers). To put this in perspective, a typical human hair is 80 microns in diameter. Since common dust particles are many microns in diameter, a dust particle on a chip during fabrication can easily lead to a nonfunctioning device. Integrated circuits are therefore manufactured in *clean rooms* containing ten or less particles per cubic foot. To maintain this cleanliness, filtered air continuously flows down from the ceiling, through the room and out small holes in the floor. Everyone in the room dresses in specially designed lint free garments which so totally encase the wearer, that they leave only a small region between the nose and eyes open to the clean environment. Figure A.2 shows an operator in a clean room garment.

Fabricating an integrated circuit is accomplished by sequentially performing hundreds of operations on the chip, some of which are illustrated in Figure A.1. Instead of performing these operations on one chip at a time, the processes are performed on large, thin, round wafers of silicon up to 200 mm (8 inches) in diameter, such as the one being held by the operator in Figure A.2. Each of these wafers contains many chips of the same design in a regular rectangular array, as shown on the wafer in Figure A.2. When finished, the wafers are sawed apart, separating them into the individual chips.

Perhaps the most critical process performed on the wafer during manufacturing is *photolithography*, or *lithography*. This process is similar to black and white photography except there are no half tones. Lithography is used to repeatedly form patterns on the wafer that are aligned to previously formed patterns. This process eventually forms all the transistors and the necessary circuitry. For example, lithography is used early in the process to form the regions of appropriately doped silicon. First, liquid *photoresist* (or simply *resist*) is dripped onto the center of a rapidly spinning wafer. Much as in carnival spin-art, the resist coats the wafer and dries to form a thin, hard, uniform film on the polished silicon surface. Since resist is a photosensitive material, the coated wafer then acts much like photographic print paper. Light is projected onto the wafer through a patterned *mask or reticle*, to create the pattern with the desired doped regions on the resist. Because the instrument that projects this pattern does so by sequentially projecting and stepping from one chip position on the wafer to the next, the instrument is frequently called a *stepper*. "Exposing" the wafer thus and dipping it into a developing solution, results in "patterned" resist. The thin plastic-like resist remains intact over (unexposed) regions that must remain undoped and it is not present over (exposed) areas requiring doping. Appropriately chosen impurity atoms are then *implanted* into the silicon surface by placing the wafer perpendicular to a beam of accelerated, ionized dopant atoms. The resist mask prevents implantation where it is present on the surface by capturing the dopant before it penetrates into the silicon surface. After implantation the resist is removed along with the imbedded impurity atoms using a *plasma* and a *chemical bath*, leaving the doped silicon pattern on the wafer. Critical parameters for implantation are average aerial density and uniformity of impurity atom density, depth of the implanted atoms and the alignment of patterns to one

Figure A.2. *Operator in a clean room garment holding a round wafer 8 inches in diameter. The clean room garment reduces particle contamination on the wafer during integrated circuit manufacture. There are several hundred integrated circuits (the small rectangles) on the wafer.*

another. Implants are typically followed at some point by a *diffusion* step to activate and drive the impurities to specified depths, and by high temperature *annealing* to repair the wafer surface after the implantation. Critical parameters here are average temperature and uniformity of temperature across the wafer.

After the doped regions are formed with a series of patterned implants, the transistors and the wiring are formed with sequential layers of patterned insulators and conductors. Unlike the patterned implants, the patterned insulators and conductors are formed by first depositing thin films of the material on the whole wafer, depositing the photoresist, then exposing and developing it. This time the resist is left over regions where the films are meant to remain. Placing the wafers in an atmosphere that *etches* the film results in the resist pattern transferring into the film because of its masking effect. Early etch processes simply used *chemical baths* of acid to etch the films. Unfortunately this technique results in film feature shapes smaller than the mask because the liquid etchant also acts laterally undercutting the mask edge. More modern etch processes use chemically active *plasmas*, ionized gases, to etch films. Ions traveling through the plasma bombard the films and prevent mask undercutting. Critical parameters for these processes are feature size, pattern alignment, and unintended etching of underlying material.

While many materials can be patterned using lithography and etch processes, a few materials have broad application. For example, *silicon dioxide* is used extensively as an electrical insulator. This material is formed in three ways. First, parts of the silicon wafer itself can be oxidized by heating the wafer in an oxygen ambient. This process is called *oxidation*. Second, silicon dioxide can be formed on the wafer by heating it in an ambient gas containing silicon and oxygen. This process is called *chemical vapor deposition (CVD)* or *low pressure chemical vapor deposition (LPCVD)*. Third, the CVD process can be forced to occur at reduced temperature by adding a plasma to the ambient. This process is called *plasma enhanced CVD (PECVD)*. In general, chemical vapor deposition is used to react gases with heat or plasma to form a vapor that deposits a film on the wafer surface. The film can be a conductor such as a metal, a semiconductor such as silicon, or an insulator such as silicon dioxide.

Metals such as aluminum are deposited using yet another plasma process called *sputtering*. In this *metallization* process, metal from the *target* is transferred to the wafer through an argon plasma. For sputtering and all other deposition processes, important parameters are average film thickness, uniformity of thickness, and film properties such as conductivity.

As the minimum device feature size decreases with each new technology, greater planarity across a chip is required for the lithographic process. *Chemical mechanical polishing (CMP)* is used to smooth the contours of the wafer surface by holding the wafer face down in a rotating carrier and pressing it onto a rotating polishing pad that is covered with a chemical slurry.

It is important that semiconductor processes not introduce *particles* onto the wafer surface, since they can lead to defects. One defect in an integrated circuit containing millions of circuit elements can cause it to malfunction and be scrapped. A measure of the quality of an integrated circuit manufacturing line is its *yield*. That is, the fraction of total chips processed which result in good final product. Yields for new technologies can be relatively low, but some well established facilities can yield over 90%.

The sequences of, 1) applying then patterning a layer of resist, performing an operation and then removing the resist or, 2) first depositing a film, applying then patterning resist and etching the pattern into the film, are repeated ten to twenty times during the manufacturing process. Optimizing each process and then continuously monitoring and controlling it are crucial to a successful manufacturing effort.

The wafer fabrication process, from beginning to end, can last from a few weeks to a few months, depending on the complexity of the technology. After the wafers are finished processing, individual chips are electrically tested and the wafer is sawed apart. Good chips are then *packaged* by mounting them in small plastic holders or packages. Contact points on the chips are then connected to metal pins in the packages used to wire them to their applications.

Because of the many wafers processed in modern integrated circuit factories and the many sequential processes necessary to complete a wafer, statistical methods are frequently used to develop, improve and analyze both individual processes and their integration into the whole manufacturing sequence. This book gives many examples of how statistical methods have been used to improve integrated circuit manufacturing. Hopefully it will be useful to engineers and scientists with a cursory knowledge of statistics, as well as to statisticians with a cursory knowledge of engineering and science.

REFERENCES

Two books on integrated circuit manufacturing are:
> Wolf, Stanley and Richard N. Tauber, *Silicon Processing for the VLSI Era*, Sunset Beach, California: Lattice Press, 1986.
> Sze, S.M. *VLSI Technology*, second edition, New York: McGraw-Hill, 1988.

Five books on integrated circuit manufacturing are:
> Pierret, R.F., *Semiconductor Fundamentals*, second edition, Reading, Massachusetts: Addison-Wesley Publishing Co., 1988.
> Wolf, Stanley and Richard N. Tauber, *Silicon Processing for the VLSI Era*, Sunset Beach, California: Lattice Press, 1986.
> Sze, S.M., *VLSI Technology*, second edition, New York: McGraw-Hill, 1988.
> Van Zant, Peter, *Microchip Fabrication*, third edition, New York: McGraw-Hill, 1997.
> Ghandhi, Sorab K., *VLSI Fabrication Principles*, *Silicon & Gallium Arsenide,* Second Edition, New York: John Wiley & Sons, 1994

ACKNOWLEDGMENTS

Special thanks go to Carmen Gullón and Francisco Gallego from the training department in the Madrid semiconductor manufacturing plant of Lucent Technologies for Figure A1 on semiconductor processing. SEMATECH is gratefully acknowledged for the picture in Figure A2 of an operator in clean room garb.

BIOGRAPHY

Dr. Ronald Schutz has been working in the semiconductor industry for seventeen years. During that time he has managed semiconductor research at Bell Laboratories, semiconductor joint development projects at SEMATECH, and development/manufacturing engineering at Lucent Technologies. He is currently a development manager with Siemens Components at the DRAM Development Alliance in East Fishkill, New York. Ron has published over thirty papers and had ten patents awarded to him.

GLOSSARY OF SELECTED STATISTICAL TERMS

Veronica Czitrom

The first section of this glossary describes graphs frequently used for data analysis, with reference to both straightforward and unusual uses in the case studies. The second section defines some common statistical terms, such as population, sample, normal distribution, t distribution, uniformity, variance, confidence interval, and hypothesis test. The third section describes analysis of variance (ANOVA) and related topics.

GRAPHS

This section describes the following types of graphs:

x-y graphs	capability graphs	main effects graphs
scatter plots	bar graphs	interaction graphs
trend charts	Pareto charts	contour plots
control charts	box plots	three-dimensional graphs
dot diagrams	multi-vari plots	normal probability plots
histograms		

An *x-y graph* is a graph of two variables, one on the horizontal or x axis, and the other on the vertical or y axis, that helps visualize the relationship between the variables. As illustrated in Figure G.1a, an x-y graph can show individual observations, or a function such as a line or a curve. Many types of graphs are derived from x-y graphs.

A *scatter plot* is an x-y graph that shows individual observations as a function of two variables, such as the individual observations shown in Figure G.1a. A scatter plot helps visualize the degree of correlation (strength of the relationship) between two variables. It is important to note that correlation does not imply causality, since the values of two variables may be correlated without changes in one variable *causing* changes in the other variable. If the horizontal x-axis of a scatter plot is an independent variable and the vertical y-axis is a dependent variable (response), the x-y graph helps visualize how, and whether, the independent variable affects the response. Reece and Shenasa (Chapter 30, Figure 30.5) use a scatter plot to show how a sinusoidal pattern in the response was initially missed due to a sparse sampling plan. Buckner et al. (Chapter 16, Figures 16.5 and 16.11) show a scatter plot of the residuals of a regression model as a function of the predicted values from a designed experiment.

A *trend chart*, or *time trend chart*, graphs data in chronological order. A trend chart is a scatter plot in which the horizontal axis is time. Buckner et al. (Chapter 16, Figures 16.2 and 16.8) use two trend charts that indicate the time at which each data point was collected (and the corresponding histograms) to display data from a unimodal distribution and from a bimodal distribution. Freeny and Lai (Figure 18.7) use a trend chart to illustrate a downward trend in the residuals of removal rate from a chemical mechanical polishing process.

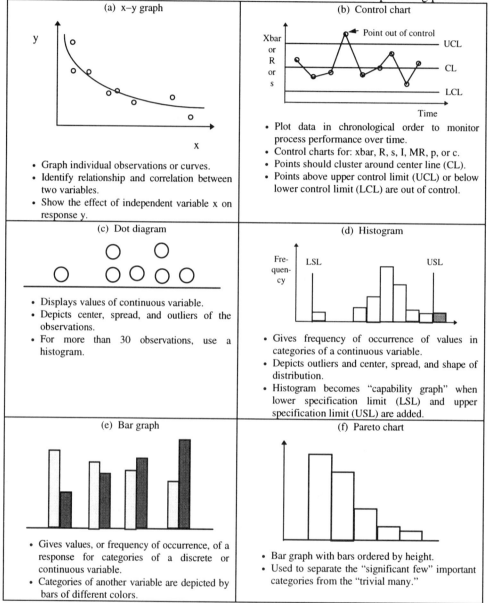

Figure G.1. *Graphs:* (a) *x–y graph*, (b) *control chart*, (c) *dot diagram*, (d) *histogram*, (e) *bar graph, and* (f) *Pareto chart*.

A ***control chart*** plots data in chronological order to monitor process performance over time. A control chart is a trend chart in which the points are connected by lines, and to which three lines are added (see Figure G.1b): a center line (CL) around which the observations are expected to cluster, an upper control limit (UCL) above which few observations are expected to fall, and a lower control limit (LCL) below which few observations are expected to fall. A process is said to be out of (statistical) control if the points on a control chart exhibit unlikely behavior, such as a point above the upper control limit, a point below the lower control limit, or eight consecutive points on one side of the center line (three of the Western Electric rules). The fourth observation in the control chart in Figure G.1b is out of control because it is above the upper control limit. Different types of control charts are used to plot different types of data, and different characteristics of the data. For continuous data, $\bar{x}$ control charts plot averages of groups of observations, R control charts plot ranges of groups of observations, s control charts plot standard deviations of groups of observations, and I control charts plot individual observations. Control charts are described in greater detail in Chapter 21. Ray Marr (Chapter 22, Figure 22.1) used a control chart to identify a process drift over time. Lynch and Markle (Chapter 7, Figures 7.4 and 7.5) used control charts to identify large reductions in variability.

A *dot diagram*, or *dot plot*, displays observations from a continuous variable as "dots" above a horizontal axis, as shown in Figure G.1c. A dot plot can be used to visualize the location, spread, and distribution of the observations, and to identify outliers. A dot diagram is used when there are less than 30 or so observations, and a histogram is used when there are more than 30 or so observations. Pankratz (Chapter 3, Figure 3.2) illustrates five carbon levels, and the fact that more samples are at low carbon levels, using a dot plot.

A *histogram* gives the number, percentage, or fraction, of occurrences (frequency distribution) that fall in different categories of a continuous variable. Figure G.1d shows a histogram. A histogram is useful for detecting outliers and for visualizing the location or central tendency of the observations, the scatter or spread in the observations, and the shape or distribution of the observations. The distribution can have different shapes, such as symmetric, non-symmetric, skewed to the right or to the left (long-tailed to the right or to the left, respectively), unimodal (one "peak"), or multimodal (more than one "peak"). The histogram in Figure G.1d shows a fairly symmetric distribution, with one outlier on the left (uncharacteristically small observation). The continuous variable on the horizontal axis is usually divided into between 7 and 12 categories, so that the histogram has between 7 and 12 bars. A histogram can be used to visually judge whether a set of observations appears to come from a normal distribution. A histogram is used if there are more than 30 or so observations, and a dot plot is used if there are less than 30 or so observations. Lynch and Markle (Chapter 7, Figures 7.7 and 7.8) use histograms to illustrate a major reduction in variability. Buckner et al. (Chapter 16, Figures 16.2 and 16.8) display data in trend charts that indicate the time at which each data point was collected, and also display the same data as histograms that show the location, spread, and distribution of the data as well as outliers. Their histogram and time plot in Figure 16.2 reveal a unimodal distribution and an outlier, and their histogram and time plot in Figure 16.8 reveal a bimodal distribution and an outlier.

A *capability graph* is a histogram to which the upper and lower process specification limits have been added. The histogram in Figure G.1d is really a capability graph, since it includes the specification limits. The last bar on the right indicates observation(s) outside of specifications. A capability graph gives a visual assessment of the capability of a process to meet specifications. A process is capable if the observations have reduced variability compared to the distance between the specifications, and if the observations are centered around the specifications. The process depicted in Figure G.1d is not very capable, since the natural variability of the observations is almost the same as the distance between specifications, and the distribution is not well centered around the specifications. A normal distribution (with the mean and the standard deviation of the observations) can be overlaid on the histogram. Canning and Green (Chapter 26, Figures 26.4, 26.5, and 26.6) use capability graphs to understand the capability of three key processes.

A *bar graph*, or *bar chart*, gives, as heights of bars, the magnitude of a response (such as the number of dollars spent), or the number or percentage of occurrences (frequency distribution) that fall in different categories of an independent variable. The categories can be a partition of a continuous variable, or the discrete values of a categorical variable. The categories of a second independent variable can be represented by bars of different colors or shadings. The bar graph in Figure G.1e gives a response in the vertical direction, four categories of the independent variable represented by the horizontal axis, and two categories (light and dark shades for the bars) for an additional independent variable. Lynch and Markle (Chapter 7, Figure 7.8) display a major reduction in variability with a bar graph of the magnitude of four variance components, with each variance component before and after a major process change.

A *Pareto chart*, or *Pareto diagram*, is a bar graph with bars ordered by height. Figure G.1f shows a Pareto chart. Usually, a few of the tallest bars, which correspond to the most frequent categories, account for most of the height of the bars. For this reason, a Pareto chart is said to separate the "significant few" categories from the "trivial many". Canning and Green (Chapter 26, Figure 26.14) use a Pareto chart to identify the most frequent reasons for reliability assists. Buckner et al. (Chapter 16, Figures 16.3 and 16.9) use Pareto charts to identify the most important factors and interactions in a designed experiment.

A *box plot*, or *box and whisker plot*, simultaneously displays the location and spread of groups of data, their distribution, and outliers. The vertical axis is the response, and the horizontal axis gives groups or categories of an independent variable. Figure G.2a shows a box plot of oxide thickness on nine wafers for two processing conditions, old and new. In addition, the oxide thickness at each processing condition is classified by a second variable, etcher, and the three etchers are represented by circles, squares, and triangles. A box contains the "middle" half of the observations, going from the *first quartile* Q_1 (25% of the observations are smaller) at the bottom of the box, to the *third quartile* Q_3 (75% of the observations are smaller) at the top of the box. The line in the middle of the box indicates the *median* or *second quartile* Q_2 (half the observations are larger, half are smaller), and an asterisk often indicates the mean (average). The *interquartile range* is the height of the box (third quartile minus first quartile), and is the "range" of the middle 50% of the data. The "whisker" below the box can extend to the smallest observation, and the whisker above the box can extend to the largest observation. Alternatively, the whisker below the box can extend to the smallest observation that is not farther than 1.5 times the interquartile range below the box, and the whisker above the box can extend to the largest observation that is not farther than 1.5 times the interquartile range above the box, in which case points below the lower whisker and above the higher whisker are *outliers*, and are plotted individually. The individual observations may be displayed inside each box as in Figure G.2a, or the boxes may be shown without the individual observations. The median and the mean of a box are two indicators of the location of the "middle" of the data, and the box indicates where the middle 50% of the data is. The width of a box (interquartile range) gives the dispersion or spread of the middle 50% of the data, and the distance from whisker to whisker gives the dispersion (range) of the data if there are no outliers. For the nine observations in the box on the right hand side in Figure G.2a, the median goes through an observation represented by a triangle (four observations are smaller, four are larger), the lower quartile goes through another triangle (two observations are smaller, six are larger), and the upper quartile goes through a square (six observations are smaller, two observations are larger). A box and whiskers that are not symmetric indicate a distribution that is not symmetric. For example, a box where both the bottom whisker and the bottom half of the box are much shorter than the top whisker and the top half of the box indicate a skewed distribution, with many small observations and a few larger ones, and a histogram with a long tail to the right. Buckner et al. (Chapter 4, Figure 4.1) identified a gauge warm-up effect using box plots graphed in chronological order. Freeny and Lai (Chapter 18, Figure 18.2) give boxplots for three grouping variables.

A *multiple variable plot*, or *multi-vari plot,* simultaneously displays the data as a function of several variables. A multi-vari plot gives an excellent graphical representation of variance components. For example, the multi-vari plot in Figure G.2b represents data from two furnace runs (the two groups of circles and lines), with three wafers (three groups of circles) in each furnace run, and five sites on each wafer (each circle is the value of the response at one wafer site). The lines in each furnace run go from the average of the first wafer to the average of the second wafer, and from the average of the second wafer to the average of the third wafer. The spread of the circles depicts site-to-site variability, the changes in the lines (connecting wafer averages) depict wafer-to-wafer variability, and the differences between the groups of circles and lines depicts run-to-run variability. In Figure G.2b, one observation on each wafer is lower than the rest; it was found to be from the site at the center of the wafer, which had lower values than the edge of the wafer due to a radial pattern on the wafer. The figure depicts a fair amount of site-to-site variability (the circles are fairly spread out), very little wafer-to-wafer variability (the lines are almost horizontal), and some run-to-run variability (the first group of observations is slightly higher than the second). This visual assessment of site-to-site, wafer-to-wafer, and run-to-run variability can be quantified using analysis of variance, as indicated in the last section of this glossary. The multi-vari plot presented in Figure G.2b is similar to the boxplot in Figure G.2a in that both classify the observations by two variables. Kahn and Baczkowski (Chapter 10, Figure 10.3) use a multiple variable graph different from the one in Figure G.2b to visualize two fixed effects and three variance components.

Four important types of graphs for analyzing the results of a designed experiment (see Part 3) are main effects graphs, interaction graphs, contour plots, and three-dimensional graphs. Main effects graphs and interaction graphs are the main graphical analysis tools for full factorial and fractional factorial designs. They are intuitive and easy to understand, they display most (if not all) the results of the experiment, and can lead to significant process improvement without further numerical analysis. Contour plots, and to a lesser extent three-dimensional graphs, are the main graphical analysis tools for second-order designs with factors at three or more levels. Figures G.2c, G.2d, G.2e, and G.2f all correspond to the same experimental design in two factors (temperature and pressure) at two levels each in four runs. The four graphs show how the same response, yield, looks using the different types of graphs.

(a) Box plot 2 processing conditions, 3 etchers, 18 wafers 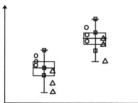 • Displays outliers, and center, spread, and distribution, for groups of data. • The box contains the "middle" 50% of the data (25% larger, 25% smaller).	(b) Multi-vari plot 2 furnace runs, 3 wafers/run, 5 sites/wafer Site-to-site variability = circles, wafer-to-wafer variability = lines 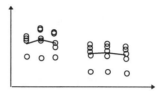 • Displays location and spread of response as a function of two variables. • Particularly useful for nested designs.
(c) Main effects graph Main effects of pressure and temperature on yield 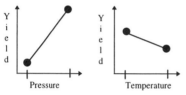 • Used for full factorial and fractional factorial designed experiments. • Depicts effect of a factor on the response. • Connects average response at factor levels.	(d) Interaction graph Interaction between temperature and pressure 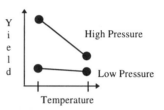 • Used in designed experiments. • Depicts effect of one factor on a response at different levels of another factor. • The factors interact if the lines are not parallel.
(e) Contour plot Yield as function of pressure and temperature 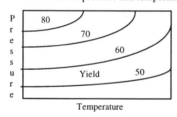 • Gives lines of constant value (contour lines) of a continuous response as a function of two continuous variables. • Ovelapping contour plots can be used to make trade-offs between multiple responses. • Topographical maps are contour plots of height above sea level.	(f) Three-dimensional graph Yield as a function of pressure and temperature 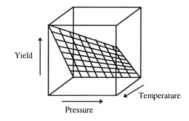 • Depicts surface of values of a continuous response as a function of two continuous variables. • Equivalent to a contour plot.

Figure G.2. *Graphs:* (a) *The box plot and* (b) *the multi-vari plot present three variables each. The other four graphs* ((c) *main effects graph,* (d) *interaction graph,* (e) *contour map, and* (f) *three-dimensional graph) give visual representations of the results of the same designed experiment.*

A *main effects plot*, or *main effects graph*, for a factor in a full or fractional factorial experimental design gives the average of all the values of the response at each level of the factor, with the averages connected by straight lines. Figure G.2c shows the main effects plots for two factors (pressure and temperature) at two levels each, for the response yield. Yield increases with pressure, and decreases slightly with temperature. It is useful to add the run (treatment combination) averages at each level of the factor to the main effects plot to visualize the spread of the observations at each level of the factor. Lewis et al. (Chapter 27, Appendices F and G) give main effects plots for two responses and eight factors.

An *interaction plot*, or *interaction graph*, for two factors in a full or fractional factorial experimental design gives the average of the values of the response at each level of one factor (connected by straight lines), for each level of another factor. Figure G.2d gives the interaction graph between temperature and pressure, for the response yield. If the lines in the interaction plot are not parallel the two factors interact, and if the lines in the interaction plot are parallel or almost parallel the two factors do not interact. In Figure G.2d temperature and pressure interact, since the effect of temperature on yield is different at different levels of pressure: yield is insensitive to temperature at low pressure, and yield decreases with temperature at high pressure. An equivalent interaction graph, which may look different but contains the same information, is obtained by switching the roles of the two factors and placing pressure on the horizontal axis and indexing the curves by temperature. The main effects graph of temperature in Figure G.2c can be obtained from the interaction graph in Figure G.2d by averaging the values of yield at the low and high temperatures. When the interaction between two factors is large, the main effects lose physical significance. The main effects graph in Figure G.2c indicates that yield decreases with temperature; the interaction graph in Figure G.2d indicates that yield is insensitive to temperature when pressure is at its low level, while yield decreases steeply when pressure is at its high level. Barnett et al. (Chapter 17, Figures 17.3 to 17.6) give several interaction plots for a designed experiment with factors at two levels each, and show that the corresponding contour plots provide the same information.

A *contour plot*, or *contour map*, or *contour graph*, is used to represent three continuous variables. In a designed experiment, two factors correspond to the horizontal and vertical directions of a contour plot, and the contour lines correspond to points along which the response has a constant value. Figure G.2e shows a contour plot of yield as a function of pressure and temperature. At low temperature and high pressure yield is around 90, and at high temperature and high pressure yield is around 63. For low pressure yield has a constant value around 40 for all temperatures, and for high pressure yield decreases from 90 to 63 with increasing temperature. This information agrees with the interaction graph in Figure G.2d. Topographical maps used by miners and hikers are contour plots where the contour lines indicate geographical locations with the same height above sea level. A contour plot can be used to make trade-offs between multiple responses by overlapping the contour lines for the different responses on the same contour plot. Preuninger et al. (Chapter 19, Figures 19.3 to 19.6) use overlapping contour plots for two responses to make trade-offs between the responses in a designed experiment. Buckner et al. (Chapter 16, Figure 16.6) give wafer maps, which are contour plots of the values of a response on the surface of a wafer. Buckner et al. (Chapter 15, Figure 15.1) show parallel contour lines for the responses stress and uniformity, which indicate the absence of a pressure by argon flow interaction for both responses; the vertical contours of stress indicate that pressure has an effect on stress while argon flow does not. Barnett et al. (Chapter 17, Figures 17.4 and 17.6) show how a saddle point (with the response decreasing towards two opposite corners of the contour plot and increasing towards the other two corners) corresponds to an interaction plot where one line has a positive slope and the other line has a negative slope.

A *three-dimensional graph* gives a three-dimensional view of a continuous response as a function of two continuous variables. In a designed experiment, two factors are shown in the plane at the base of the figure, and the response is depicted as a surface above the plane. Figure G.2f shows a three-dimensional graph corresponding to the contour plot in Figure G.2e. Hurwitz and Spagon (Chaper 9, Figure 9.2) give a wafer map of wafer thickness above the wafer surface.

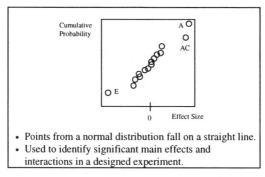

Figure G.3. *Normal probability plot.*

A *normal probability plot* is used to determine whether a set of independent observations with the same variance follows a normal distribution, by judging whether the observations fall approximately on a straight line in the normal probability plot. Observations that do not fall on a straight line may indicate outliers, a skewed (non-symmetric) distribution, or other non-normal characteristics. In a designed experiment, a normal probability plot can be used to help identify significant effects as points that are far from a line around zero where most of the effects cluster, and to check residuals for normality and for outliers. Figure G.3 shows the normal probability plot of the effects of a designed experiment. Three effects, main effects A and E and interaction AC, stand out as significantly different from noise. A normal probability plot has the grid along one axis ruled so that the cumulative normal distribution function plots as a straight line against values of the random variable. A broken line on a normal probability plot can indicate a "bad" (uncharacteristic) observation. A *half-normal probability plot* uses the absolute values of the observations. Hall and Carpenter (Chapter 12, Figure 12.1) test a response for normality using a normal probability plot. Lewis et al. (Chapter 27, Figures 27.4 and 27.5) identify significant factors in a designed experiment using normal probability plots of the factor effects. Freeny and Lai (Chapter 18, Figures 18.3, 18.4, and 18.6) use half-normal probability plots to identify significant effects and to check residuals for normality.

References for graphs

Introductory statistics books usually include descriptions of graphs. Three references are:

Box, George E.P., William G. Hunter, and Stewart J. Hunter. *Statistics for Experimenters*. New York, NY: John Wiley and Sons, Inc., 1978.
Daniel, Wayne W. *Essentials of Business Statistics*. Boston, MA: Houghton Mifflin Co., 1988.
Chambers, John M., William S. Cleveland, Beat Kleiner, and Paul A. Tukey, *Graphical Methods for Data Analysis*, Pacific Grove, California: Wadsworth & Brooks/Cole, 1983.

Three superb books on visual display of information are:

Tufte, Edward R., *The Visual Display of Quantitative Information*, Cheshire, Connecticut: Graphics Press, 1983.
Tufte, Edward R., *Envisioning Information*, Cheshire, Connecticut: Graphics Press, 1990.
Tufte, Edward R., *Visual Explanations*, Cheshire, Connecticut: Graphics Press, 1997.

STATISTICAL TERMS

This section introduces the following basic statistical concepts:
- population and sample;
- distributions: Normal, t, and Poisson;
- statistics: mean, variance, standard deviation, uniformity;
- confidence intervals and hypothesis testing.

Population and sample

A *(statistical) population* is the collection of all possible observations of interest. A *sample* from a population is the set of observations that is actually obtained, and is a subset of the population. A sample is usually assumed to be a *random sample*, one in which all elements in the population have the same likelihood of being selected, so some segment of the population (e.g., large observations) is not favored over others. A population is quantitative if the observations are numerical, and a population is qualitative if the observations are attributes (such as type of defect and type of equipment). The population of oxide thickness measurements is quantitative and consists of all the measurements that could be taken, and the 12 measurements of oxide thickness that Mary took yesterday are a sample from that population.

Distributions

We will describe three distributions, the normal distribution, the t distribution, and the Poisson distribution. These are theoretical distributions that correspond to a population. The shape of the distribution of a sample can be displayed using a histogram. The normal and t distributions correspond to continuous numerical variables that take on real values, and the Poisson distribution corresponds to a discrete numerical value that takes on non-negative integer values (0, 1, 2, ...).

The *normal distribution*, or *Gaussian distribution*, is perhaps the most frequently encountered distribution in engineering applications. It corresponds to measurements of continuous variables such as thickness, resistivity, and stress. The distribution is named after K.F. Gauss, who used it in the nineteenth century to describe the propagation of errors in astronomical data. The normal distribution is bell-shaped or mound-shaped and symmetric, and it corresponds to a continuous random variable that can take on any real value. The normal distribution is really a family of distributions, identified by two parameters: the *mean* μ which locates the middle of the distribution, and the *variance* σ^2 which is a measure of the spread of the distribution. The formula for the density function of a random variable X that follows a normal distribution with mean μ and variance σ^2 is given by

$$f(x) = \frac{1}{\sigma\sqrt{2\pi}} \exp\left[-\frac{(x-\mu)^2}{2\sigma^2}\right].$$

Areas under the curve of this density function represent probabilities. The area under the curve to the right of a value x is the probability that a value of the normal random variable will be as large or larger than the value x. The *standard normal random variable Z* has mean $\mu = 0$ and variance $\sigma^2 = 1$. The normal random variables X (with mean μ and variance σ^2) and Z (with mean 0 and variance 1) are connected by the formula Z = (X - μ)/σ, or equivalently X = μ + Zσ. The probability that a standard normal random variable is larger than a specified number is given in tables, which are then used to compute probabilities for a normal distribution with mean μ and variance σ^2. A histogram or a normal probability plot can be used to visually check if the observations in a sample appear to come from a normal distribution. The Shapiro-Wilk test is a numerical test for normality for independent observations in a sample. The population standard deviation σ (square root of the variance σ^2) is *sigma*, for the greek letter σ. The "Six Sigma" quality initiative in industry is an effort to improve quality by reducing variability. 99.73 % of the normal distribution is between $\mu - 3\sigma$ and $\mu + 3\sigma$ (within three standard deviations from the mean), 95.44% of the normal distribution is between $\mu - 2\sigma$ and $\mu + 2\sigma$, and 68.26% of the normal distribution is between $\mu - \sigma$ and $\mu + \sigma$. For a sample from a normal distribution, 99.74% of the observations are expected to fall within three standard deviations of the mean.

The ***t distribution*** *(or Student distribution) with* ν *degrees of freedom, or* t_ν is similar in shape to the normal distribution, except that it has "thicker" tails, that is, it has higher variability or spread. In addition, the t distribution tends to the normal distribution as the number of degrees of freedom ν gets large. The t distribution is in reality a family of distributions, one for each value of ν, where ν = 1, 2, 3, ... is a positive integer. The probability that a t random variable with ν degrees of freedom is larger than a specified number is given in tables. Consider the problem of estimating the population mean μ by the sample mean $\bar{x}$ computed from a random sample of size n. By the central limit theorem, $\bar{x}$ has an approximately normal distribution with mean μ and variance σ^2/n when n is large enough (over 30, say), so that Z = ($\bar{x} - \mu$)/($\sigma/\sqrt{n}$) has an approximately standard normal distribution. If the population variance σ^2 is unknown and is estimated by the sample variance s^2, then T = ($\bar{x} - \mu$)/($s/\sqrt{n}$) has approximately a t distribution with n - 1 degrees of freedom. The t distribution is used for confidence intervals and hypothesis testing for a population mean μ. In design of experiments, the population means of interest are the mean effects of the factors, which are tested for statistical significance. The *t statistic* T corresponding to a coefficient in a regression is an observation from a t distribution. The *p value* corresponding to the t statistic T is the probability that a value as large (in absolute value) as the t statistic or even larger could have been obtained just by chance, and is the area under the density function of the t distribution in the two tails of the distribution, to the left of (smaller than) -|T| and to the right of (larger than) |T|.

The *Poisson distribution* is used to count the number of times unlikely independent events (such as traffic accidents) occur over a specified period of time. It is also used to count the number of times unlikely independent events (such as radioactive particle decay) occur over a specified volume or area. The Poisson distribution would seem like the ideal distribution for particle counts in semiconductor manufacturing: the number of particles on a wafer is of interest, and the event of the appearance of each particle is unlikely. The reason the Poisson distribution is not a good approximation for particle counts is that particles tend to cluster (e.g., from scratches), so that the appearance of individual particles are not independent events. In general, the discrete *Poisson random variable* X counts the total number (0, 1, 2, ...) of independent events that occur in a unit, when the probability of an individual event is small and constant. The distribution of the Poisson random variable X is given by

$$P(x) = \lambda^x e^{-\lambda}/x!, \quad x = 0, 1, 2, ...$$

where the parameter λ is the expected number of occurrences of the event in the unit, and $x! = 1 \cdot 2 \cdot 3 \cdot ... \cdot x$ is the factorial of x.

Statistics

This section defines two types of statistics:
- Measures of the "middle" (or location, or central tendency), of the data: mean (average), and median;
- Measures of the spread (or variability, or dispersion) of the data: range, variance, standard deviation, and non-uniformity.

The description of the box plot in the last section includes a discussion of measures of central tendency and of spread. Outliers are described at the end of the section.

The mean of a sample of observations or of a population is a measure of central tendency, of the location of the "middle". The *(sample) mean*, or *(sample) average*, $\bar{x}$ of a sample of n observations $x_1, x_2, ... , x_n$, is the sum of all the observations divided by the number of observations n:

$$\bar{x} = \Sigma\, x_i/n.$$

The *population mean* μ is usually unknown. The sample mean $\bar{x}$ is the usual estimate of the "true" or "theoretical" population mean μ.

The *median* of a sample of observations is a value such that half of the observations are larger, and half are smaller. To find the sample mean, order the observations from smallest to largest. For an odd number of observations, the median is the "middle" observation. For an even number of observations, the median is the average of the two "middle" observations. The median is less sensitive to outliers than the mean, which is affected by unusually large or small observations.

The *range* of a sample of observations is the difference between the largest and the smallest observations. The range is used as a measure of the spread of the observations for small samples, say ten or less observations. The range is very easy to compute, but is very sensitive to outliers. The variance, standard deviation, and non-uniformity are less sensitive to outliers than the range (they are computed using all the observations), and are usually used when a sample has ten or more observations.

The variance of a sample or of a population is a measure of dispersion, spread, or variability. The *(sample) variance* s^2 of a sample of n observations $x_1, x_2, \ldots, x_n$ is given by

$$s^2 = \frac{\sum (x_i - \bar{x})^2}{n - 1},$$

where $\bar{x}$ is the average of the observations and the summation Σ is over the n observations. The variance is the sum of the squares of the distances between the observations and the mean $\bar{x}$, divided by n - 1. Since the numerator of the variance is a sum of squares, and the denominator is greater than zero if the sample has two or more observations, the sample variance is greater than or equal to zero. Sometimes the denominator of s^2 is given as n instead of n-1, and calculators often give two values of s^2: the "sample" variance with n - 1 in the denominator, and the "population" variance with n in the denominator. The *population variance* σ^2 is a measure of the variability or spread of the distribution of the population. The population variance σ^2 has a constant value that is usually not known, and is estimated by the sample variance s^2. Analysis of variance can give negative estimates of variances, which are usually set to zero.

The standard deviation of a set of observations or of a distribution is a measure of dispersion, spread, or variability. The *(sample) standard deviation* s is the positive square root of the sample variance s^2. The standard deviation has the same units as the observations (e.g., volts) while the variance has the square of the units of the observations (e.g., volts^2), which often does not have physical meaning. The *population standard deviation* σ is the positive square root of the population variance σ^2. The population standard deviation is *sigma*, for the greek letter σ. Standard deviations (for example, those corresponding to different components of variance) cannot be added directly to give a total standard deviation, but variances can be added to give a total variance. To obtain a total standard deviation, take the squares of the component standard deviations (which gives their variances), add them, and take the square root of the sum.

Non-uniformity, usually called *uniformity* (although this is a misnomer), or *coefficient of variation*, is a measure of variability or lack of homogeneity. Its definition is not standardized. The usual engineering definition of non-uniformity is $(s/\bar{x})100\%$, which expresses the standard deviation s as a percentage of the mean $\bar{x}$ (or equivalently $s/\bar{x}$, which expresses the standard deviation as a fraction of the mean). Since the formula for non-uniformity combines the standard deviation (measure of variability) and the mean (measure of central tendency), from the statistical point of view it is often better to study the standard deviation (or the variance) as a measure of variability instead of the non-uniformity, although the results of the analysis will usually be the same. Sometimes uniformity is defined as $(3s/\bar{x})100\%$ [or $3s/\bar{x}$], as $R/\bar{x}$, as the standard deviation s, as the variance s^2, or as some other measure of variability.

An *outlier* is an observation that is much larger or much smaller than most of the observations, or is in some way very different from the majority of the observations. Outliers can bias the results of a numerical analysis. It is usually wise to have reasonable engineering grounds to remove an outlier from a data set. Sometimes it is precisely the unusual observations that are of greatest interest, such as the process settings that give an unusually low wafer contamination. The data can be analyzed twice, once with outliers and once without them, to see whether the conclusions derived from both analyses differ. Czitrom and Reece (Chapter 8, Figure 8.3) found nine observations that had unusually high oxide thickness that was traced to a likely film residue on a wafer, and eight observations with unusually low thickness due to a faulty gauge.

Confidence Intervals and Hypothesis Testing

This section considers some concepts relating to confidence intervals and hypothesis testing, and the connection between them. The next section, on analysis of variance, considers hypothesis testing for regression analysis in greater depth.

A *confidence interval* for a parameter is an interval of numbers which attempts to include the true (population) value of the parameter, with a specified level of confidence. The confidence interval for a parameter is determined from data in a sample, and its formula depends on the parameter considered. Examples of parameters of interest are the population mean, the population variance, and the difference between two population means. The parameter is estimated by a single value (point estimate), and the confidence interval is built around that single value to take sampling variability into account. For example, the population mean μ from a normal population with unknown standard deviation is estimated by the sample mean $\bar{x}$, and the confidence interval goes from $\bar{x} - t_{n-1,\alpha/2} s/\sqrt{n}$ to $\bar{x} + t_{n-1,\alpha/2} s/\sqrt{n}$, or briefly

$$\bar{x} \pm t_{n-1,\alpha/2} s/\sqrt{n},$$

where $\bar{x}$ is the mean and s is the standard deviation of a random sample of size n, $t_{n-1,\alpha/2}$ is the value of student's t statistic with n-1 degrees of freedom such that the probability that the random variable T is greater than $t_{n-1,\alpha/2}$ is $\alpha/2$, and where $(1-\alpha)100\%$ is the level of confidence. The $\pm t_{n-1,\alpha/2} s/\sqrt{n}$ values of the confidence interval are built around $\bar{x}$ to take into account the fact that different samples give different values of $\bar{x}$ (sampling variability). The user selects the *level of confidence (1-α)100%* (or the *level of significance* α) that the true value of the parameter will actually be in the *(1-α)100% confidence interval* constructed from the sample data. If the user wants to be 95% confident that the confidence interval will contain the true value of the parameter, the user constructs a 95% confidence interval. Equivalently, the user selects the risk (level of significance) he/she is willing to take that the confidence interval does *not* contain the parameter. The level of confidence is the percentage of instances that the true value of the parameter will be in a confidence interval for repeated constructions of the confidence interval from different samples. Typical levels of confidence used in engineering applications are 90%, 95%, and 99% (which correspond to values of α of 0.1, 0.05, and 0.01 respectively). A 100% confidence interval is useless. A 100% confidence interval for the mean is the interval between $-\infty$ and ∞, which only says that the mean is a number (which indeed is true 100% of the time). There are trade-offs between the level of confidence, the sample size, and the confidence interval width. For a given level of confidence, the narrower we want the width of the confidence interval to be, the larger the sample size has to be. For a given sample size, the more confident we want to be that the true value of the parameter actually lies in the confidence interval, the wider the confidence interval has to be (to make sure the interval "catches" the true value). For a given width of the confidence interval, the more confident we want to be that the true value of the parameter lies in the confidence interval, the larger the sample size needs to be.

A *hypothesis test* on a population parameter such as the mean μ is performed to determine the plausibility of a *(null) hypothesis* compared to an *alternative hypothesis*. Three common hypothesis tests are

1. The null hypothesis is that the population mean μ equals a specific value μ_0, versus the alternative hypothesis that it does not:

$$H_0: \mu = \mu_0,$$
$$H_a: \mu \neq \mu_0.$$

2. The null hypothesis is that the population mean μ is less than or equal to a specific value μ_0, versus the alternative hypothesis that it is greater:

$$H_0: \mu \leq \mu_0,$$
$$H_a: \mu > \mu_0.$$

3. The null hypothesis is that the population mean μ is greater than or equal to a specific value μ_0, versus the alternative hypothesis that it is less:

$$H_0: \mu \geq \mu_0,$$
$$H_a: \mu < \mu_0.$$

To determine whether the null hypothesis is true, the results from a sample are compared to the results that would be expected if the hypothesis were true. For the first hypothesis test, the test statistic $T=(\bar{x}-\mu_0)\sqrt{n}/s$ is compared to the value $t_{n-1,\alpha/2}$ from a table of the t distribution, and the null hypothesis is rejected if and only if the absolute value of T is greater than $t_{n-1,\alpha/2}$.

Two types of mistakes can be made in hypothesis testing: rejecting the null hypothesis when it is true (*type I error*), and failing to reject the null hypothesis when it is false (*type II error*). The *probability* α of making a type I error is selected by the user, and the *probability* β of making a type II error can be reduced by using a larger sample to test the hypothesis. For a given sample size, if the probability of making a type I error is decreased (increased), then the probability of making a type II error increases (decreases). Hypotheses that are frequently tested are that a factor does not affect the response (changes in the input do not affect the output), that a group of factors in a designed experiment do not affect the response, that two or more treatments (e.g., machines or seeds) are not different, that a treatment is larger than a specified amount (e.g., the tensile strength is greater than a specific number), that one treatment has a higher mean than another treatment, and that the variances from two populations are equal.

Confidence intervals and hypothesis tests are related. For example, the hypothesis that the true value of the mean μ has a specific value μ_0 is rejected if, and only if, μ_0 is *not* in the corresponding confidence interval (equivalently, we fail to reject the hypothesis if, and only if, μ_0 is in the confidence interval). The probability α of a type I error in this hypothesis test corresponds to a level of significance α for the confidence interval. Note that only one confidence interval corresponds to the hypothesis test for an arbitrary value of μ_0 (which can take on an infinite number of values), so that a confidence interval gives more information than a hypothesis test. For this reason, it is usually better to give a confidence interval than to make a hypothesis test when making a statement about a parameter. On the other hand, to determine whether a specific factor affects a response in a designed experiment, it is usually more convenient to do a hypothesis test using analysis of variance than to use confidence intervals.

References for statistical terms:

> Box, George E.P., William G. Hunter, and Stewart J. Hunter, *Statistics for Experimenters*, New York, NY: John Wiley and Sons, Inc., 1978.
> Mendenhall, William, and Richard L. Scheaffer, *Mathematical Statistics with Applications*, North Scituate, Massachusetts, 1973.

ANALYSIS OF VARIANCE (ANOVA)

Introduction

***Analysis of variance** (ANOVA)* partitions the total variability observed in a response into components corresponding to different sources of variation. Analysis of variance is used in the book to test a source of variation for statistical significance (e.g., to determine whether a change in temperature affects the response), or to estimate the variability corresponding to a source of variation (eg., to estimate variance components such as site-to-site variability and wafer-to-wafer variability).

Analysis of variance is used throughout the book. In a gauge study (Part 1), sources of variation are tested for significance, and variance components are estimated. In a passive data collection (Part 2), variance components such as site-to-site variability, wafer-to-wafer variability, and lot-to-lot variability, are estimated. In a designed experiment (Part 3), sources of variation such as factors and interactions are tested for significance, and components of variance such as site-to-site variability are used as responses. In statistical process control (SPC, Part 4), variance components are monitored for stability.

We will first define some basic analysis of variance concepts, and then we will consider how an ANOVA table is used to test sources of variation for statistical significance, and how an ANOVA table is used to estimate variance components.

Basic Concepts

The sampling plan used to collect the data defines ***factors***, or *variables*. If the data in a gauge study includes several operators and cassette types, then operator and cassette type are factors. Examples of factors in a passive data collection are furnace run, wafer location in the furnace, and site on a wafer. Examples of factors in a designed experiment are temperature, pressure, and gas flow. During the experiment to collect the data on the response, each factor has to be at two or more levels (it is necessary to have at least two levels of temperature to determine whether changes in temperature affect the response, and it is necessary to have at least two lots to estimate lot-to-lot variability). A factor can be fixed or random. A factor is *fixed* if the main interest is in whether the response changes when the level of the factor changes, and the levels of the factor are the specific ones of interest. In a designed experiment temperature is a fixed factor, since it is of interest to know whether a change in temperature affects the response, when temperature changes from

300°C to 350°C. A factor is *random* if the main interest is in the *variability* in the response caused by variability in the factor, and the levels of the factor considered in the data are representative of all possible factor levels. In a passive data collection lot is a random factor, since it is of interest to know the variability between lots, and the particular wafer lots considered in a given sample are representative of the lots of interest. If the factor "machine" is at five levels (there are five machines), machine is a fixed factor if interest is in whether those particular five machines are different, and machine is a random factor if interest is in variability between machines and the five machines are representative of all machines.

A *model* corresponds to a data set, and one or more terms in the model correspond to each factor. For example, a model for the response sheet resistance as a function of the fixed factor "operator" is given by

$$\text{Sheet resistance} = \mu + \text{Operator}_i + \varepsilon,$$

where μ is a constant, Operator_i is a constant term corresponding to the ith operator, ε is a random error term associated with each observation and assumed to have a normal distribution with mean 0 and variance σ^2, and the observations are independent. If "operator" is a random factor, then Operator_i is a random variable with a normal distribution with mean 0 and variance $\sigma^2_{\text{Operator}}$. In design of experiments, the model for continuous fixed factors is usually expressed in terms of coefficients and coded variables x_i that take on values between -1 and +1 at the low and high values of the factor respectively (see the introduction to Part 3). For example, a model for the response thickness as a function of temperature and pressure is

$$\text{Thickness} = \beta_0 + \beta_1 \, x_{\text{Temp}} + \beta_2 \, x_{\text{Press}} + \beta_{12} \, x_{\text{Temp}} \, x_{\text{Press}} + \varepsilon,$$

where all the β's are constant coefficients that are estimated from the data, the term $\beta_1 \, x_{\text{Temp}}$ corresponds to the main effect of temperature, and the term $\beta_{12} \, x_{\text{Temp}} \, x_{\text{Press}}$ corresponds to the interaction between temperature and pressure. A term in the model corresponding to a fixed factor is a constant. A term in the model corresponding to a random factor is a random variable with a mean and a variance, and a variance component is an estimate of the variance. A *fixed effects model* has fixed factors only, a *random effects model* has random factors only, and a *mixed model* has both fixed and random factors. The interaction between a fixed factor and a random factor is considered random. One or more terms in the model, such as linear and quadratic terms, can correspond to one factor. The assumptions underlying the model are used to test whether the terms in the model corresponding to fixed factors are statistically significant, and to estimate the variance components of the random factors.

In the next section we will consider the analysis of variance for a fixed effects model, where the main objective is to test which terms should be in the model, that is, which terms are statistically significant (different from zero) and hence which factors affect the response. In the last section we consider analysis of variance for a random effects model, where the main objective is to find the variance components.

The ANOVA Table: Testing Sources of Variation for Significance

The main computations in an analysis of variance are presented in an *analysis of variance (ANOVA) table*. We will introduce analysis of variance for fixed effects using the example in Table G.1, from a gauge study by Buckner et al.

Table G.1. *ANOVA table for a gauge study. Six fixed effects (operator, cassette type, slot loading, prior film, wafer handler, and probe conditioning) are sources of variation that affect the response, sheet resistance. From Table 2.6a by Buckner et al., Chapter 2.*

Source of Variation	Degrees of Freedom	Sum of Squares	Mean Square	F Value	p value
Operator	2	0.00951250	0.00475625	1.50	0.2788
Cassette Type	1	0.00640000	0.00640000	2.02	0.1926
Slot Loading	1	0.00062500	0.00062500	0.20	0.6683
Prior Film	1	0.00090000	0.00090000	0.28	0.6081
Wafer handler	1	0.00062500	0.00062500	0.20	0.6683
Probe Conditioning	1	0.00122500	0.00122500	0.39	0.5509
Residual	8	0.02528750	0.00316094		
Total	15	0.04457500			

The first column in an ANOVA table corresponds to the *sources of variation*. The *total variability* in the observations is partitioned into the variability corresponding to the factors (terms in the model), and the variability that remains after the model is fitted to the data is the *residual* or *error* source of variation that will be used to estimate σ^2. In Table G.1, the six factors in the gauge study (operator, cassette type, slot loading, prior film, wafer handler, and probe conditioning) are sources of variation.

Another column in the ANOVA table gives the *sums of squares* corresponding to the sources of variation. The *total (corrected) sum of squares SST* is the total variability in the observations. If the N observations are $y_1, y_2, ..., y_N$ and their average is $\bar{y}$, the total sum of squares is $SST = \Sigma(y_i - \bar{y})^2$. The total sum of squares is the sum of the squares of the differences between each observation and the average of all the observations, that is, the sum of the squares of the deviations of the observations from their average. The total sum of squares is the numerator of the sample variance of all the observations. The total sum of squares is partitioned into sums of squares corresponding to each source of variation, namely the terms in the model, and the *residual* or *error sum of squares SSE*. SSE is given by $\Sigma(y_i - \hat{y}_i)^2$, which is the sum of the squares of the residuals $e_i = y_i - \hat{y}_i$, where y_i is the ith observation and $\hat{y}_i$ is the estimate of the ith observation using the model. The residual sum of squares SSE is the sum of the squares of the distances between the observations and the estimates of the observations given by the model, that is, the residual sum of squares SSE is the variability in the observations that is not explained by the model. In Table G.1, the sum of the six sums of squares corresponding to the six factors, plus the residual sum of squares, gives the total sum of squares 0.04457500. For a 2^k full factorial or 2^{k-p} fractional factorial experimental design with factors at two levels each coded as -1 and +1, the sum of squares corresponding to a factor is the square of the coefficient in the model multiplied by the number of observations N (or equivalently, the square of the main effect of the factor multiplied by N/4).

The number of *degrees of freedom (df)* for a source of variation represents the number of independent terms in the corresponding sum of squares. The number of degrees of freedom corresponding to the total sum of squares is the total number of observations minus one, N-1. There are only N-1 "independent" terms in the total sum of squares $\Sigma(y_i-\bar{y})^2$, since by the definition of the average $\Sigma(y_i-\bar{y})$ is identically equal to zero. N-1 is the denominator of the sample variance of all the observations. The N-1 degrees of freedom are partitioned among the sources of variation. In Table G.1, the total degrees of freedom are N-1=15, which means that the analysis of variance corresponds to N=16 observations. The sum of the number of degrees of freedom corresponding to the six factors (2+1+1+1+1+1=7), plus the 8 degrees of freedom corresponding to the residual, give the 15 total degrees of freedom. For a categorical factor, the number of degrees of freedom is the number of levels of the factor minus one. The categorical factor "operator" in Table G.1 has three levels (the operators were Larry, Curly, and Moe, see Table 2.2 of Chapter 2), so "operator" has 3-1=2 degrees of freedom. The other five factors are also categorical and are at two levels each, so they have 2-1=1 degrees of freedom each. In a designed experiment, continuous or discrete factors at two levels have 2-1=1 degrees of freedom. Continuous factors at three levels can be separated into linear and quadratic terms, with one degree of freedom each.

Another column in the ANOVA table gives mean squares. The **mean square** corresponding to a source of variation is its sum of squares divided by its degrees of freedom. This quotient essentially gives an "average" sum of squares per degree of freedom. In Table G.1, the sum of squares for operator 0.0095125, divided by the 2 degrees of freedom for operator, give the mean square for operator 0.00475625. The *mean square error*, or *residual mean square*, is an estimate of the variance σ^2 corresponding to the "noise" in the observations, or natural variability, that has not been attributed to one of the sources of variation. In Table G.1, the estimate of σ^2 is 0.00316094. The mean square error is the aggregate of estimates of the variance σ^2 computed from replicates at the same experimental conditions, as well as terms that are not currently in the model. If there are replicates, the residual sum of squares can be divided into a *pure error* component corresponding to the replicates, and *lack of fit* that is used to check whether the model appears to be missing important terms, such as quadratic terms. Measurement error is included in the residual mean square. The total mean square (which is the variance of all the observations) is not computed because it is not used in the analysis of variance.

The ***F statistic***, or *F value*, corresponding to a source of variation is the ratio of the mean square of the source of variation divided by the residual mean square. In Table G.1, the mean square for operator 0.00475625 divided by the residual mean square 0.00316094 gives the F value for operator 1.50. The F statistic can be thought of as a "signal-to-noise ratio" that compares the mean square corresponding to a source of variation to the mean square error corresponding to random variability. If the F statistic is much larger than one, the source of variation contains variability beyond that which can be attributed to random variation, the source of variation is statistically significant, and the factor or factors corresponding to the source of variation have an effect on the response. On the other hand, if the F statistic is close to one, the source of variation has variability comparable to random variation, the source of variation is not statistically significant, the term can be removed from the model, and the factor or factors do not affect the response. The source of variation can be tested for statistical significance by comparing the F statistic to a value in a table for the F distribution with the degrees of freedom for the source of variation for the numerator and the degrees of freedom for the residual for the denominator, although it is easier to test for significance using the p value in the next column of the ANOVA table. The F statistic corresponding to a fixed factor tests the null hypothesis that the term in the model has a coefficient of zero (that is, the term should not be in the model because the source of variation has no effect on the response), against the alternative hypothesis that the term in the model has a coefficient different from zero.

The *p value*, *significance probability*, or *Prob>F*, is used to test an individual source of variation for statistical significance, to determine whether it has an effect on the response. The p value corresponding to a fixed factor is the probability of seeing an effect that large just by chance if the source of variation really has no effect on the response. In Table G.1, there is a probability of 0.2788 (28 chances in 100) of seeing the effect of operator as large as it is if operator really does not have an effect on sheet resistance. The p value is computed from the F value as the area under the curve of the F distribution to the right of the F value. In Table G.1, 0.2788 is the area under the curve of the F distribution to the right of 1.50. A value of 0.05 or 0.1 is often used as a threshold, so that sources of variation that have p values less than 0.05 or 0.1 are statistically significant (the fixed factor appears to affect the response) and are kept in the model. Factor effects that have p values greater than 0.05 or 0.1 are not considered statistically significant and are removed from the model. All the p values in Table G.1 are larger than 0.05, so none of the factors are statistically significant. When all the factors in an analysis of variance are statistically significant or none of the factors are statistically significant, it is wise to question whether the estimate of error (residual or error mean square) is too small or too large. For example, the estimate of error can be too small if replications are taken one after the other without randomization, so the mean square error is underestimated because it does not include the variability of going from one experimental set-up to another. Buckner et al. (Chapter 2) remark that the estimate of the residual error was probably too large because they performed the experiment over five days and hence captured many sources of variation in addition to those in the table. This agrees with the fact that none of the factors in Table G.1 are significant. This highlights the importance of having a good estimate of variability to test the significance of sources of variation in an analysis of variance.

The next step in the analysis of the data is to remove the least significant factor from the model, either slot loading or wafer handler in Table G.1, and repeat the analysis of variance. This will change the residual sum of squares, and hence the F values and the p values. Terms are removed from the model until only statistically significant terms remain.

Table G.2. *ANOVA Table with the six factors in Table G.1 combined into one "Model" term. From Table 2.6a by Buckner et al., Chapter 2.*

Source of Variation	Degrees of Freedom	Sum of Squares	Mean Square	F Value	p value
Model	7	0.01928750	0.00275536	0.87	0.5654
Residual	8	0.02528750	0.00316094		
Total	15	0.04457500			

Table G.2 gives a summary of the analysis of variance in Table G.1. Table G.2 pools the sources of variation corresponding to the six factors in the gauge study into a single source of variation due to the regression model. Equivalently, we can think of the "model" source of variation in Table G.2 as being partitioned into the six sources of variation given in Table G.1. The sums of squares corresponding to the factors in the model are added into a *regression sum of squares SSR*, or *model sum of squares*. The sums of squares for the factors in Table G.1 add up to the sum of squares for the model 0.01928750 in Table G.2, and the residual and total sums of squares are the same in both tables. The degrees of freedom corresponding to the factors in Table G.1 add up to the 7 degrees of freedom for the model in Table G.2, and the residual and total degrees of freedom are the same in both tables. In Table G.2, the 7 degrees of freedom for the model and the 8 degrees of freedom for the residual add up to the 15 total degrees of freedom, and the sum of squares for the model SSR=0.01928750 and the sum of squares for the residual SSE=0.02528750 add up to the total sum of squares SST=0.04457500. The mean square for the model 0.00275536 is the model sum of squares divided by the model degrees of freedom. The mean square error (or residual mean square) 0.00316094 that is used to estimate σ^2 is the same in Tables G.1 and G.2. In Table G.2, the F value 0.87 is the mean square for the model divided by the residual mean square. The F value in Table G.2 is used to test the entire model for significance, to determine whether the terms in the model help explain any of the variability in the model, that is, whether any of the factors in the model have an effect on the response. The p value of 0.5654 in Table G.2 is much larger than 0.05, indicating that the entire model with all the factors is not statistically significant. Again, the fact that the estimate of the mean square error is probably too large means that the p value of 0.5654 is probably larger than it should be.

In the analysis of variance in Table G.2, the total sum of squares SST is the sum of the regression (model) sum of squares SSR and the residual sum of squares SSE (SST=SSR+SSE). The *coefficient of determination* R^2 is the fraction (or percentage) of the variability in the data explained by the model, R^2=SSR/SST. In Table G.1, R^2=0.01928750/0.04457500=0.43, so that 43% of the variability in the data is explained by the model. A value of R^2 above 0.9 or 0.8 is usually considered satisfactory. The value of R^2 is usually higher (the model fits the data better) when the response is the mean than when the response is a variability (variance or uniformity) or a particle count. The *adjusted coefficient of determination, R^2 adjusted*, adjusts for the fact that as the number of terms in the model increases, R^2 increases and gets closer to 1 (or 100%) whether the model is suitable or not. R^2 adjusted is always less than R^2.

In design of experiments (Part 3), a *t statistic* and a *p value* are often given with the least squares estimate of each coefficient in the model. If the coefficient has one degree of freedom, the square of the t statistic is equal to the F statistic in the analysis of variance table, and the p values corresponding to the t statistic and to the F statistic are identical. A t statistic can only be used to test a source of variation with one degree of freedom for statistical significance, while an F statistic can be used to test a source of variation with one or more degrees of freedom for significance, so an F statistic can be used to test several sources of variation simultaneously. The sources of variation in an analysis of variance table can be grouped in different ways. For example, in a designed experiment, all the interaction terms can be grouped and tested to determine whether any interaction terms are needed in the model. If the test is statistically significant, the interaction terms are tested individually to determine which one of the interactions is the one that is needed.

Sometimes the constant term in the model is included as a source of variation, with a sum of squares of $N\bar{y}^2$. In this case, the *total sum of squares* in the ANOVA table is Σy_i^2. This total sum of squares Σy_i^2 is the sum of the total corrected sum of squares $\Sigma(y_i-\bar{y})^2$ used in Tables G.1 and G.2, and the sum of squares $N\bar{y}^2$ for the constant term. The sums of squares corresponding to the other terms in the model and the residual sum of squares remain the same.

The ANOVA Table: Estimating Variance Components

A ***variance component***, or *component of variance*, is the variance of a random factor. The standard deviation of the variance component is the square root of the variance. When estimating the variance components using an analysis of variance table, the computations of the number of degrees of freedom, sums of squares, and mean squares, are the same as those in the analysis of variance for fixed factors described in the last section. However, the mean squares are used to estimate the variance components, instead of for finding F values and p values as in the last section. We will use analysis of variance to estimate the variances of three factors for the example in Table G.3, from a passive data collection by Lynch and Markle.

Table G.3. *ANOVA table for three nested factors, run, wafer nested in run, and site nested in wafer. There are 9 runs, 6 wafers/run, and 9 sites/wafer. A column for sums of squares is not included in this table, but it was used to derive the column of mean squares; the sums of squares can be retrieved by multiplying the mean squares by their degrees of freedom. From Table 7.1 by Lynch and Markle, Chapter 7.*

Source of Variation	Degrees of Freedom	Mean Square	Expected Mean Square	Variance component	Variance component as %
Run-to-run	8	6106.5	$\sigma_s^2 + 9\sigma_w^2 + 54\sigma_r^2$	94.2	26
Wafer-to-wafer	45	1019.7	$\sigma_s^2 + 9\sigma_w^2$	94.6	26
Site-to-site	432	168.3	σ_s^2	168.3	47
Total	485			357.1	100

Lynch and Markle considered three sources of variation, runs, wafers within runs, and sites within wafers, and were interested in estimating run-to-run variability, wafer-to-wafer variability, and site-to-site variability. The three factors are nested. One factor is *nested* in another factor if the levels of the first factor are similar but not identical at different levels of the second factor. For example, wafers are nested in runs because wafer number 1 in run 1 is different from wafer number 1 in run 2 (for example, one wafer may have a thicker oxide layer than the other wafer). We will consider sites on a wafer to be nested in wafers. Lynch and Markle's example has 9 sites per wafer, 6 wafers per run, and 9 runs. Table G.3 has no residual source of variability, because the total variability is partitioned among the three sources of variation. Site-to-site variability, also known as within-wafer or across-wafer variability, includes measurement error. If it is known, the measurement error (precision) can be removed from site-to-site variability by subtraction: σ^2site (corrected) = σ^2site (calculated) - σ^2measurement error.

The variance components σ_s^2 (site-to-site), σ_w^2 (wafer-to-wafer), and σ_r^2 (run-to-run) are estimated by equating the mean squares to the *expected mean squares* in the analysis of variance Table G.3. This gives a system of three equations in three unknowns, the three variance components σ_s^2, σ_w^2, and σ_r^2. From the third equation, the *site-to-site variance component* is given by

$$\sigma_s^2 = 168.3$$

From the second equation, the *wafer-to-wafer variance component* is

$$\sigma^2_w = 1019.7/9 - \sigma^2_s/9 = 113.3 - 168.3/9 = 94.6$$

From the first equation, the *run-to-run variance component* is

$$\sigma^2_r = 6106.5/54 - \sigma^2_s/54 - \sigma^2_w/6 = 113.1 - 168.3/54 - 94.6/6 = 94.2$$

These three variance components are given in the next to the last column in Table G.3. The total variance 357.1 is the sum of the three variance components. The last column of Table G.3. gives the *variance components as a percentage* of the total variance. Site to site variability contributes almost half (47%) of the total variability, so reducing it would have the greatest impact on process improvement.

Table G.4. *Structure of oxide thickness data for 9 sites/wafer, 6 wafers/run, for one run. The last two columns give the average and the variance of the measurements on each wafer. The two values in the last row are the variance of the wafer averages $s^2_{\bar{y}}$, and the average of the wafer variances $\bar{y}_{s^2}$.*

Wafer number	Site 1	Site 2	Site 3	Site 4	Site 5	Site 6	Site 7	Site 8	Site 9	Wafer average	Wafer variance
1	y_{11}	y_{12}	y_{13}	y_{14}	y_{15}	y_{16}	y_{17}	y_{18}	y_{19}	$\bar{y}_1$	s^2_1
2	y_{21}	y_{22}	y_{23}	y_{24}	y_{25}	y_{26}	y_{27}	y_{28}	y_{29}	$\bar{y}_2$	s^2_2
3	y_{31}	y_{32}	y_{33}	y_{34}	y_{35}	y_{36}	y_{37}	y_{38}	y_{39}	$\bar{y}_3$	s^2_3
4	y_{41}	y_{42}	y_{43}	y_{44}	y_{45}	y_{46}	y_{47}	y_{48}	y_{49}	$\bar{y}_4$	s^2_4
5	y_{51}	y_{52}	y_{53}	y_{54}	y_{55}	y_{56}	y_{57}	y_{58}	y_{59}	$\bar{y}_5$	s^2_5
6	y_{61}	y_{62}	y_{63}	y_{64}	y_{65}	y_{66}	y_{67}	y_{68}	y_{69}	$\bar{y}_6$	s^2_6
										$s^2_{\bar{y}}$	$\bar{y}_{s^2}$

We will now give an intuitive derivation of the variance components, and of the degrees of freedom for the analysis of variance in Table G.3. Table G.4. shows the structure of the oxide thickness data for one run, with six wafers in the run and nine sites on each wafer. Two statistics are computed from the oxide thickness measurements at the nine sites on each wafer: the wafer average, and the wafer variance. These are shown in the last two columns of Table G.4.

First consider the site-to-site source of variability. For each wafer, an estimate of site-to-site variability is the variance of the oxide thicknesses at the nine wafer sites (the wafer variances in the last column of Table G.4). Using all wafers, an estimate of site-to-site variability is the average of the wafer variances for all 6x9=54 wafers, given by 168.3. This is the site-to-site variance component

$$\sigma^2_s = 168.3$$

which is also the mean square for site 168.3 shown in Table G.3. Each wafer variance is computed using 9 sites, and has 9-1=8 degrees of freedom (the denominator of each variance). The number of degrees of freedom for site-to-site variability using all the wafers is 54x8=432, as shown in Table G.3.

Now consider the wafer-to-wafer source of variability. For each run, an estimate of wafer-to-wafer variability is the variance of the six wafer averages ($s^2_{\bar{y}}$ in the last row of Table G.4). Using all runs, an estimate of wafer-to-wafer variability is the average of the 9 variances of wafer averages, which is 113.3 (which is also the mean square for wafer in Table G.3 divided by 9 sites on one wafer, 1019.7/9). The wafer-to-wafer variance component is

$$\sigma^2_w = 113.3 - \sigma^2_s/9 = 94.6$$

Subtracting $\sigma^2_s/9$ "corrects" the wafer-to-wafer variability 113.3 for the fact that each wafer average used to compute it contains site-to-site variability. For each run, the variance of the six wafer averages has 6-1=5 degrees of freedom (the denominator of the variance). Since there are 9 runs, there are a total of 9x5=45 degrees of freedom for wafer-to-wafer variability, as shown in Table G.3.

Finally, consider the run-to-run source of variation. There are nine runs, and we can compute the average oxide thickness for each run. An estimate of run-to-run variability is the variance of the nine run averages, which is 113.1 (which is also the mean square for run 6106.5 in Table G.3 divided by the total number of wafer sites in one run, 9x6=54). The run-to-run variance component

$$\sigma^2_r = 113.1 - \sigma^2_s/54 - \sigma^2_w/6 = 94.2$$

Subtracting the term $\sigma^2_s/54 + \sigma^2_w/6$ "corrects" the run-to-run variability 113.1 for the fact that each run average used to compute it contains both wafer-to-wafer variability and site-to-site variability. Since there are nine wafer runs, the run-to-run source of variability has 9-1=8 degrees of freedom, as shown in Table G.3.

When variance components are estimated by equating the expected mean squares to the mean squares and solving for the variance components, the estimate of a variance component may be negative. Since a variance (which is a sum of squares) cannot be negative, a negative estimate of a variance component is set to zero.

References for analysis of variance:

Box, George E.P., William G. Hunter, and Stewart J. Hunter, *Statistics for Experimenters*, New York, NY: John Wiley and Sons, Inc., 1978.

Montgomery, Douglas C. *Design and Analysis of Experiments,* 4th ed. New York, NY: John Wiley and Sons, Inc., 1997.

Statistical Topic by Chapter

The following table shows where each statistical topic can be found within each section and each chapter.

	\multicolumn{22}{c	}{CHAPTER NUMBER}		
	Gauge					PDC								DOE							SPC				Reliability				C	G
TOPIC	1	2	3	4	5	6	7	8	9	10	11	12	13	14	15	16	17	18	19	20	21	22	23	24	25	26	27	28	30	G
ANOVA	x	x			x	x	x	x	x	x	x	x		x	x	x	x	x	x	x									x	x
Assist																										x			x	x
Availablility																									x	x	x			
Calibration of gauge	x		x																						x		x			
Capability	x			x	x	x	x														x		x		x	x			x	x
Control charts	x		x			x	x					x									x	x	x	x	x				x	x
Individuals							x														x		x	x					x	
MR																					x		x						x	
R			x																		x		x						x	
s						x															x			x					x	
xbar			x			x															x	x	x	x					x	
Drift over time	x			x		x	x							x			x				x	x			x					x
Designed experiments		x	x		x			x		x	x			x	x	x	x	x	x	x						x			x	
Center points														x	x	x	x		x										x	
Central composite														x	x			x											x	
D-optimal														x															x	
Folding over														x		x	x													
Fractional factorial		x												x		x	x										x		x	
Full factorial								x		x				x	x			x	x	x										
Nested					x							x		x					x										x	
Robust														x				x	x								x			
Split-plot								x		x				x	x					x							x			
Three-quarter														x			x													
Gauge studies	x	x	x	x	x																								x	
Interaction plot									x					x		x				x							x			x
Marathon																									x	x			x	
P/T, gauge	x	x		x	x																								x	
Passive data collection						x	x	x	x	x	x	x	x		x							x	x		x				x	
Precision of gauge	x	x	x	x	x																								x	
Preventive maintenance																									x		x			
Regression		x	x							x				x	x	x	x	x	x	x						x			x	x
Reliability																									x	x	x	x	x	
Repeatability of gauge	x			x	x																								x	
Reproducibility of gauge	x			x	x																								x	
Residual analysis															x		x		x										x	
Risk											x	x																		
Sampling	x			x		x		x			x	x		x							x		x		x				x	x
SPC	x		x			x	x														x	x	x	x	x					x
Time	x			x		x	x				x	x	x	x				x			x	x			x	x	x	x		x
Trade-offs														x	x		x	x	x							x				
Variance components	x				x	x	x	x	x	x	x	x		x										x		x			x	x

INDEX

Accuracy of gauge, 5
Addelman design, 12, 16
Adjusted coefficient of determination R^2, 191, 496
Advantages of designed experiments, 185
Age replacement policies, 403–404, 406–412, 421–423
  availability for, 407–412, 422, 423
Air flow in clean room, 146
Alarm, false, 302
Alias, 180
  chain, 180, 236, 239, 249
    breaking, 249
Alignment, 134–136, 283–285, 365, 366, 368, 472
  clear field, 285
  dark field, 285, 371
  marks, 369
Alternative hypothesis, 490
Analysis of variance, 491–499; *see* ANOVA
Anneal, 473
ANOVA, 190–192, 481, 491–499
  in DOE, 190, 192, 204–207, 218–222, 240, 243–244, 255, 259, 280, 288–291, 293–295, 372–375, 461
  in gauge study, 15, 49–55
  in PDC, 66, 76–79, 94–100, 105–111, 122–128, 136–137, 147–151, 434–435
  in reliability, 372–375
  table, 493–497
Array
  control, 182, 286, 390
  inner, 182, 390
  noise, 182, 286, 390
  orthogonal, 182
  outer, 182, 390
  product, 182, 391
Assignable causes of variability, 74–75, 300
Assists, 360, 383, 387, 388, 389

mean time between, 360
mean wafers between, 388, 396
Attribute data, 303
Augmenting an experimental design, 235, 245
Availability,
  equipment, 360, 404, 405, 407–412, 414–420, 422–423
  for age-replacement policy, 407–412, 422–423
  for condition-based policy, 414–421, 423
  ratio, 405
Average, 487; *also see* mean
Axial point, 181

Baczkowski, Carole, 65, 67, 68, 115–132, 172, 189, 481
Balanced design, 174
Bar graph, 478, 480
  examples, 65, 79, 99, 435
Barnett, Joel, 188, 194, 195, 235–250, 483, 484
Batch process,
  control limits for, 303, 307, 309, 337, 339, 342–344, 354–356
  correlation in, 347
Bell-shaped distribution, 485
Between-wafer variability, 497–499; *also see* variance components
Bias of gauge, 5
Bimodal distribution, 222, 224, 479
Binomial distribution, 307
Blasko, Joseph, 188, 194, 196, 265–281, 483
Block, 94, 122, 175
Blocking, 94, 122–125, 175, 241
Bonding integrated circuits, 49
Box-Behnken design, 180, 182, 183
Box plot, 480, 482
  examples, 43, 65, 92–94, 150, 190, 256, 376–381, 432

Broadbent, Eliot, 187, 193, 195, 213–234, 477, 479, 480, 483
Buckner, James, 4, 5, 6, 7, 9–17, 39–44, 172, 173–176, 178, 179, 187, 193, 194, 195, 189, 199–211, 213–234, 477, 479, 480, 481, 483, 493–497
Bulls eye, 109–110, 213
Bunny suit, 116, 120, 472, 473

Calibration, of gauge, 5, 19, 20, 28, 34
 curve, 28, 33–37
 interval, 34
 limits, 28, 33–37
 model, 22, 33–37
Cammenga, David, 173–176, 178, 179, 187, 194, 195, 199–211, 483
Canning, John, 300, 361, 362, 365–385, 480
Cant, 109–110
Cantell, Brenda, 188, 194, 196, 283–296
Capability
 analysis, 310, 328–335, 371, 435–436, 462
 gauge, 5, 30, 39, 40, 42, 47–48, 56–59, 451
 graph, 478, 479
  examples, 66, 310, 328–334, 371–374, 436, 454
 index
  C_p, 310, 436, 440, 462
  C_{pk}, 56–59, 66, 310, 328–335, 371–374, 436
 in passive data collection, 66, 436, 462
 process, 66, 75–76, 310, 325, 328–335, 440, 462, 479
 reliability, 366, 371, 382
 utilization, 382
Carpenter, Steven, 4, 65, 67, 68, 146–154, 484
Categorical
 factor, 174
 response, 174
c control chart, 304, 308
CD, 368
Cdf, 407
Censoring times, 407
Center
 line, 300, 478
 point, 179, 203, 218, 238, 441
Central
 composite design, 178, 180, 181, 205, 266, 271, 455
 limit theorem, 305, 341, 486

Characterization of process, xxiii
Chemical
 bath, 470–472, 474
 mechanical polishing, 106, 251, 252, 474
 vapor deposition (CVD), 39, 40, 199, 201, 213, 215, 429, 474
  low pressure (LPCVD), 474
  plasma enhanced (PECVD), 316, 328, 474
Chin, Barry, 6, 7, 9–17, 39–44, 172, 187, 189, 193, 195, 213–234, 477, 479, 480, 481, 483, 493–497
Chip, 469, 473
 memory, 389
 packaging, 475
Chi-square distribution, 151–153
Circuit,
 integrated, 469–471, 473
 manufacturing, ix, xvii
 printed board, 325–327, 469
CL, 300, 478
Clean room, 472
 air flow in, 146
 garments in, 116, 120, 472, 473
CMP, 106, 251, 252, 474
Coefficient
 of determination R^2, 190, 496
  adjusted, 191, 496
 of variation, 488
 model, 107, 112, 177, 204, 206, 220, 225, 445, 458
Common cause variation, 300
Completion time, 165
Condition-based replacement policies, 403–404, 413–421, 423
 availability for, 414–421, 423
Conductor, 474
Confidence
 interval, 34–37, 76, 83–85, 245–246, 489
  relationship to hypothesis test, 490
 level, 489
Confirmation
 experiment, 194, 245
 run, 175, 452
Conforming items, 303
Confounding, 180, 235, 239, 254, 390
 resolution III, 180
 resolution IV, 181, 236, 238
 resolution V, 181
Connections, electrical, 201, 470, 471
Contamination,

carbon, 19
particles, 90, 115–117, 145, 308, 320, 473, 474
Continuous
　data, 303
　factor, 174
　response, 173
Contour plot, 481–483
　examples, 205, 206, 223, 227, 232, 233, 243–245, 275–277, 451, 452, 464, 465
Contrast, 105–106, 108–111
Control
　array, 182, 286, 390
　factor, 182, 272, 286, 390
　limits, 19, 301, 319–320
　　in presence of time trend, 315–323
　　for batch process, 303, 307, 309, 337, 339, 342–344, 354–356
　　lower, 301, 310, 319, 479
　　1-sigma, 302
　　statistical, 301, 319
　　3-sigma, 302
　　too narrow, 303, 309, 337, 340
　　too wide, 303, 317
　　2-sigma, 302
　　upper, 300, 478
　　workable, 337
　out of, 79, 300, 301, 340, 479
　parameter, 182, 272, 286, 390
　statistical, 300, 479
Control chart, 299–309, 478
　c, 304, 308
　center line, 300, 478
　control limits, 19, 301, 319–320, *See also* control limits
　count, *see* c control chart
　for batch process, 303, 307, 309, 337, 339, 342–344, 347, 354–356
　fraction nonconforming, *see* p control chart
　i, 75–76, 303, 306, 327–334, 340, 344–346, 354–356, 434
　individuals, *see* i control chart
　in PDC, 65, 74–75, 299, 316, 327
　moving range, *see* MR control chart
　MR, 303, 306–307, 327–334
　nonconforming, *see* c control chart
　p, 304, 307
　R, 26, 303, 304, 327–334
　range, *see* R control chart
　s, 74–75, 303, 305, 346, 354–356, 434
　standard deviation, *see* s control chart

　x, *see* i control chart
　$\bar{x}$, 26, 74–75, 303–305, 315–325, 327–334, 340, 346, 354–356, 434
　xbar, *see* $\bar{x}$ control chart
　zone, 302, 328
COO, 361
Correlation, 342, 347
Cost of ownership model, 361
Count control chart, 304, 308
Covariate, 189
C_p capability index, 310, 436, 440, 462
C_{pk} capability index, 56–59, 66, 310, 328–335, 371–372, 436
Criss-cross design, 184
Critical dimension, 368
Crossed
　design, 182, 286
　factor, 112, 183
Crowder, Stephen, 363, 403–424
Cumulative distribution function, 407
Current, electrical, 469–471
CVD, 39, 40, 199, 201, 213, 215, 429, 474
Cycle, equipment, 360, 405
Czitrom, Veronica, xxi–xxvii, 3–7, 4, 63–69, 65, 66, 67, 87–104, 171–198, 172, 188, 189, 194, 195, 235–250, 299–314, 359–364, 477–499, 483, 484, 489

Data,
　attribute, 303
　continuous, 303
　intervals, 303
　measurement, 303
　variables, 303
Defective, 304, 307, 308
Defining relation, 180, 239, 247–248, 390
Degradation of equipment, 404
Degrees of freedom, 84, 486, 493
Deposition,
　chemical vapor (CVD), 39, 40, 199, 201, 213, 215, 429, 474
　low pressure (LPCVD), 474
　plasma enhanced (PECVD), 316, 328, 474
Design, 174
　Addelman, 12, 16
　augmenting, 235, 245
　balanced, 174
　Box-Behnken, 180, 182, 183

central composite, 178, 180, 181, 205, 266, 271, 455
criss-cross, 184
crossed, 182, 286
D-optimal, 182, 183, 442
experimental, 174
factor, 182
foldover, 182, 183, 218
fractional factorial, 12–13, 178–180, 218, 238, 266, 388, 390, 441
   2^{k-p}, 179
full factorial, 91, 95, 117, 122, 174, 178–180, 203, 254, 272, 286
   2^k, 179
hierarchical, 183, see nested
matrix, 174
mixture, 184
nested, 48–55, 98, 123, 136–137, 181, 183, 434, 440, 497
orthogonal, 174, 176, 186
Plackett-Burman, 182, 184
resolution, 180
response surface, 175, 182, 183, 199
robust, 175, 178, 181, 182, 266, 271, 283, 286, 388, 391
rotatable, 183
RSM, 175, 182, 183, 199
second-order, 175
split-plot, 94–95, 122–128, 173, 182, 184, 287–288, 391
split-split plot, 184, 288, 292–295
split-unit, 184, 287–288; *also see* split-plot
strip-plot, 182, 184, 292
three-quarter, 181, 184, 235, 239, 247
Designed experiment, xxiii, xxv, 171–296
  advantages of, 185
  graphical analysis, 184
  hidden replication in, 12, 176, 185
  in gauge study, 12, 19, 24
  independence of effects, 186
  orthogonality of, 174, 176, 186
  practical considerations in, 186
  precision of, 185
  projection property of, 185
  regression, 204, 206, 218–225, 240–244, 255–262, 273–278, 280, 289–295, 392, 444–452, 457–461
  reliability, 387
  resources for, 185
  SPC, 299

variance components in, 189, 439–440, 455
Design of experiments (DOE), 171–296; *see* design and designed experiment
Destructive measurement, 47–49, 51–57
Detection limit of measurement, 19, 28–30
df, 84, 486, 493
Deterministic model, 177
Differences, successive, 318, 321
Diffusion, 473
Discrete
  factor, 174
  response, 173
Distribution,
  bell-shaped, 485
  bimodal, 222, 224, 479
  binomial, 307
  chi-square, 151–153
  F, 192
  frequency, 479
  Gaussian, 485
  lognormal, 159–162
    shape, 160
  multimodal, 479
  normal, 159–163, 479, 485–486
    shape, 160
    standard, 486
  Poisson, 308, 487
  skewed, 479
  symmetric, 479
  t, 486
  unimodal, 218, 219, 479
  Weibull, 159–163
    shape, 160
DOE, *see* design and designed experiment
Doping, 469, 471, 472
D-optimal design, 182, 183, 442
Dot plot, 479
  examples, 24, 478
Downtime, equipment, 360
Drain of transistor, 469, 471
Drift, 43, 251, 254, 261, 315–323, 412, 415

Effect
  confounded, 180, 235, 239, 254, 390
  fixed, 97, 112, 125, 491, 492
  interaction, 123, 176, 190, 219–226, 235, 242–246, 289, 392
  main, 123, 176, 190, 392
  random, 97, 112, 125, 492

EIP, 359
Electrical
 connections, 201, 470, 471
 current, 469–471
 tests, 475
Environmental factor, 182
Equipment
 availability, 360, 404, 405, 407–412, 414–420, 422–423
 cycle, 360, 405
 degradation, 404
 dependent uptime, 361
 development program, 359
 downtime, 360
 failure, 360, 383, 404, 405
 improvement project, 359
 maintainability, 361
 qualification, xxiii
 reliability, 359–424, 360, 365–367, 387
 utilization, 361
 wear, 404
Equivalence of wafers, 88
Error,
 difference, 317
 independence of, 23
 mean square, 494
 measurement, 4; *also see* precision
  removal of, 98, 147
 pure, 494
 root-half-mean-square-successive, 317
 root mean square, 315–317
 standard, 191, 255
 sum of squares, 493, 495, 496
 term, 177
 types I and II, 147, 303, 490
Estimate, 343
 least squares, 177, 190
Estimated value, 177
Estimator efficiency, 322
Etching, 71–72, 236–237, 265, 267, 471, 472, 474
Evolutionary operation, 182, 184
EVOP, 182, 184
Expected
 mean squares, 50–54, 97, 497
 time, 407
Experiment
 confirmation, 194, 245
 designed, *see* designed experiments
 one-factor-at-a-time, 172
 screening, 174, 218, 235, 270, 390, 437

sensitivity, 175
 sequential, 175, 202, 218
Experimental
 design, 174
 region, 174
 unit, 174, 287, 288
Exploratory data analysis, 189

Factor, 173, 174
 categorical, 174
 coding, 177
 continuous, 174
 control, 182, 272, 286, 390
 crossed, 112, 183
 design, 182
 discrete, 174
 environmental, 182
 fixed, 97, 112, 125, 491, 492
 level, 174
 nested, 112, 123, 181, 183, 434, 497
 noise, 182, 272, 286, 390
 orthogonal, 174, 176
 random, 97, 112, 125, 492
 setting, 174
 space, 174
Factorial experiment, *see* full factorial and fractional factorial designs
Failures
 equipment, 360, 383, 404, 405
 mean time between, 360, 387, 388, 397, 407
 time to, 405, 407
F-distribution, 192
Feed rate limited regime, 214, 217, 228
Filter for air, 146
First-order model, 177
Fit,
 of model, 190, 213
 lack of, 494
Fitted value, 177
Fixed
 effects, 97, 112, 125, 491, 492
 factor, 97, 112, 125, 491, 492
Flow rate limited regime, 217
Foldover design, 182, 183, 218; *see also* semifolding
Fraction nonconforming control chart, 304, 307
Fractional factorial design, 12–13, 178–180, 218, 238, 266, 388, 390, 441
 2^{k-p}, 179
F-ratio, 192
Freedom, degrees of, 84, 486, 493

Freeny, Anne, 188, 194, 196, 251–264, 477, 481, 484
Frequency distribution, 479
F-statistic, 192, 494
Full factorial design, 91, 95, 117, 122, 174, 178–180, 203, 254, 272, 286
 2^k, 179
F-value, 192, 494

Gadson, William, 188, 194, 196, 283–296
Gate of transistor, 265, 267, 469–471
Gauge
 accuracy, 5
 ANOVA, 15, 49–55
 bias, 5
 calibration, 5, 19, 20, 28, 33–37
 capability, 5, 30, 39, 40, 42, 47–49, 56–59, 451
 destructive testing, 47–49, 51–57
 precision, 4, 9, 14, 29, 39, 40, 48, 52, 57, 432
 P/T, 5, 10, 14, 39, 42, 48, 432–433
 reliability, 49, 57, 359
 repeatability, 4, 39–42, 47–55, 431
 reproducibility, 4, 39–42, 49–55, 431
 SPC, 6, 19, 26, 30, 299
 stability, 5
 studies, xxii, xxiii, 3–59, 431–433
 variance components in, 49–55
Gaussian distribution, 485
Generator, 180, 239
Graphical data analysis; *also see* plots
 for designed experiment, 189
 for passive data collection, 65
 for SPC, 309
Graphs, 65, 189, 477–485
 bar, 478, 480
  examples, 65, 79, 99, 435
 box, 480, 482
  examples, 43, 65, 92–94, 150, 190, 256, 376–381, 432
 capability, 478, 479
  examples, 66, 310, 328–334, 372–374, 436, 454
 contour, 481–483
  examples, 205, 206, 223, 227, 232, 233, 243–245, 275–277, 451, 452, 464, 465
 control charts, 300–309, 478
  examples, 26, 65, 74–75, 301, 318, 329–334, 340, 346, 356, 434
 dot, 479
  examples, 24, 478
 half-normal probability plot, 484
  examples, 192, 255, 257, 258
 histograms, 478, 479
  examples, 65, 76, 77, 219, 224
 interaction plot, 190, 482, 483
  examples, 121, 221, 226, 242, 246, 289, 290, 395, 451, 462, 481
 main effect plot, 190, 481–483
  examples, 123, 392, 400, 401
 multivariate, 482
  examples, 65, 122, 481
 normal probability plot, 484
  examples, 65, 148, 161, 162, 192, 255, 393, 394, 447, 459
 Pareto, 478, 480
  examples, 65, 190, 220, 225, 383
 residual, 478
  examples, 222, 226, 260, 261, 447, 448, 458–460, 478
 scatter, 477, 478
  examples, 27, 30, 139, 420, 436
 snapshot, 65, 163–165
 three-dimensional, 482, 484
  examples, 109, 481
 time trend, 477
  examples, 5, 43, 219, 224, 261, 318, 412, 415, 418
 x-y graph, 477, 478
Green, Kent, 300, 361, 362, 365–385, 480
Green, Todd, 6, 7, 39–44, 481
Group, SPC focus, 338, 341
Growth model, 361
GRR, 48, 49, *see* gauge studies

Half-normal probability plot, 484,
 examples, 192, 255, 257, 258
Hall, Kathryn, 4, 65, 67, 68, 146–154, 484
Hazard function, 410
Hegemann, Victor, 4, 7, 47–59
Henri, Jon, 6, 7, 9–17, 39–44, 172, 187, 189, 193, 195, 213–234, 477, 479, 480, 483, 481, 493–497
Hidden replication, 12, 176, 185
Hierarchical design, 183; *also see* nested design
Histogram, 478, 479
 examples, 65, 76, 77, 219, 224
Hole, 469
Horrell, Karen, xxi–xxvii

Huang, Richard, 187, 193, 195, 213–234, 477, 479, 480, 483
Hurwitz, Arnon, 66, 67, 105–114, 172, 189, 484
Hutchens, Craig, 172, 188, 362, 387–401, 483, 484
Hypothesis
 alternative, 490
 null, 490
 test, 147, 152, 191, 303, 490
  relationship to confidence interval, 490

i control chart, 75–76, 303, 306, 327–334, 344–346, 354–356, 434
Implant, 472
Implanter, 416
Independence of
 effect estimates, 186
 errors, 23
Individuals control chart, 75–76, 303, 306, 327–334, 340, 344–346, 354–356, 434
Inner array, 182, 390
Input variable, 173
Insulator, 474
Integrated circuit, 469–471, 473
 bonding, 49
 manufacturing, ix, xvii
Interaction, 176, 235
 effect, 123, 176, 190, 219–226, 235, 242–246, 289, 392
 plot, 190, 482, 483
  examples, 121, 221, 226, 242, 246, 289, 290, 395, 451, 462, 481
Interquartile range, 480
Intervals
 calibration, 34
 confidence, 34–37, 76, 83–85, 245–246, 489
  relationship to hypothesis test, 490
 data, 303
 specification, 5, 310
Inverse regression model, 22, 29, 33
IRONMAN, xxvi
Ion implanter, 416

JDP, 359, 365
John, Peter W.M., 188, 194, 195, 235–250, 483, 484
Joint development project, 359, 365
Joshi, Madhukar, 308, 309, 312, 313, 337–357

Kahn, William, 65, 67, 68, 115–132, 172, 189, 481
Kinetic regime, 214, 217, 228–229
Knowledge,
 prior, 202, 217
 theoretical, 199, 209–210
Kook, Taeho, 188, 194, 196, 265–281, 483

Laboratory for device development, 155
Lack of fit, 494
Lai, Warren, 188, 194, 196, 251–264, 477, 481, 484
Lambert, Diane, 65, 67, 68, 155–165
Landwehr, James, 65, 67, 68, 155–165
Lattice, 182
LCL, 301, 319, 479
Least squares estimates, 177, 190
León, Ramón, 188, 194, 195, 235–250, 483, 484
Level,
 of confidence, 489
 of factor, 174
 of significance, 489
Lewis, Donald, 172, 188, 362, 387–401, 483, 484
Limit
 calibration, 28, 33–37
 control, 19, 301, 319–320; *also see* control limits
 probability, 302
 sigma, 302
 SPC, 301
 specification, 48, 56, 309–310, 338, 479
  lower, 5, 48, 66, 310, 338
  one-sided, 48, 56, 58–59
  upper, 5, 48, 66, 310
 statistical, 301, 319
Linear
 model, 177
 term, 177
Lithography, 134, 340, 360, 472
Liu, K.C., 4, 7, 47–59
Location, measures of, 4, 487
Lognormal distribution, 159–162
 shape, 160
Low pressure chemical vapor deposition (LPCVD), 474
Lower
 control limit, 301, 310, 319, 479
 specification limit, 5, 48, 66, 310, 338
LPCVD, 474

LSL, 5, 48, 66, 310, 338
Lynch, Richard, 65, 66, 67, 71–85, 172, 189, 300, 479, 480

Main effect, 123, 176, 190, 392
  plot, 190, 481–483
    examples, 123, 392, 400, 401
Maintainability of equipment, 361
Maintenance, 360
  emergency, 404
  preventive, 383, 403–423, 405
  scheduled, 403
  unscheduled, 403
Management support, 337
Marathon, xxvi, 359, 361, 365, 366, 384, 365–385, 454
Markle, Richard, 65, 66, 67, 71–85, 172, 189, 300, 479, 480
Marr, Ray, 308, 313, 315–324, 479
Mask, 470, 472
Mass flow controller, 409
Matrix for experimental design, 174
Mean
  population, 159, 487
  sample, 159, 487
  square, 50–54, 97, 494
    error, 494
    expected, 50–54, 97, 497
    residual, 494
  standard deviation of sample, 341
  time between assists, 360
  time between failures, 360, 387, 388, 397, 407
  time to repair, 361, 367, 383
  wafers between assist, 388, 396
  wafer between interrupt, 383
Measurement
  data, 303
  destructive, 47–49, 51–57
  detection limit, 19, 28–30
  error, 4, *see* precision
    removal of, 98, 147
  studies, see gauge studies
  system, 3
Median, 480, 487
Meester, Steven, 188, 194, 196, 265–281, 483
Memory
  chips, 389
  repair, 387, 389
Metallization, 201, 215, 470, 471, 474
Microbalance, 39
Micron, 472

Mitchell, Teresa, 4, 7, 47–59, 172, 189
Mixed effects model, 112, 492
Mixture experiment, 184
Model, 107, 112, 173, 176–177, 492
  calibration, 22, 33–37
  coefficients, 107, 112, 177, 204, 206, 220, 225, 445, 458
  cost of ownership, 361
  deterministic, 177
  first-order, 177
  fit, 190, 213
  fixed effects, 112, 492
  growth, 361
  hierarchical, *see* nested
  inverse regression, 22, 29, 33
  linear, 177
  mixed effects, 112, 492
  nested, 55, 77, 83, 181, 347, 355–356
  polynomial, 176
  quadratic, 177, 273
  random effects, 112, 492
  regression, 176, 190, 206, 444, 457
  second-order, 177, 273
  statistical, 177
  sum of squares, 496
  uniformity, 213, 214
  Weibull, 409
Monnig, Kenneth, 187, 193, 195, 213–234, 477, 479, 480, 483
Moving range, 306, 321
  control chart, 303, 306, 327–334
MR control chart, 303, 306, 327–334
MTBA, 360
MTBF, 360, 387, 388, 397
MTTR, 361
Multimodal distribution, 479
Multivariate plot, 482
  examples, 65, 122, 481
MWBA, 388
MWBI, 383

Nested
  design, 48–56, 98, 123, 136–137, 181, 183, 434, 440, 497
  factor, 112, 123, 181, 183, 434, 497
  model, 55, 77, 83, 181, 347, 355–356
Nitride process, 315, 316
Noise
  array, 182, 286, 390
  factor, 182, 272, 286, 390
  parameter, 182, 272, 286, 390
Nonconforming
  item, 303, 307, 308

fraction, control chart, 304, 307
Nonconformities control chart, 304, 308
Nondefective, 303
Non-linear term, 177
Non-uniformity, 213, 214, 232–233, 238, 488
Normal
   distribution, 159–163, 479, 485–486
     shape, 160
   probability plot, 484
     examples, 65, 148, 161, 162, 192, 255, 393, 394, 447, 459
   standard, 486
Normality, test of, 193, 486
n-type silicon, 469, 471
Null hypothesis, 490

Observational study, *see* passive data collection
One-factor-at-a-time experiment, 172
Optical recognition system, 389
Optimization, 446, 450, 462
Order of experimental runs, 175, 263
Orthogonal
   array, 182
   design, 174, 176, 186
   factor, 174, 176
   polynomial, 254
Outer array, 182, 390
Outlier, 42, 193, 315, 316, 320, 480, 488
Output variable, 173
Overlay, 365, 368, *see also* alignment
Oxidation process, 72, 89, 470, 471, 474

Packaging chips, 475
Pankratz, Peter, 5, 6, 19–37, 172, 300, 479
Parameter,
   control, 182, 272, 286, 390
   noise, 182, 272, 286, 390
Pareto chart, 478, 480
   examples, 65, 190, 220, 225, 383
Particles, 90, 115–117, 146, 308, 320, 473, 474
Particulate contamination, *see* particles
Passive data collection, xxii, xxiv, 63–165, 315, 316, 327, 433, 452
   graphical data analysis, 65
   in designed experiment, 200, 208, 440
   process stability, 65, 74–75
   process capability, 66, 436, 462
   reliability, 359, 366
   sampling plan, 64, 66, 70

SPC, 65, 74–75, 299, 316, 327
   variance components, 76–79, 90, 94–100, 107–111, 123–128, 136–140, 147–151, 434–435
p control chart, 304, 307
PDC, *see* passive data collection
PECVD, 316, 328, 474
Pepper, Dwayne, 66, 68, 133–144
Photolithographic process, 134–136, 340, 367, 472
Photoresist, 470–472
Plackett-Burman design, 182, 184
Planarization, 106, 251
Plasma, 474
   enhanced chemical vapor deposition, 316, 328, 474
   etching, 71–72, 265, 267, 471, 472, 474
Plots, *see* graphs
PM, 383, 405, 403–423
Point in experimental design, 174
Poisson, 308, 487
Polarity, 285
Polynomial, 254
   orthogonal, 254
   regression model, 176
Polysilicon, 470
Population, 485
   mean, 159, 487
   standard deviation, 159, 488
   variance, 159, 488
Ppm, 311
Practical significance, 191
Precision,
   designed experiment, 185
   gauge, 4, 9, 14, 29, 39, 40, 48, 52, 57, 432
   increase by sampling, 5, 20, 30
   to tolerance ratio, 5, 10, 14, 39, 42, 48, 432–433
Predicted value, 177
Prediction, 176, 190
Predictor variable, 173
Preventive maintenance, 383, 405, 403–423
Preuninger, Fred, 188, 194, 196, 265–281, 483
Printed circuit board, 325–327, 469
Probability
   limit, 302
   plot
     half-normal, 192, 255, 257, 258, 484

normal, 65, 148, 161, 162, 192, 255, 393, 394, 447, 459, 484
significance, 191, 495
Process characterization, xxiii
Product array, 182, 391
Productivity, 365, 367
Projection property of designed experiment, 185
P/T, 5, 10, 14, 39, 42, 48, 432–433
p-type silicon, 469
Pure error, 494
p-value, 191, 192, 486, 495, 496

QFD, 326
Quadratic model, 177, 273
Qualification
  of equipment, xxiii
  plan, ix, xii, xxi–xxvii, 39, 365–367, 384, 427, 429
Quality function deployment, 326
Qual plan, *see* qualification plan
Quartile,
  first, 480
  second, 480
  third, 480
Queue time, 158, 165

RAM, 360, 361
Ramírez, José, 188, 194, 196, 283–296
Random
  effect, 97, 112, 125, 492
  effects model, 112, 492
  factor, 97, 112, 125, 492
  order, 175
  sample, 485
Randomization restrictions, 203, 292
Range, 487
  control chart, 26, 303, 304, 327–334
  moving, 306, 321
    control chart, 303, 306–307, 321, 327–334
  interquartile, 480
Rational subgroup, 302, 304
R control chart, 26, 303, 304, 327–334
Reece, Jack, 4, 65, 66, 67, 87–104, 172, 189, 427–428, 429–468, 477, 489
Regime,
  feed rate limited, 214, 217, 228
  flow rate limited, 217
  kinetic, 214, 217, 228–229
Region, experimental, 174
Registration
  in photolithography, 134–136

for printed circuit boards, 326
Regression,
  in designed experiment, 204, 206, 218–225, 240–244, 255–262, 273–278, 280, 289–295, 392, 444–452, 457–461
  in gauge study, 23, 29, 33–37
  in passive data collection, 107–112
  model, 176, 190, 206, 444, 457
    a-priori, 190
    inverse, 22, 29, 33
  polynomial, 176
  stepwise, 190, 204, 241–242, 457
  sum of squares, 496
Reliability, xxiii, xxv
  equipment, 359–424, 360, 365–367, 387
  gauge, 49, 57, 359
  in SPC, 300, 359
  in passive data collection, 359, 366
  variance components in, 372–375
Repair, mean time to, 361, 367, 383
Repeatability of gauge, 4, 39–42, 47–55, 431
Replacement policies,
  age, 403–404, 406–412, 421–423
  condition-based, 403–404, 413–421, 423
Replication, 175
  hidden, 12, 176, 185
Reproducibility of gauge, 4, 39–42, 49–55, 431
Residual, 177, 193, 445
  analysis, 193, 260–261, 445, 458–460
  mean square, 494
  plot, 222, 226, 260, 261, 447, 448, 459–460, 478
  studentized, 193
  sum of squares, 493, 495, 496
  variability, 493
Resist, 470–472
Resolution of experimental design, 180
  III, 180
  IV, 181, 236, 238
  V, 181
Resources for designed experiment, 185
Response, 173, 189
  categorical, 174
  continuous, 173
  discrete, 173
  prediction of, 190
  surface, 183
    design, 175, 182, 183, 199

methodology, 175, 182–183, 199
Reticle, 134–136, 283–285, 470, 472
RHMSSD, 317
Risk, 139–142, 146–147
RMSE, 315–317
Robust design, 175, 178, 181, 182, 266, 271, 283, 286, 388, 391
Root-half-mean-square-successive-difference, 317
Root mean square error, 315–317
Rotatable design, 183
R^2, 190, 496
R^2 adjusted, 191, 496
RSM, 175, 182, 183, 199
Rule
  run, 301
  Western Electric, 302, 340, 479
Run, 174
  confirmation, 175, 452
  exploratory, 175
  marathon, xxvi, 359, 361, 365, 366, 384, 365–385, 454
  order, 175, 263
  rule, 301
  time, 159
Run-to-run variability, 497–499; *also see* variance components

Sample, 485
  mean, 159, 487
  random, 485
  size
    calculation, 12, 152
    effective, 337, 338, 340–347
    reduction, 133, 136, 147
  standard deviation, 341, 488
  variance, 159, 488
Sampling
  increasing precision, 5, 20, 30
  plan
    and C_{pk}, 310, 433–437
    and variance components, 64
    for designed experiment, 186
    for passive data collection, 64, 66, 70
  skip-lot, 133
  systematic, 73
Scace, Robert I., xviii
Scanning electron microscope, 267, 269
Scheduled maintenance, 403
Schutz, Ronald, 469–475
s control chart, 74–75, 303, 305, 346, 354–356, 434

Scatter plot, 477, 478
  examples, 27, 30, 139, 420, 436
Screening
  experiment, 174, 218, 235, 270, 390, 437
  for outliers, 316, 320
Second-order
  experimental design, 175
  model, 177, 273
SEM, 267, 269
SEMATECH, ix, xix
  qualification plan, ix, xii, xxi–xxvii, 39, 365–367, 384, 427, 429
Semiconductor, 469
  manufacturing, ix, xvii, 469–475
SEMI E10 standards, 359
Semifolding, 184, 236, 239, 247
Sensitivity experiment, 175
Sequential experimentation, 175, 202, 218
Setting, of factor, 174
Shapiro-Wilk test for normality, 486
Shenasa, Mohsen, 172, 189, 427–428, 429–468, 477
Shewhart control charts, 300, *see* control charts
Shift, 315, 319, 320
Shyu, Ming-Jen, 65, 67, 68, 155–165
Sigma, 486, 488
  limit, 302
  six, 310, 325, 486
Significance,
  level, 489
  practical, 191
  probability, 191, 495
  statistical 175, 191, 492, 495
Silicon, 469
  dioxide, 470, 471, 474
  n-type, 469, 471
  p-type, 469
Site-to-site variability, 497–499; *also see* variance components
Six sigma, 310, 325, 486
Skewed distribution, 479
Skip limit, 138–139
Skip-lot sampling plan, 133
Smith, Joseph, 172, 188, 362, 387–401, 483, 484
Snapshot plot, 65, 163–165
Sorell, Mark, 187, 193, 195, 213–234, 477, 479, 480, 483
Source of
  transistor, 469, 471

variation, 493; *also see* ANOVA
Space, factor, 174
Spagon, Patrick, 66, 67, 105–114, 172, 189, 484
SPC, xxiii, xxv, 299–357
   implementation, 337, 347–350
   in designed experiment, 299
   in gauge study, 6, 19, 25–26, 30, 299
   in passive data collection, 65, 74–75, 299, 316, 327
   in reliability, 300, 359
   limit, 301
   variance components in, 330
Special cause of variability, 300, 339
Specification, 309–310, 339
   interval, 5, 310
   limit, 48, 56, 309–310, 338, 479
      lower, 5, 48, 66, 310, 338
      one-sided, 48, 56, 58–59
      upper, 5, 48, 66, 310
Spectrometer, 19, 21
Spencer, William J., ix
Split-plot design, 94–95, 122–128, 173, 182, 184, 286–288, 391
Split-split-plot design, 184, 288, 292–295
Split-unit design, 184, 286–288; *also see* split-plot design
Sprague, Kimberley, 308, 309, 312, 313, 337–357
Sputtering, 474
Stability
   gauge, 5
   process, 65, 74–75
   reliability, 375
Standard deviation, 341
   control chart, 74–75, 303, 305, 346, 354–356, 434
   estimate of, 317, 488
   of sample mean, 341
   population, 159, 488
   sample, 341, 488
Standard
   error, 191, 255
   normal random variable, 486
   SEMI E10, 359
Star point, 181
State variable, 413, 415, 421
Statistical
   control, 300, 479
      limits, 301, 319
      out of, 79, 300, 301, 340, 479
   model, 177
      limit, 301
   process control, *see* SPC
   significance, 175, 191, 492, 495
Stepper, 134, 284, 365, 472
Stepwise regression, 190, 204, 241–242, 457
Strip-plot design, 182, 184, 292
Subgroup, rational, 302, 304
Successive differences, 318, 321
Sum of squares, 493, 461
   error, 493, 495, 496
   model, 496
   regression, 496
   residual, 493, 495, 496
   total, 493, 495, 496
Surface mount technology, 325–327
Switching by transistor, 469, 471
Symmetric distribution, 479
Systematic sampling, 73

Taguchi, Genichi, 182, 272, 391
Target, 474
t distribution, 486
Term
   error, 177
   linear, 177
   non-linear, 177
Test of
   hypothesis, 147, 152, 191, 303, 490
      relationship with confidence interval, 490
   normality, 193, 486
Three-dimensional graphs, 482, 484
   examples, 109, 481
Three-quarter design, 181, 184, 236, 239, 247
Throughput, 144, 366, 367
Time,
   between assists, 360
   between failures, 360, 387, 388, 397, 407
   censoring, 407
   completion, 165
   drift, 43, 251, 254, 261, 315–323, 412, 415
   graphs of, 5, 43, 219, 224, 261, 318, 412, 415, 418
   expected, 407
   non-scheduled, 360
   operations, 360
   queue, 158–159, 165
   repair, 361, 367, 383
   run, 159

sequence, 43
to failure, 405, 407
trend 5, 43, 219, 224, 261, 318, 412, 415, 418, 477
total, 360
unproductive, 382
warm-up effect, 39, 40, 42
TiN, 199
Tolerance, 5, 83–85
Total
sum of squares, 493, 495, 496
variability, 493
Tradeoff, 193, 199, 207, 251, 266, 272, 275, 291, 483
Transformation, 119, 192, 240, 396
Transistor
as switch, 469, 471
creation of, 470–471
drain, 469, 471
gate, 265, 267, 469–471
source, 469, 471
Treatment combination, 174
t-statistic, 191, 486, 496
tungsten, 40, 199, 201, 213, 215
t-value, 191, 496
2^k full factorial design, 179
2^{k-p} fractional factorial design, 179
Type I error, 147, 303, 490
Type II error, 147, 303, 490

UCL, 300, 478
Uniformity, 213, 214, 232–233, 238, 488
signed, 214, 219
Unimodal distribution, 218, 219, 479
Unit, experimental, 174, 286–288
Upper
control limit, 300, 478
specification limit, 5, 48, 66, 310
Uptime
equipment, 360
dependent, 361
expected, 407
operational, 361
supplier dependent, 361, 367
USL, 5, 48, 66, 310, 338
Utilization
capability, 382
of equipment, 361
operational, 361, 382
total, 361

Variability
assignable cause of, 74–75, 300

chance cause of, 300
common cause of, 300
error, 493
natural, 175
process, 300
residual, 493
sources of, 493; *also see* ANOVA and variance components
special cause of, 300, 339
total, 493
Variable,
categorical, 174
continuous, 174, 303
input, 173
output, 173
predictor, 173
response, 173
state, 413, 415, 421
Variables data, 303
Variance
estimate of, 494
sample, 159, 488
population, 159, 488
Variance component, 64, 66, 481, 497–499
estimation of, 50–54, 97
in DOE, 189, 439–440, 455
in gauge study, 49–55
in PDC, 76–79, 90, 94–100, 107–111, 123–128, 136–140, 147–151, 434–435
in reliability, 372–375
in SPC, 330
sampling plan, 64
Variation
assignable cause, 74–75, 300
chance cause, 300
common cause, 300
special cause, 300, 339
Venor, Kevin, 187, 193, 195, 213–234, 477, 479, 480, 483
Very large scale integration, 472
VLSI, 472

Wafer, 472, 473
between assist, 388, 396
between interrupt, 383
equivalence, 88
planarization, 106, 251, 252
recycled, 89–90
virgin, 90
Wafer-to-wafer variability, 497–499; *also see* variance components

Warm-up effect, 39, 40, 42
Watson, Ricky, 3, 308, 309, 312, 313, 325–335
Wear of equipment, 404
Weber, Ann, 173–176, 178, 179, 187, 194, 195, 199–211, 483
Weibull distribution, 159–162
    model, 409
    shape, 160
Western Electric rules, 302, 340, 479
Whisker of boxplot, 480
Wire, 469–471, 474
Within-wafer variability, 497–499; *also see* variance components

x control chart, 75–76, 303, 306, 327–334, 340, 344–346, 354–356, 434
$\bar{x}$ control chart, 26, 74–75, 303–305, 315–325, 327–334, 340, 346, 354–356, 434
xbar control chart, *see* $\bar{x}$ control chart
x-y graph, 477, 478

Yield, 474
    loss, 49, 54–59, 142, 213, 215

Zone, control chart, 302, 328